BIOTECHNOLOGY IN AGRICULTURE, INDUSTRY AND MEDICINE

LIGNIN

PROPERTIES AND APPLICATIONS IN BIOTECHNOLOGY AND BIOENERGY

BIOTECHNOLOGY IN AGRICULTURE, INDUSTRY AND MEDICINE

Additional books in this series can be found on Nova's website
under the Series tab.

Additional E-books in this series can be found on Nova's website
under the E-books tab.

BIOCHEMISTRY RESEARCH TRENDS

Additional books in this series can be found on Nova's website
under the Series tab.

Additional E-books in this series can be found on Nova's website
under the E-books tab.

BIOTECHNOLOGY IN AGRICULTURE, INDUSTRY AND MEDICINE

LIGNIN

PROPERTIES AND APPLICATIONS IN BIOTECHNOLOGY AND BIOENERGY

RYAN J. PATERSON
EDITOR

Nova Science Publishers, Inc.
New York

Library of Congress Cataloging-in-Publication Data

Lignin : properties and applications in biotechnology and bioenergy / Ryan J. Paterson.
    p. cm. -- (Biotechnology in agriculture, industry and medicine biochemistry research trends)
  Includes index.
  ISBN 978-1-61122-907-3 (hardcover)
  1. Lignin--Biotechnology. I. Paterson, Ryan J.
  TP248.65.L54L534 2011
  660.6--dc22
                        2010044709

Published by Nova Science Publishers, Inc. † New York

CONTENTS

PREFACE

As a biopolymer, lignin is unusual because of its heterogeneity and lack of a defined primary structure. Its most commonly noted function is the support through strengthening of wood (xylem cells) in trees. This book presents current research in the study of lignin, including the biotechnological applications of lignin; the isolation of lignin from alkaline pulping liquors; biobleaching of paper pulp with lignin degrading enzymes; lingin and funal pathogenesis; plastic moldable lignin and lignin valorization into polyurethanes.

Chapter 1 - Lignin is the second most abundant natural renewable resource and aromatic (phenolic) polymer on Earth. Lignocellulosic biomass is mainly comprised of cellulose, hemicellulose and lignin that are strongly intermeshed and chemically bonded by non-covalent forces and covalent cross-linkages. Lignin is an aromatic polymer synthesized from phenylpropanoid precursors. The complex structure of lignocellulose in plants forms a protective barrier to cell destruction by microorganisms. Various pre-treatment methods have been applied for the proper utilization of the valuable carbon sources from the lignocellulosic biomass. A smaller group of filamentous fungi, namely white-rot fungi have the ability to break down lignin, the most recalcitrant component of plant cell walls. Large amount of lignin is derived as waste product from wood processing in pulp and paper industry and is burned to generate energy and recover chemicals, which represents an important environmental problem.

Researchers around the globe have been trying hard to crack the key behind the complex structure of lignocellulosic biomass. Genetic engineering opens new avenues for producing tailor-made plant products with improved structural properties. Many new target genes of both plant and microbial origin are increasingly becoming available for designing plants suitable for downstream processing. Till date, not much research has been carried out to utilize lignin, a gigantic reservoir of carbon having promising applications. The lack of value-added applications from lignin is mainly caused by the heterogeneity, odor and color problems of lignin-based products. Only a small proportion of lignin is currently used in biotechnological applications, such as an adhesive, tanning agent, adsorbent of acids, metal ions, dyes, bile acids, organic pollutants, cholesterol, surfactants, pesticides and phenols, bio(polymer) additives, as a precursor for chars and activated carbon. Lignin has also been used as an antioxidant, such as in cosmetic formulations, an application of particular interest in relation to human health. The immense reservoir of carbon should be increasingly exploited in the development of economically viable biotechnological products within the context of sustainable development.

This chapter deals with the brief introduction on lignocellulosic biomass structure, different delignification strategies, types, properties and various compounds which are produced from lignin degradation. This chapter also discusses the potential applications of lignin in industrial biotechnology in the context of sustainable alternatives to non-renewable products.

Chapter 2 – This chapter reviews the hydrothermal processing of biomass, based on treatments with water or steam. Hydrothermolysis of lignocellulosic materials enables the fractionation and the selective extraction of bioactive compounds from plants. Along some fractionation processes, lignin is depolymerized, and the dissolved fragments are incorporated to the liquid phase. Upon pressurized water extraction (in which water remains in liquid state), free or polymeric phenolic compounds present in the lignocellulosic feedstocks can be solubilized at improved yields respect to conventional extraction, due to the changes in solvent properties and to the participation of autocatalytic hydrolysis reactions. The properties of both depolymerized lignin fractions and benzoic and hydroxycinnamic acids are considered, with particular emphasis on recent and emerging applications in the cosmetic and pharmaceutical industries. For comparative purposes, additional information concerning lignins extracted by traditional methods is presented.

Chapter 3 – The development of commercially viable "green products" based on natural resources for both matrices and reinforcements for a wide range of applications is on the rise. Biodegradable plastics made from renewable resources have received increased attention. The driving force for this interest arises from sustainability gains and environmental amelioration provided through a reduced dependence on petroleum reserves, increased disposal options, and lower levels of greenhouse gases. This development has occured in spite of acknowledged challenges related with material properties, recycling, and cost. Most of the interest in biodegradable plastics is aimed at developing low cost composites that are economical in high-volume applications. Applications considered most relevant include packaging and consumer products. Lignin is one of the most abundant natural raw materials available on earth, after cellulose. Lignin research comes out of the tradition of organic chemistry. The applied RandD has primarily been driven by the needs for knowledge in relation to pulping and bleaching. Lignin recovery and utilization research has been highly empirical and raw material-based rather than consumer-oriented. The 300 million dollar lignin business today is dominated by products from sulphite-pulping liquors, originally based on the need to solve a pollution problem. It is now a chemical business associated to the pulp mill. The products are primarily directed to the process industry, but increasingly higher value functional products are produced. The lignin business is faced with a high cost of RandD and has to carefully select which markets to serve. There are presently several new fields of lignin application that are too complex and competitive and require synergistic effort to prove their viability for uses in fields like coatings and adhesives. The use of lignocellulosics as fillers and reinforcements in thermoplastics has been gaining acceptance in commodity plastics applications in recent years. Another major advancement in engineered lignin is in product and performance enhancement. Advanced engineered lignin based composites are currently being developed that will simultaneously meet the diverse needs of users for high-performance and economical commodity products. These engineered lignin biocomposites will provide advanced performance, durability, value, service-life, and utility while at the same time being a fully sustainable technology.

The focus of this review is to highlight the state of knowledge in processing, manufacturing, characterization, and the challenges in the integration lignin based coatings and adhesives for their potential applications in various fields.

Chapter 4 – The reduction in the emission of gaseous pollutants, the diversification of energy sources, and the dependence on the volatile oil market, all make bio-refinery a good candidate for the production of consumer goods and energy. The valorization of lignin is central to the implementation of the bio-refinery as an alternative to the traditional petrochemical industry. The bio-refinery is the comprehensive use of the total lignocellulosic biomass through the production of high added value products.

Lignin is a polyphenolic polymer with very complex branching structure. Its aromatic structure makes it the only natural polymer with these features. Hence, the huge interest to obtain lignin with homogeneous properties to head production for chemicals, fuels, and/or energy.

There are several treatments applicable for lignocellulosic materials and the fractionation for its major components, cellulose, hemicellulose, and lignin. These treatments are, in general, more focused to achieve good quality solid fraction (cellulose) and pursue the delignification regardless of how it affects lignin structure. However, the different extraction processes strongly affect the structure of the lignin removed, making it difficult to produce homogeneous fractions with industrial interest.

The separation of the liquid fraction after the treatment of the raw material may be a solution to the problem of heterogeneity.

The ultra filtration is an efficient separation process that allows the good separation and purification of lignin to dissolve in the pulping liquor. This method also allows differentiated lignin fractions with diverse physical-chemical characteristics.

Ultra filtration separation method has been compared with another simple method, selective precipitation; in order to establish the effect of the separation process on the properties of lignin fractions produced by both methods. Selective precipitation is a very low energy consumption and easy separation method that intends to solve the no-homogeneity of the lignin samples.

Lignin is known to be a natural antioxidant. This capacity is influenced by the properties of the lignin, mainly by the phenol content. As previously mentioned the separation method can reach homogenous fractions and so, enhance their antioxidant capacity. For this purpose, ultra filtrated fractions were studied to know how the final properties of the different fractions affect the antioxidant capacity of the lignin.

This work presents the results of the study and provides a comparison for the physical-chemical characterization of different fractions of lignin obtained by two methods of separation, ultra filtration and selective precipitation. Finally, an antioxidant capacity test was done to confirm the effect of the ultra filtration on the lignin properties.

Chapter 5 – Lignin is the second most abundant polymer in nature and a major by-product of the paper industry. Crude lignin is also generated as a waste stream in the organosolv delignification process and the steam explosion process for cellulosic ethanol production. There is an increasing interest in utilizing lignin as a potential raw material for the chemical industry in the past two decades, due to the its immense quantities produced annually and the depletion of fossil fuels as well as the environmental concerns associated with the use of fossil resources. Lignin is an amorphous macromolecule of three phenyl-propanols i.e., p-hydroxyl-phenyl propanol, guaiacyl-propanol and syringyl-propanol. These phenyl-propanols are linked mainly by condensed linkages (e.g., 5-5 and β-1 linkages), and more dominantly by ether linkages (e.g., O-4 and O-4) between the three main lignin building blocks. This macromolecule can decompose/degrade into oligomeric and monomeric phenolic compounds via thermochemical technologies (such as pyrolysis, hydrothermal liquefaction, and hydrolysis) and some biological processes. Lignin is thus attractive to many industries such as the manufacture of phenol-formaldehyde (PF) resins and adhesives, as the bio-phenols derived from lignin could potentially substitute for petroleum-based phenol in the resin synthesis. This chapter provides an overview of chemistry of PF resin synthesis, liquefaction of lignocellulosic materials and production of bio-oil-based PF resins, and the recent advances in lignin extraction, modification and de-polymerization.

Chapter 6 – Biomass has been important resource because of its renewable and carbon neutral characteristics. Steam gasification is a very promising technology for energy conversion and hydrogen production with high efficiency. In steam gasification of biomass, initially pyrolysis takes place, producing volatiles (tar and gases) and solid residue (char). Then, steam reacts with the residual char and tar, producing gases such as H_2, CO and CO_2. Moreover, a large amount of tar is evolved in biomass gasification. Tar causes blockages and corrosion of pipes, and also reduces the overall thermal efficiency of the process [Bridgewater, 1995; Devi et al., 2003]. It is, therefore, imperative to rapidly convert the char and tar into gases to achieve high overall efficiency of steam gasification of biomass.

Chapter 7 – Alkaline pulping is the most widely used pulping method for the pulping of lignocelluloses. The spent liquor generated, called the black liquor, contains almost all the lignin present in the original raw material. Most of this lignin finds its way, as a fuel, to a power boiler for captive power generation. Methods, however, exist for the isolation of lignin from the black liquor. Commercial isolation of lignin from spent pulping liquors amounts to about 1 million tonnes annually, which is less than 2% of the total technical potential. A huge opportunity, thus, exists for tapping this potential to put lignin to better uses and niche applications. Precipitation of lignin by acidification is the current most practiced separation method. Another method is by membrane separation in its different forms. Recently, some electrochemical methods have been used for the isolation of lignin from black liquor. This chapter provides an insight into the different methods for the separation of lignin from alkaline pulping liquors.

Chapter 8 – Taking into consideration the advantages of using natural materials as biosorbents in the cleaning and/or clearing wastewater technologies, the specialists' attention has been turned to the development of some new treatment methods, as a part of the environmental biotechnologies. These include, among others, biosorption. A special attention is focused on lignocellulosic materials, due to their fundamental characteristics, such as accessibility and low prices, mechanical resistance, high porosity and specific surface area, hydrophilic character that ensures a rapid sorption rate, tolerance to biological adsorbed solid layers, easy functionalization, the possibility of be used in different forms (particles of different dimensions, fibbers, filters, textile materials for cloths) and regimes (discontinuous and continuous processes).

The paper is a review about their researches and comparisons with literature regarding different types of industrial wasted lignocellulose and lignin materials with sorptive properties that were used into biosorption processes of wastewater treatment.

Chapter 9 – Lignin is a three dimensional natural plant biopolymer formed by radical coupling of hydroxycinnamyl subunits called monolignols mainly p-coumaryl, sinapyl and coniferyl alcohols and creates together with hemicelluloses, a glueing matrix for cellulose microfibrils in tracheary elements and fibers of higher plants. It is the second most abundant component of the cell wall of vascular plants and it protects cellulose towards hydrolytic attack by saprophytic and pathogenic microbes.

Though important for plants, the removal of lignin represents a key step for carbon recycling in land ecosystems. Its removal is also considered as a fundamental issue for the optimal agroindustrial utilization of plant biomass. Lignin polymer is considered to be recalcitrant towards degradation by chemical and biological means. This owes to its molecular architecture in which different non-phenolic phenylpropanoid units form a complex 3D mesh linked by several ether and carbon-carbon bonds. A few microbial species are known for their ligninolytic potential. These ligninolytic microbes exhibit a unique strategy for lignin degradation, which is based on unspecific one-electron oxidation of the benzenic rings catalyzed by synergistic action of extracellular haemperoxidases and peroxide-generating oxidases.

Chapter 10 – Lignocellulosic biomass is present in very large amounts as a result of world-wide pulp and paper processes. Lignin is a major potential renewable, non-fossil source of aromatic and cyclohexyl compounds. It is a feasible raw material for many valuable substances such as activated carbon, vanillin, vanillic acid, dispersing agents, ion-exchange agents, polymer fillers, binding agents for the production of fibre boards, artificial fertilizers and complexing agents. The utilization of lignin can benefit both the green/renewable chemistry and forestry industries. Various methods, including chemical, biological, photochemical and electrochemical techniques have been explored for the modification of lignin to produce value-added products. Each of these advanced approaches are reviewed in this chapter. The chemical treatment of lignin, which leads to the generation of lignin based epoxy resin and activated carbon is also examined. Lignin modification using enzymes in combination with specific chemical mediators is summarized. The photochemical alteration and electrochemical oxidation of lignin for the generation of vanillin, vanillic acid and other value-added products are discussed. In addition, the effectiveness of combining photochemical and electrochemical approaches is evaluated.

Chapter 11 – The present chapter deals with pyrolysis characteristics of different kinds of lignins, focusing on an industrial raw lignin arising from the Kraft pulping process, a commercial alkali lignin, and Klason lignins lab-isolated from two lignocellulosic biomasses with different lignin contents (27 and 57 wt%), emerging from the processing of agro-industrial products. Characterization of the lignins includes determination of ash content, elemental composition, Fourier-transform infrared (FT-IR) spectra, and surface morphological features by scanning electronic microscopy (SEM). Pyrolysis characteristics of the different lignin samples as well as of the whole biomasses from which they are obtained, in the case of Klason lignins, are comparatively examined by non-isothermal thermogravimetric analysis from room temperature up to 1000 °C. In order to investigate possible effects of mineral matter inherently present in the industrial Kraft lignin on its pyrolytic behaviour, pyrolysis characteristics are also determined using samples prior subjected to demineralization by a mild acid treatment. The industrial Kraft lignin possesses the highest contents of ash (16 wt%) and elemental carbon (62.2 wt%) among all the investigated raw lignins. A similar pyrolytic behavior is found for the industrial raw Kraft lignin and the commercial alkali one. It is characterized by differentiated thermal degradation domains, as evidenced by three peaks in reaction rate profiles, successively attributable to moisture evolution, primary and secondary pyrolysis, with progressive increase in temperature. Mineral matter reduction of the industrial Kraft lignin induces some structural changes, as suggested by SEM images and FT-IR spectra, and noticeable modifications in its pyrolytic behavior, leading to shift primary pyrolysis to higher temperatures, to increase the maximum primary pyrolysis rate, and to inhibit secondary pyrolysis. Pyrolysis characteristics for binary mixtures composed of equal proportions of the commercial alkali lignin and polyethylene in powder form, as a representative major polymeric waste of massive post-consumed plastics, are also examined following some current research trends towards alternative energy generation based on the advantageously favourable environmental nature of bio-resources and the higher energy content of synthetic polymers. No interactions between the lignin and polyethylene are found, the pyrolytic behaviour of the mixtures arising from independent thermal degradation of the individual constituents. On the other hand, the two Klason lignins separated from sawdust of *Aspidosperma australe* wood and nutshells from *Bertholletia excelsa*, exhibit different ash contents and elemental compositions as well as noticeable differences in their pyrolysis characteristics depending on the biomass source and with respect to the pyrolytic behaviour of the untreated parent biomasses. Compared to the raw alkali lignins, the Klason lignins do not seem to undergo secondary pyrolysis and are more resistant to thermal degradation, likely due to more condensed chemical structures related to the method applied for isolation. Overall, the results highlight the marked influence of both the botanical origin of the bio-resource and extraction method used to obtain the lignins on their physicochemical characteristics and pyrolytic behaviour.

Chapter 12 – Lignin, one of the most abundant natural polymers, is expected to play in the near future an important role as raw material for the world's biobased economy for the production of bioproducts and biofuels. The pulping processes currently used in the paper industry produce a degraded lignin employed in low-added value utilizations and energy production. However, among the various pretreatment methods currently studied for the production of pulp and/or ethanol, the organosolv processes seem to be very promising. Organosolv processes use either low-boiling solvents (e.g., methanol, ethanol, acetone), which can be easily recovered by distillation, or high-boiling solvents (e.g., ethyleneglycol, ethanolamine), which can be used at a low pressure. These procedures not only produce a cellulose-rich pulp but also large amount of high-quality lignins which are relatively pure, primarily unaltered and less condensed than other pretreatment lignins. This review presents the progress of organosolv pretreatment of lignocellulosic biomass for the production of high-quality lignins. Impacts of the process conditions on the chemical structure of the recovered lignin fractions and on delignification mechanisms are exposed. Recent utilizations of organosolv lignins for the production of materials (e.g. biodegradable polymers and adhesives) are given.

Chapter 13 – Paper pulp bleaching, the most polluting and expensive step on pulp production, consists on removal and decolouration of residual lignin after wood cooking process. Pulp bleaching is performed during several steps using chemical reagents with chlorine species generally involved. Environmental restrictions forced pulp industry to implant ECF (elemental chlorine free) and TCF (totally chlorine free) bleaching processes. Enzymes involved in natural degradation of lignin have been proposed as possible biobleaching agents: peroxidases and laccases can modify the structure of lignin and improve the subsequent chemical steps. The introduction of an enzymatic stage has been studied in several bleaching sequences with different results depending on operational conditions, kind of pulp to be bleached, enzyme used, bleaching sequence employed, etc. A growing interest exists regarding the application of laccase-mediator systems: couples of the enzyme laccase and low molecular weight compounds able to increase the oxidation power of the enzyme and to diffuse through places where the enzyme cannot access due to its large size. Laccases are commercially available in high quantities and their bleaching capability was already demonstrated. The main limitations of laccase application for lignin removal on pulp are the cost of mediator, the effectiveness of some of the tested mediators, and the suitability of enzyme treatments on mill equipment. Lignin modification with these enzymes and their application for pulp biobleaching has been studied for several authors, but more knowledge is needed to get industrial implementation.

Chapter 14 – Lignin is the second most abundant molecule in nature. The evolution of the lignin pathway allowed plants to conquer dry lands, and as a consequence became essential for the continental success of life as a whole. It is a relevant component of the lignocellulosic biomass which includes, wood, straw, bagasse and other materials that can be used as a renewable resource. Lignin was the first source of bio-energy that men used as firewood and then the main compound as humus in the agricultural revolution. Its biosynthesis results from a free radical polymerization of phenolic compounds. It is composed with various proportions of three main phenylpropanoids: coniferyl, sinapyl and hydroxycoumaryl alcohols as well as other minor compounds. Once completely assembled within cell walls making it impermeable and confers physical and chemical resistance to plant cells and, ultimately, the entire plant. Lignin has a high caloric content and might be an important compound in new generation biofuels industry, both directly by pyrolysis using energy to produce energy play the mill and indirectly by its thermochemical conversion in bio-oil and other derivatives. The research on lignocellulosic biomass focuses mainly on the processes of conversion of ordinary biomass. On the other hand, breeding, genetic engineering and crop management of plants are also important research areas devoted to improve plant biomass for industrial applications. Both biomass processing and plant production require accurate determination of lignin content and characteristics. In addition, little is known about the changes resulting from genetic modification in glyphosate-resistant (GR) crops in terms of secondary metabolism of plants, in particular lignin biosynthesis. Regarding these effects, some pesticides currently used in weed management, including glyphosate, can directly affect the synthesis of secondary compounds. Even though GR crops are resistant to glyphosate, recent reports suggest that glyphosate, or its main metabolite aminomethylphosphonic acid (AMPA), can decrease lignin content in these crops. Therefore, further research should be conducted to evaluate the different rates of glyphosate in GR crops. Such research is important to evaluate glyphosate effects on GR crop physiology and nutrition, and associated lignin production. In this work, we review: (1) the general properties and chemical composition of lignin; (2) the new uses under study and proposed uses for industry; (3) the potential for processing lignin biomass to provide bioenergy and the contribution to a new generation of biofuels, and (4) the potential threat to its production using current weed management in cropping systems.

Chapter 15 – Lignin is the second most abundant natural aromatic polymer after cellulose in terrestrial ecosystems and represents nearly 30% of the organic carbon sequestered in the biosphere (Boudet et al. 2003). Lignins are interlinked with cellulose and hemicellulose conferring structural strength, rigidity and impermeability to the woody cell wall, while providing natural resistance against chemical or microbial attack and environmental stresses

(Foster et al. 2010). Additionally, lignin waterproofs the cell wall thus enabling transport of water and solutes through the vascular system.

Chapter 16 – Environmental degradation, such as drought and salinity stresses, is a major factor in limiting plant growth and productivity, will become more severe and widespread in the world. To overcome the environmental stress, genetic engineering in woody plants needs to be implemented. The adaptation of plants to environmental stress is controlled by cascades of molecular networks. For woody plant species, the effects of longer periods of stress need to be considered to regard the actual tolerance. This chapter focuses on the molecular mechanism of abiotic stress responses of woody plants. The basis of genetic engineering for enhanced biomass and stress tolerance in woody plants will also summarized in this chapter.

Chapter 17 – The accumulation of lignin is one of the important plant defense mechanisms against pathogens and wound. In this work the mechanisms of local lignification of pathogen infected zones are discussed. The importance of the structurally functional organization of some peroxidase isoforms, promoting the phenolic compound polymerization with participation of reactive oxygen species (ROS) on a pathogenic fungi mycelium is considered. The analysis of acetylation degree of pathogen polysaccharides in the induced plant defense and its role in the local lignification of infection zones was carried out. The post infectious accumulation analysis of lignin in wheat plants has shown that resistant plant tissues increased activity of lignin synthesis enzymes. The authors consider this effect is associated with highest activity of pathogen polysaccharide-specific apoplastic peroxidase. For example, an anionic peroxidase in wheat plants is characterized by property to contact with the chitin of pathogenic fungi. Analysis of the amino acid chain of some peroxidases of *Arabidopsis*, wheat, zucchini, corn and rice homologous zone allows allocating isoforms genes from a large amount of genes. In order to prove the polysaccharide-specificity of allocated peroxidases to the site of a gene presumably coding polysaccharide-specific domain of peroxidase have been picked up and designed primers which we have used for an estimation of gene expression of wheat anionic peroxidase gene under infection by fungus disease agents. Sequencing results, received from peroxidase DNA and cDNA of some plants have proved the accuracy of choosing primers. The anionic peroxidase gene expression or repression repeatedly increased or decreased under infection with fungus pathogens and the influence of signal molecules and elicitors. Thus, a comparative estimation of the lignin content, activity of lignification enzymes, ability to sorption on chitin of plant peroxidases, definition of their immunochemical affinity, a homology of molecular structure of polysaccharide-specific zone have allowed to assume, that local lignin accumulation under fungal pathogenesis shows the universality via of activation of polysaccharide-specific peroxidases.

Chapter 18 – Some batch sorption experiments were carried out to remove Methylene Blue cationic dye from aqueous systems using industrial lignin as a low cost sorbent. The solution's pH, amount of industrial lignin, contact time, initial dye concentration and temperature were the studied operating variables.

To establish the most suited type of sorption mechanism to describe the dye retaining onto the solid sorbent, the data were analyzed using the Langmuir, Freundlich and Dubinin-Radushkevich models for the sorption isotherms. The results of this experimental study indicate that the tested solid material has a moderate capacity for dye molecules uptake.

Chapter 19 – The effects of chemical modification of lignin on the moldability of this material were examined. Softwood kraft lignin was chemically modified with benzyl chloride under alkaline conditions at various mole ratios of benzyl chloride to wood hydroxyl groups (ratio = 1-3) at a reaction time of 8 h. The extent of benzylation was assessed by weight gain, and Fourier transform infrared (FTIR) and solid state ^{13}C nuclear magnetic resonance (NMR) spectroscopies. FTIR spectroscopy revealed the reduction of lignin hydroxyl group bands, an increase in aromatic bands and an increase in acryl and alkyl ether bands, which were consistent with etherification. The thermal and flexural properties of the benzylated lignin were assessed by differential scanning calorimetry (DSC), dynamic rheometry, and dynamic mechanical analysis. The results from DSC were consistent with data from rheometry. Results have also shown that the benzylated lignin thermal transition temperature and mechanical properties can be manipulated by the extent of benzylation.

Chapter 20 – Lignins are complex phenolic polymers occurring in higher plant tissues and are the second most abundant terrestrial polymer after cellulose. Due to their very complex structure, lignins are amorphous polymers with rather limited industrial use. One of the uses of lignin is production of biofuels. Fuels are generally those made from non-edible lignocellulosic biomass. These biofuels have some clear advantages. Plants can be bred for energy characteristics, and not for food, and a larger fraction of the plant can be converted to fuel. Lignocellulosic crops can be grown on poor quality land, requiring fewer fertilizers. There are substantial energy and environment benefits primarily due to greater biomass usability per unit of land area. Within the bioenergy sector, biotechnology, and in particular genetic engineering, has the potential to be applied to agricultural production - to optimize the productivity of biomass; to raise the ceiling of potential yield per hectare; to modify crops to enhance their conversion to fuels - and to the biomass conversion process, for example by developing more effective enzymes for the downstream processing of biofuels. It has become possible to process lignocellulose at high substrate levels and the enzyme performance has been improved. Also the cost of enzymes has been reduced. Genetic research into dedicated energy crops and manufacturing processes is still at an early stage.

Chapter 21 – In recent years, considerable effort has been devoted to the thermochemical transformation of agricultural, biomass and industrial residues in order to generate energy, chemicals or activated carbons. Extensive research has been focused on the production of activated carbons from waste materials of agricultural origin, mostly based on lignocellulosic materials. The control of the activated carbon porous texture requires controlling each stage in the activation process. In the case of physical activation, the efficiency of the prior carbonization of the raw material is a key step in reducing the quantity of disorganized carbon inside the char and consequently the extent of subsequent activation required. Similarly, in the case of chemical activation, economic and environmental constraints make it imperative to reduce the amount of activating agent, energy and water needed for subsequent washing as far as possible. This can be carried out by optimizing the corresponding heat treatments. Within the framework of the valorization of agricultural and food industry waste, several recent studies have investigated lignocellulosic precursors for the production of activated carbon. During the carbonization of lignocellulosic precursors, hemicellulose, cellulose and lignin (spatially distributed within the matter) decompose at different rates and within distinct temperature ranges. The differences in reactivity of these three components during carbonization, as well as competition between the reactions and thermal effects accompanying their decompositions make the study of carbonization complex. At this stage of the process, the porosity of the adsorbent is not fully developed. Pyrolysis products such as tars have to be released from the char by means of the subsequent activation step in order to fully open the potential porosity of the material. However, the activation procedure usually induces heterogeneity in the pore size distribution due to the competition between diffusion and chemical phenomena. Moreover, the thermal effect of each reaction induces enhancement or inhibition of other kinetics. Improving the carbonization procedure (in terms of heating rates, plateau,...) can therefore reduce the amount of activation agent needed, lead to narrowing the pore size distribution and reduce the cost of the whole development process. The objective of our study is to optimize the initial carbonization stage of the lignocellulosic precursors. The thermal decomposition of the three major lignocellulosic components (hemicellulose, cellulose and lignin) was first studied separately by thermogravimetry to predict their respective contributions in terms of weight fraction and carbon production in the obtained char. The decomposition kinetics of these lignocellulosic compounds were then studied within both a synthetic blend and natural materials, namely coconut shell and plum stones. A model composed of three independent chemical kinetics was validated for various particle sizes and heat treatment rates. The use of a particular technique is proposed to reduce the number of adjustable parameters needed in the model from 15 to 3.

In: Lignin
Editor: Ryan J. Paterson

ISBN 978-1-61122-907-3
© 2012 Nova Science Publishers, Inc.

Chapter 1

BIOTECHNOLOGICAL APPLICATIONS OF LIGNIN: "SUSTAINABLE ALTERNATIVE TO NON-RENEWABLE MATERIALS"

G. S. Dhillon,[1] F. Gassara,[1] S. K. Brar,[1][*] and M. Verma[2]

[1]INRS-ETE, Université du Québec, 490, Rue de la Couronne,
Québec, Canada
[2]Institut de recherche et de développement en agroenvironnement Inc. (IRDA),
2700 rue Einstein, Québec, Canada

ABSTRACT

Lignin is the second most abundant natural renewable resource and aromatic (phenolic) polymer on Earth. Lignocellulosic biomass is mainly comprised of cellulose, hemicellulose and lignin that are strongly intermeshed and chemically bonded by non-covalent forces and covalent cross-linkages. Lignin is an aromatic polymer synthesized from phenylpropanoid precursors. The complex structure of lignocellulose in plants forms a protective barrier to cell destruction by microorganisms. Various pre-treatment methods have been applied for the proper utilization of the valuable carbon sources from the lignocellulosic biomass. A smaller group of filamentous fungi, namely white-rot fungi have the ability to break down lignin, the most recalcitrant component of plant cell walls. Large amount of lignin is derived as waste product from wood processing in pulp and paper industry and is burned to generate energy and recover chemicals, which represents an important environmental problem.

Researchers around the globe have been trying hard to crack the key behind the complex structure of lignocellulosic biomass. Genetic engineering opens new avenues for producing tailor-made plant products with improved structural properties. Many new target genes of both plant and microbial origin are increasingly becoming available for designing plants suitable for downstream processing. Till date, not much research has been carried out to utilize lignin, a gigantic reservoir of carbon having promising applications. The lack of value-added applications from lignin is mainly caused by the heterogeneity, odor and color problems of lignin-based products. Only a small proportion

---

[*] Phone: 1 418 654 3116; Fax: 1 418 654 2600; E-mail: satinder.brar@ete.inrs.ca.

of lignin is currently used in biotechnological applications, such as an adhesive, tanning agent, adsorbent of acids, metal ions, dyes, bile acids, organic pollutants, cholesterol, surfactants, pesticides and phenols, bio(polymer) additives, as a precursor for chars and activated carbon. Lignin has also been used as an antioxidant, such as in cosmetic formulations, an application of particular interest in relation to human health. The immense reservoir of carbon should be increasingly exploited in the development of economically viable biotechnological products within the context of sustainable development.

This chapter deals with the brief introduction on lignocellulosic biomass structure, different delignification strategies, types, properties and various compounds which are produced from lignin degradation. This chapter also discusses the potential applications of lignin in industrial biotechnology in the context of sustainable alternatives to non-renewable products.

Keywords: Lignin; lignocellulose; delignification; non-renewable biomass; biotechnological applications.

INTRODUCTION

Lignin is the most abundant natural non-carbohydrate organic compound in fibrous materials and second most abundant biopolymer on earth. Lignins are complex phenolic polymers obtained from abundant and renewable resources, such as trees, plants and agricultural crops. Woody plants represent a complex polymeric composition in which lignin, cellulose and hemicellulose are the main components. They are present in huge amounts in large variety of foods, such as cereal brands. Structurally, lignin is a three-dimensional phenylpropanoid polymer mainly linked by ether bonds between monomeric phenylpropane units most of which are not readily hydrolyzable (McCrady, 1991). Lignin is a constituent of the cell wall of various cell types of plants, such as wood fibres, vessels, and tracheids. It constitutes 20–30% of the total weight of wood. In lignocellulosic biomass, lignin encrusts as an amorphic mass of the cellulose fibres, which gives the lignified cell wall high mechanical strength and enhance the resistance to microbial degradation.

The importance of lignin in plants should be considered from different aspects, i.e. its role in plant development, contribution to mechanical strength and protection from dreadful conditions (Walker, 1975). From the biomass utilization point of view, lignin has always been considered as an important barricade which hinders polysaccharide utilization (Van Soest, 1994). Industrial lignins are by-products from the pulp and paper industry, as well as from various other biomass-based industries. However, some properties of lignin, such as their non-toxic nature, potential of high value, extremely versatile performance, being inexpensive and available in large amounts make them increasingly important in many industrial applications due to which their uses have expanded into hundreds of applications (Boudet et al., 2003; Bajpai, 2004; Eddy and Yang, 2005).

About 50 million tons of lignin is being produced annually as by-product of paper production processes (Dam et al., Lignin applications, www.biomassandbioenergy.nl). The lignin represents the global potential of bioenergy. The lignin possesses highly reactive locations that can be modified through a selection of chemical, physical and enzymatic reactions, which gives enormous potential for their exploitation as industrial raw materials. Further, enhancement of the amount of functional groups, such as phenolic hydroxyl groups in lignin can be achieved by chemical reduction using sodium dithionite (Dence and Reeve, 1996).

STRUCTURE OF LIGNIN

Lignins are very complex natural polymers with many random couplings, due to which their exact chemical structure is not known. However, lignins are classified as three-dimensional amorphous polymers composed of methoxylated phenylpropane structures. The most important chemical functional groups in lignin include the hydroxyl, methoxyl, carbonyl and carboxyl groups in various numbers and proportion, depending on origin and extraction processes (Gosselink et al., 2004a, b). Owing to these biochemical properties which can be modified according to its use leads to its multifarious applications.

Lignin is very unusual as compared to other abundant natural polymers due to their low degree of order and high degree of heterogeneity in structure (Sederoff et al., 1999). The heterogeneity of lignin is caused by variations in the polymer composition, size, cross-linking and functional groups due to differences in raw material, pulping and isolation conditions (Gosselink et al., 2004b). In general, lignin is made up of three primary precursors, i.e. trans-coniferyl, trans-sinapyl and trans-*p*-coumaryl alcohols (Sarkanen and Ludwic, 1971). Differences exist in molecular composition and linkage type between the phenylpropane monomers (*p*-hydroxyphenyl, guaiacyl, and syringyl units) derived from coniferyl, sinapyl and coumaryl alcohol precursors, respectively. Differences in lignin composition are not only found between plants of different genetic origin, but also between different tissues of the same plant. Difference in structure of lignins exists mainly due to the presence of three alcohol units in different proportions as given below:

- In softwood lignins, the structural elements are mainly derived from approximately 80% coniferyl, 14% *p*-coumaryl and 6% sinapyl alcohols. In most of the softwood lignins, coiferyl alcohols make upto 95% of total elements.
- Hardwood lignins are made up of of 56% coniferyl alcohol, 40% sinapyl alcohol and 4% *p*-coumaryl alcohol. In hardwoods and dicotyl crops like flax and hemp various ratios of coniferyl/sinapyl have been observed.
- Lignins derived from grass and cereals straws are rich in *p*-coumaryl units (Jung and Fahey, 1983).

Lignification (lignin polymerization/formation) in plants results in an almost random series of bonding and a very complex structure due to lack of enzymatic control (Jung and Fahey, 1983). The susceptibility of lignin to chemical/enzymatic disruption is attributed to the existence of strong carbon-carbon (C-C) and ether (C-O-C) bonds in its structure (Harkin, 1973).

Sarkanen and Ludwic (1971), classified lignins into two main groups, namely guaiacyl and guaiacyl-syringyl lignin. Majority of lignin derived from the gymnosperm is typically guaiacyl lignins, although they contain small amounts (<1.5%) of syringyl units and a rather lower proportion of *p*-hydroxyphenyl units.

However, dicotyledon and grass lignins are true guaiacyl-syringyl lignins (Schwartz et al., 1989). The syringyl content of woody angiosperm lignins varies between 20 and 60%. In herbaceous angiosperms (dicotyledons), the range can be from 10 to 65%. According to Zadrazil (1984) and Khazaal et al. (1993b), a white-root fungus efficiently degrades lignin in the presence of oxygen. However, under anaerobic conditions, the lignin is poorly degraded (Kirk and Farrell, 1987).

TYPES OF LIGNIN

The physical and chemical properties of lignin differ depending on the extraction technology. Lignins are classified into different types according to extraction process.

- Lignosulfonates (also called lignin sulfonates and sulphite lignins) are products of sulfite pulping. Lignosulfonates are very useful owing to their dispersing, binding, complexing and emulsifying properties. In 1880s, lignosulfonates were first industrially used in leather tanning and dye baths and since then, the applications have been extended to food products, serving as emulsifiers in animal feed and as a raw material in the production of vanillin.
- Kraft lignins (also called sulfate lignins) are obtained from the kraft pulping process.

The commercial lignin is divided into two major categories: 1) conventional or sulphur containing lignins, such as kraft lignin and lignosulfonates and; 2) non-sulphur containing lignins obtained from different processes, most of which are not yet commercially implemented on industrial scale, such as soda lignins, organosolv lignins, steam explosion lignins, hydrolysis lignins (mostly from biofuel plants), and oxygen delignification lignins (Mansouri et al., 2006).

Lignosulfonates are available in large quantities around 1 million tonnes of solids per year and kraft lignins are found in moderate quantities around 100,000 tonnes of solids per year (Gosselink et al. 2004).

Industrially used conventional lignins are mainly obtained from soft woods. In non-sulphur lignins, only the soda lignin has some potential for industrial availability whereas hydrolyzed lignin obtained from bioethanol production plants provide important new opportunities for industrial level production replacing transportation fuels of non-renewable nature fossil fuels. They can be directly obtained from woods and non-woods.

In spite of their abundance, only a minimal part of the available lignins is commercialized. To increase industrial use of lignins, it is necessary not only to improve lignin extraction processes, but also to determine the chemical composition and structure of lignins; to characterize their chemical reactivity and functional properties; and to develop new applications for them (Boeriu et al., 2004). Further, there has been an increasing trend in the potential health application of lignins.

DIFFERENT SOURCES OF LIGNIN

Lignin is a by-product of pulping and bleaching processes and it represents an important environmental problem. Most of the lignin extracted from lignocellulosic materials for cellulose and paper production in modern pulp mills is burned to generate energy and recover chemicals. The lack of other value-added applications is mainly caused by the heterogeneity, odour and colour problems of lignin-based products. The common technologies in pulp industries to extract lignin have been focused on the optimum cellulose yield and maximum degradation of lignin (Bonini et al., 2008).

APPLICATIONS OF LIGNIN

In recent years, a great deal of research was devoted to the development of lignin-containing polymeric materials and other value-added products. Lignins are amorphous polymers with rather limited industrial use due to their very complex structure. They are usually seen as waste products of pulp and paper industries. Large quantity of lignin is produced as side product from the biofuel plants during processing of lignocellulosic biomass to bio-ethanol.

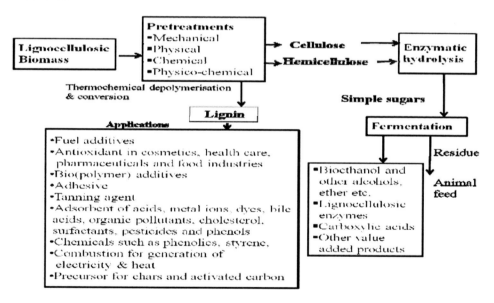

Figure 1. Flowchart showing the utilization of lignocellulosic substrates and various uses of lignin.

They are often used as fuel for the energy balance of the process. There is a very small market for the large quantities of industrial lignins produced by paper and pulp industry. Currently, only 2% of the lignin produced was used in valuable industrial processes, rest is being burned to generate energy and to recover chemicals. The existing markets consist of very low value products (Gossilink et al., 2004 b). The lignin comprises of different functional groups, depending upon the genetic origin of lignin: a) ethers of various types, primary and secondary hydroxyl groups, carboxyl and ester functions, among others and; b) applied pulping and sequences and extraction processes, such as the materials derived from the sulphite and kraft processes, lignosulphonates and kraft lignins contain sulphur-containing groups, such as thiols and sulphuric acids. Efforts have been mostly dedicated to develop new techniques for lignin modification which will soon give rise to technically more useful materials having broad range of applications. Various applications of lignin are presented in Figure 1.

ACTIVATED CARBON (ACs)

ACs have a very porous structure with a large internal surface area ranging from 500 to 2000 $m^2 g^{-1}$ and owing to this property, they possess good adsorption capacities towards various substances. They have found application in removal of a wide variety of pollutants including organic and inorganic substances from the liquid or gaseous phases (McKay et al., 1985; Lalezary et al., 1986; Urano et al., 1991; Pirbazari et al., 1992; Carrott et al., 1991; LeCloirec et al., 1997; Hu et al., 1998; Walker and Weatherley, 1999; Gabaldo'n et al., 2000; 2005; 2006a; 2006b; Malhas et al., 2002;). Among the applications of these materials, the most important is their adsorption of dissolved species in the wastewater such as dyes, chlorinated or nitrated compounds, phenols, and aromatic surfactants, among other substances. Activated carbon exists in two types: powdered activated carbon (PAC) and, in particular, granular activated carbon (GAC). PAC is a material that is constituted by small particles (diameter less than 0.2 mm). This structure gives PAC a large external surface area and small diffusional resistance, leading to high rate of adsorption. Compared to PAC, GAC is comprised of larger carbon particles, up to about 5 mm in diameter and smaller external surface areas. This structure allow GAC to adsorb gases and vapors and facilitate their use in fixed bed filtration systems due to the fact that the granular form is more adaptable to continuous contact. Moreover, in contrast to PAC, its regeneration is often impossible due to the difficulty of separating powdered adsorbent from the fluid, and the GAC can be easily regenerated.

Activated carbon (PAC and GAC) is produced using different precursors such as bagasse (Ahmedna et al., 2000); scrap tires and saw dust (Hamadi et la., 2001); almond, pecan, English walnut, black walnut and macadamia nut (Toles et al., 1998); pistachio (Wartelle et al., 2001); hazelnut shells (Kobya 2004); rice husk (Yalem and Sevine, 2000) and rice bran (Akhtar et al., 2005).

The selection of precursors essentially determines the range of adsorptive, physical, and chemical properties that can be attained in the activated carbon products. Depending on the requirement of adsorbents, one can design an activated carbon by choosing appropriate precursors and optimizing the carbonization and the activation process conditions. The designed carbon can be tailored to the surface area, pore size distribution, and surface chemistry needs of application.

Cheap raw materials have been recommended for the preparation of activated carbons. Trials directed towards the preparation of activated carbons from agricultural by-products have been widely carried out over recent decades as these materials often find no other economic use even in undeveloped countries where they have been used for many years as fuels in rural areas; and because of the absence of an economic use, these materials often represent a solid pollutant as far as the environment is concerned.

Nevertheless, the reuse of waste to produce activated carbon is a good option to protect the environment and produce value-added products. Lignin is the second most abundant natural waste after cellulose which is generally used for its fuel value (Hayashi et al., 2000). Therefore, lignin was used as precursor to produce different higher value products, such as activated carbon in order to apply it in wastewater treatment and in agricultural application as fertilizer. The use of lignin as precursor to produce activated carbon was justified by many published studies that showed that lignin can be a good adsorbent over a wide concentration range. However, the uptake of adsorbates by lignin depends on the nature of adsorbate and the adsorption mechanism is often unknown: it might occur by simple adsorption or by a combination of processes (Lalvani et al., 2000; Crist et al., 2004; Carrot and Carrot, 2007) as the surface area of natural lignin is inferior to 10 m^2/g (Pérez et al., 2006). It was shown that the uptake varied linearly with the solute/lignin ratio and that the adsorption was irreversible with the hydroxides and partially reversible with the acids (Wedekind and Garee, 1928a; 1928b). Many authors have showed that lignin is able to adsorb different adsorbates, such as dyes, bile acids, cholesterol, surfactants, pesticides and phenols (Dizhbite et al., 1999; Ludvık and Zuman, 2000; Cu (II) (Merdy et al., 2002; Acemioglu et al., 2003; Sciban and Klasnja. 2004; Allen et al., 2005; Mohan et al., 2006; Van Beinum et al., 2006), Pb(II) (Srivastava et al., 1994; Lalvani et al., 1997), Zn(II), Sr(II) (Crist et al., 2002; 2003). The sorption capacity of lignin on different adsorbates was highly variable in the literature. These authors found that the sorption capacity depended on the kind of adsorbates and increased with increasing pH and with temperature and suggested that the high adsorption capacity was due to polyhydric phenols and other functional groups on the lignin surface.

It was shown that the binding strength of lignin to metallic ions was found to be in the order Pb(II) > Cu(II) > Zn(II) > Cd(II) > Ca(II) > Sr(II) (Crist et al., 2002, 2003). It was also shown that the adsorption capacities of non-activated lignin were lower than activated carbon for aqueous phase metal ion removal (Mussato et al., 2010). However, the adsorption capacity can be improved by means of an activation procedure (Suhas et al., 2007).

PRODUCTION OF AC FROM LIGNIN

Lignin can be used as precursor to produce activated carbon as it is rich in carbon and it is able to sorb many molecules (Srivastava et al., 1994; Dizhbite et al., 1999; Lalvani et al., 2000; Ludvı´k and Zuman, 2000; Demirbas, 2004; Basso et al., 2004; Allen et al., 2005). The production of activated carbon from lignin can be carried out by activation. The objective of activation is to increase the diameter of pores created during carbonization of lignin by two methods: physical activation and chemical activation.

Physical Activation

The lignin can be transformed to activated carbon by physical activation. The process is generally carried out in a two-step scheme, and lignin is developed into AC using gazes. The first step of physical activation is the carbonization, which involves the formation of a nonporous char by pyrolysis of lignin at temperatures in the range 600–900 °C in an inert, usually nitrogen, atmosphere. The second step of physical activation is activation during which the char comes in contact with an oxidizing gas, such as CO_2 or steam, in the temperature range 600–1200 °C. During activation, the more disorganized carbon is removed and a micropore structure is formed.

Many research works have reported production of activated carbon from lignin by physical activation. Rodrıguez-Mirasol et al. (1993) produced activated form of Eucalyptus kraft lignin by the activation of the material at 850 °C for 20 h in CO_2 atmosphere. The activation produces ACs having an appreciable surface area of 1853 m^2 g^{-1} with a micropore volume of 0.57 cm^3 g^{-1}. The appreciable surface area was obtained with longer time of activation. This was due to the low reactivity of lignin. The activated carbon was prepared from eucalyptus kraft lignin in CO_2 atmosphere at 1073 K activation temperature (Luis et al., 2007). The surface properties of the ACs play a significant role when these materials are used for adsorption from liquid phases. The physico-chemical properties and the surface chemical structure of the ACs were studied by means of N_2 adsorption experiments, elemental analysis, X-ray photoelectron spectroscopy (XPS), and X-ray photoelectron spectroscopy (TPD). The XPS and TPD spectra of ACs suggested the presence of aromatic rings and oxygenated functional groups in the surface material. Adsorption isotherms data were fitted to the Freundlich and Fritz-Schlu¨nder equations. The values of ΔH, ΔG, and ΔS were calculated, and these indicated that the process was exothermic in nature in all the examined cases.

The effects of temperature and steam treatment on the structural formation of carbons from hydrolytic lignin (Baklanova et al. (2003) and cotton lignin (Perezdrienko et al., 2001) have also been studied. The results showed that the increase in temperature of carbonization hydrolytic lignin caused slight changes in the BET surface area and the average micropore that decreased with the carbonization temperature increases. It was also shown that carbonization temperatures of 600–700 °C could be used for producing microporous char materials with a minimum average size of micropores using argon (Baklanova et al., 2003). Moreover, it was found that steam activation of the carbonized samples produced ACs having surface area as high as 865 m^2 g^{-1}, with a corresponding micropore volume of 0.37 cm^3 g^{-1} (Baklanova et al., 2003). The variation of carbonization temperature of cotton lignin from 400 to 800 °C makes the concentration of the acidic surface centers lower. An additional formation of carboxyl groups resulted in 40% combustion loss as shown with the activation of carbonized carbons while using superheated steam at 800 °C (Perezdrienko et al., 2001). Further, it was suggested that lignin carbons containing only acidic functional groups, or those containing both acidic and basic centers could be simultaneously prepared by varying the combustion loss in activation of lignin carbon (Perezdrienko et al., 2001).

The activated carbon was also prepared from thermo-compressed and virgin fir wood by physical activation with CO_2. The physical activation caused the structural alteration which enhanced the enlargement of micropores leading to their degradation and the formation of mesopores (Khesami et al., 2007).

Moreover, activated carbons have also been prepared by physical activation of commercial by-product kraft lignin (as received, after de-ashing and after impregnation with NaCl) and natural cork in CO_2 atmosphere (Carott et al., 2008). The results obtained showed that the presence of natural inorganic impurities increased the reactivity of lignin significantly and also decreased the values for the micropore volume and widened the micropore size. Pure de-ashed lignin is known to be non-reactive but allows microporous activated carbons to be obtained which have very narrow micropore width of ~0.5-0.6nm. When the de-ashed lignin is impregnated with NaCl, similar micropore volumes and width can be obtained but in a considerably shorter time (~30min instead of ~8h) which would result in a considerable energy saving and is therefore a promising procedure for the production of microporous activated carbons from low-cost kraft lignin.

Chemical Activation

Activated carbon from lignin was also produced by chemical activation (Gonzalez-Serrano et al., 1997; Hayashi et al., 2000; Zou and Han., 2001; Fierro et al., 2003b; Fierro et al., 2006; Gonzalez-Serrano et al., 2004; Fierro et al., 2007; Khesami et al., 2007). Chemical activation consisted of impregnation of lignin with chemicals, such as H_3PO_4, KOH or NaOH followed by heating under a nitrogen flow at temperatures in the range 450– 900°C, depending on the chemical used. During chemical activation, carbonization and activation proceed simultaneously. The materials produced by this process often possess higher micropore volumes, wider micropore sizes and they are generally used for the adsorption in liquid phase.

A commercially available Kraft lignin was chemically activated with two alkaline hydroxides, viz NaOH and KOH, using different preparation conditions (Fierro et al., 2007). The activation was made at various temperatures, hydroxide to lignin mass ratios, activation times, flow rates of inert gas, and heating rates. The resultant activated carbons were characterized in terms of BET surface area, total, micro and meso-pore volumes, average pore width, carbon yield and packing density. The influence of each parameter of the synthesis on the properties of the activated carbon was discussed, and the action of each hydroxide was methodically compared. The results showed that irrespective of the preparation conditions, KOH leads to the most microporous materials having surface areas and micropore volumes typically 1.5 and 1.2 times higher than those obtained with NaOH, which is in agreement with some early works. However, the surface areas and the micropore volumes obtained were much higher than previous studies, ranging up to 3100 m^2/g and 1.5 cm^3/g, respectively, using KOH. The study of preparation parameter influences the properties of the final materials showing insight into the activation mechanisms. The results of lignin chemically activated with hydroxides were compared with those obtained with anthracites in order to look for similarities and differences.

The activated carbon was also prepared from thermo-compressed and virgin fir wood by physical activation with KOH. The chemical activation conferred to activated carbon a heterogeneous and excessively microporous nature. Moreover, the coupling of thermo compression chemical activation resulted in a satisfactory yield (23%), a high surface area (> 1700 m^2/g) and a good adsorption capacity for two different pollutants in aqueous solution (methylene blue and phenol). Activated carbon prepared from thermocompressed wood exhibited a higher adsorption capacity for the two pollutants than commercial activated carbon (Khesami et al., 2007).

Activated carbon was also produced by fast pre-carbonization of cornstalk lignin in a fluidized bed followed by K_2CO_3 activation (Yong et al., 2010). The results showed that the product was essentially microporous carbon whose Brunauer-Emmett-Teller surface area and pore volume when the carbon was activated at 800°C were 1410 m²/g and 0.77 mL/g, respectively. The potential usefulness of the resulting carbons for removal of phenol from water and their subsequent bio-regeneration capabilities were also investigated. The kinetics study showed that all the carbons exhibited a fast adsorption rate and the carbon activated at 800°C had the largest amount of phenol adsorbed due to its greater specific surface area and pore volume. The adsorption isotherms by applying the Langmuir method showed that the monolayer adsorption capacity of carbon activated at 800°C could reach 110.9 mg/g.

The activated carbon was prepared from fast pre-carbonization of reed lignin in the fluidized-bed followed by K_2CO_3 activation (Yong et al., 2006). The impregnation ratio was ζ = 1:1 in this research. The effect of activation temperature upon the BET specific surface area and specific pore volume of the carbon was also studied. It was found that when activation temperature was increased (below 800°C), the pores were opened and widened. Above 800 °C, the excess widening of pores led to the decrease of BET surface area and micropore volume. The BET specific surface area and specific pore volume of the carbon activated at 800 °C were s = 1217 m² g⁻¹ and v = 0.65 ml g⁻¹, respectively. The potential application of carbon activated at 800 °C for removal of pollutants was also studied. The Langmuir monolayer adsorption capacity q_m of this activated carbon of adsorbing phenol, nitrobenzene, and Cr (VI) can reach 136.2, 393.7 and 52.6 mg g⁻¹, respectively. The experimental results showed good adsorption capacity.

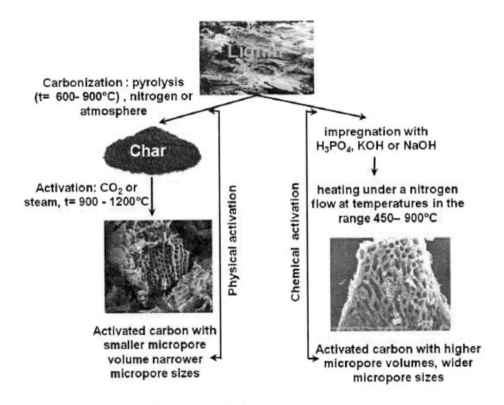

Figure 2. Production of activated carbon from lignin.

Different types of activated carbon were prepared by chemical activation of brewer's spent grain (BSG) lignin using H_3PO_4 at various acid/lignin ratios (1, 2, or 3 g/g) and carbonization temperatures (300, 450, or 600 °C), according to a 22 full-factorial design (Mussatto et al., 2010). The activated carbon formed was characterized with regard to their surface area, pore volume, and pore size distribution. The resulting materials were then used for detoxification of BSG, hemicellulosic hydrolysate (a mixture of sugars, phenolic compounds, metallic ions, among other compounds). BSG carbons presented BET surface areas between 33 and 692 m^2/g, and micro- and mesopores with volumes between 0.058 and 0.453 cm^3/g. The activated carbon already formed showed high capacity for adsorption of metallic ions, mainly nickel, iron, chromium, and silicon. These sorbents were also able to reduce the concentration of phenolic compounds and color. Thus, activated carbons produced from BSG lignin had characteristics similar to those commercially found and demonstrated high adsorption capacity.

COMPOSITES PRODUCTION

Another principal application of lignin is its use as precursor to produce biocomposites. Different composites have been reported in literature using lignin as precursors (Thielemans et al., 2001; Kadla et al., 2002; Chen et al., 2009; El-Wakil, 2009;).

Polyesters and vinyl esters were investigated along with lignin as precursor by Thielemans et al. (2001). In this study, the solubility of different lignins (pine kraft, hardwood, ethoxylated, and maleinated) was determined in different resin systems (acrylated epoxidized soybean oil, hydroxylated soybean oil, soy oil monoglyceride, and a commercial vinyl ester) to know the compatibility of lignin with the resin systems that were used. Further, the use of lignin as a filler was also studied. An increase in the glass-transition temperature was noticed, and the modulus at 20°C decreased due to the plasticizing effect of lignin. The modification of lignin was carried out to improve its effect on the matrix properties by adding double bond functionality, thus making it possible to incorporate the lignin molecule in the resin through free-radical polymerization (Thielemans et al., 2001). Modified lignin was introduced in several resins by reaction with maleic anhydride and epoxidized soybean oil and was tested for its effect on the solubility, glass-transition temperature, and modulus. The solubility of lignin in styrene-containing resins as well as the chemical incorporation of lignin in the resin was improved by the modification of lignin. The treatment of surfaces of natural hemp fibers by lignin was also investigated to utilize lignin's natural affinity for cellulosic fibers. The treatment was carried out to cure the surface defects on the natural fibers and increase the bonding strength between the resin and fiber. The improvement shown during the treatment depended on the amount of lignin covering the fibers.

Others polymers, such as carbon fibers have been produced from a commercially available kraft lignin, without any chemical modification, by thermal spinning followed by carbonization. To produce carbon fibers, kraft lignin was pre-treated thermally under vacuum to produce a fusible lignin with excellent spinnability to form a fine filament. Consequently, lignin with poly(ethylene oxide) (PEO) was blended to facilitate fiber spinning, but at PEO levels greater than 5%, the blends could not be stabilized without the individual fibers fusing together. Carbon fibers produced had an overall yield of 45%. The tensile strength and modulus increased with decreasing fiber diameter, and are comparable to those of much smaller diameter carbon fibers produced from phenolated exploded lignins. In view of the mechanical properties of tensile strength of 400–550 MPa and modulus 30–60 GPa shown by kraft lignin (Kadla et al., 2002). kraft lignin should be further investigated as a precursor for general grade carbon fibers.

Furthermore, biodegradable composites were produced using lignin as precursor in combination with others materials (wheat gluten) (El-Wakil, 2009). These bio-composites were prepared, characterized and the addition of silica to the aforementioned composites was studied in order to improve the thermal and mechanical properties of these biocomposites. Moreover, the effect of wheat gluten percent and the extent of its modification on the blends properties were investigated via diametric tensile strength, thermo-mechanical analysis (TMA), scanning electron microscope (SEM), thickness swelling and thermogravimetric analysis (TGA) (El-Wakil, 2009). According to the results of this study, a significant improvement in the diametric tensile strength, thickness swelling, uniformity in the fracture surface, and the shift of glass transition temperature (Tg) towards higher values were shown with the increase in wheat gluten percent and its modification extent. These results reflect the enhancement of interaction between alkali lignin and wheat gluten. Alkali lignin/wheat gluten blends filled with silica possessed distinguishable characteristics and improved diametric tensile strength, low thermal expansion, and high Tg. Interestingly, TMA results showed that high dimensional stability against heating (thermal expansion percent) could be obtained using 60% wheat gluten modified with 15% sodium silicate and filled with 10% silica. This sample showed the highest Tg and the lowest thickness swelling in addition to smooth, uniform, and glossy surface as seen from the SEM images and TMA charts.

In addition to the composite lignin/wheat gluten, biodegradable composite films based on chitosan and lignin with varied composition were produced by Chen et al. (2010) *via* the solution-casting technique. This study provided a simple and cheap way to prepare fully biodegradable chitosan/lignin composites, which could be used as packaging films or wound dressings. The composites formed were analyzed by FT-IR that showed the existence of hydrogen bonding between chitosan and lignin. Moreover, SEM images of these composites showed that lignin could be well dispersed in chitosan when the lignin content is below 20 wt% due to the strong interfacial interaction (Chen et al., 2010). An improvement of tensile strength, storage modulus, thermal degradation temperature and glass transition temperature of chitosan was shown by adding lignin. The improvement was due to strong interaction between lignin and chitosan further aided by good dispersion. All these studies showed that lignin can be used as precursor to produce different composites useful in many fields, such us packaging. However, the properties of these composites produced from lignin should be studied in more details to improve them further and make them more useful for applications.

FUELS PRODUCTION

Ligninocellulosic wastes are used to produce biofuels particularly ethanol and buthanol (Akin-osanaiye et al., 2005; Mtui and Nakamura, 2005; Sjöde et al., 2007; Okuda et al., 2007; Li et al., 2007; Cara et al., 2008; Sørensen et al., 2008).The basic steps of production include pre-treatment, saccharification, fermentation, and distillation. Pre-treatment is designed to help separate cellulose, hemicellulose and lignin so that the complex carbohydrate molecules constituting the cellulose and hemicellulose can be broken down by enzyme catalyzed hydrolysis (water addition) into their constituent simple sugars. Cellulose is a crystalline lattice of long chains of glucose (6-carbon) sugar molecules. Its crystalline nature makes it difficult to unbundle into simple sugars, but once unbundled, the sugar molecules are easily fermented to ethanol using well-known microorganisms, and some microorganisms for fermentation to butanol are also known. Hemicellulose consists of polymers of 5-carbon sugars and is relatively easily broken down into its constituent sugars, such as xylose and pentose. However, fermentation of 5-carbon sugars is more challenging than that of 6-carbon sugars. Some relatively recently developed microorganisms are able to ferment 5-carbon sugars to ethanol (Aden et al., 2002; Jeffries, 2006). Lignin consists of phenols, which for practical purposes are not fermentable. However, lignin can be recovered and utilized as a fuel to provide process heat and electricity at an alcohol production facility. This lignin separated from ligninocellulosic waste can be converted into bio-oil with a low O/C ratio by pyrolysis in the presence of formic acid and an alcohol (Gellerstedt et al., 2008). A one-step pyrolysis of lignins at about 380°C in the presence of formic acid and an alcohol, such as ethanol degrade this natural waste completely to a liquid bio-oil (Kleinert et al., 2008). Fourier transform infrared spectroscopy (FTIR), nuclear magnetic resonance (NMR), electrospray ionization mass spectrometry (ESI-MS), and size-exclusion chromatography (SEC) demonstrated that lignin is completely degraded by pyrolysis (Gellerstedt et al., 2008). The bio-oil formed has a low-molecular-mass distribution with a preponderance of aliphatic hydrocarbon structures. The results of this study showed that phenolic compounds present in this oil also contain carboxyl groups. Hence, lignin can be used as precursor to produce a bio-oil by pyrolysis. However, the bio-oil is formed contain aliphatic hydrocarbon and aromatic compounds. Thus, the optimization of pyrolysis conditions and more analyses is required to obtain a good bio-oil with fewer amounts of aromatic compounds.

LIGNINOLYTIC ENZYMES PRODUCTION

Many agro-industrial wastes containing lignin, such us bagasse, banana waste, wood, corncob, corn, wheat bran and wheat straw were used to produce ligninolytic enzymes (Pradeep et al., 2002; Reddy et al., 2002; Tychanowicz et al., 2006; D'Souza et al., 2006; Levin et al, 2008). These enzymes were used in many biotechnological application, such us bioremediation (Wesenberg et al. 2003; Akhtar et al. 2005a, b; Mohan et al. 2005; Husain, 2006, Husain and Husain, 2008), juice clarification (Giovanelli et Ravasini, 1993; Maier et al., 1994), and dye removal (Vaithanomsat et al., 2010). These enzymes were produced using different types of reactor that are presented in Table 1.

Lignin is a very complex molecule, so that its degradation is very difficult. Lignin resists attack by most microorganisms; even anaerobic processes cannot attack the aromatic rings at all, and aerobic breakdown of lignin is slow. In nature, only basidiomycetes white-rot fungi are able to degrade lignin efficiently by solublization and mineralization (Kirk et al, 1975 b, Kirk et al., 1976). Lignin biodegradation by white-rot fungi is an oxidative process and phenol oxidases are the key enzymes (Kuhad et al., 1997; Leonowicsz et al., 1999; Rabinovich et al., 2004). The enzymes responsible for the degradation of lignin are lignin peroxidases (EC1.11.1.14) (LiP), manganese peroxidases (EC 1.11.1.13) (MnP), versatile peroxidase(VP; EC 1.11.1.16), and laccases (EC 1.10.3.2).

Table 1. Use of lignin rich wastes to produce ligninolytic enzymes

Support	Microorganism	Type of reactor	Enzyme	Production
Bagasse	*Pleurotus ostreatus, Phanerochaete chrysosporium*		Laccase, MnP, LiP	772 U/g of laccase 656 U/g of Lip 982 U/g of MnP
Banana waste	*P. ostreatus, Pleurotus sajor-caju*	1000-ml conical flasks	Laccase, LiP	1.7106 U/mg of laccase 0.1632 U/mg of LiP
Corn	*Pleurotus pulmonarius*	250-ml Erlenmeyer flasks	Laccase	180 U/ml of laccase
Olive mill wastewater	P. tigrinus	stirred-tank reactor the air-lift reactor	Laccase, MnP	$4600+-/98 \text{ U l}^{-1}$ Of laccase $410+-/22$ of MnP
Wheat bran	P. pulmonarius	250 ml Erlenmeyer flasks	Laccase	8,600 U/g substrate
Wheat bran	Fomes sclerodermeus		Laccase, MnP	6.3 U g^{-1} of MnP 270 U g^{-1} of laccase
Wheat straw	P. chrysosporium	250-ml Erlenmeyer flasks	LiP, MnP	2600 U/L LiP 1375 U/L MnP
Wheat straw	P. ostreatus	100-mL conical flasks	Laccase	69,4 U/L de laccase
Wheat straw	P. pulmonarius		Laccase	8,600 U/g substrate
Wood	Bjerkandera sp. strain BOS55	Erlenmeyer flasks	LiP, MnP	$660 \text{nmol ml}^{-1} \text{ min.}^{-1}$ of LiP $1320 \text{nmol ml}^{-1} \text{ min.}^{-1}$ of MnP
Wood	Ceriporiopsis subvermispora	Erlenmeyer flasks	MnP, laccase, Beta-glucosidase, xylinase	9600 U/l of xylinase 126 U/L of laccase 50 U/L of perxidase 260 U/L of Beta-glucosidase
Poplar wood	*Trametes trogii*	100-ml Erlenmeyer flasks	Endoxylanase Laccase MnP	780 Ug−1 of endoxylanase 901 Ug−1 of laccase 20Ug−1 of MnP
barley bran	*Trametes versicolor*	Immersion reactor	Laccase	600 U/L of laccase
barley bran	*Trametes versicolor*	Expanded-bed reactor	Laccase	600 U/L of laccase
barley bran	*Trametes versicolor*	Tray reactor	Laccase	3500 U/L of laccase
cornob	P. chrysosporium	Packed-bed	MnP, LiP, laccase	(200 U/l of MnP 300 U/l of LiP 200U/L of laccase

These enzymes are produced by white-rot fungi especially *Botrytis cinerea, P. chrysosporium, Stropharia coronilla, P. ostreatus* and *Trametes spp.* (Howard et al., 2003; Martinez et al., 2004). The process of lignin degradation is enhanced by the cooperative action of several accessory enzymes, which may include glyoxal oxidase (EC 1.2.3.5), aryl alcohol oxidase (veratryl alcohol oxidase; EC 1.1.3.7), pyranose 2-oxidase (glucose 1-oxidase; EC 1.1.3.4), cellobiose/quinone oxidoreductase (EC 1.1.5.1), cellobiose dehydrogenase (EC 1.1.99.18),bacterial and fungal feruloyl and *p*-coumaroyl esterases [Mester et al., 2004, Baldrian, 2005; Martinez et al., 2005; Kersten and Cullen, 2007]. Bacterial and fungal feruloyl and p-coumaroyl esterases are relatively novel enzymes capable of releasing feruloyl and p-coumaroyl compounds and play an important role in biodegradation of recalcitrant cell walls in grasses (Kuhad et al. 1997).

LiP and MnP oxidizes the substrate by two consecutive one-electron oxidation steps with intermediate cation radical formation. LiP and MnP were discovered in the mid-1980s in *P.chrysosporium* and described as true lignases because of their high redox potential value (Gold et al., 2000; Martínez, 2002). LiP degrades non-phenolic lignin units (up to 90% of the polymer), whereas MnP generates Mn^{3+}, which acts as a diffusible oxidizer in phenolic or non-phenolic lignin units via lipid peroxidation reactions (Jensen et al., 1996; Cullen and Kersten, 2004). Laccases are blue copper oxidases that catalyze the one-electron oxidation of phenolics and other electron-rich substrates (Hammel, 1997). Laccase, MnP and LiP produced by white-rot fungi oxidize the lignin polymer and generate aromatic radicals. Lignin degradation is carried out by different non-enzymatic reactions, including C-4-ether breakdown (b), aromatic ring cleavage, Cα–Cβ breakdown, and demethoxylation. The aromatic aldehydes released from Cα–Cβ breakdown of lignin from the substrates for H_2O_2 generation by AAO in cyclic redox reactions involving AAD. Phenoxy radicals from C4-ether breakdown can repolymerize the lignin polymer if they are not first reduced by oxidases to phenolic compounds. The phenolic compounds formed can be again oxidized by laccases or peroxidases. Phenoxy radicals can undergo Cα–Cβ breakdown giving rise to p-quinones. Quinones contribute to oxygen activation in redox cycling reactions involving oxygen activation in redox cycling reactions with QR, laccases, and peroxidases. This results in reduction of the ferric iron present in wood, either by superoxide cation radical or directly by the semiquinone radicals, and its re-oxidation with concomitant reduction of H_2O_2 to a hydroxyl free radical. The latter is a very mobile and strong oxidizer that can initiate the attack on lignin in the initial stages of wood decay, when the small size of pores in the still-intact cell wall prevents the penetration of ligninolytic enzymes. Subsequently, lignin degradation proceeds by oxidative attack of the enzymes described above. In the final steps, simple products from lignin degradation penetrate in the fungal hyphae and are incorporated into intracellular catabolic routes (Martínez et al., 2005). Fungal feruloyl and p-coumaroyl esterases are capable of releasing feruloyl and p-coumaroyl units and play an important role in biodegradation of recalcitrant cell walls in grasses (Kuhad et al., 1997). These enzymes act synergistically with xylanases to disrupt the hemicellulose-lignin association, without mineralization of lignin *per se* (Borneman et al., 1990; Fillingham et al., 1999). Therefore, hemicelluloses degradation is required before the efficient lignin removal can actually start. In *P. chrysosporium*, a co-metabolizable carbon source is essential for lignin degradation (Kirk et al., 1976), and it is produced in response to nitrogen starvation (Keyser et al., 1978). This indicates that the ligninolytic system is formed as part of secondary metabolism in this organism. Carbohydrate starvation likewise leads to a rapid but transient onset of ligninolytic activity (Jeffries, 1987, 1994). High oxygen levels ameliorate lignin biodegradation through the production of hydrogen peroxide as the extracellular oxidant and the subsequent induction of ligninolytic activity (Kirk and Farell, 1987; Faison and Kirt, 1983; Kirk and Cullen, 1998). Hydrogen peroxide is finally derived from the co-metabolism of cellulose and/or hemicellulose (Jeffries et al., 1987).

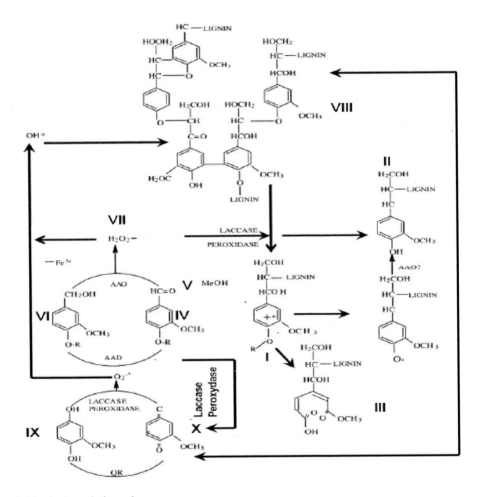

Figure3. Lignin degradation scheme.

POTENTIAL HEALTH BENEFITS OF LIGNIN

In recent years, interest in the physiological role of bioactive compounds present in plants has increased dramatically particularly in relation to human health. Lignin is among one of the major components of dietary fibre. Lignin is a complex hydrophobic molecule that typically occurs in cell walls with heteroxylans. Various potential health benefits have been credited for different lignin properties, such as their high activity in binding sodium salts of cholic acid, antitumoral, antiviral and immunopotentiating activity; as well as, antibacterial and antiparasitic action (Mitjans and Vinardell, 2005, Vinardell et al., 2008). Lignins have been demonstrated as antioxidants and effective free radical scavengers that stabilize the reactions induced by oxygen and its radical species (Sakagami et al., 1992; Dizhbite et al., 2004; Ugartondo et al., 2008; 2009). These properties allow the use of lignins for cosmetics formulations, such as topical applications. Vinardell et al. (2008) have studied the potential antioxidant action of industrial lignins obtained from different sources, such as bagasse, lignosulfonate, curan and steam explosion and their potential eye and skin irritation due to their possible use in topical applications. They successfully established that all the lignins studied are not harmful to eyes and skin, which opens up new avenues of potential use in cosmetics and topical formulations. Ugartondo et al. (2009) conducted a study to evaluate whether lignins prevent H_2O_2-induced lipid peroxidation. They investigated the protective effect of industrial lignins obtained from different sources against H_2O_2-induced oxidation in normal human red blood cells (RBCs).They incubated the RBCs with different amount of lignins, challenged with H_2O_2 and analyzed for lipid peroxidation. From this study, they concluded that the inhibitory effects of different lignins against lipid peroxidation were notably related to their molecular weights. Bagasse with low molecular weight proved to be the greatest antioxidant whereas lignosulfonate with the highest molecular weight was the lowest antioxidant lignin.

As a major component in dietary fibre, lignins can inhibit the activity of enzymes related to the generation of superoxide anion radicals and obstruct the growth and viability of cancer cells (Lu et al., 1998). Promising applications of lignins exist in the cosmetic, pharmaceutical and food processing industries, owing to the antioxidant activity of lignins, their capacity to reduce the production of radicals, and ability to stabilize reactions induced by oxygen and its radical species. The anti-human immunodeficiency viral activity of lignins, such as lignosulfonate extracted after pulping processes, lignin-like substances extracted from lignocellulosics and anticancer activity of lignin F have also been previously reported (Suzuki et al., 1989; Ichimura et al., 1998; Jiang et al., 2001).

ANTI-NEURODEGENERATIVE EFFECTS OF BIOACTIVE DERIVATIVES FROM LIGNIN

The oxidative stress and inflammation are implicated in the neurodegenerative diseases. Neuronal toxicity is mediated and stimulated by reactive oxygen species (ROS) or reactive nitrogen species (RNS) and thus the oxidative stress accelerates the cell death of neurons in neurodegenerative disorders (Conrad et al., 2000; Kidd, 2000). In fact, such neuronal cell death has been found to be attenuated by antioxidants and free radical scavengers (Nagai et al., 2002; Lee et al., 2003). Recent studies by Ugartondo et al. (2009) have shown that lignins possess inhibitory effects against H_2O_2-induced lipid peroxidation. Lignins have been attracting increasing interest due to their antioxidant, anti-inflammatory and anti-carcinogenic properties. In an effort for protection against neurodegenerative diseases, Akao et al. (2004) had proposed a new therapeutic strategy to protect neurons from cell death by attenuating the apoptotic signal transduction. In their study, they converted lignin, a durable aromatic network polymer into highly active lignophenol derivative with newly developed phase-separation technique. These bioactive lignophenol derivatives were found to show the potent neuroprotective function against oxidative stress. Among the compounds investigated, the lignocresol derivative from bamboo (lig-8) possessed the most potent neuroprotective activity against H_2O_2-induced apoptosis in human neuroblastoma cell line SH-SY5Y by preventing the caspase-3 activation either via caspase-8 or caspase-9. Moreover, it was observed that lig-8 exerted the anti-apoptotic activity by inhibiting dissipation of the mitochondrial membrane permeability transition induced by H_2O_2 or by the peripheral benzodiazepine receptor ligand PK11195. Lig-8 was also proved to be having high antioxidant activity in the cells exposed to H_2O_2, as assessed by flow cytometry using 5-(and-6)-chloromethyl-2`7`-dichlorodihydrofluorescein diacetate and in vitro reactive oxygen species-scavenging potency. Depending on their results, they suggested that lig-8 is a promising neuroprotector, which influences the signaling pathway of neuronal cell death and it would be helpful to delay the progress of neurodegenerative diseases.

Soil Reclamation

The agricultural land tenure is facing a problem of reduction in the resources of humus in the soil and lack of high quality fertilizers. Conventional organic fertilizers used for retaining humus in the soil, such as manure, bird dung and peat cannot essentially fill up the deficit humus in soil. So it is necessary to find some alternative sources of organic raw materials, including the secondary waste materials. In this effort, Kapustina et al. (2006) demonstrated the potential use of hydrolysis lignin (HL), a waste product of hydrolysis yeast production process in agriculture. HL is potentially used as a component of organic fertilizers and organo-mineral mixtures. The HL had been found to be involved in humus formation and the effect of its complex forming properties on the regime of plant nourishment and on the soil microflora is well understood. The application of HL into soil improves its structure, increases its absorption capacity resulting in increased fertility. The humification of HL in soil occurs at very slow pace in comparison with usual organic fertilizers due to which its utilization can give a prolonged positive effect. Dumitru et al. (2003) observed that the introduction of lignin in sandy soil had a favourable effect on the growth of *Phaseolus vulgaris* plantlets, stimulating the growth rate and the quantity of green and dry biomass, as well as productivity of seeds. Similarly, Balas and Popa (2007) also showed that the supplementation of lignin at different doses produced stimulatory effects on the number of established *Lycopersicon esculentum* plantlets. The growth speed and height of plantlets was also found to be accelerated. Popa et al. (2005) also obtained similar observations stating that the presence of lignin had a beneficial effect on the development of plantlets.

A study conducted by Smidt et al. (2008) proved that lignin can be used for the formation of humic substances during composting. In their study, they used different lignin sources for co-composting with biogenic waste materials comprising yard and kitchen waste free of leftovers. Generally, the organic waste is processed by biological aerobic treatment to get composts for different purposes. They concluded that addition of lignin can induce a synergistic effect in composts with regard to humic acid formation, concentration depending on the source of input lignin added. During composting, humic substances are built up, so it is considered as a humification technology. Humic substances are considered as advantageous for soils to maintain aggregate stability and therefore help to prevent soil erosion. The humic acids formed, especially humic acid by co-composting, represents a stable fraction of compost organic matter which is mainly responsible for the water and nutrient holding capacity and availability. Owing to their dark color they have a regulating effect on soil temperature.

Humification contributes to the carbon sequestration, promising environmental benefit in the context of global warming. The content of the humic acid that can be obtained during the composting process depends on the chemical composition of input material. Lignin moieties are known to be a suitable provider of phenolic compounds required for humification processes (Zeichmann, 1994).

Kapustina et al. (2006) successfully demonstrated the use of HL for rehabilitation of radionuclides polluted soil. Among various kinds of HL, such as neutralized, acid and lignin from dumps; neutralized lignin from dumps showed much better absorption (40-58%) of ^{137}Cs and ^{90}Sr than the acid lignin. The absorbed ions did not desorb and were present in non-exchangeable form. HL from dumps is found to be the most promising material for use as a component of land reclamation sorbents and organomineral mixtures due to its physicochemical properties, such as increased sorption and fixation capacity.

Biopolymers Blended with Kraft Lignin

The industrial utilization of starch as food packaging is hindered by its swelling and partial dissolution in moist environments. Chemical modification of starch has been proved to be an effective way to reduce its water affinity but at the risk of cost and biodegradability (Sagar and Merrill, 1995). Another potential approach is to blend starch with hydrophobic compounds. Technical lignins available as by-products from the pulp and paper industry have already been integrated into synthetic thermoplastics, either in their native state or after grafting with compatible molecules (Kosikowa et al., 1993; Oliveira and Glasser, 1994). Baumberger et al. (1998) reported the influence of kraft lignin on some properties of thermally moulded starchy films. Rials et al. (1989) and Muzzarelli and Ilari (1994) have already demonstrated the properties of lignin based materials including cellulose or chitosan.

Lignins as Allelochemicals

Different organic compounds, known as allelochemicals are released from plants and microbes are known to affect the growth and different functions of receiving species. The allelochemicals exert their effects on patterns in vegetation communities, plant succession, seed preservation, germination of fungal spores, the nitrogen cycle mutuality association, crop productivity and plant defence mechanisms. Allelopathy is tightly associated with competition for resources and stress from diseases, temperature extremes, moisture deficit and herbicides. Allelochemical production is increased under the effect of such stress responses and their action of allelochemical is also accentuated. The allelopathic inhibition usually results from a combination of allelochemicals which affect various physiological processes in the receiving plant or microorganism (Einhelling, 1995).

Allelochemicals transmit from one higher plant to another in a terrestrial community through different forms, such as volatiles, aqueous leachates or various exudates. Volatiles may travel through the atmosphere from a donor plant to a receiving species. The volatile compounds are also adsorbed on soil particles and solubilized in the soil solution. Water-soluble allelochemicals leached from shoot tissue into the soil matrix and exudates from roots are a regular occurrence. Hence, spatial movements of allelochemicals can be up to a short distance. The soil acts as allelochemical pool, from where the roots of a receiving plant take up allelochemicals or lipid-soluble compounds adsorbed on soil particles and they can partition directly into root tissue. Plant residues decomposing in the soil will result in localized regions of higher allelochemical concentrations and the impact of allelochemicals in the soil on a receiving plant often depends on the chance encounters of the root system with such region which are rich in allelochemicals. Mostly, the allelopathic agents reported from higher plants are secondary compounds that are formed from either the acetate or shikimate pathway, or their chemical skeleton comes from a combination of these two origins as presented in Figure 4.

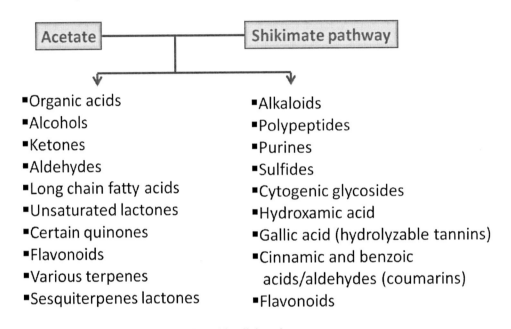

Figure 4. Different allelochemicals implicated in allelopathy.

Popa et al. (2008) demonstrated the promising application of lignin and polyphenols as allelochemicals. They suggested using these compounds to enhance crop production and develop a sustainable agriculture, including weed and pest control through crop rotations, residue management and various approaches in biocontrol.

They concluded that both the lignins used in this study; flax lignin and ammonium lignosulphonate have a biostimulating effect on mitotic division, in the radicular meristems of *P. Vulgaris.*They mentioned the possibility of induction of this process in case of flax lignin, as a result of the improvement of micro media conditions at plant root level, correlated with the beneficial influence of lignin on the flora present in soil. This effect was found to be doubled by the protective character, especially of ammonium lignosulphonate, materialized in the diminution of the anatelophases frequency at lower level against the control sample. Exciting prospective is to implement these allelochemicals as herbicides, pesticides and growth stimulants.

Chelating Complexes

Sena-Martins et al. (2008) thoroughly reviewed different applications of lignin. They described the use of lignin in chelating complexes which have potential use in bioremediation of heavy metals. In one study, Goncalves and Benar (2001) treated a lignin obtained from sugarcane acetosolv pulping with a polyphenoloxidase enzyme which resulted in increased number of hydroxyl and carbonyl groups. Thus, the chelating ability of treated lignin was improved and was 110% higher than in the non-treated acetosolv lignin. The chelating complexes obtained after enzyme treatment can be either used for removing heavy metals in pulps or to treat effluents containing heavy metals.

Treatment of Porous Materials

Porous articles can be treated with an enzyme catalyst, which forms macromolecule with a phenolic and/or an aromatic amine compound in their vicinity. This process results in increased properties, such as strength, flame resistance, antibacterial and antiseptic properties, adhesiveness, chemical agent-slow-releasing properties, colouring properties, dimension stability, crack resistance, deodorizing properties, humidity controlling properties, moisture conditioning properties, surface smoothness, ion exchangeability, among others. Echigo and Ohno. (2003), in their patent described the process in which porous particles can be treated with an enzyme catalyst, which can form macromolecules by reacting with a phenolic and/or an aromatic amine compounds inside them including lignin and lignosulfonic acid. These phenolic and aromatic amine compounds can be oxidized by peroxidase enzymes.

Lignin-Based Coatings and Paintings

Lignin-based compositions can be used for the preparation of coatings by enzymatic polymerization, which can be used for protection, beautification or treatment of lignocellulosics, wood or paper-based products. This technique allows coating of object using an environmentally safe process comprising: a) enzymes, such as catechol oxidase, laccase and peroxidises and; b) lignin derivatives, such as lignosulfonate, kraft and organosolv lignins. Lignin polymerization can be accelerated by the addition of laccase mediator ABTS (2,2`-azino-bis, 3-ethylbenzthiazoline-6-sulfonic acid) and lignin co-polymerization agents in the case of waterproof coatings (Bolle and Aehle, 2001; Sena-Martins et al., 2008).

Similarly, enzymatic polymerization of lignin-based compositions can be performed to prepare paintings, including protective paintings. In this process, a mixture containing a solution of lignin (lignosulfonate, kraft and organosolv lignins) plus a dye or a pigment, and an enzyme (catechol oxidase, laccase and peroxidase) is incubated under conditions and sufficient incubation time is given to reach the required viscosity. The resulting mixture is applied on an article and it is subject to proper conditions and time to form a painting on its surface (Bolle and Aehle, 2001; Sena-Martins et al., 2008). Park et al. (2008) worked to develop lignin-based coatings and composites. They demonstrated that the lignin-phenol formaldehyde (PF) resin films are effective water-barrier coatings for cardboard substrates. In their initial study, they focused on the lignin purification method to obtain material with uniform thermal properties. They performed thermal and rheological tests with different wt% of lignin substitution into PF resins in order to optimize the lignin-based resin for applications in coatings and composite materials.

Lignin-Based Polymer-Template Complexes

A lignosulfonate can be used for the formation of polymer-template complex via enzymatic polymerization. In this method, the reaction mixture, containing a monomer, a template (a micelle, a borate-containing electrolyte, or a lignosulfonate) and an enzyme (catechol oxidase, laccase and peroxidase) is incubated under appropriate conditions to allow the monomer to align along the template and polymerize, making a polymer-template complex. The so formed complex possesses exceptional electrical and optical stability, water solubility and processibility. Owing to these properties, it can be used for wide range of applications, such as light-weight energy storage devices e.g. rechargeable batteries, electrolytic capacitors, anti-static and anti-corrosive coatings for smart windows, and biological sensors (Samuelson et al., 2004). The utilization of lignin as a template for phenolic polymer synthesis has also potential applications in the production of polymeric dispersants, soil conditioning agents, phenolic resins or adhesives, and laminates, among others.

Lignin containing phenolic resins can be prepared by various methods (Tock et al, 1987; Matt and Doucet, 1988). Majority of these methods employ formaldehyde to hydroxymethylate lignin and /or phenol in order to cross-link the polymer. However, these lignin phenol resins are as toxic as formaldehyde itself. The enzymatic polymerization of phenols has proven to be promising alternative for the synthesis of phenolic resins. In the presence of H_2O_2, peroxidases catalyze the oxidation of phenols that eventually give rise to high molecular weight polymers (Xu et al, 1995; Nicel and Wright, 1997). Blinkovsky and Dorkick (1993) and Popp et al. (1993) described the copolymerization of phenols with kraft lignin that has been performed in aqueous-organic solvent mixtures catalyzed by horseradish peroxidase. The enzyme horseradish peroxidase, when encapsulated in reverse micelles, was able to catalyze the polymerization of phenolic compounds and aromatic amines (Rao et al., 1993; Liu et al., 1999).

Lignin-Based Biodispersants

Recently, a lignin-based formulation for control of microbial populations in industrial water circuits, such as in paper machine recirculating (''white'') waters was developed by an Austrian company (Bioconsult Gessellschaft fuer Biotechnologie mbH, Hallein, Austria) (Oberkofler, 2001). Generally, various types of biocides are added to prevent slime deposition in the white water of paper machines. The total volume of slime (dry matter per liquid volumc) and slime solids content in percent are crucial parameters. The slime solids content is linked to the ease of re-suspension of deposited slime via the shear forces of circulating fluid when the turbulence of the flow is adjusted. Non-wood soda lignin based formulations proved to be highly effective, which is currently used in many paper machines in Europe, where it is replacing toxic and less environmental friendly products (Lora and Glasser, 2002).

Lignin-Based Adhesives

The use of lignin in the adhesive sub-sector is likely to be in the fibreboard production. Lund et al. (2000) and Felby et al. (2002; 2004) have demonstrated the technology in which the *in situ* cross-linking of lignin via laccases (oxidoreductases) were used at the pilot plant level to provide boards comparable to those produced using urea-formaldehyde adhesives. Lignin-based adhesives market is likely to expand due to the utilization of safer sources over the traditional chemical usage in fibreboard formation.

Polyolefins (PO)

Lignin feedstock is likely to find potential applications in the polyolyfin subsector. Lignin is integrated via polymer blends and UV stabilization as reflected in the studies of Cazacu et al. (2004a, 2004b) and Gosselink et al. (2004), respectively. Incorporation of lignin in PO provides significant benefits related to environment and renders it more susceptible to biodegradation as compared to plastics and PO-based products which are recalcitrant to biodegradation. Variation of the degree of lignin incorporation is reported to be accompanied by the increase in degree and rate of biodegradation (Rusu and Tudorachi, 1999; Tudorachi et al., 2000 a, 2000b).

The global market for PO is vast with polyethylene (PE) and polypropylene (PP) representing approximately 60% of all the thermoplastics produced and is considered as having large economic value. Low density polyethylene continues to be a large global market with an estimated value of $19 billion in 2001 whereas global demand for all types of PE is likely to exceed by 63.4 tonnes in 2006. PP resins are also growing at faster pace with the consumption of 38 M tonnes in 2005 representing a value of about $42 billion (Anon, 2004). It is most likely that there will be enormous growth opportunities predicted for PP to replace paper, metal, wood, glass and natural fibres in developing countries with their probable shift from chiefly agrarian to industrial economy. This market is likely to expand especially by the blend polymer approach leading to products with unique properties directly attributed to lignin (Stewart, 2008).

CONCLUSIONS AND FUTURE OUTLOOK

Due to their renewable character and abundance in nature, plant biomass is an important resource for energy, food and useful chemicals. The aim of economically feasible biomass processing is to achieve a complete utilization of various biomass components.

Various processes were developed for utilization of cellulose and hemicelluloses fractions of lignocellulosic biomass. However, not much interest has been shown towards efficient utilization of lignin due to its recalcitrant structure. Lignins are natural products widely distributed in the plant kingdom with multiple chemical and biophysical functionalities which can be utilized in many possible ways. Conventionally, lignin and its allied compounds are used in fabrication of adsorbents, biocomposites, polymers, among others. Lignins and lignin-derived compounds are not currently used in the para-pharmaceutical or cosmetic industries. They could, however, represent potential alternative to a number of plant polyphenol-containing extracts being currently used. The potential health benefits and endless industrial applications of lignins make them a promising field of research. There is an urgent need to have detailed characterization and properties of the different lignins due to wide range of heterogeneity in lignins. This will pave the possible ways to determine their behaviour in different applications, and the ways in which their structures influence their biological properties.

The research emphasis will be focussed on value-added utilization of the lignin extracted during ethanol production from lignocellulosic biomass. A substantial amount of the fossil oil that is utilized today for production of certain chemicals can be competitively replaced by alternative processes based on lignin as raw material. The research outcomes will significantly broaden the understanding of various aspects of lignin and its potential applications. The development of high-value products from lignin will achieve complete utilization of biomass and improve the economics of the lignocellulosic biomass processing. Hence, the great challenge to scientific community is to find new applications for lignin.

ACKNOWLEDGMENTS

The authors are sincerely thankful to the Natural Sciences and Engineering Research Council of Canada (Discovery Grant 355254), FQRNT (ENC 125216) and MAPAQ (No. 809051) for financial support. The views or opinions expressed in this article are those of the authors.

REFERENCES

Acemiog˘lu, B., Samil, A., Alma, M.H., Gundogan, R., 2003. Copper(II) removal from aqueous solution by organosolv lignin and its recovery. *J. Appl. Polymer Sci.* 89, 1537–1541.

Aden, A., Ruth, M., Ibsen, K., Jechura, J., Neeves, K., Sheehan, J., Wallace, B., Montague, L., Slayton, A., Lukas, J., 2002. Lignocellulosic biomass to ethanol process design and economics utilizing co-current dilute acid prehydrolysis and enzymatic hydrolysis for corn stover. NREL/TP-510-32438, National Renewable Energy Laboratory, Golden, CO.

Ahmedna, M., Marshall, W.E., Rao, R.M., 2000. Surface Properties of granular activated carbons from agricultural byproducts and their effects on raw sugar decolorization, *Bioresource Tech* . 71, 103-112.

Akao, Y., Seki, N., Nakagawa, Y., Yi, H., Matsumoto, K., Ito, Y., Ito, K., Funaoka, M., Maruyama, W., Naoi, M., Nozawa, Y. (2004). A higly bioactive lignophenol derivative from bamboo lignin exhibits a potent activity to suppress apoptosis induced by oxidative stress in human neuroblastoma SH-SY5Y cells. *Bioorganic and Medicinal Chemistry.* 12, 4791-4801.

Akhtar,M., Bhanger, M. I., Iqbal , S., Hasany, S. M., 2005. Efficiency of rice bran for the removal of selected organics from water kinetics and thermodynamics investigation", *J. Agri. Food. Chem.* 53, 8655-8662.

Akhtar, S., Khan, A.A., Husain, Q., 2005a. Potential of immobilized bitter gourd (Momordica charantia) peroxidases in the decolorization and removal of textile dyes from polluted wastewater and dyeing effluent. *Chemosphere*. 60, 291–301.

Akhtar, S., Khan, A.A., Husain, Q., 2005b. Partially purified bitter gourd (Momordica charantia) peroxidase catalyzed decolorization of textile and other industrially important. *Biores Technol*. 96, 1804–1811.

Akin-osanaiye, B. C., Nzelibe, H.C., Agbaji, A.S., 2005. Production of ethanol from Carica papaya (pawpaw) agro waste: effect of saccharification and different treatments on ethanol yield. *Afric. J. Biotechnol*. 4(7), 657-659.

Allen, S.J., Koumanova, B., Kircheva, Z., Nenkova, S., 2005. Adsorption of 2-nitrophenol by technical hydrolysis lignin: kinetics, mass transfer, and equilibrium studies. *Ind. Eng. Chem. Res*. 44, 2281–2287.

Anon. (2004). Global commodity price reporting and intelligence: Phenol (Europe). http://www.icislor.com.

Avellar, B.K., Glasser, W.G., 1998. Steam-assisted biomass fractionation. I. Process considerations and economic evaluation. *Biomass Bioenerg.,* 205–218.

Bajpai, P. (2004). Biological bleaching of chemical pulps. Crit. Rev. Biotechnol. 24, 1-58.

Baklanova, O.N., Plaksin, G.V., Drozdov, V.A., Duplyakin, V.K., Chesnokov, N.V., Kuznetsov, B.N., 2003. Preparation of microporous sorbents from cedar nutshells and hydrolytic lignin. *Carbon* 42, 1793–1800.

Balas, A. and Popa, V. I. (2007). The influence of natural aromatic compounds on the development of *Lycopersicon esculentum* plantlets. *Bioresources*. 2(3), 363-370.

Baldrian, P., 2005. Fungal laccases - occurrence and properties. *FEMS Microbiology Reviews*. 30, 215–242,

Baldrian, P., Valášková, V., Merhautová ,V., Gabriel, J., 2005. Degradation of Lignocellulose by *Pleurotus ostreatus* in the Presence of Copper, Manganese, *Lead and Zinc. Res. Microbiol*. 156, 670-676.

Basso, M.C., Cerrella, E.G., Cukierman, A.L., 2004. Cadmium uptake by lignocellulosic materials: effect of lignin content. *Sep. Sci. Technol*. 39, 1163–1175.

Baumberger, S., Lapierre, C., Monties, B. and Della Valle, G. (1998). Use of kraft lignin as filler for starch films. *Polymer Degradation and Stability*. 59, 213-211.

Boeriu, C.G., Bravo, D., Gosselink, R.J.A., van Dam, J.E.G., 2004. Characterization of structure dependent functional properties of lignin with infrared spectroscopy. *Ind. Crops Prod*. 20, 205–218.

Bolle, R., Aehle, W. (2000). Lignin-based paints. *Patent* 6,072,015, 1-8.

Bolle, R., Aehle, W. (2001). Lignin-based coatings. *US. Patent* 6,217,942 B1, 1-6.

Borneman,W.S., Hartley, R.D., Morrison, W.H., Akin, D.E., Ljungdahl, L.G., 1990. Feruloyl and p-coumaroyl esterase from anaerobic fungi in relation to plant cell wall degradation. *Appl. Microbiol. Biotechnol*. 3, 345–51.

Boudet, AM., Kajita, S., Grima-Pettenati, J., Goffner, D. (2003). Lignins and lignocellulosics: a better control of synthesis for new improved uses. *Trends Plant Sci*. 8, 576-81.

Cara, C., Ruiza, E., Ballesteros, M., Manzanares, P., Negro, M.J., Castro, E., 2008. Production of fuel ethanol from steam-explosion pretreated olive tree pruning. *Fuel*. 87(6), 692-700.

Carlo Bonini, Maurizio D'Auria, Pasqua Di Maggio, Rachele Ferri. 2008. Characterization and degradation of lignin from steam explosion of pine and corn stalk of lignin: The role of superoxide ion and ozone. *Industrial crops and products*. 27, 182–188.

Carrott, P.J.M., Carrott, M.M.L.R., Roberts, R.A., 1991. Physical adsorption of gases by microporous carbons. *Colloid. Surf.* 58,385–400.

Carrott, P.J.M., Carrott, M.M.L.R., Moura˜o, P.A.M., Lima, R.P., 2003. Preparation of activated carbons from cork by physical activation in carbon dioxide. *Adsorp. Sci. Technol*. 21, 669–681.

Carrott, P.J.M., Moura˜o, P.A.M., Carrott, M.M.L.R., Gonc, alves, E.M., 2005. Separating surface and solvent effects and the notion of critical adsorption energy in the adsorption of phenolic compounds by activated carbons. *Langmuir*. 21, 11863–11869.

Carrott, P.J.M., Carrott, M.M.L.R., Mourao, P.A.M., 2006a. Pore size control in activated carbons obtained by pyrolysis under different conditions of chemically impregnated cork. *J. Anal. Appl. Pyrolysis*.75, 120–127.

Carrott, P.J.M., Moura˜o, P.A.M., Carrott, M.M.L.R., 2006b. Controlling the micropore size of activated carbons for the treatment of fuels and combustion gases. *Appl. Surf. Sci.* 252, 5953–5956.

Carrot, S., Carrot M., 2007. Lignin- from natural adsorbent to activated carbon: A review. *Bioresour. Technol*. 98, 2301-2312.

Cazacu, G., Mihaies, M., Pascu, C., Profire, L., Kowarskik, A. L., Vasile, C. (2004a). Polyolyfin/lignosulfonate blends. *Macromol. Meter. Eng*. 289, 880-889.

Cazacu, G., Pascu, M. C., Profire, L., Kowarskik, A. L., Mihaies, M., Vasile, C. (2004b). Lignin role in a complex polyolyfin blend. *Ind. Crops prod*. 20, 261-273.

Chen, L., Tang, C-Y., Ning, N-Y., Wang,C-Y, Fu, Q., Zhang, Q., 2010. Preparation and properties of chitosan/lignin composites films. *Chinese Journal of Polymer Science*. 27, 739−746.

Conrad, C.C., Marshall, P. L., Talent, J. M., Malakowsky, C. A., Choi, I., Gracy, R. W. (2000). Oxidized proteins inAlzheimer's plasma. *Biochem. Biophy. Res. Commun.* 275, 678-68.

Cotoruel, L. M., Marques, M.d., Rodrı´guez-Mirasol, J., Cordero, T., 2007. Adsorption of Aromatic Compounds on Activated Carbons from Lignin: Equilibrium and Thermodynamic Study. *Ind. Eng. Chem. Res*. 46, 4982-4990.

Couto, S.R., Herrera, J.L.T. 2006. Industrial and biotechnological applications of laccases: A review. *Biotechno Adv*. 24, 500-513.

Crist, R.H., Martin, J.R., Crist, D.R., 2002. Heavy metal uptake by lignin: comparison of biotic ligand models with an ion-exchange process. *Environ. Sci. Technol*. 36, 1485–1490.

Crist, D.R., Crist, R.H., Martin, J.R., 2003. A new process for toxic metal uptake by a kraft Lignin. *J. Chem. Technol. Biotechnol*. 78, 199–202.

Crist, R. H., Martin, J. R., Crist, R. D.,2004. Use of a Novel Formulation of Kraft Lignin for Toxic Metal Removal from Process Waters. *Separation Science and Technology*. 39 (7), 1535-1545.

Cullen, D., Kersten, P.J., 2004. Enzymology and molecular biology of lignin degradation. In: Brambl R, Marzluf GA, editors. The Mycota III. Biochemistry and molecular biologyBerlin-Heidelberg: Springer-Verlag, p. 249–73.

Dam, J. V., Gosselink, R., Jong, E. D. Lignin applications. Agrotechnology and food innovations. (www.biomassandbioenergy.nl), Cited 20th Aug, 2010.

Demirbas, A., 2004. Adsorption of lead and cadmium ions in aqueous solutions onto modified lignin from alkali glyccrol dclignication. *J. Ilazard. Mater.* B109, 221–226.

Dence, C. W., Reeve, D. W. (1996).Pulp bleaching: Principles and practise, Tappi press, Atlanta Georgia.

Dizhbite, T., Zakis, G., Kizima, A., Lazareva, E., Rossinskaya, G., Jurkjane, V., Telysheva, G., Viesturs, U., 1999. Lignin – a useful bioresource for the production of sorption-active materials. *Bioresour. Technol.* 67, 221–228.

Dizhbite, T., Telysheva, G., Jurkjane, V., Viesturs, U., 2004. Characterization of the radical scavenging activity of lignins-natural antioxidants. *Bioresource Technol.* 95, 309–317.

DSouza, D.T., Tiwari, R., Sah, A.K., Raghukumar, C., 2006. Enhanced production of laccase by a marine fungus during treatment of colored effluents and synthetic dyes. *Enzyme Microb Technol.* 38,504–511.

Dumitru, M., Popa, V. I., Obreja, I., Campeanu, M. M. (2003).The influence of some lignin products on the metabolic processes of plants. Buletinul I. P., XLIX (LIII), 1-2, 109-118.

Echigo, T., Ohnu, R. (2003). Composition for treating porous articles, treatment method, and use thereof. *US Patent* 0,017,565 A1, 1-22.

El-Wakil, N.A., 2009. Use of Lignin Strengthened with Modified Wheat Gluten in Biodegradable Composites. *Journal of Applied Polymer Science.* 113, 793–801.

Einhelling, F. A. (1995). Allelopathy: current status and future goals. In: Inderjit, K. M. M., Einhelling, F. A. (Eds.), Allelopathy: Organisms, Processes and applications, ACS, symposium series no. 582. American chemical society, Washington D.C, 1-7.

Felby, C., Hassingboe, J., Lund, M. (2002). Pilot-scale production of fireboards made by laccase oxidized wood fibres: board properties and evidence for cross-linking of lignin. *Enzyme Microb. Technol.* 31, 736-741.

Faison, B.L, Kirk, T.K., 1983 Relationship between lignin degradation and production of reduced oxygen species by Phanerochaete chrysosporium. *Appl. Environ. Microbiol.* 46,1140–11455.

Felby, C., Thygesen, L. G., Sanad, S., Barsberg, S. (2004). Native lignin for bonding of fibre boards - evaluation of bonding mechanisms in boards made from laccase-treated fibres of beech (*Fagus sylvatica*) *Ind. Crops Prod.* 20, 81-189.

Fenice, M., Giovannozzi-Sermanni, G., Federici, F., D'Annibale, A., 2003. Submerged and solid-state production of laccase and manganese-peroxidase by *Panus tigrinus* on olive-mill wastewater-based media. *J. Biotechnol.* 100, 77–85.

Ferraz, A., Cordova, A.M., Machuca, A., 2003. Wood biodegradation and enzyme production by Ceriporiopsis subvermispora during solid-state fermentation of Eucalyptus grandis. *Enzyme Microb. Technol.* 32,59–65.

Fierro, V., Torne, V., Montane´, D., Salvado´ , J., 2003b. Activated carbons prepared from kraft lignin by phosphoric acid impregnation. In: Proc. Carbon 2003, Oviedo, Spain.

Fierro, V., Torné-Fernández, V., Celzard, A., 2006. Kraft lignin as a precursor for microporous activated carbons prepared by impregnation with orthophosphoric acid: synthesis and textural characterisation. *Micropor. Mesopor. Mater.* 92, 243–250.

Fillingham, I.., Kroon, P.A., Williamson, G., Gilbert, H.J., Hazlewood, G.P., 1999. A modular cinnamoyl ester hydrolase from the anaerobic fungus Piromyces equi acts synergistically with xylanase and is part of a multiprotein cellulose-binding cellulase-hemicellulase complex. *Biochem. J.* 343:215–24.

Fujian, X., Hongzhang, C., Zuohu, L., 2001. Solid-state production of lignin peroxidase (LiP) and manganese peroxidase (MnP) by *Phanerochaete chrysosporium* using steam-exploded straw as substrate. *Biores. Technol.* **80**, 149–151.

Gabaldon, C., Marzal, P., Seco, A., Gonzalez, J.A., 2000. Cadmium and copper removal by a granular activated carbon in laboratory column systems. *Sep. Sci. Technol.* 35, 1039–1053.

Gellerstedt,G., Li, J., Eide, I., Kleinert,M., Barth.T., 2008. Chemical Structures Present in Biofuel Obtained from Lignin. *Energy and Fuels.* 22, 4240–4244.

Giovanellil, G., Ravasini, G., 1993. Apple juice stabilization by combined enzyme membrane filtration process. *Lebensm.-Wiss. Technol.* 26, 1–7.

Gold, M.H., Youngs, H.L., Gelpke, M.D., 2000. Manganese peroxidase. *Met Ions Biol Syst.* 37:559–86.

Gonclaves, A.R., Benar, P. (2001). Hydroxymethylation and oxidation of organosolv lignins and utilization of the products. *Bioresour.Technol.* 79(2), 103-11.

Gonzalez-Serrano, E., Cordero, T., Rodriguez-Mirasol, J., Rodriguez, J.J., 1997. Development of porosity upon chemical activation of kraft lignin with ZnCl2. *Ind. Eng. Chem. Res.* 36, 4832–4838.

Gonzalez-Serrano, E., Cordero, T., Rodriguez-Mirasol, J., Cotoruelo, L., Rodriguez, J.J., 2004. Removal of water pollutants with activated carbons prepared from H3PO4 activation of lignin from kraft black liquors. *Water Res.* 38, 3043–3050.

Gosselink, R.J.A., de Jong, E., Guran, B., Aba¨cherli, A., 2004a. Coordination network for lignin-standardisation, production and applications adapted to market requeriments (EUROLIGNIN). *Ind. Crops Prod.* 20, 121–129.

Gosselink, R.J.A., Snijder, M.H.B., Kranenbarg, A., Keijsers, E.R.P., de Jong, E., Stigsson, L.L., 2004b. Characterisation and application of NovaFiber lignin. *Ind. Crops Prod.* 20 (2), 191–203.

Hamadi, N.K., Chen, X.D., Farid, M.M., Lu, M.G., 2001.Adsorption kinetics for the removal of chromium (VI) from aqueous solution by adsorbents derived from used tyres and saw dust. *Chem. Eng. J.* 84, 95-105.

Hammel, K. E., Jensen, Jr., K. A., Mozuch, M. D., Landucci, L. L., Tien, M., and Pease, E. A. (1993). Ligninolysis by a purified lignin peroxidase. Journal of Biological Chemistry, 268, 12274–12281, Medline.

Hammel., K.E., 1997. Fungal degradation of lignin. In: Cadisch G, Giller KE, editors. Plant litter quality and decomposition. CAB-International. 33–46.

Hartley, R.D. (1978) The lignin fraction of plant cell walls. *Am. J. Clin. Nutr.* 31, S90-S93.

Hayashi, J., Kazehaya, A., Muroyama, K., Watkinson, A.P., 2000. Preparation of activated carbon from lignin by chemical activation. *Carbon.* 38, 1873–1878.

Howard, R.L., Abotsi, E., Jansen van Rensburg, E.L., Howard, S., 2003. Lignocellulose biotechnology: issues of bioconversion and enzyme production. *Afr. J. Biotechnol.* 2, 602–19.

Hu, J., Aizawa, T., Ookubo, Y., Morita, T., Magara, Y., 1998. Adsorptive characteristics of ionogenic aromatic pesticides in water on powdered activated carbon. *Water Res.* 32, 2593–2600.

Husain, Q., 2006. Potential applications of the oxidoreductive enzymes in the decolorization and detoxification of textile and other synthetic dyes from polluted water: a review. *Crit. Rev. Biotechnol.* 60,201–221.

Husain. M., Husain, Q., 2008. Applications of redox mediators in the treatment of organic pollutants by using oxidoreductive enzymes: a review. *Crit. Rev. Environ. Sci. Technol.* 38,1–41.

Ichimura, T., Watanable, O., Maruyama, S. (1998). Inhibition of HIV-1 protease by water soluble lignin-like substance from an edible mushroom *Fuscoporia oblique. Biosci. Biotechnol. Biochem.* 62, 575-577.

Jahan, M.S., Chowdhury, D.A., Isalm, M.K., 2006. Characterization and evaluation of golpata fronds as pulping raw materials. *Bioresource Technol.* 97, 401–406.

Jahan, M.S., Chowdhury, D.A., Isalm, M.K., Moeiz, S.M., 2007. Characterization of lignin isolated from some nonwood available in Bangladesh. *Bioresource Technol.* 98, 465–469.

Jeffries, T.W., 2006. Engineering yeasts for xylose metabolism. *Current Opinion in Biotechnology.* 17(3),320–326.

Jeffries, T.W., 1987. Physical, chemical and biochemical considerations in the biological degradation of wood. In: Kennedy JF, Phillips OG, William AP, editors. Wood and cellulosics: industrial utilization biotechnology, structure and propietaries. Chichester, West Sussex, England: Ellis Harwood Ltd.; 1987. Chapter 24.

Jensen, J., BaoW, K.A., Kawai, S., Srebotnik, E., Hammel, H.E., 1996. Manganese-dependent cleavage of nonphenolic lignin structure by Ceriporipsis subvermispora in the absence of lignin peroxidase. *Appl. Environ. Microbiol.* 62:3679–86.

Jiang, Y., Satoh, K., Aratsu, C., Kobayashi, N., Unten, S., Kakuta, H., Kikuchi, H., Nishikawa, H., Ochiai, K., Sakagami, H. (2001). Combination effect of lignin F and natural products. *Anticancer Research.* 21(2A), 965-970.

Jung, H-J. and Fahey, G.C.Jr. (1983) Interactions among Phenolic monomers and *in vitro* fermentation. *J. of Dairy Sci..* 66, 1255.

Junhong Liu, Yuan Weiping, Tao Lo. 1999. Copolymerization of lignin with cresol catalyzed by peroxidase in reversed micellar systems. *Electronic Journal of Biotechnology*, 2(2), 82-87.

Kadla, J.F., Kubo, S., Venditti, R.A., Gilbert, R.D., Compere, A.L., Griffith, W., 2002. Lignin-based carbon fibers for composite fiber applications. *Carbon.* 40, 2913–2920.

Kapustina, IB., Moskalchuk, LN., Matushonok, TG., Pozylova, NM. And Khololovich, ME. (2006). Investigation of hydrolysis lignin for the purpose of its possible use as a land reclamation sorbent for rehabilitation of soil polluted with radionuclides. *Chemistry for Sustainable Development,* 14, 13-18.

Kersten, P., Cullen, D., 2007. Extracellular oxidative systems of the lignin-degrading Basidiomycete Phanerochaete chrysosporium. *Forest Genet Biol.* 2007,44:77–87.

Keyser. P., Kirt, T.K., Zeikus, J.G., 1978. Ligninolytic enzyme system of Phanerochaete chrysosporium: synthesized in the absence of lignin in response to nitrogen starvation. *J. Bacteriol.* 135:790–7.

Khazaal K., Owen A.P.D., Palmer, J. and Harvey, P. (1993 b) Treatment of barley straw with ligninase: effect on activity and fate of the enzyme shortly after being added to straw. *Animal Feed Sci. and Technology.* 41, 15-21.

Khesami, L., Ould- Dris, A., Capard, R., 2007. Copressed wood activated carbon. *Bioresources.* 2(2), 193-209.

Kidd, P. M. (2000). Parkinson's disease as multifactorial oxidative neurodegeneration: implications for integrative management.*Altern. Med. Rev.* 5(6), 502-29.

Kirk, T. K., Connors, W. J., Zeikus, G., 1976. Requirements for a growth substrate during lignin decomposition by two wood-rotting fungi. *Applied and Environmental Microbiology.* 32, 192–194.

Kirk, S.H. and Farrell, R.L. (1987) Enzymatic "combustion": the microbial degradation of lignin. *Ann. Rev. Microb.* 41. 465-505.

Kirk, T.K., Farell, R.L., 1987. Enzymatic "combustion": the microbial degradation of lignin. *Annu. Rev. Microbiol.* 41:465–505.

Kirk, T.K., Cullen, D., 1998. Enzymology and molecular genetics of wood degradation by whiterot fungi. In: Young RA, Akhtar M, editors. Environmentally friendly technologies for the pulp and paper industry. New York: John Wiley and Sons; 1998. p. 273–308.

Kleinert, M.;,Barth, T.,2008 .*Energy Fuels*, 22, 1371–1379.

Kobya, M.,2004. Adsorption kinetics and equlibrium studies of Cr(VI) by hazelnut shell activated carbon, *Ads. Science. Tech* . vol 22 .

Kosikowa, B., Demianova, V. and Kacurakova, M., J. (1993). Sulfur-free lignins as composites of polypropylene films *Appl. Polym. Sci.*, 47, 1065-1073.

Kuhad, R.C., Singh, A., Ericsson, K.E. L., 1997. Microorganisms and enzymes involved in the degradation of plant fiber cell walls. *Adv Biochem Eng Biotechnol.* 57,45–125

Lalvani, S.B., Wiltowski, T.S., Murphy, D., Lalvani, L.S., 1997. Metal removal from process water by lignin. *Environ. Technol.* 18, 1163–1168.

Lalvani, S.B., Hu"bner, A., Wiltowski, T.S., 2000. Chromium adsorption by lignin. Energy *Sources* 22, 45–56.

LeCloirec, P., Brasquet, C., Subrenat, E., 1997. Adsorption onto fibrous activated carbon: applications to water treatment. *Energy. Fuels* 11, 331–336.

Lee, S. R., Im, K. J., Suh, S. I., Jung, J. G. (2003). *Protective effect of green tea polyphenol (-)-epigallocatechin gallate and other antioxidants on lipid peroxidation in gerbil brain homogenates. Phytother Res.* 17, 206-*209.*

Leonowicsz, A., Matuszewska, A., Luterek, J., Ziegenhagen, D.,Wojtas-Wasilewska, M., Cho, N.S., 1999. Biodegradation of lignin by white rot fungi. *Fungal Genet Biol* . 27, 175–85.

Levin, L., Herrmann, C., Victor Papinutti, L., 2008. Optimization of lignocellulolytic enzyme production by the white-rot fungus Trametes trogii in solid-state fermentation using response surface methodology. *Bioch. Eng. J.* 39, 207-214.

Li, A., Antizar-Ladislao, B., Khraisheh, M., 2007. Bioconversion of municipal solid waste to glucose for bio-ethanol production. *Bioprocess Biosystems Eng.* 30(3), 189-196.

Lora, J. H. And Glasser, W. G. (2002). Recent industrial applications of lignin: A sustainable alternative to Nonrenewable materials. *Journal of polymers and the environment.* 10 (1/2), 39-48.

Lu, F.J., Chu, L.H., Gau, R.J., 1998. Free radical-scavenging properties of lignin. Nutr. Cancer 30, 31–38. Mitjans, M., Vinardell, M.P., 2005. Biological activity and health benefits of lignans and lignins. *Trends Comp. Biochem. Physiol.* 11, 55–62.

Ludvık, J., Zuman, P., 2000. Adsorption of 1,2,4-triazine pesticides metamitron and metribuzin on lignin. *Microchem. J.* 64, 15–20.

Lund, M., Hassingboe, J., Felby, C., (Sep.2000). Oxidoreductase catalyzed bonding of wood fibres. In: Proceedings of the 6th European workshop on Lignocellulosics and Pulp. Bordeaux, France, 113-116.

Maier, G., Frei, M., Wucherpfennig, K., Dietrich, H., Ritter, G., 1994. Innovative processes for production of ultrafiltrated apple juices and concentrates. *Fruit Process.* 5, 134–138.

Malhas, A.N., Abuknesha, R.A., Price, R.G., 2002. Removal of detergents from protein extracts using activated charcoal prior to immunological analysis. *J. Immunol. Methods* 264, 37–43.

Martínez, A.T., 2002. Molecular biology and structure-function of lignin-degrading heme peroxidases. *Enzyme Microb Technol.* 30, 425–32.

Martinez, G., Larrondo, N., Putman, N., Gelpke, M.D.S., Huang, K., Chapman, J., 2004. Genome sequence of the lignocellulose degrading fungus Phanerochaete chrysosporium strain RP78. *Nature Biotechnol.* 22,1–6.

Martinez, A. T., Speranza, M., Ruiz-Duenas, F. J., Ferreira, P., Camarero, S., Guillen, F., 2005. Biodegradation of lignocellulosics: microbial, chemical, and enzymatic aspects of the fungal attack of lignin. *International Microbiology.* 8, 195–204, Medline.

Mester, T., Varela, E., Tien, M., 2004. Wood degradation by brown-rot and white-rot fungi. The Mycota II: genetics and biotechnology (2nd edition). Springer-Verlag Berlin Heidelberg.

McCrady, E. (1991). The nature of lignin. *Alkaline paper advocate.* 4(4), 1-4.

McKay, G., Bino, M.J., Altameni, A.R., 1985. The adsorption of various pollutants from aqueous solutions onto activated carbon. *Water Res.* 19, 491–495.

Merdy, P., Guillon, E., Aplincourt, M., Dumonceau, J., Vezin, H., 2002. Copper sorption on a straw lignin: experiments and EPR characterization. *J. Colloid Interf.* Sci. 245, 24–31.

Mohan, S.V., Karthikeyan, J., 1997. Removal of lignin and tannin colour from aqueous solution by adsorption onto activated charcoal. *Environ. Pollut.* 97, 183–187.

Mohan, S.V., Prasad, K.K., Rao, N.C., Sarma, P.N., 2005. Acid azo dye degradation by free and immobilized horseradish peroxidase catalyzed process. *Chemosphere.* 58,1097–1105.

Mtui, G., Nakamura, Y., 2005. Bioconversion of lignocellulosic waste from selected dumping sites in Dar es Salaam, *Tanzania Biodegradation.* 16(6), 493-499.

Mussatto, S. I., Fernandes, M., Rocha, G.J.M., Órfão, J.J.M., Teixeira, J.A., Roberto. I. C., 2010. Production, characterization and application of activated carbon from brewer's spent grain lignin. *Bioresour. Technol.* 101, 2450–2457.

Muzzarelli, R. and Ilari, P. (1994). Chitosans carrying the meth- oxypheyl function typical of lignin. *Carbohydr. Polym.*, 23, 155-160.

Nagai, K., Jiang, M. H., Hada, J., Nagata, T., Yajima, Y., Yamamoto, S., Nishizaki, T. (2002). (-)-Epigallocatechin gallate protects against NO stress-induced neuronal damage after ischemia by acting as an anti-oxidant. *Brain Res.* 956, 319-322.

Oberkofler, J. (2001). Sulphur-free lignin and derivatives thereof for reducing the formation of slime and deposits in industrial plants. WO 01/68530 A2.

Okuda, N., Ninomiya, K., Takao, M., Katakura, Y., Shioya, S., 2007. Microaeration enhances productivity of bioethanol from hydrolysate of waste house wood using ethanologenic *Escherichia coli* KO11. *J. Biosci. Bioeng.* 103(4), 350-357.

Oliveira, W. and Glasser, W. G., J. (1994). Multiphase materials with lignin. *Wood Chem. Techn.*, 14(i), 119-126.

Papinutti, V.L., Diorio, L.A., Forchiassin, F., 2003. Production of laccase and manganese peroxidase by *Fomes sclerodermeus* grown on wheat bran. *J. Ind. Microbiol. Biotechnol.* **30**: 157–160.

Park, Y., Doherty, W.O.S., Halley, P. J. (2008). Developing lignin-based resin coatings and composites. *Industrial crops and products*. 27, 163-167.

Pérez, N., Rincón, G., Delgado, L., González, N. (2006). Use of biopolymers for the removal of heavy metals produced by the oil industry-A feasibility study. Adsorption 12: 279–286.

Perezdrienko, I.V., Molodozhenyuk, T.B., Shermatov, B.E., Yunusov, M.P., 2001. Effect of carbonisation temperature and activation on structural formation of active lignin carbons. Russian *J. Appl. Chem.* 74, 1650–1652.

Pirbazari, M., Badriyha, B.N., Kim, S.H., Miltner, R.J., 1992. Evaluating GAC adsorbers for the removal of PCBs and toxaphene. *J. Am. Water Works Assoc.* 84, 83–90.

Popa, Vl., Dumitru, M., Volf, I., Anghel, N. (2005). Lignin and polyphenols as allelochemicals. Proceedings of the seventh forum of international Lignin Institute, Barcelona, 69-70.

Popa, Vl., Dumitru, M., Volf, I., Anghel, N. (2008). Lignin and polyphenols as allelochemicals. *Industrial Crops and Products*, 27, 144-49.

Pradeep, V., Datta, M., 2002. Production of ligninolytic enzymes for dye decolorization by cocultivation of white-rot fungi *Pleurotus ostreatus and Phanerochaete chrysosporium* under solid-state fermentation. *Applied Biochemistry and Biotechnology*. 109,102-103.

Rabinovich, M.L., Bolobova, A.V., Vasil'chenko 2004. Fungal decomposition of natural aromatic structures and xenobiotics: a review. *Appl. Biochem. Microbiol.* 40,1–17.

Reddy, G.V., Babu, P.R., Komaraiah, P., Roy, K.R.R.M, Kothari, I.L., 2003. Utilization of banana waste for the production of lignolytic and cellulolytic enzymes by solid substrate fermentation using two *Pleurotus* species (*P ostreatus and P. sajor-caju*). *Process Biochem.* **38**, 1457–1462.

Reddy, N., Yang, Y. (2005). Biofibres from agricultural byproducts for industrial applications. *Trends Biotechnol.*, 23, 22-27.

Rials, T.G., and W.G. Glasser. (1989). Multiphase materials with lignin. IV. Blends of hydroxypropyl cellulose with lignin. *J. Appl. Polymer Sci.* 37, 2399-2415.

Rodriguez-Mirasol, J., Cordero, T., Rodrı´guez, J.J., 1993. Activated carbons from CO2 partial gasification of eucalyptus kraft lignin. *Energy Fuels* 7, 133–138.

Rusu, M., Tudorachi, N. (1999). Biodegradable composite materials based on polyethylene and natural polymers I. Mechanical and thermal properties. *J. Polym. Eng.* 19, 355-369.

Sagar, A. and Merrill, E., J. (1995). Properties of Fatty-Acid Esters of Starch. *J. Appl. Polym. Sci.*, 58, 1647-1656.

Sakagami, H., Kohno, S., Takeda, M., Nakamura, K., Nomoto, K., Ueno, I., Kanegasaki, S., Naoe, T., Kawazoe, Y., 1992. O_2-scavenging activity of lignins, tannins and PSK. *Anticancer Res.* 12, 1995–2000.

Samuelson, L., Bruno, F., Tripathy, S., Nagaranjan, R., Kumar, J., Liu, W. (2004). Enzymatic polymerization. *US Patent*. 0,233,46 A1, 1-8.

Sarkanen, K.V. and Ludwic, C.H. (1971). Lignins occurrence, formation, structure and reactions. Wiley-Interscience, New York.

Schwartz, P.B.; Youngs, V.L. and Shelton, D.R. (1989) Isolation and characterisation of lignin from hard red spring wheat bran. *Cereal Chemistry*. 66 No.4, 289-295.

Sciban, M., Klasnja, M., 2004. Study of the adsorption of copper(II) from water onto wood sawdust, pulp and lignin. *Adsorp. Sci. Technol.* 22, 195–206.

Sederoff, R. R., Mackay, J. J., Ralph, J., Hatfield, R. D. (1999). Unexpected variation in lignin. *Curr. Opi. Plant Biol.* 2, 145-52.

Sena-Martins, G., Almeida-Vara, E., Duarte, J.C. (2008). Eco-friendly new products from enzymatically modified industrial lignins. *Industrial crops and products.* 27, 189-195.

Sjöde, A., Alriksson, B., Jönsson, L.J., Nilvebrant, N.O., 2007. The potential in bioethanol production from waste fiber sludges in pulp mill-based biorefineries. *J. Appl. Biochem. Biotechnol.* 140(1-12), 327-337.

Smidt, E., Meissl, K., Schmutzer, M., Hinterstoisser, B. (2008). Co-composting of lignin to build up humic substances-Strategies in waste management to improve compost quality. *Industrial crops and products*, 27, 196-201.

Srivastava, S.K., Singh, A.K., Sharma, A., 1994. Studies on the uptake of lead and zinc by lignin obtained from black liquor – a paper-industry waste material. *Environ. Technol.* 15, 353–361.

Stewart, D. (2008). Lignin as a base material for materials applications: Chemistry, application and economics. *Industrial crops and products.* 27, 202-207.

Suhas, Carrott, P.J.M., Ribeiro Carrott, M.M.L., 2007. Lignin – from natural adsorbent to activated carbon: a review. *Bioresour. Technol.* 98, 2301–2312.

Sun, Y., Zhang, J-P., Yang,G., Li, Z-H., 2006. Removal of Pollutantswith Activated Carbon Produced from K2CO3 Activation of Lignin From Reed Black Liquors. *Chem. Biochem. Eng. Q.* 20 (4) 429–435.

Suzuki, H., Tochikura, T. S., Iiyama, K., Yamazaki, S., Yamamoto, N., Toda, S., (1989). Lignosulfonate, a water solubilised lignin from the waste liquor of the pulping process, inhibits the ninfectivity and cytopathic effects of human immunodeficiency virus in vivo. *Agric. Biol. Chem.* 53, 3369-3372.

Thielemans, W., Can, E., Morye, S. S., Wool R. P., 2002. Novel Applications of Lignin in Composite Materials. *Journal of Applied Polymer Science*, 83, 323–331.

Toles, C.A., Marshall, W.E., John, M.M., 1998.Phosphoric acid activation of nutshells for metals and organic remediation process optimization. *J. Chem. Technol. Biotechnol.* 72, 255-263.

Tudorachi, N., Rusu, M., Cascaval, N. C., Constantin, L., Rugina, V. (2000a). Biodegradable polymeric materials I. Polyethylene-natural polymer blends. *Cell. Chem. Technol.* 34, 101-111.

Tudorachi, N., Cascaval, N. C., Rusu, M. (2000b). Biodegradable polymer blends based on Polyethylene and natural polymer : degradation in soil. *J. Polym. Eng.* 20, 287-304.

Tychanowicz, G.K., De Souza, D.F., Souza, C. G. M, Kadowaki, M.K., Peralta, R.M., 2006. Copper improves the production of laccase by the white-rot fungus *Pleurotus pulmonarius* in solid state fermentation. *Brazil Arch Biol Technol.* 49,699–704.

Ugartondo, V., Mitjans, M., Vinardell, M. P. (2008). Comparative antioxidant and cytotoxic effects of lignins from different sources. *Bioresour. Technol.* 99(14), 6683-6687.

Ugartondo, V., Mitjans, M., Vinardell, M. P. (2009). Applicability of lignins from different sources as antioxidants based on the protective effects on lipid peroxidation induced by oxygen radicals. *Industrial crops and products*, 30, 184-187.

Urano, K., Yamamoto, E., Tonegawa, M., Fujie, K., 1991. Adsorption of chlorinated organic compounds on activated carbon from water. *Water. Res.* 25, 1459–1464.

Vaithanomsat, P., Apiwatanapiwat,W., Petchoy,O., . Chedchant, J., 2010. Production of Ligninolytic Enzymes by White-Rot Fungus *Datronia* sp. KAPI0039 and Their Application for Reactive Dye Removal. International Journal of Chemical Engineering. Article ID 162504, 6 pages doi:10.1155/2010/162504.

Vinardell, MP., Ugartondo, V., Mitjans, M. (2008). Potential applications of antioxidant lignins from different sources. *Industrial crops and products*, 27, 220-223.

Van Soest, P.J. (1994) Nutrition ecology of the ruminants. 2nd edition. Cornel University Press. 156-176.

Van Beinum, W., Beulke, S., Brown, C.D., 2006. Pesticide sorption and desorption by lignin described by an intraparticle diffusion model. *Environ. Sci. Technol.* 40, 494–500.

Walker, G.M., Weatherley, L.R., 1999. Kinetics of acid dye adsorption on GAC. *Water Res.* 33, 1895–1899.

Wartelle, L. H ., Marshall, W.E., 2001. Nutshell as granular activated carbon physical chemical and adsorptive properties," Journal of chemical technology and biotechnology. 76, 451-455.

Wedekind, E., Garee, G., 1928a. Sorptive power of lignin. Z. *Angew Chem.* 41, 107.

Wedekind, E., Garee, G., 1928b. *Colloidal nature of liginic acid. Kolloid* – Z 44, 202.

Wesenberg, D., Kyriakides, I., Agathos, S.N, 2003. White-rot fungi and their enzymes for the treatment of industrial dye effluents. *Biotechnol Adv.* 22,161–187.

Yalem,N., Sevine, V., 2000. Studies of the surface area and porosity of actrivated carbon prepared from rice husk. *Carbon.* 38, 1943-1945.

Zadrazil, F. (1984) Microbial conversion of lignocellulose into feed. In Straw and Other Fibrous By-Products as Feeds, ed. F. Sundst☐l and E. Owen. Elsvier, 223 Amsterdam, chap. 9, 276-279.

Zou, Y., Han, B-X., 2001. Preparation of activated carbons from Chinese coal and hydrolysis lignin. *Adsorp. Sci. Technol.* 19, 59–72.

Ziechmann, W. (1994).Humic substances. A study about their theory and reality. B1-Wissenchaftsverlag, Mannheim, Leipzig, Wein, Zurich.

In: Lignin
Editor: Ryan J. Paterson

ISBN 978-1-61122-907-3
© 2012 Nova Science Publishers, Inc.

Chapter 2

BIOLOGICAL PROPERTIES AND APPLICATIONS OF PHENOLICS FROM HYDROTHERMAL PROCESSING

Enma Conde[1]*, Andrés Moure[1], Herminia Domínguez[2] and Juan C. Parajó[3]
[1]University of Vigo, Spain
Departamento de Enxeñería Química.
[2]Facultade de Ciencias de Ourense. Universidade de Vigo.
[3]Edificio Politécnico. As Lagoas. 32004 Ourense. Spain

ABSTRACT

This chapter reviews the hydrothermal processing of biomass, based on treatments with water or steam. Hydrothermolysis of lignocellulosic materials enables the fractionation and the selective extraction of bioactive compounds from plants. Along some fractionation processes, lignin is depolymerized, and the dissolved fragments are incorporated to the liquid phase. Upon pressurized water extraction (in which water remains in liquid state), free or polymeric phenolic compounds present in the lignocellulosic feedstocks can be solubilized at improved yields respect to conventional extraction, due to the changes in solvent properties and to the participation of autocatalytic hydrolysis reactions. The properties of both depolymerized lignin fractions and benzoic and hydroxycinnamic acids are considered, with particular emphasis on recent and emerging applications in the cosmetic and pharmaceutical industries. For comparative purposes, additional information concerning lignins extracted by traditional methods is presented.

1. INTRODUCTION

Vegetal biomass from herbaceous crops, forest or lignocellulosic wastes is a potential resource for energy, food and chemicals; and can provide a sustainable alternative to the oil based economy. Particular attention deserves the rational use of forest and agricultural wastes, due to their abundance, renewable character, environmental benefit and economic implications.

Wastes of lignocellulosic nature can be processed to achieve an integral utilization. For this purpose, an efficient fractionation of the major feedstock polymers (hemicellulose, lignin, and cellulose) and their further transformation into valuable products is needed.

The conversion of vegetal biomass into materials, chemicals and energy shows aspects related to the cracking of petroleum. The development of a sustainable industry based on renewable resources and green processes, oriented to the manufacture of safe and environmentally compatible products, avoiding hazardous and polluting chemicals, is in the scope of the "biorefinery" concept. Industrial biorefinery relies on the green chemistry principles, and includes biotransformations with enzymes or microorganisms to produce food and non-food products, including agro-industrial intermediates. Whereas conventional biorefineries have simple materials, chemicals and biofuels as major products, the second generation biorefineries are expected to produce sustainable chemicals that could be the base of valuable molecules, chemical intermediates, advanced materials and second-generation biofuels (Clark, 2007).

Fractionation of plant biomass can be carried out by physical operations (i.e., mechanical comminution), physicochemical transformations (i.e., steam explosion), chemical reactions (i.e., ozonolysis, acid hydrolysis, alkaline hydrolysis, or organosolv delignification) and biological treatments (i.e., enzymatic and microbial transformations).

The major drawbacks of conventional fractionation operations include the use of chemicals, generation of sugar degradation products and waste streams. In processes based on the utilization of acids, the corrosion of extraction vessels, the degradation of hemicellulose into monomers, and the formation of furfural are major problems. Alkaline extraction of hemicelluloses requires lower temperature and pressure and causes hydrolysis of the ester linkages, but environmental concerns and reagent recovery costs limit their potential. The development of hydrothermal autocatalyzed processes for fractionation of lignocellulosic materials should be encouraged, based on their environmental and operational advantages.

Lignin behaves as a natural free radical scavenger, owing to its aromatic structure and functional groups. Its complex structure is highly influenced by the plant resource and by the extraction and separation method. The antioxidant capacity of fractions solubilized during hydrothermal processing, as well as their properties and applications, are revised in this chapter. Additional information on related products coming from other processes is also presented for comparison.

2. LIGNOCELLULOSIC MATERIALS, LIGNIN, AND LIGNIN PRECURSORS

Lignocellulosic materials (LCM) can be fractionated in their major constituents (cellulose, hemicelluloses and lignin). Cellulose and hemicelluloses are polysaccharides that form the cell wall, and most industrial processes have been developed for their utilization. Lignin is the second most abundant organic natural compound in the plant kingdom (after cellulose), and usually accounts for 20-40% of the dry matter of LCM. Lignin provides rigidity, confers structural support, impermeability and resistance, and acts as a barrier against microbial attacks to the cell wall polysaccharides, protecting against pests and diseases.

Chemically, lignin is very stable, and shows a complex amorphous, polymeric, disordered, and aromatic structure formed by random polymerization of three precursors: trans-coniferyl, trans-sinapyl and trans-*p*-coumaryl alcohols, with ether-linked aromatic rings. The structure and monomer distribution pattern of this polymer depends on factors such as sampling method, plant family, physiological state, location and cultivation conditions (temperature, light intensity, water availability, latitude, harvest and storage periods), and extraction processes (Gosselink *et al.*, 2004). The predominant groups differ depending on the origin of the lignin: gymnosperm lignins contain mainly guaiacyl groups, woody angiosperm lignins contain guaiacyl-syringyl groups and lignins from grasses contain guaiacyl-syringyl-*p*-hydroxyphenyl groups. Lignin contains considerable amounts of phenolic and hydroxybenzoic acids, which play a major role in the covalent linkage of polysaccharides (arabinoxylans, xyloglucans, pectins) with lignin by ester and ether bonds.

Lignin can be obtained from trees, agricultural crops, and industrial byproducts from LCM processing (such as pulp or biofuel facilities). Lignins present a variety of functional groups, depending on the type of feedstock and the separation method. Lignin is rarely isolated in a pure state, and is usually associated to cellulose and hemicelluloses, i.e. commercial lignins contain 2–8% carbohydrates, which play a crucial role on the reactivity and properties of the polymer. Purified lignin is commercially available as a by-product of the pulp industry, and consists of low molecular weight mono-phenolic fragments with biological properties different from those of native lignin. Different pulping processes yield diverse types of purified lignin. In sulfite pulping, lignin is converted into lignosulphonates, which present sulphur-containing groups.

Sulfate (or *kraft* lignins) are obtained by delignification in media made up of sodium hydroxide and sodium sulphide. The Alcell process, based on the utilization of aqueous ethanol at temperatures in the range 185 - 195 °C, disrupts the structure of native lignin, yielding phenolic fragments of low molecular weight and enhanced hydrophobicity. Lignins with different properties are obtained using other delignification methods and raw materials (for example, non-woody raw materials). The emergence of the cellulosic bio-ethanol industry is expected to increase the production of lignin (Wang *et al.*, 2009). The industrial use of lignins requires improvements in lignin extraction, whereas new knowledge on the chemical composition, reactivity and structure of lignins will enable new applications (Boeriu *et al.*, 2004).

In pulp industries and in tree biorefineries, lignin is more often a problem than an opportunity, and is typically used as a solid fuel (Clark, 2007). Lignin from industrial processes is non-toxic, inexpensive and largely available. Approximately 50–100 million tonnes of lignin are annually separated from wood in pulp industries, but less than 3% of this material is processed into value-added products. Lignin is the richest natural source of aromatic compounds, which are basic structures for the development of synthetic precursors suitable for manufacturing products currently obtained from oil (Clark, 2007; García *et al.*, 2010). The use of these lignins as feedstock for novel chemicals is an important research field (Boudet *et al.*, 2003), and the possibility of increasing the value-added applications by improving the lignin purity of the commercial lignins has been suggested (Gosselink *et al.*, 2004; Sena-Martins *et al.*, 2008). The highly reactive locations can be modified through chemical, physical and/or enzymatic reactions. Biotechnological modifications for upgrading industrial lignins have been revised (Sena-Martins *et al.*, 2008).

Hydroxycinnamates are incorporated into the lignin polymer during lignification, resulting in extensive crosslinking between lignins and polysaccharides. Hydroxycinnamic acids are structural components of plant cell walls and therefore of the diet. The research on hydroxycinnamates, already revised by Kroon and Williamson (1999), has increased recently due to their biological activities and potential health benefits. The epidemiological evidence linking high intake of fruit and vegetables with reduced risk of chronic diseases has drawn much attention, since there are evidences of protection against free radicals and reactive oxygen species involved in aging and degenerative diseases. Dietary hydroxycinnamates consist predominantly of ferulic acid (abundant in an insoluble form in cereal brans) and caffeic acid. It has been suggested that the intake of hydroxycinnamates could exceed that of the flavonoids, and that they could provide the majority of the antioxidant activity in many diets.

3. HYDROTHERMAL PROCESSES FOR EXTRACTION AND REACTION

The interest in more environmentally friendly physicochemical methods to utilize biomass has favoured the research on hydrothermal technologies, based on the application of hot, pressurized water or steam. These technologies have been claimed to mimic nature, since they played an important role during formation of fossil fuels (Jin and Enomoto, 2009). Some of the technological, economic, and environmental advantages over conventional extraction technologies include selectivity, cleanliness and ease of product recovery, enabling cost savings on both raw material and energy.

The hot pressurized water is highly reactive and can be used either as an agent for extraction or to achieve the hydrolytic transformation of monomers and polymers. Water in subcritical state (temperature in the range 100–374 °C and pressure <22.1 MPa) is environmentally safe, non-toxic, and presents unique solvent and transport properties, including decreased dielectric constant, surface tension, and viscosity with increasing temperature. It has unique properties, being a good solvent for ionic species at ambient pressure and temperature, and for nonpolar solutes near the critical temperature. As a polar fluid, water dissolves polar compounds more readily than nonpolar compounds but the operational conditions of the subcritical extraction can be tuned to optimize the composition of the extract, since a temperature increase decreases the dielectric constant up to values similar to the ones of methanol or acetonitrile, improving the extractability of nonpolar compounds.

On the other hand, hot pressurized water has a high reactivity. The dissociation constant of water at 200–300 °C is three orders of magnitude greater than at ambient temperature, a fact increasing the reactivity of subcritical water with certain organic compounds. The reactions are commonly summarized as "hydrolysis reactions" which are catalyzed by acids, or may arise from simply hydrothermal transformations. The unique properties of hot and subcritical water led to the investigation of specific reactions for production of chemicals, degradation of pollutants, or the formation of particles. These potential applications have been comprehensively revised in a recently work by Brunner (2009). Therefore, subcritical water is a promising solvent and a suitable reaction medium for biomass conversion, enabling the manufacture of products to be used in the food, plastic and fine chemical industries (Wiboonsirikul and Adachi, 2008; Cheng *et al.*, 2009; Asghari and Yoshida, 2010).

Some other common alternatives terms for hydrolytic reactions, catalyzed by the acids derived from the raw material, are noncatalytic hydrothermolysis, hot-compressed water, or subcritical water treatments. The steam explosion technology is often considered as a type of hydrothermal processing. In steam explosion, defibration is caused by a sudden pressure drop of the reaction media (obtained by steaming the sample at high temperature and pressure) to atmospheric pressure.

The reactions involved in hydrothermal processing are autocatalytic: they start with hydronium ions from water autoionization, and their progress is favoured by the *in situ* generation of organic acids (such as formic and levulinic acid from sugar-degradation products, phenolic acids from hemicellulose substituents and acetic acid from acetyl groups). pH drops along the hydrothermal processes of extraction and reaction: for example, Amidon and Liu (2009) reported changes in extraction rate (initially slow and progressively faster) as the pH fell during the extraction. Some pH profiles measured during hydrothermal treatments are shown in Figure 1.

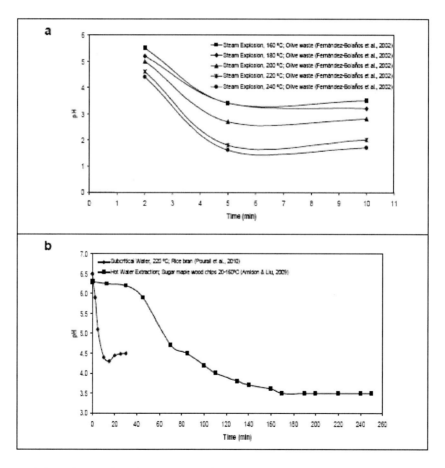

Figure 1. Variation of pH during biomass processing by a) steam explosion, b) pressurized hot water extraction.

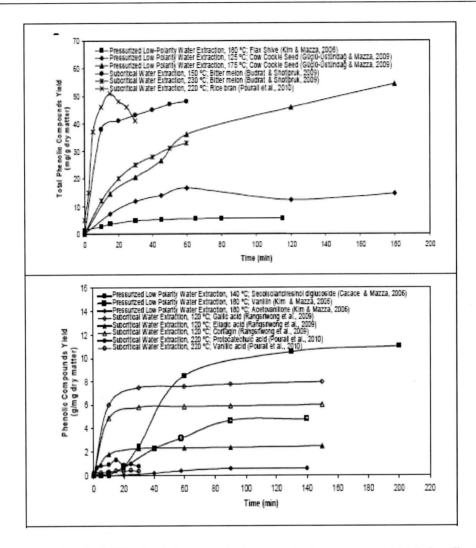

Figure 2. Kinetics of solubilization during presurized water extraction processes: a) total phenolics, b) selected phenolics.

Along hydrothermal processing of LCM, water-soluble extractives are removed from solid phase, hemicelluloses react to yield soluble oligomeric sugars, monosaccharides, and sugar-degradation products (furfural and hydroxymethylfurfural), whereas some cell wall-linked phenolic compounds can also be solubilized (Felizón *et al.*, 2000; Garrote *et al.*, 2004; Rodríguez *et al.*, 2008). The major phenolic compounds, derived from partial depolymerization of lignin and lignin–hemicellulose linkages, identified in liquors from acidic and autohydrolytic treatments of LCM, have been compiled (Garrote *et al.*, 2004). The structures of depolymerized lignin fraction included *p*-substituted phenols, hydroxybenzoic acids (gallic, vanillic, syringic, 4-hydroxybenzoic, protocatechuic, 3-hydroxybenzoic), aldehydes (syringaldehyde and sinapaldehyde), acetophenones and carboxylic acids. Ferulic and *p*-coumaric acids (the most abundant hydroxycinnamic acids) are linked to arabinoxylans or pectins through ester bonds. These lignin-derived compounds can have commercial value in relation to different applications as final products or as intermediate in chemical and biotechnological synthesis, i.e. the hydroxybenzoic and hydroxycinnamic acids. Figure 2 shows the kinetics of solubilization of some phenolic compounds from different materials.

When hot pressurized water is used for fractionation of lignocellulosic biomass, depolymerization and solubilization of hemicelluloses and acid-soluble lignin take place. The lignin obtained from different hydrothermal (or autohydrolysis) processing can be recovered, concentrated and purified for several purposes. In autohydrolysis, the cleavage of carbohydrate-lignin bonds increases the lignin extractability in subsequent extractions with aqueous alkali or organic solvents. However, the maximum lignin extractability is achieved only in a narrow range of autohydrolysis conditions, due to lignin repolymerization, which has negative effects on lignin properties and reactivity. The increased values of acid-insoluble lignin determined in LCM treated under harsh conditions can be ascribed to condensation reactions of dissolved lignin, sugar degradation products and phenolic extractives. Both lignin depolymerization and repolymerization reactions may occurr simultaneously during autohydrolysis (Lora and Wayman, 1980; Robert et al., 1988; Garrote et al., 2004; Lechinsky et al., 2009). The presence of sodium hydroxide during autohydrolysis can suppress acidolysis-related reactions (Li and Gellerstedt, 2008), as neutralization of acidic products avoids the acid-catalysed lignin reactions, allowing the isolation of lignins with higher content of carboxylic acid groups, lower average molecular weight, light color and preserving some of the native β-O-4 linkages.

It has been proposed that carbonium-ion-initiated repolymerization occurs during autohydrolysis, and that improved lignin extractability can be achieved in the presence of carbonium ion scavengers (Wayman and Lora, 1978). The presence of a reactive phenol in autohydrolysis media enables an efficient lignin depolymerization and a large incorporation of the phenol into the lignin matrix without any repolymerization. Autohydrolysis in the presence of 2-naphthol proceeds with a comprehensive depolymerization of lignin, since both acidolysis and homolytical cleavage of β-O-4 structures would be operative, and the competing lignin–lignin condensation would be suppressed, resulting in higher contents of phenolic hydroxyl groups and incorporated naphthalene rings (Li and Gellerstedt, 2008). Despite the almost quantitative lignin extraction after autohydrolysis in the presence of 2-naphthol, it has been suggested that this is a petroleum-derived product, and that more environmentally friendly agents should be used.

Several studies reported on the application of subcritical water as a green and environmentally friendly reagent for the hydrolytic conversion of LCM. In the following paragraphs, representative examples of phenolics solubilization by hydrothermal processing are commented. They are grouped in those aimed at fractionating the raw material and in those aimed at extracting bioactive compounds.

3.1. Fractionation Treatments

Steam explosion of olive stones solubilized most hemicelluloses and a small part of lignin, and the lignin isolated by acid precipitation and further alkaline dissolution was a de-etherified product (with extensive cleavage of the β-aryl ether linkage) potentially suitable as a fuel, as a polymer precursor, as a resin or as a therapeutic agent (Fernández-Bolaños et al., 1999; Rodríguez et al., 2008). Bamboo lignin is characterized by the presence of ester- and/or ether-linked p-coumaric acid to lignin. The yield of p-hydroxybenzaldehyde decreased after steam explosion, probably because the severe operational conditions caused the cleavage of ester and ether bonds between lignin and polysaccharides in bamboo cell walls, and lignin was depolymerized due to the cleavage of ether linkages. Alkaline nitrobenzene oxidation of bamboo lignin gave vanillin, syringaldehyde and *p*-hydroxybenzaldehyde as major products (Shao *et al.*, 2008). High yields of phenolic hydroxyl groups were detected in lignin prepared from steam-exploded pulp, since the *p*-coumaric acid esterified to lignin side chain was cleaved by steam explosion. *Eucalyptus globulus* wood lignin obtained from autohydrolyzed samples recovered by ethyl acetate extraction showed low contents of aliphatic hydroxyl groups and β-O-4 structures, and an increased proportion of phenolic hydroxyl groups. In lignins isolated from samples treated under severe autohydrolysis conditions, the contents of β-O-4 structures were half of the ones determined for native lignin. Fragmentation reactions dominated over condensation reactions, except for the low-molecular-weight lignin dissolved in autohydrolysis liquors, which was highly susceptible toward acid-catalyzed condensation (Leschinsky *et al.*, 2008 and 2009).

The susceptibility of LCM toward lignin solubilization depends on the raw material considered: for example, whereas most lignin was extracted from bamboo and chinquapin at low severity, a significant amount of lignin from Japan cedar remained insoluble (Ando *et al.*, 2000). *Pinus* species have a polyphenol-rich bark which has been widely studied. Proanthocyanidins in pine bark, composed of flavan-3-ol subunits linked mainly through C4-C8 bonds, have a number of physiological activities and are employed as a dietary supplement in Europe. A comparison of pine bark extraction yields using boiling water has been reported (Ku *et al.*, 2007). Degraded lignin was isolated from Japanese red pine wood by treatments with subcritical water in the presence and absence of phosphate buffer (Asghari and Yoshida, 2010). The effect of severity of steam explosion of *Lespedeza* stalks was probably connected to the balance between lignin depolymerization and repolymerization reactions. The steric hindrance of the methoxyl group at the C-5 position could explain that less syringyl units than guaiacyl units condensed during steam explosion, resulting in higher syringyl / guaiacyl ratios (Wang *et al.*, 2010).

3.2. Extraction

The growing interest in bioactive compounds derived from plants, due to their health-promoting effects, has incentivated the search for efficient extraction techniques. Conventional solid-liquid extraction using organic solvents (such as ethanol, methanol, ethyl acetate, or acetone) can be expensive and lack selectivity, and their environmental effects can become restrictive, as more strict regulations limits their use in food and pharmaceutical industries. Alternative extractions processes based on the use of pressurized safe non toxic solvents have been developed (Pronyk and Mazza, 2009).

The hot-water extraction of tobacco stalk yielded coniferyl and sinapyl alcohols, showing maximum concentrations of both alcohols under defined conditions, and decreased yields under harsh conditions, possibly due oxidation reactions. These alcohols were also identified (at lower levels) in methanol extracts, but not in chloroform or acetone extracts, due to the different capacities of the various solvents to hydrolyze the covalently bound coniferyl and sinapyl alcohols (Andersen *et al.*, 1980). These phenolic lignin monomers might contribute to flavor, aroma, and other desirable organoleptic properties of tobacco and tobacco smoke, but lignin might also be a precursor of phenols with undesirable health implications.

Flaxseed is a rich source of the lignan formed by two cinnamic acid residues and a 2,3-dibenzylbutane nucleus, the most abundant being secoisolariciresinol diglucoside, whose metabolites may provide health benefits due to their weak oestrogenic or anti-oestrogenic effects, antioxidant activity, ability to induce phase 2 proteins and/or inhibit the activity of certain enzymes. Additionaly, they may protect against cardiovascular diseases and metabolic syndrome by reducing lipid and glucose concentrations, lowering blood pressure, and decreasing oxidative stress and inflammation. Flax lignans may also reduce cancer risk by preventing pre-cancerous cellular changes and by reducing angiogenesis and metastasis. The potential health benefits have been reviewed (Adolphe *et al.*, 2010). In defatted flaxseed extracts, secoisolariciresinol diglucoside exists in oligomeric form, largely in ester linkages to 3-hydroxy-3-methylglutaric acid and cinnamic acid and with other phenolic compounds also present in glucosidic form. Pressurized low polarity water extraction of flaxseed produced maximum amounts of lignans and proteins at 160°C and 5.2 MPa. The secoisolariciresinol diglucoside yield increased sharply at temperatures higher than 120 °C, although thermal degradation was observed at 160 °C (Cacace and Mazza, 2006). The water extraction of phenolic compounds (*p*-hydroxybenzaldehyde, vanillic acid, syringic acid, vanillin, acetovanillone, and ferulic acid) from flax stalks has been proposed (Kim and Mazza, 2006).

Subcritical water extraction (at 25-200 °C) selectively extracted bioactive phenolic compounds from rosemary leaves, leading to fractions activity comparable to those achieved by Supercritical Fluid Extraction. The more polar compounds (scutellarein, rosmanol, rosmanol isomer, and genkwanin) were extracted at low temperature, whereas operation at higher temperatures increased the solubility of less polar compounds (such as carnosol, rosmadial, carnosic acid, methyl carnosate, rosmanol, epirosmanol, epirosmanol methyl ether, carnosol isomer, rosmanol isomer, and some flavonoids such as cirsimaritin, scutellarein, and genkwanin) (Ibáñez *et al.*, 2003). Carnosic acid, one of the most active compounds from rosemary, started to be extracted at 100 °C, and its yield increased with temperature. Despite the oxidative unstability of carnosic acid, it was the major compound in extracts obtained at 200 °C (accounting for more than 70% of the dissolved material). The absence of decomposition at this temperature could be due to the fact that the water employed was deoxygenated, and that no exposure to air occurred along extraction (Ibáñez *et al.*, 2003).

Subcritical water extraction of rice bran at 100-360 °C yielded phenolic compounds (caffeic, ferulic, gallic, gentisic, *p*-coumaric, *p*-hydroxybenzoic, protocatechuic, sinapic, syringic, vanillic acids and vanillin). Both phenolic content and antioxidant activity first increased with temperature but the phenolic compounds may be decomposed under severe conditions (Pourali *et al.*, 2010). The increase in phenolics yield observed at temperatures up to 220 °C was attributed to the cleavage of ester and/or ether bonds in lignin/phenolic–carbohydrate complexes of rice bran, and also to the more favourable solvating conditions of subcritical water. The optimal temperature depended on the compound considered (i.e. 230 °C for protocatechuic and 295 °C for vanillic acid); whereas longer residence times and/or higher temperatures promoted decomposition reactions under subcritical water conditions (Pourali *et al.*, 2010). Hydrothermal treatments have been proposed to release ferulic acid from wheat bran. Rose and Inglett (2010) developed a two-stage hydrothermal processing (at 130 °C to solubilize starch and pentose sugars, and at 170-220 °C to release soluble hemicellulose-derived products, including ferulic acid as esterified ferulate).

Under conditions of subcritical water (250 - 350 °C, 20 MPa) and short reaction times, the lignin macromolecular structure of switchgrass (*P. virgatum*) can be significantly broken down, yielding lignin-derived compounds (such as 3-ethylphenol, ethylguaiacol and guaiacol) suitable as precursors for aromatic chemicals. The possibility of using subcritical water to obtain functional products of high value in the chemical industry has been recently pointed out (Chen *et al.*, 2009).

Pressurized low-polarity water (PLPW) extraction of flax shive was proposed for the extraction of phenolic compounds (including *p*-hydroxybenzaldehyde, vanillic acid, syringic acid, vanillin, acetovanillone, and ferulic acid). The extraction of all phenolic compounds, except ferulic acid, increased with temperature and NaOH concentration, yielding up to 5.8 g of phenolics /kg of flax shive at 230.5 °C in the presence of 0.63 M NaOH (Kim and Mazza, 2006). Fractionation of flax shive into cellulose, hemicellulose, and lignin has been accomplished in two-stage extraction process using water and aqueous ammonia (Buranov and Mazza, 2007, 2009 and 2010).

During the first stage, deionized water at 190 °C and 5.2 MPa solubilized 84% of hemicellulose and 32% of lignin; and the subsequent aqueous ammonia treatment at 200 °C removed more than 77% of lignin and 95% hemicelluloses. The two-stage (water/alkaline) processing enabled the sequential extraction and separation of hemicellulose and lignin, and yielded a nearly pure cellulose solid residue with a negligible concentration of furfural (Kim and Mazza, 2009).

The liquefaction of empty palm fruit bunch in subcritical water (270 °C, 20 bar, 20 min) has been assayed in the presence of alkalis, whose reactivity was in the order K_2CO_3 > KOH > NaOH. Treatments resulted in up to 65.6% lignin solubilization, in comparison with 24.3% in the absence of alkalis. The major products were phenols and esters: the highest phenolic yield was quite sensitive to the operating conditions, and the maximum value was achieved with K_2CO_3; whereas NaOH processing yielded the maximum concentration of esters. Since esters can found applications as biodiesel, hot compressed water processing present potential for biofuels, with advantages respect to pyrolysis in terms of expected lower post-processing costs (Akhtar *et al.*, 2010).

Microwave processing enables a rapid and uniform heating of samples, and can be employed in technologies dealing with hot compressed water as an alternative to electrical or steam heating. Fast (2 min) extraction of polysaccharides and polyphenols from tea residues has been reached with microwave heating, and this technology has been scaled up to the semi-industrial scale (Tsubaki *et al.*, 2008). Ultrasonic irradiation of sugar-cane bagasse has also been employed to obtain lignin fractions having the same primary structure, composition, and similar or slightly higher purity than products obtained in the absence of ultrasounds (Sun *et al.*, 2004).

3.3. Chemical and Biotechnological Transformation of Phenolic Compounds

Lignin fractions can be transformed either by chemical or biotechnological means with the aim of producing compounds demanded by the food, health, cosmetic and pharmaceutical industries. The production of aldehydes (vanillin, syringaldehyde, hydroxybenzaldehyde) from lignin has been addressed by alkaline air oxidation of partially sulfonated lignin from spent sulfite liquors, by aqueous alkaline treatment, and by oxidation of lignin in the presence of metal catalysts (Wu *et al.*, 1994). Alkaline air oxidation of spent sulfite liquors to produce vanillin also leads to syringaldehyde, a product suitable for the manufacture of 3,4,5-trimethoxybenzaldehyde, a building block for the preparation of specific drugs (Wu *et al.*, 1994). Chemical oxidation with molecular oxygen of kraft black liquor from *Pinus pinaster* wood was more efficient than enzymatic oxidation and than a combined chemical - enzymatic oxidation (Mathias *et al.*, 1995). Alkaline nitrobenzene oxidation of bamboo lignin gave vanillin, syringaldehyde and *p*-hydroxybenzaldehyde as major products (Shao *et al.*, 2008). Singlet oxygen mediated degradation of lignin can also be used to obtain fine chemicals. Lignins exhibit different reactivity toward singlet oxygen depending on their nature: those with only guaiacyl units are more easily oxidized than those containing both guaiacyl and syringyl units. The beech steam-exploded lignin irradiated in the presence of oxygen produced vanillin, sinapyl alcohol, 4-hydroxy- 3,5-dimethoxybenzaldehyde, and 2,4-dioctylphenol (Bentivenga *et al.*, 1999 and 2000). Photochemical degradation of steam-exploded lignin from *Pinus brutia* was proposed to obtain antioxidant phenolics (including trans-sinapyl alcohol, 4-hydroxy-3,5-dimethoxybenzaldehyde, 4-hydroxy-3,5-dimethoxyphenylacetone, 4-hydroxy- 3-methoxybenzaldehyde, cis-sinapyl alcohol, and sinapyl aldehyde) (Bonini *et al*, 2002). Agricultural residues are also possible raw materials for the production of valuable chemicals from lignin. Sugarcane bagasse is a source of vanillin and *p*-coumaric, ferulic, and syringic acids (Arni *et al.*, 2010); and ferulic, *p*-coumaric and gallic acids have been obtained from vine shoots (Max *et al.*, 2010). The recovery of ferulic acid, coumaric acid, vanillic acid, and vanillin from agricultural residues, such as wheat and oilseed flax straws, was enhanced by enzymatic treatments with feruloyl esterase and xylanase (Tapin *et al.*, 2006). Ferulic acid can be naturally extracted from agroindustrial and forestal wastes either by an enzymic treatment or from clean technologies coupled with biotechnological methods, and it can be used as a substrate for the biotecnological production of vanillin (Falconnier *et al.*, 1994; Walton *et al.*, 2000; Priefert *et al.*, 2001), or for value-added chemicals, i.e. 4-vinyl guaiacol (Yang *et al.*, 2009).

Microbial biotransformations can be employed for the conversion of renewable resources into commercial products. The wide range of activities, the high chemo-, regio- and enantio-selectivity, the mild operation conditions and the ecological nature of products are important advantages over chemical synthesis. The potential biocatalytic transformations of ferulic acid have been revised (Rosazza *et al.*, 1995; Lomascolo *et al.*, 1999; Serra *et al.*, 2005; Mathew and Abraham, 2006). Biotransformation of cinnamic and ferulic acids into commercially valuable fine chemicals such as natural flavours has received continued attention (Tsujiyama and Ueno, 2008). Vanillin and benzaldehyde are among the most important aromatic flavour compounds used in the food and perfume industries, and can be obtained as intermediates of the microbial degradation of cinnamic and ferulic acid, respectively. In these degradative pathways also appear benzoic and vanillic acids, an important starting molecule in the synthesis of oxygenated aromatic chemicals such as vanillin (Brunati *et al.*, 2004).

4. BIOLOGICALLY RELEVANT FUNCTIONS

The interest in bioactive plant compounds has increased in the last decades, promoted by studies on their physiological role in issues related to human health. The potential health applications of lignins are attracting interest, based on their diverse chemical and biological functionalities, including antioxidant capacity as well as antitumoral, antiviral, immunopotentiating, antiparasite and antibacterial activities (Mitjans and Vinardell, 2005). Lignin has been proposed as an alternative to plant phenolic rich extracts currently used in the para-pharmaceutical or cosmetic industries, and its dietary importance as a part of the fiber complex might be more important than previously considered. Additionally, low molecular weight phenolic monomers of lignin may possess biological effects not characteristic of native lignin. The mono-phenolic fragments of purified lignin may have beneficial effects on productivity of farm animals, safety of animal products and environment. Some antioxidant assays reported to confirm the antioxidant activity of lignin precursors are summarized in Table 1, and other biological activities are summarized in Table 2.

Table 1. Some assays employed to measure the *in vitro* antioxidant activity of lignin model compounds

Antioxidant test	Reference	Lignin-model compounds / Phenol and catechol derivatives
ABTS radical scavenging	Villaño et al., 2005	Caffeic acid, p-coumaric acid , ferulic acid, gallic acid, gentisic acid, protocatechuic acid, tyrosol, vanillic acid
	Gülçin, 2006	Caffeic acid
	Loo et al., 2008	Catechol, 3-methoxycatechol, syringol
	Tabart et al., 2009	Catechin, ellagic acid, gallic acid
	Bountagkidou et al., 2010	p-Hydroxybenzaldehyde, protocatechuic aldehyde, salicylaldehyde, syringaldehyde, vanillin, p-hydroxybenzoic acid, protocatechuic acid, salicylic acid, syringic acid, vanillic acid
	von Gadow et al., 1997	Caffeic acid, p-coumaric acid, ferulic acid, p-hydroxybenzoic acid, protocatechuic acid, syringic acid, vanillic acid
	Kim et al., 2006	Caffeic acid, trans-cinnamic acid, p-coumaric acid, ferulic acid, p-hydroxybenzoic acid, salicylic acid, syringic acid, vanillic acid
	Terpinc et al., 2010	Caffeic acid, chlorogenic acid, p-coumaric acid, ferulic acid, protocatechuic acid, rosmarinic acid, sinapic acid, umbellic acid
DPPH radical scavenging	Brand-Williams et al., 1995	Caffeic acid, coumaric acid, eugenol, ferulic acid, gallic acid, gentisic acid, guaiacol, isoeugenol , phenol, protocatechuic acid, γ-resorcylic acid, vanillic acid, vanillin, zingerone,
	von Gadow et al., 1997	Caffeic acid, p-coumaric acid, ferulic acid, p-hydroxybenzoic acid, protocatechuic acid, syringic acid, vanillic acid
	Lee et al., 2004	Coniferyl alcohol, ferulic acid
	Dizhbite et al., 2004	4-Allyl-2-methoxyphenol, coniferyl alcohol, coniferyl aldehyde, eugenol, guaiacol, guaiacyl propanol-1, guaiacyl propanone-1, isoeugenol, propyl guaiacol, syringol
	Villaño et al., 2005	Caffeic acid, p-coumaric acid, ferulic acid, gallic acid, gentisic acid, protocatechuic acid, tyrosol, vanillic acid
	Gülçin, 2006	Caffeic acid
	Loo et al., 2008	Catechol, 3-methoxycatechol, syringol
	Tabart et al., 2009	Catechin, ellagic acid, gallic acid

Antioxidant test	Reference	Lignin-model compounds / Phenol and catechol derivatives
	Bountagkidou et al., 2010	p-Hydroxybenzaldehyde, protocatechuic aldehyde, salicylaldehyde, syringaldehyde, vanillin, p-hydroxybenzoic acid, protocatechuic acid, salicylic acid, syringic acid, vanillic acid
	Terpinc et al., 2010	Caffeic acid, chlorogenic acid, p-coumaric acid, ferulic acid, protocatechuic acid, rosmarinic acid, sinapic acid, umbellic acid
Ferric reducing antioxidant power	Loo et al., 2008	Catechol, 3-methoxycatechol, syringol
Ferric thiocyanate method	Gülçin, 2006	Caffeic acid
Ferrous metal ions chelating	Gülçin, 2006	Caffeic acid
Hydroxyl radical scavenging	Zou et al., 2002	Sinapic acid
Lipid oxidation inhibition	Zou et al., 2002	Sinapic acid
	Prasad et al., 2007	Ferulic acid
	Sudheer et al., 2008	Ferulic acid
Phosphatidylcholine liposome oxidation	Bountagkidou et al., 2010	p-Hydroxybenzaldehyde, protocatechuic aldehyde, salicylaldehyde, syringaldehyde, vanillin, p-hydroxybenzoic acid, protocatechuic acid, salicylic acid, syringic acid, vanillic acid
Nitric oxide radical scavenging	Lee et al., 2004	Coniferyl alcohol, ferulic acid
Nitrogen monoxide scavenging	Zou et al., 2002	Sinapic acid
Oxygen radical absorbance capacity	Villaño et al., 2005	Caffeic acid, p-coumaric acid, ferulic acid, gallic acid, gentisic acid, protocatechuic acid, tyrosol, vanillic acid
	Tabart et al., 2009	Catechin, ellagic acid, gallic acid
Peroxyl radical scavenging	Barclay et al., 1997	4-Allyl-2,6 dimethoxyphenol, coniferyl alcohol, coniferyl aldehyde, eugenol, isoeugenol, 4-propylguaiacol
	Bountagkidou et al., 2010	p-hydroxybenzaldehyde, protocatechuic aldehyde, salicylaldehyde, syringaldehyde, vanillin, p-hydroxybenzoic acid, protocatechuic acid, salicylic acid, syringic acid, vanillic acid
Peroxynitrite scavenging	Zou et al., 2002	Sinapic acid
Phosphomolybdenum method	Loo et al., 2008	Catechol, 3-methoxycatechol, syringol
Radical chain-breaking	Itagaki et al., 2009	Ferulic acid
Rancimat	von Gadow et al., 1997	Protocatechuic acid, caffeic acid, p-hydroxybenzoic acid, p-coumaric acid, ferulic acid, syringic acid, and vanillic acid
	Bountagkidou et al., 2010	p-Hydroxybenzaldehyde, protocatechuic aldehyde, salicylaldehyde, syringaldehyde, vanillin, p-hydroxybenzoic acid, protocatechuic acid, salicylic acid, syringic acid, vanillic acid
Reducing power	Gülçin, 2006	Caffeic acid
Superoxide radical scavenging	Zou et al., 2002	Sinapic acid

Table 1. (Continued)

Antioxidant test	Reference	Lignin-model compounds / Phenol and catechol derivatives
	Lee et al., 2004	Coniferyl alcohol, ferulic acid
	Gülçin, 2006	Caffeic acid
	Itagaki et al., 2009	Ferulic acid
	Tabart et al., 2009	Catechin, ellagic acid, gallic acid
	Terpinc et al., 2010	Caffeic acid, chlorogenic acid, p-coumaric acid, ferulic acid, protocatechuic acid, rosmarinic acid, sinapic acid, umbellic acid
Thiobarbituric acid	Itagaki et al., 2009	Ferulic acid
Total reactive antioxidant potential	Zou et al., 2002	Sinapic acid

Table 2. Examples of biological activities of model lignin monomers and precursors

Biological activity	Reference	Lignin-model compounds/Phenol derivatives
Antiangiogenic	Lim et al., 2008	Vanillin
Anticancer	Mori et al., 1999	Ferulic acid
	Feng et al., 2005	Chlorogenic acid
	Lee et al., 2008	3,4-Dihydroxybenzaldehyde
Antigenotoxic	Ferguson et al., 2005	p-Coumaric acid, ferulic acid
Antihemolytic	Tabart et al., 2009	Catechin, ellagic acid, gallic acid
Antiinflamatory	Lim et al., 2008	Vanillin
	Sudheer et al., 2008	Ferulic acid
	Itoh et al., 2009	Syringic, vanillic acid
Antimicrobial	Barber et al., 2000	Caffeic acid, coniferaldehyde, coniferyl alcohol, p-coumaraldehyde, p-coumaric acid, p-coumaryl alcohol, eugenol, ferulic acid, sinapaldehyde, sinapic acid, sinapyl alcohol
Antimutagenic	King et al., 2007	Cinnamaldehyde, vanillin
Antinociceptive	Lim et al., 2008	Vanillin
Antiproliferative	Kampa et al., 2004	Caffeic acid, 3,4-dihydroxy-phenylacetic acid, ferulic acid, protocatechuic acid, sinapic acid, syringic acid
Antiulcer	Barros et al., 2008	Caffeic acid, cinnamic acid, p-coumaric acid, ferulic acid
Anxiolytic	Yoon et al., 2007	Sinapic acid
Apoptotic	Kampa et al., 2004	Caffeic acid, 3,4-dihydroxy-phenylacetic acid, ferulic acid, protocatechuic acid, sinapic acid, syringic acid
	Yin et al., 2009	Protocatechuic acid
DNA protector	Sudheer et al., 2008	Ferulic acid
LDL oxidation inhibitor	Cheng et al., 2007	Caffeic acid, chlorogenic acid, p-coumaric acid, ferulic acid, sinapic acid
Photoprotective	Saija et al., 2000	Caffeic acid, ferulic acid
Skin protector	Kim, 2007	Gallic acid
UV-B–induced oxidative stress inhibitor	Prasad et al., 2007	Ferulic acid
Xanthine oxidase inhibitor	Itagaki et al., 2009	Ferulic acid

4.1. Antioxidant Activity

The controversial toxicological studies about the safety of synthetic antioxidants and the consumers´ preferences for natural additives have boosted the research on natural antioxidants. These compounds exert beneficial healthy actions additional to the protective action against oxidation, including protecting from reactive species. The reactive oxygen species (ROS) are generated through metabolic processes in biological systems, but their uncontrolled production can be involved in cancer, atheroscleosis and degenerative diseases associated with aging. Lipid oxidation, initiated by free radicals, light and metal ions, progresses by a radical chain reaction and ends by secondary nonradical compounds. The action of antioxidants can occur by a number of mechanisms, including scavenging of the species responsible for oxidation initiation, interruption of the chain reaction propagation, synergistic increase of the antioxidant activity, as reducing agents or metal chelators, or by cooperative actions. Therefore, a number of tests have been used to measure the *in vitro* antioxidant activity (Antolovich *et al.*, 2002; Sánchez-Moreno, 2002; Huang *et al.*, 2005; Prior *et al.* 2005; Roginsky and Lissi, 2005, Frankel and Finley, 2008; Magalhães *et al.*, 2008; Karadag *et al.*, 2009). However, the *in vivo* relevance can not be extrapolated from these tests (Haenen *et al.*, 2006; Fernández-Panchón *et al.*, 2008), as it is also affected by aspects such as absorption, metabolism and excretion. One method widely used is the DPPH (1,1-diphenyl-2-picrylhydrazyl) radical scavenging; it offers good correlation with the linoleate oxidation and the ABTS$^+$ radical cation scavenging (Brand-Williams *et al.*, 1995; Huang *et al.*, 2005), and the values could predict the activity of lignin in other redox systems (Dizhbite *et al.*, 2004). This spectrophotometric procedure was used for both polymeric and low molecular weight substrates with some modification (Brand-Williams *et al.*, 1995). Operation in inert argon atmosphere was proposed to prevent autooxidation of lignin (Dizhbite *et al.*, 2004). In order to characterize the antioxidant capacity of a product, an extract or a fraction of them, the utilization of several assays is recommended; Table 3 summarizes the *in vitro* tests reported with hydrothermall processed plant biomass.

Literature has reported on the activiy of some lignin precursors and related monomeric and dimeric structures (Kasprzycka-Guttman and Odzeniak, 1994; Barclay *et al.*, 1997; Dizhbite *et al.*, 2004), commercial kraft lignin (Dizhbite *et al.*, 2004), alkali lignin (Lu *et al.*, 1998; Satoh *et al.*, 1998; Dizhbite *et al.*, 2004), and lignin isolates at the laboratory scale (Gregorová *et al.*, 2007). Antioxidant activity has been reported for acid soluble lignin fractions released upon mild acid hydrolysis (Dizhbite *et al.*, 2004; González *et al.*, 2004; Cruz *et al.*, 2005; Moure *et al.*, 2007) or autohydrolysis (Garrote *et al.*, 2003 and 2008; Moure *et al.*, 2005). Steam explosion results in the solubilization of phenolic compounds (Felizón *et al.*, 2000; Fernández-Bolaños *et al.*, 1998) with antioxidant activity (Castro *et al.*, 2008; Conde *et al.*, 2008). The thermal stability of the ethyl acetate solubles from acid hydrolysis was higher than for synthetic antioxidants (Cruz *et al.*, 2007).

The antiradical activity of lignin isolates varies with the type and origin of raw lignocellulosic material, the isolation method and the further purification stages. The antioxidant activity of lignins is influenced by aspects such as the molecular weight distribution, functionalization, π-conjugation, heterogeneity and presence of carbohydrates (Barclay *et al.*, 1997; Dizhbite *et al.*, 2004; Ugartondo *et al.*, 2008). Studies on the structure-activity relationships are available for different lignins and lignin model compounds (Barclay *et al.*, 1997; Dizhbite *et al.*, 2004; Ugartondo *et al.*, 2008), although the adscription of the antioxidant efficiency to specific components is more difficult than in the case of structurally simpler phenolics (Sakagami *et al.*, 2005). Further study is needed to understand the relationship between the chemical structure of the lignins and their antioxidant activities (Ugartondo *et al.*, 2008).

Table 3. Examples of hydrothermal treatments of vegetal biomass. A) processes performed for biomass fractionation and antioxidant assays reported for the ethyl acetate or for the hot water soluble compounds, B) processes for the extraction of bioactive compounds and antioxidant assays reported for the aqueous extract

Hydrothermal treatment	Plant biomass	Antioxidant assay	Reference
A) Fractionation			
Autohydrolysis	*Castanea sativa* burs	ABTS, βC, DPPH, FRAP, RP, ORAC	Conde *et al.*, 2011
	Eucalyptus globulus wood	DPPH	Garrote *et al.*, 2003
	Eucalyptus globulus wood	βC, DPPH	González *et al.*, 2004
	Eucalyptus globulus wood	ABTS, βC, DPPH, FRAP, RP, ORAC	Conde *et al.*, 2011
	Hordeum vulgare husks	ABTS, DPPH, FRAP, RP	Conde *et al.*, 2008
	Hordeum vulgare husks	DPPH	Garrote *et al.*, 2008
	Miscanthus sinensis	DPPH	García *et al.*, 2010
	Olea europea pruning	ABTS, βC, DPPH, FRAP, RP,	Conde *et al.*, 2009
	Pinus radiata wood	ABTS, βC, DPPH, FRAP, HOSC, RP	Moure *et al.*, 2005
	Prunus amygdalus shells	ABTS, βC, DPPH, FRAP, RP, ORAC	Conde *et al.*, 2011
	Vitis vinifera pomace	ABTS, βC, DPPH, FRAP, RP, ORAC	Conde *et al.*, 2011
	Winemaking waste solids	DPPH	Cruz *et al.*, 2004
	Zea mays cobs	DPPH	Garrote *et al.*, 2003
	Zea mays cobs	ABTS, βC, DPPH, FRAP, RP, ORAC	Conde *et al.*, 2011
Steam Explosion	*Olea europea* pruning	ABTS, βC, DPPH, FRAP, RP	Conde *et al.*, 2009
	Olea europea wood	ABTS, βC, DPPH, FRAP, RP	Castro *et al.*, 2008
	Sasa palmata leaves	DPPH	Kurosumi *et al.*, 2007
B) Extraction			
Hot Water Extraction	*Acacia confusa* leaves	ABTS, RP, SRS	Tung *et al.*, 2009

Hydrothermal treatment	Plant biomass	Antioxidant assay	Reference
	Hyphaene thebaica fruits	DPPH, FRAP, HOSC, RP, SRS	Hsu *et al.*, 2006
	Crataegus cuneata	HOSC, SRS	Satoh *et al.*, 1998
Pressurized Hot Water Extraction	*Peumus boldus* leaves	ABTS, DPPH	Del Valle *et al.*, 2005
	Morinda citrifolia roots	ABTS, DPPH	Pongnaravane *et al.*, 2006
Pressurized Low Polarity Water Extraction	*Saponaria vaccaria* seed	ABTS, DPPH, FRAP	Güçlü-Üstündağ and Mazza, 2009
Subcritical Water	*Betula pendula* bark	DPPH	Co *et al.*, 2009
Extraction	*Eucalyptus grandis* leaves	NOSC	Kulkarni *et al.*, 2008
	Terminalia chebula Retz. fruits	ABTS	Rangsriwong *et al.*, 2009

ABTS, 2,2,′-azinobis(3-ethylbenzothiozoline-6-sulfonate) radical scavenging; βC, β-Carotene bleaching; DPPH, (α, α-diphenyl-β-picrylhydrazyl) radical scavenging; FRAP, Ferric reducing antioxidant power; RP, Reducing power; ORAC, Oxygen radical absorbance capacity; HOSC, Hydroxyl radical scavenging capacity; SRS, Superoxide radical scavenging; NOSC, Peroxynitrite radical scavenging capacity

Antioxidant activity has been reported for lignins available from commercial sources or from hydrothermal processing. The radical scavenging capacity of the methanol soluble fraction of kraft lignin was higher than that of the whole kraft lignin sample, and similar to the ones of δ-tocopherol and ascorbic acid (Brand-Williams *et al.*, 1995; Dizhbite *et al.*, 2004). Lignins from pulping exhibited higher antioxidant activity than Trolox (Košíková *et al.*, 2006). Cacao husk lignin fractions, obtained by precipitation with acid and ethanol synergistically enhanced the superoxide anion and hydroxyl radical scavenging activity of vitamin C (Sakagami *et al.*, 2008). A process consisting on steam explosion (1 min, 250 °C) followed by hot water and methanol extractions was used for separating the antioxidant compounds from *S. palmata* leaf, resulting in a product yield of 22% with 14% the radical scavenging capacity of butylated hydroxyanisole (BHA) (Kurosumi *et al.*, 2007).

The hot water extraction causes hydrolytic reactions, enabling for example the solubilization of catechins and theaflavins (Tsubaki *et al.*, 2008). Hot water *Pinus radiata* bark extracts exhibited a potent DPPH radical scavenging activity, ascribed to the presence of flavan-3-ols and an entire series of procyanidins from monomers to longer polymers. Monomeric components such as taxifolin, quercetin, protocatechuic acid, ferulic acid, caffeic acid, etc., are also present in these extracts (Ku *et al.*, 2007). Autohydrolysis of *Pinus pinaster* wood yielded antioxidants with activity comparable to synthetic compounds (Moure *et al.*, 2005). The subcritical water extraction of rosemary yielded extracts with higher DPPH radical scavenging capacity than those extracted by supercritical fluids. The activity was due to carnosic acid, carnosol, rosmanol and rosmanol-related polar compounds. Increasing temperatures improved the yields of all compounds except carnosic acid, which was preferentially extracted at 200 °C (Ibáñez *et al.*, 2003). Hydrothermal processing of wheat bran released soluble products containing esterified ferulate, with antioxidant activity comparable to Trolox (Rose and Inglett, 2010). The refining procedures also affect the structure and properties of lignin isolates, although this aspect has been scarcely studied. The antiradical capacity of *Miscanthus sinensis* lignin depended on the fractionation treatments (soda, organosolv or autohydrolysis). Ultrafiltration with ceramic membranes of different molecular weight cut-offs (10 and 15 kDa) allowed the production of lignin fractions with different antioxidant capacity (with advantage for products obtained with the highest cut-off) (García *et al.*, 2010).

Recent studies reported on the antioxidant efficiency in biological systems, based on hemolysis protection (Ugartondo *et al.*, 2008; Vinardell *et al.*, 2008) or hydrogen peroxide induced oxidation in human red blood cells (Ugartondo *et al.*, 2009). Industrial lignins of different types (lignosulfonate, *kraft*, and steam explosion) protected cell membranes against lipid peroxidation according to their molecular weights: low molecular weight lignins were the most effective radical scavengers, with a behavior similar to that of epicatechin (Ugartondo *et al.*, 2009). It has been suggested that the action of dietary fiber against colon cancer may be partly determined by the free radical-scavenging ability of lignin (Lu *et al.*, 1998), and by the medicinal efficacy of plant extracts (Satoh *et al.*, 1998).

4.2. Antimutagenic and Anticarcinogenic Activity

The antioxidant properties may explain only a part of the antitumor effects caused by dietary phenolics, since other modulatory actions could be responsible for interfering in the development of malignant tumors (Ren *et al.*, 2003; Bonfili *et al.*, 2008; Pan *et al.*, 2008). Potential cancer chemopreventive agents can act by several major mechanisms, including inhibiting phase I and phase II enzymes, scavenging DNA reactive agents, suppressing the over-expression of pro-oxidant enzymes, modulating hormone homeostasis, suppressing hyper-cell proliferation induced by carcinogens, inducing apoptosis, counteracting angiogenesis, and/or inhibiting certain phenotypic expressions of preneoplastic and neoplastic cells (Ren *et al.*, 2003). Some studies with lignin are available in this field, although flavonoids have received more attention in recent years (Bonfili *et al.*, 2008).

Lignin can protect living organisms against different genotoxic compounds, avoiding the drawbacks of synthetic compounds. The potential medicinal applications of industrial lignins from pulping could be based on the protective effect on DNA, and make them potential antimutagenic and anticarcinogenic agents. Endogenous oxidative damage of DNA is also considered as an important etiologic factor in the development of chronic diseases. The biologically modified lignin decreased the level of DNA strand breaks of mammalian cells exposed to oxidative treatment with hydrogen peroxide, as well as to the alkylating damage with N-methyl-N,-nitro-N-nitrosoguanidine (Košíková *et al.*, 2006). Water-soluble, sulfur-free lignin obtained by fractionation of hardwood hydrolyzate did not increase the level of DNA damage substantially, whereas other lignins (obtained by oxidation of water-soluble sulfur-free lignin, or by extraction of oxidized lignin) increased the oxidative damage to DNA (Slamenova *et al.*, 2000). The water-soluble sulfur-free lignin obtained by fractionation of hardwood hydrolyzate reduced DNA lesions induced by H_2O_2 or visible light-excited methylene blue in rat testicular cells and rat peripheral blood lymphocytes both *in vitro* and *ex vivo* (Lábaj *et al.*, 2004). Literature has been reported on the *in vitro* protective effect of lignin on rat hepatocyte DNA, and on *ex vivo* experiments with lignin fed rats: in both cases, lignin significantly reduced the direct DNA strand breaks induced by H_2O_2, the alkali-labile sites of DNA, and the oxidative DNA lesions induced by visible light-excited methylene blue (Lábaj *et al.*, 2007).

Lignin byproducts from concentrated spruce kraft liquors, modified chemically with diluted sulfuric acid or biologically by *Geotrichum klebahnii* and *Sporobolomyces rosens*, showed protective actions against model carcinogens in hamster cells, in lignin-fed rats and in human colon carcinoma cells (Košíková *et al.*, 2002 an 2009). The protective effect of *kraft* lignins from spruce or beech on DNA damage could be due to the reactions of the mutagenic agents with lignin hydroxyl groups, and also correlates with the adsorption affinity to bind mutagenic N-nitrosamines and to the cross-linking density. Lignins obtained from *kraft* pulping inhibited mutagenicity and provided *in vitro* DNA-protective effect against oxidative stress induced by H_2O_2 and visible light in mice lymphocytes and testicular cells. This behavior could not be ascribed just to the antioxidant properties, as probably lignin is partially digested by gut microflora to yield lignans, and these compounds are responsible for the decreased DNA damage (Košiková and Lábaj, 2009). Lignin biopolymers were reported to protect human hepatoma cells and human colonic cells from the genotoxicity of an effective anti-HIV drug (3'-azido-3'-dideoxythymidine) (Slameňová *et al.*, 2006).

Alkali-lignin can inhibit the activity of enzymes related to the generation of superoxide anion radicals and limit the growth and viability of cancer cells (Lu *et al.*, 1998). Lignin-related structures extracted with NaOH from pine cone and precipitated with acid or with ethanol (containing complexes with sugars or polysaccharides) showed antitumor activity on mice cells (Sakagami *et al.*, 1991).

4.3. Antimicrobial and Antiviral Activity of Lignin-Derived Compounds

Antimicrobial drugs are losing their effectiveness due to the evolution of pathogen resistance. This antibiotic-resistance has led to some countries to ban the use of antibiotics for certain animal-related applications. Natural products could be the base of alternative, effective and less toxic antimicrobials: in particular, some plant polyphenolics possess strong anti-bacterial, anti-fungal and anti-parasitic activities, and could be employed as an alternative to in-ration antibiotics (Wang *et al.*, 2009).

The antimicrobial properties of lignin fragments are well recognized, although the variety of assays employed to assess the antimicrobial activity does not easily allow a direct comparison between studies. The mechanisms of antibacterial action vary among phenolic compounds, and the mode of action of natural complex mixtures of phenolic compounds can not be predicted. The antimicrobial action of phenolics has been related with the cell membrane damage and subsequent lysis of bacteria and release of cell contents, with inactivation of cellular enzymes, modification of membrane permeability, and decreases in electron transport and nutrients uptake; and could interfere with the metabolic synthesis of macromolecules and nucleic acids.

Phenolic compounds naturally found in plants and spices show antimicrobial activity and, although they are less active than synthetic molecules, their utilization at higher doses is possible. The antimicrobial action of low-molecular-weight lignin related compounds against bacteria, yeast-like organisms and protozoa yeast has been reported (Zemek *et al.*, 1987). The mutagenesis leading to antibiotic resistance has to be limited (Birosová *et al.*, 2005). In this field, vanillin and lignin reduced the frequency of mutations leading to ciprofloxacin resistance in *Salmonella typhimurium*; whereas lignin was more effective than vanillin as an inhibitor of spontaneous and induced mutations leading to gentamicin resistance (Birosová *et al.*, 2005). Antimicrobial effects of lignin fragments and precursors have been compiled (Baurhoo *et al.*, 2008; Barber *et al.*, 2000). The antimicrobial activity of simple phenolic compounds with chemical structures related to those present in hydrolyzates from mild-acid based processes have been summarized (Garrote *et al.*, 2004). Several of the intermediates involved in the general phenylpropanoid pathway and lignin specific pathway are naturally occurring hydroxycinnamates with antimicrobial activity against yeasts and bacteria. Hydroxycinnamaldehydes possess antifungal and antibacterial activities, whereas the hydroxycinnamyl alcohols possessed little antimicrobial activity (Barber *et al.*, 2000). Lignin-related structures complexed with sugars or polysaccharides extracted with NaOH from pine cone induced antimicrobial activity against *Staphylococcus aureus*, *Escherichia coli*, *Pseudomonas aeruginosa*, *Klebsiella pneumoniae* and *Candida albicans*, and induced antiparasite activity against *Hymenolepis nana* in mice (Sakagami *et al.*, 1991).

In animal nutrition, native lignin is mostly regarded as a barrier to nutrient digestibility. The strong carbon–carbon and ether linkages make lignin resistant to degradation, limiting digestion of forages and decreasing their nutritional value. Although lignosulphonates are useful feed pellet binders, the interest in lignin as a biological feed additive in animal nutrition is quite recent.

Lignin is not digested by monogastric animals, but the rumen bacterial flora facilitates degradation of benzyl ether linkages of lignin polymers. In contrast to native lignin, purified lignin does not represent a barrier to digestion in monogastric animals. Several *in vitro* and *in vivo* studies have demonstrated the antimicrobial properties of purified lignin fragments (Baurhoo *et al.*, 2008). Data on the productivity increase and health benefits derived from the utilization of purified lignin in farm animals non-fed with antibiotics was reviewed (Baurhoo *et al.*, 2008). The antimicrobial activities of semipurified industrial lignin byproducts from lignocellulosic biorefineries suggest that these phenolics may have value as feed additives for ruminants and broilers. Several studies confirmed that lignin can be a potential alternative to antibiotics as growth promoters in broilers, based on the prebiotic action of purified lignin. Prebiotics are non digestible components of diet able of selectively stimulating the growth or metabolic activity of beneficial bacteria, limiting intestinal colonization of pathogens (Gibson and Roberfroid, 1995). The dietary inclusion of purified lignin improved chickens weight gain and feed efficiency, and reduced the concentrations of volatile fatty acids in the ceca and large intestine (Ricke *et al.*, 1982). Lignin reduced the intestinal translocation of pathogenic bacteria following burn injury in rats and inhibited the *in vitro* growth of *E. coli*, *Staphylococcus aureus*, and *Pseudomonas sp.* (Nelson *et al.*, 1994). Lignin from the Alcell pulping process, composed of low molecular weight phenolic fragments, has been reported to improve growth performance of veal calves and to inhibit the *in vitro* growth of *Escherichia coli* (Phillip *et al.*, 2000). Supplementation with purified lignin improved the feed efficiency, but did not affect fecal shedding of *E. coli* by lambs (Wang *et al.*, 2009). Birds fed with lignin had increased population of beneficial bacteria in the ceca, increased villi height and number of goblet cells in the jejunum, and lower population of *E. coli* in the litter than birds fed with antibiotics (Baurhoo *et al.*, 2007a). Feeding lignin at low levels could be a dietary strategy to reduce the *E. coli* load in the gut and litter of broilers, being an alternative to the use of antibiotics as growth promoters in poultry production (Baurhoo *et al.*, 2007b). Alcell lignin increased the intestinal concentrations of *Lactobacilli* and *Bifidobacteria* in broiler chickens, and may exert health benefits in monogastric animals, improving the morphological structures in intestines, inhibiting the growth of pathogenic enteric bacteria, and promoting the safety of these products to humans (Baurhoo *et al.*, 2007b; Baurhoo *et al.*, 2008).

Based on the scarce literature available, it can be seen that animal responses to purified lignin depend on dosage, animal species and source and type of the lignin. Due to the diversity of feedstocks and processing methods, further research is needed in order to establish optimum dosages of bioactive lignin products leading to health and welfare of animals, safety of animal products, and increased production benefits (Baurhoo *et al.*, 2008), and to assess their effects as feed additives for ruminant animals (Wang *et al.*, 2009).

Natural lignified material and polymers from dehydrogenation reactions inhibited the cytopathic effect of human immunodeficiency virus in human myelogenous leukemic cell lines, as well as in MT-4 and MOLT-4 cells (Kunisada *et al.*, 1992). These authors also reported that myeloperoxidase might not be involved in the anti-HIV activity induced by lignins. The anti-HIV potential was reported for a synthetic lignin, obtained by dehydrogenation of *p*-coumaric acid. The serum preparations collected after intravenous injection of this lignin into mice inhibited the cytopathogenicity of an HIV-1 strain (Shimizu *et al.*, 1993). Carboxylated lignins, based on a 4-hydroxy cinnamic acid scaffold synthesized by enzymatic oxidative coupling inhibited herpes simplex virus-1 entry into mammalian cells, and were more potent than sulfated lignins, which are macromolecular mimetics of heparan sulfate lignins (Thakkar *et al.*, 2010).

Cacao husk lignin fractions obtained by acid and ethanol precipitation, showed anti-human immunodeficiency virus activity comparable to that of conventional anti-HIV compounds, as well as anti-influenza virus activity (Sakagami *et al.*, 2008). Alkali-soluble lignin isolated from *Pinus nigra* seed cones showed anti-HIV-1 activity (Eberhard and Young, 1996). Lignin-carbohydrate complexes prepared from pine cones showed potent antiviral activity (Sakagami *et al.*, 2005). The majority of the patients treated with pine cone lignin and ascorbic acid reduced both severity and recurrence of herpes simplex virus type 1 symptoms, usually treated with antiviral drugs and natural remedies (López *et al.*, 2009). Lignin from pine cone extract showed antiviral activity and granulocyte iodination stimulating activity comparable to those of commercial alkali-lignin. However, the pine cone extract, with a lignin-like polyphenolic skeleton and molecular matrix of unknown components, had higher immunopotentiating activities than commercial alkali-lignin, such as antitumor and antimicrobial activities and splenocyte stimulating activity (Sakagami *et al.*, 1989). Lignin structures extracted with NaOH from pine cone and precipitated with acid or with ethanol showed antiviral activity against the virus of human immunodeficiency, herpes simplex and influenza. The *in vivo* immunopotentiating activity was ascribed to polysaccharide conjugated lignins, since no activity was observed for acid-hydrolyzed fractions. The antiviral activity of lignins prepared by polymerization of phenylpropanoid units was comparable to that of the undecomposed counterparts of the pine cone extract (Sakagami *et al.*, 1991).

Immunological and antiviral activities have been reported for the water-soluble lignin of an extract of *Lentinus edodes* mycelia culture medium, consisting of a highly condensed and polycarboxylated structure solubilized by this organims from bagasse (Suzuki *et al.*, 1990).

These authors reported the activation of murine macrophages, proliferation of bone marrow cells, and *in vitro* inhibition of the replication of human immunodeficiency virus.

The potential activity of lignin-based antioxidant, antimicrobial and antiviral formulations for topical use has been assessed by cytotoxicity evaluation, which confirmed that lignins can be used in cosmetics and pharmaceutical preparations without harmful effects in human cells. Cosmetic and pharmaceutical applications of lignins require absence of skin irritation. The cytotoxic effects of the lignins on the cell membrane integrity of human keratinocyte HaCaT and murine fibroblast 3T3 cells were measured. The effective antioxidant concentrations were smaller than the cytotoxic ones, confirming their safe use as antioxidants suited for topical applications (Ugartondo *et al.*, 2008). The information ensuring that lignins are safe for normal body cells at the concentrations employed in formulations is relevant for commercial applications based on antioxidant and therapeutic properties.

5. COMMERCIAL INTEREST AND APPLICATIONS

Commercical applications can be based on biological activities (including antioxidant activity), but also applications related to polymers, cosmetics, pharmaceuticals and food processing industries can be highlighted. Stability studies showed that solutions of industrial lignins remained stable with no alteration after 60 days at room temperature without any special storage condition and after 2 h of exposure to ultraviolet irradiation (Ugartondo *et al.*, 2008).

The amount and variety of functional groups of lignins make them compatible additives for polymer manufacture, and enables their modification to yield materials with special properties. The industrial incorporation of sulfur-free, water-insoluble lignins into polymeric systems (such as automotive brakes, wood panel products, biodispersants, polyurethane foams, and epoxy resins for printed circuit boards) has been reviewed (Lora and Glasser, 2002). Lignin from wood pulping is suitable as an stabilizer against thermooxidative aging of polypropylene composites and styrene-butadiene vulcanizates, avoiding the use of commercial additives or enhancing their antioxidant potency (Košiková and Lábaj, 2009; Košiková *et al.*, 2006; Košíková and Sláviková, 2010). Optically transparent polypropylene films with appropriate mechanical properties can only be manufactured with a limited lignin load, as the proportion of lignin could limit the rheological and strenght properties of polyolefin composites (Košiková *et al.*, 2001). The oxidative stabilizing effect of wheat straw and wheat bran lignins (Pouteau *et al.*, 2003), *kraft* lignin (Pouteau *et al.*, 2003; Fernándes *et al.*, 2006), and commercial hydrolytic lignin (Canetti *et al.*, 2006) on diverse natural or synthetic polymers has been reported. As an example, the stabilization efficiency of lignin in virgin and recycled polypropylene is comparable to that of a commercial product (Gregorová *et al.*, 2007). Other advantages derived from its incorporation into plastics are related to the absence of health hazards and to the environmental benefits derived from its biodegradable nature. Lignins (for example, from wet-mill ethanol plants) can be used as a performance enhancer of asphalts. In blends with asphalts, lignins act as a filler and improve the intermediate and low-temperature properties of the binders, allowing to delay long-term aging (McCready and Williams, 2008).

CONCLUSION

Lignin is an underutilized polymer present in the cell walls of plants, which can be separated from the raw materials by chemical processing. When industrial wastes and byproducts from agricultural, forestry or industrial activities are subjected to fractionation, the environmental impact caused by the unutilized feedstocks is alleviated. The potential of pressurized water as a solvent and as an autocatalytic reaction media has received increasing interest, and has been object of extensive research. Purified lignin has been proposed as a natural feed additive (with antimicrobial and prebiotic effects), as an antioxidant and as a therapeutic agent for its high affinity toward mutagenic and carcinogenic compounds. Phenolic compounds produced by hydrolysis of lignin are promising chemicals due to their growing demand in the food, fragrance, and polymer industries. The heterogeneity of lignins requires a detailed structural characterization, as a key information for the assessment of their biological properties. The development of optimized extraction processes and the physico-chemical and biological characterization of different lignins and lignin-derived fractions will promote the industrial applications of lignins.

REFERENCES

Adolphe, J. L., Whiting, S. J., Juurlink, B. H. J., Thorpe, L. U., and Alcorn, J. (2010). Health effects with consumption of the flax lignan secoisolariciresinol diglucoside. *British Journal of Nutrition*, 103, 929-938.

Akhtar, J., Kuang, S. K., and Amin, N. S. (2010). Liquefaction of empty palm fruit bunch (EPFB) in alkaline hot compressed water. *Renewable Energy*, 35, 1220-1227.

Amidon, T. E., and Liu, S. (2009). Water-based woody biorefinery. *Biotechnology Advances*, 27, 542-550.

Andersen, R. A., Vaughn, T. H., and Kasperbauer, M. J. (1980). Coniferyl and sinapyl alcohols: Major phenylpropanoids released in hot water extracts of tobacco and alfalfa. *Journal of Agricultural and Food Chemistry*, 28, 427-432.

Ando, H., Sakaki, T., Kokusho, T., Shibata, M., Uemura, Y., Hatate, Y. (2000). Decomposition behavior of plant biomass in hot-compressed water. *Industrial and Engineering Chemistry Research*, 39, 3688-3693.

Antolovich, M., Prenzler, P. D., Patsalides, E., McDonald, S., and Robards, K. (2002). Methods for testing antioxidant activity. *Analyst*, 127, 183-198.

Arni, S. A., Drake, A. F., Borghi, M. D., and Converti, A. (2010). Study of aromatic compounds derived from sugarcane bagasse: II. Effect of concentration. *Chemical Engineering and Technology*, 33, 523-531.

Asghari, F. S., and Yoshida, H. (2010). Conversion of Japanese red pine wood (*Pinus densiflora*) into valuable chemicals under subcritical water conditions. *Carbohydrate Research*, 345, 124-131.

Barber, M. S., McConnell, V. S, and Decaux, B. S. (2000). Antimicrobial intermediates of the general phenylpropanoid and lignin specific pathways. *Phytochemistry*, 54, 53-56.

Barclay, L. R. C., Xi, F., and Norris, J. Q. (1997). Antioxidant properties of phenolic lignin model compounds. *Journal of Wood Chemistry and Technology*, 17, 73-90.

Barros, M. P., Lemos, M., Maistro, E. L., Leite, M. F., Sousa, J. P. B., Bastos, J. K., and Andrade, S. F. (2008). Evaluation of antiulcer activity of the main phenolic acids found in Brazilian Green Propolis. *Journal of Ethnopharmacology*, 120, 372-377.

Baurhoo, B., Letellier, A., Zhao, X., and Ruiz-Feria, C. A. (2007a). Cecal populations of lactobacilli and bifidobacteria and *Escherichia coli* populations after *in vivo Escherichia coli* challenge in birds fed diets with purified lignin or mannanoligosaccharides. *Poultry Science*, 86, 2509-2515.

Baurhoo, B., Phillip L., and Ruiz-Feria, C. A. (2007b). Effects of purified lignin and mannan oligosaccharides on intestinal integrity and microbial populations in the ceca and litter of broiler chickens. *Poultry Science*, 86, 1070-1078.

Baurhoo, B., Ruiz-Feria, C. A., and Zhao, X. (2008). Purified lignin: Nutritional and health impacts on farm animals-A review. *Animal Feed Science and Technology*, 144, 175-184.

Bentivenga, G., Bonini, C., D'Auria, M., and De Bona, A. (1999). Singlet oxygen degradation of lignin: A GC-MS study on the residual products of the singlet oxygen degradation of a steam exploded lignin from beech. *Journal of Photochemistry and Photobiology A: Chemistry*, 128, 139-143.

Bentivenga, G., Bonini, C., D'Auria, M., DeBona, A., and Mauriello, G. (2000). Fine chemicals from singlet-oxygen-mediated degradation of lignin - A GC/MS study at different irradiation times on a steam-exploded lignin. *Journal of Photochemistry and Photobiology A: Chemistry*, 135, 203-206.

Birosová, L., Mikulásová, M., and Chromá, M. (2005). *The effect of phenolic and polyphenolic compounds on the development of drug resistance.* Biomedical papers of the Medical Faculty of the University Palacký, Olomouc, Czechoslovakia, 149, 405-407.

Boeriu, C. G., Bravo, D., Gosselink, R. J. A., and van Dam, J. E. G. (2004). Characterisation of structure-dependent functional properties of lignin with infrared spectroscopy. *Industrial Crops and Products*, 20, 205-218.

Bonfili, L., Cecarini, V., Amici, M., Cuccioloni, M., Angeletti, M., Keller, J. N., and Eleuteri, A. M. (2008). Natural polyphenols as proteasome modulators and their role as anti-cancer compounds. *FEBS Journal*, 275, 5512-5526.

Bonini, C., D' Auria, M., and Ferri, R. (2002). Singlet oxygen mediated degradation of lignin - Isolation of oxidation products from steam-exploded lignin from pine. *Photochemical and Photobiological Sciences*, 1, 570-573.

Boudet, A. M., Kajita, S., Grima-Pettenati, J., and Goffner, D. (2003). Lignins and lignocellulosics: a better control of synthesis for new improved uses. *Trends in Plant Science*, 8, 576-581.

Bountagkidou, O. G., Ordoudi, S. A., and Tsimidou, M. Z. (2010). Structure–antioxidant activity relationship study of natural hydroxybenzaldehydes using *in vitro* assays. *Food Research International*, 43, 2014-2019.

Brand-Williams, W., Cuvelier, M.E. and Berset, C. (1995). Use of a free radical method to evaluate antioxidant activity. LWT - *Food Science and Technology*, 28, 25-30.

Brunati, M., Marinelli, F., Bertolini, C., Gandolfi, R., Daffonchio, D., and Molinari, F. (2004). Biotransformations of cinnamic and ferulic acid with *Actinomycetes*. *Enzyme and Microbial Technology*, 34, 3-9.

Brunner, G. (2009). Near critical and supercritical water. Part I. Hydrolytic and hydrothermal processes. *The Journal of Supercritical Fluids*, 47, 373-381.

Budrat, P. and Shotipruk, A. (2009). Enhanced recovery of phenolic compounds from bitter melon (*Momordica charantia*) by subcritical water extraction. *Separation and Purification Technology*, 66, 125-129.

Buranov, A. U., and Mazza, G. (2007). Fractionation of flax shives by water and aqueous ammonia treatment in a pressurized low-polarity water extractor. *Journal of Agricultural and Food Chemistry*, 55, 8548-8555.

Buranov, A.U., and Mazza, G. (2009). Extraction and purification of ferulic acid and vanillin in flax shives and wheat bran by alkaline-hydrolysis and pressurized solvents. *Food Chemistry*, 115, 1542–1548.

Buranov, A.U., and Mazza, G. (2010). Extraction and characterization of hemicelluloses from flax shives by different methods. *Carbohydrate Polymers*, 79, 17-25.

Cacace, J. E., and Mazza, G. (2006). Pressurized low polarity water extraction of lignans from whole flaxseed. *Journal of Food Engineering*, 77, 1087-1095.

Canetti, M., Bertini, F., De Chirico, A., and Audisio, G. (2006). Thermal degradation behaviour of isotactic polypropylene blended with lignin. *Polymer Degradation and Stability*, 91, 494–498.

Castro, E., Conde, E., Moure, A., Falqué, E., Cara, C., Ruíz, E., and Domínguez, H. (2008). Antioxidant activity of liquors from steam explosion of *Olea europea* wood. *Wood Science and Technology*, 42, 579-592.

Cheng, J-C., Dai, F., Zhou, B., Yang, L., and Liu, Z-L. (2007). Antioxidant activity of hydroxycinnamic acid derivatives in human low density lipoprotein: Mechanism and structure-activity relationship. *Food Chemistry*, 104, 132-139.

Cheng, L., Ye, X. P., He, R., and Liu, S. (2009). Investigation of rapid conversion of switchgrass in subcritical water. *Fuel Processing Technology*, 90, 301-311.

Clark, J. H. (2007). Green chemistry for the second generation biorefinery - Sustainable chemical manufacturing based on biomass. *Journal of Chemical Technology and Biotechnology*, 82, 603-609.

Co, M., Koskela, P., Eklund-Åkergren, P., Srinivas, K., King, J.W., Sjöberg, P.J.R. and Turner, C. (2009). Pressurized liquid extraction of betulin and antioxidants from birch bark. *Green Chemistry*, 11, 668-674.

Conde, E., Cara, C., Moure, A., Ruíz, E., Castro, E., and Domínguez, H. (2009). Antioxidant activity of the phenolic compounds released by hydrothermal treatments of olive tree pruning. *Food Chemistry*, 114, 806-812.

Conde, E., Moure, A., Domínguez, H., and Parajó, J. C. (2008). Fractionation of antioxidants from autohydrolysis of barley husks. *Journal of Agricultural and Food Chemistry*, 56, 10651-10659.

Conde, E., Moure, A., Domínguez, H., and Parajó, J. C. (2011). Production of antioxidants by non-isothermal autohydrolysis of lignocellulosic wastes. LWT - *Food Science and Technology*, 44, 436-442.

Cruz, J. M., Conde, E., Domínguez, H., and Parajó, J. C. (2007). Thermal stability of antioxidants obtained from wood and industrial wastes. *Food Chemistry*, 100, 1059-1064.

Cruz, J. M., Domínguez, H., and Parajó, J. C. (2004). Assessment of the production of antioxidants from winemaking waste solids. *Journal of Agricultural and Food Chemistry*, 52, 5612-5620.

Cruz, J. M., Domínguez, H., and Parajó, J. C. (2005). Anti-oxidant activity of isolates from acid hydrolysates of *Eucalyptus globulus* wood. *Food Chemistry*, 90, 503-511.

Del Valle, J. M., Rogalinski, T., Zetzl, C. and Brunner, G. (2005). Extraction of boldo (*Peumus boldus* M.) leaves with supercritical CO_2 and hot pressurized water. *Food Research International*, 38, 203-213.

Dizhbite, T., Telysheva, G., Jurkjane, V., and Viesturs, U. (2004). Characterization of the radical scavenging activity of lignins - Natural antioxidants. *Bioresource Technology*, 95, 309-317.

Eberhardt, T. L., and Young, R. A. (1996). Assessment of the anti-HIV activity of a pine cone isolate. *Planta Medica*, 62, 63-65.

Falconnier, B., Lapierre, C., Lesage-Meessen, L., Yonnet, G., Brunerie, P., Colonna-Ceccaldi, B., Corrieu, G., and Asther, M. (1994). Vanillin as a product of ferulic acid bioconversion by the white-rot fungus *Pycnoporus cinnabarinus* 1-937: identification of metabolic pathways. *Journal of Biotechnology*, 37, 123-132.

Felizón, B., Fernández-Bolaños, J., Heredia, A., and Guillén, R. (2000). Steam-explosion pre-treatment of olive cake. *Journal of the American Oil Chemists´ Society*, 77, 15-22.

Feng, R., Lu, Y., Bowman, L. L., Qian, Y., Castranova, V., and Ding, M. (2005). Inhibition of activator protein-1, NF-kappaB, and MAPKs and induction of phase 2 detoxifying enzyme activity by chlorogenic acid. *Journal of Biological Chemistry*, 280, 27888–27895.

Ferguson, L. R., Zhu, S-T., and Harris, P. J. (2005). Antioxidant and antigenotoxic effects of plant cell wall hydroxycinnamic acids in cultured HT-29 cells. *Molecular Nutrition and Food Research*, 49, 585-593.

Fernándes, D.M., Winkler Hechenleitner, A.A., Job, A.E., Radovanocic, E. and Gómez Pineda, E.A. (2006). Thermal and photochemical stability of poly (vinyl alcohol)/modified lignin blends. *Polymer Degradability Stability*, 91, 1192–1201.

Fernández-Bolaños, J., Felizón, B., Brenes, M., Guillen, R., and Heredia, A. (1998). Hydroxytyrosol and tyrosol as the main compounds found in the phenolic fraction of steam-exploded olive stones. *Journal of the American Oil Chemists´ Society*, 75, 1643-1649.

Fernández-Bolaños, J., Felizón, B., Heredia, A., Guillén, R., and Jiménez, A. (1999). Characterization of the lignin obtained by alkaline delignification and of the cellulose residue from steam-exploded olive stones. *Bioresource Technology*, 68, 121-132.

Fernández-Bolaños, J., Rodríguez, G., Rodríguez, R., Heredia, A., Guillén, R. and Jiménez, A. (2002). Production in large quantities of highly purified hydroxytyrosol from liquid-solid waste of two-phase olive oil processing or "alperujo". *Journal of Agricultural and Food Chemistry*, 50, 6804-6811.

Fernández-Panchón, M. S., Villano, D., Troncoso, A. M., and García-Parrilla, M. C. (2008). Antioxidant activity of phenolic compounds: from *in vitro* results to *in vivo* evidence. *Critical Reviews in Food Science and Nutrition*, 48, 649-671.

Frankel, E. N., and Finley, J. W. (2008). How to standardize the multiplicity of methods to evaluate natural antioxidants. *Journal of Agricultural and Food Chemistry*, 56, 4901-4908.

García, A., Toledano, A., Andrés, M. A., and Labidi, J. (2010). Study of the antioxidant capacity of *Miscanthus sinensis* lignins. *Process Biochemistry*, 45, 935-940.

Garrote, G., Cruz, J. M., Domínguez, H., and Parajó, J. C. (2003). Valorisation of waste fractions from autohydrolysis of selected lignocellulosic materials. *Journal of Chemical Technology and Biotechnology*, 78, 392-398.

Garrote, G., Cruz, J. M., Domínguez, H., and Parajó, J. C. (2008). Non-isothermal autohydrolysis of barley husks: Product distribution and antioxidant activity of ethyl acetate soluble fractions. *Journal of Food Engineering*, 84, 544-552.

Garrote, G., Cruz, J.M., Moure, A., Domínguez, H. and Parajó, J.C. (2004). Antioxidant activity of byproducts from the hydrolytic processing of selected lignocellulosic materials. *Trends in Food Science and Technology*, 15, 191-200.

Gibson, G.R. and Roberfroid, M.B. (1995). Dietary modulation of the human colonic microbiota: Introducing the concept of prebiotics. *Journal of Nutrition*, 125, 1401-1412.

González, J., Cruz, J. M., Domínguez H., and Parajó. J. C. (2004). Production of antioxidants from *Eucalyptus globulus* wood by solvent extraction of hemicellulose hydrolysates. *Food Chemistry*, 84, 243-251.

Gosselink, R.J.A., Snijder, M.H.B., Kranenbarg, A., Keijsers, E.R.P., De Jong, E., and Stigsson, L.L. (2004). Characterisation and application of NovaFiber lignin. *Industrial Crops and Products*, 20, 191-203.

Gregorova, A., Košíková, B., and Stasko, A. (2007). Radical scavenging capacity of lignin and its effect on processing stabilization of virgin and recycled polypropylene. *Journal of Applied Polymer Science*, 106, 1626-1631.

Güçlü-Üstünda, Ö. and Mazza, G. (2009) Effects of pressurized low polarity water extraction parameters on antioxidant properties and composition of cow cockle seed extracts. *Plant Foods for Human Nutrition*, 64, 32-38.

Gülçin, İ. (2006). Antioxidant activity of caffeic acid (3,4-dihydroxycinnamic acid). *Toxicology*, 217, 213-220.

Haenen, G. R. M. M., Arts, M. J. T. J., Bast, A. and Coleman, M. D (2006). Structure and activity in assessing antioxidant activity *in vitro* and *in vivo*: A critical appraisal illustrated with the flavonoids, *Environmental Toxicology and Pharmacology*, 21 (2 spec. iss.), 191-198.

Hsu, B., Coupar, I. M., and Ng, K. (2006). Antioxidant activity of hot water extract from the fruit of the Doum palm, *Hyphaene thebaica*. *Food Chemistry*, 98, 317–328.

Huang, D., Ou, B., and Prior, R. L. (2005). The chemistry behind antioxidant capacity assays. *Journal of Agricultural and Food Chemistry*, 53, 1841-1856.

Ibáñez, E., Kubátová, A., Señoráns, F. J., Cavero, S., Reglero, U., **and** Hawthorne, S. B. (2003). Subcritical water extraction of antioxidant compounds from rosemary plants. *Journal of Agricultural and Food Chemistry*, 51, 375-382.

Itagaki, S., Kurokawa, T., Nakata, C., Saito, Y., Oikawa, S., Kobayashi, M., Hirano, T., and Iseki, K. (2009). *In vitro* and *in vivo* antioxidant properties of ferulic acid: A comparative study with other natural oxidation inhibitors. *Food Chemistry*, 114, 466-471.

Itoh, A., Isoda, K., Kondoh, M., Kawase, M., Kobayashi, M., Tamesada, M., and Yagi, K. (2009). Hepatoprotective effect of syringic acid and vanillic acid on concanavalin A-induced liver injury. *Biological and Pharmaceutical Bulletin*, 32, 1215-1219.

Jin, F., and Enomoto, H. (2009). Hydrothermal conversion of biomass into value-added products: technology that mimics **nature.** *BioResources*, 4, 704-713.

Kampa, M., Alexaki, V. I., Notas, G., Nifli, A. P., Nistikaki, A., Hatzoglou, A., Bakogeorgou, E., Kouimtzoglou, E., Blekas, G., Boskou, D., Gravanis, A., and Castanas, E. (2004). Antiproliferative and apoptotic effects of selective phenolic acids on T47D human breast cancer cells: potential mechanisms of action. *Breast Cancer Research*, 6, 63-74.

Karadag, A., Ozcelik, B., and Saner, S. (2009). Review of methods to determine antioxidant capacities. *Food Analytical Methods*, 2, 41-60.

Kasprzycka-Guttman, T. and Odzeniak, D. (1994). Antioxidant properties of lignin and its fractions. *Thermochimica Acta*, 231, 161-168.

Kim, J. W., and Mazza, G. (2009). Extraction and separation of carbohydrates and phenolic compounds in flax shives with pH-controlled pressurized low polarity water. *Journal of Agricultural and Food Chemistry*, 57, 1805-1813.

Kim, J.-W., and Mazza, G. (2006). Optimization of extraction of phenolic compounds from flax shives by pressurized low-polarity water. *Journal of Agricultural and Food Chemistry*, 54, 7575-7584.

Kim, K-H., Tsao, R., Yang, R., and Cui, S. W. (2006). Phenolic acid profiles and antioxidant activities of wheat bran extracts and the effect of hydrolysis conditions. *Food Chemistry*, 95, 466-473.

Kim, Y. J. (2007). Antimelanogenic and antioxidant properties of gallic acid. *Biological and Pharmaceutical Bulletin*, 30, 1052-1055.

King, A. A., Shaughnessy, D. T., Mure, K., Leszczynska, J., Ward, W. O., Umbach, D. M., Xu, Z., Ducharm, D., Taylor, J. A., DeMarini, D. M., and Klein, C. B. (2007). Antimutagenicity of cinnamaldehyde and vanillin in human cells: Global gene expression and possible role of DNA damage and repair. *Mutation Research*, 616, 60-69.

Košíková, B. and Sláviková, E. (2010). Use of lignin products derived from wood pulping as environmentally desirable additives of polypropylene films. *Wood Research*, 55, 87-92.

Košiková, B., and Lábaj, J. (2009). Lignin-stimulated protection of polyprolylene films and DNA in cells of mice against oxidation damage. *BioResources*, 4, 805-815.

Košiková, B., Alexy, P., Mikulášová, M., and Kačík, F. (2001). Characterization of biodegradability of lignin – Polyethylene blends. *Wood Research*, 46, 31-36.

Košíková, B., Lábaj, J., Gregorová, A., and Slameňová, D. (2006). Lignin antioxidants for preventing oxidation damage of DNA and for stabilizing polymeric composites. *Holzforschung*, 60, 166-170.

Košíková, B., Slamenová, D., Mikulášová, M., Horváthová, E., and Lábaj, J. (2002). Reduction of carcinogenesis by bio-based lignin derivatives. *Biomass Bioenergy*, 23, 153-159.

Košiková, B., Sláviková, and E., Lábaj, J. (2009). Affinity of lignin preparations towards genotoxic compounds. *BioResources*, 4, 72-79.

Kroon, P. A., and Williamson, G. (1999). Hydroxycinnamates in plants and food: Current and future perspectives. *Journal of the Science of Food and Agriculture*, 79, 355-361.

Ku, C. S., Jang, J. P., and Mun, S. P. (2007). Exploitation of polyphenol-rich pine barks for potent antioxidant activity. *Journal of Wood Science*, 53, 524-528.

Kulkarni, A., Suzuki, S., and Etoh, H. (2008). Antioxidant compounds from *Eucalyptus grandis* biomass by subcritical liquid water extraction. *Journal of Wood Science*, 54, 153-157.

Kunisada, T., Sakagami, H., Takeda, M., Naoe, T., Kawazoe, Y., Ushijima, H., Muller, W. E. G., and Kitamura, T. (1992). Effect of lignins on HIV-induced cytopathogenicity and myeloperoxidase activity in human myelogenous leukemic cell lines. *Anticancer Research*, 12, 2225-2228.

Kurosumi, A., Sasaki, C., Kumada, K., Kobayashi, F., Mtui, G., and Nakamura, Y. (2007). Novel extraction method of antioxidant compounds from Sasa palmata (Bean) Nakai using steam explosion. *Process Biochemistry*, 42, 1449-1453.

Lábaj, J., Slamen□ová, D., Lazarová, M., and Kos□íková, B. (2004). Lignin-stimulated reduction of oxidative DNA lesions in testicular cells and lymphocytes of Sprague-Dawley rats in vitro and ex vivo. *Nutrition and Cancer*, 50, 198-205.

Lábaj, J., Slamenová, D., Lazarová, M., and Košíková, B. (2007). Induction of DNA-lesions in freshly isolated rat hepatocytes by different genotoxins and their reduction by lignin given either as a dietary component or *in vitro* conditions. *Nutrition and Cancer*, 57, 209-215.

Lee, B. H., Yoon, S. H., Kim, Y. S., Kim, S. K., Moon, B. J., and Bae, Y. S. (2008). Apoptotic cell death through inhibition of protein kinase CKII activity by 3,4-dihydroxybenzaldehyde purified from *Xanthium strumarium*. *Natural Product Research*, 22, 1441-1450.

Lee, J-Y., Yoon, J-W., Kim, C-T., and Lim, S-T. (2004). Antioxidant activity of phenylpropanoid esters isolated and identified from *Platycodon grandiflorum* A. DC. *Phytochemistry*, 65, 3033-3039.

Leschinsky, M., Sixta, H., and Patt, R. (2009). Detailed mass balances of the autohydrolysis of *Eucalyptus globulus* at 170°C. *BioResources,* 4, 687-703.

Leschinsky, M., Zuckerstatter, G., Weber, H. K., Patt, R., and Sixta, H. (2008). Effect of autohydrolysis of Eucalyptus globulus wood on lignin structure. Part 1: Comparison of different lignin fractions formed during water prehydrolysis. *Holzforschung*, 62, 645-652.

Li, J., and Gellerstedt, G. (2008). Improved lignin properties and reactivity by modifications in the autohydrolysis process of aspen wood. *Industrial Crops and Products,* 27, 175-181.

Lim, E. J., Kang, H. J., Jung, H. J., Song, Y. S., Lim, C. J., and Park, E. H. (2008). Anti-angiogenic, anti-inflammatory and anti-nociceptive activities of vanillin in ICR mice. *Biomolecules and Therapeutics*, 16, 132-136.

Lomascolo, A., Stentelaire, C., Asther, M., and Lesage-Meessen, L. (1999). *Basidiomycetes* as new biotechnological tools to generate natural aromatic flavours for the food industry. *Trends Biotechnology*, 17, 282-289.

Loo, A.Y., Jain, K., and Darah, I. (2008). Antioxidant activity of compounds isolated from the pyroligneous acid, *Rhizophora apiculata. Food Chemistry*, 107, 1151–1160.

López, B. S. G., Yamamoto, M., Utsumi, K., Aratsu, C., and Sakagami, H. (2009). A clinical pilot study of lignin-ascorbic acid combination treatment of herpes simplex virus. *In Vivo*, 23, 1011-1016.

Lora J. H. and Wayman, M. (1980). Autohydrolysis of aspen milled wood lignin. *Canadian Journal of Chemistry,* 58, 669–676.

Lora, J. H., and Glasser, W. G. (2002). Recent industrial applications of lignin: A sustainable alternative to nonrenewable materials. *Journal of Polymers and Environment,* 10, 39-48.

Lu, F.-J., Chu, L.-H., and Gau, R.-J. (1998). Free radical-scavenging properties of lignin. *Nutrition and Cancer*, 30, 31-38.

Magalhaes, L. M., Segundo, M. A., Reis, S., and Lima, J. L. F. C. (2008). Methodological aspects about *in vitro* evaluation of antioxidant properties. *Analytica Chimica Acta*, 613, 1-19.

Mathew, S. and Abraham, T.E. (2006). Bioconversions of ferulic acid, an hydroxycinnamic acid. *Critical Reviews in Microbiology*, 32, 115-125.

Mathias, A. L., Loprettis, M. I., and Rodrigues, A. E. (1995). Chemical and biological oxidation of *Pinus pinaster* lignin for the production of vanillin. *Journal of Chemical Technology and Biotechnology*, 64, 225-234.

Max, B., Salgado, J. M., Cortés, S., and Domínguez, J. M. (2010). Extraction of phenolic acids by alkaline hydrolysis from the solid residue obtained after prehydrolysis of trimming vine shoots. *Journal of Agricultural and Food Chemistry*, 58, 1909-1917.

McCready, N. S., and Williams, R. C. (2008). Utilization of biofuel coproducts as performance enhancers in asphalt binder. *Transportation Research Record*, 2051, 8-14.

Mitjans, M. and Vinardell, M.P. (2005). Biological activity and health benefits of lignans and lignins. *Trends in Comparative Biochemistry and Physiology*, 11, 55-62.

Mori, H., Kawabata, K., Yoshimi, N., Tanaka, T., Murakami, T., Okada, T., and Murai, H. (1999). Chemopreventive effects of ferulic acid on oral and rice germ on large bowel carcinogenesis. *Anticancer Research*, 19, 3775-8.

Moure, A., Domínguez, and H., Parajó, J. C. (2005). Antioxidant activity of liquors from aqueous treatments of *Pinus radiata* wood. *Wood Science and Technology*, 39, 129-139.

Moure, A., Pazos, M., Medina, I., Domínguez, H., and Parajó, J. C. 2007. Antioxidant activity of extracts produced by solvent extraction of almond shells acid hydrolysates. *Food Chemistry*, 101, 193-201.

Nelson, J.L., Alexander, J.W., Gianotti, L., Chalk, C.L., Pyles, T. (1994). Influence of dietary fiber on microbial growth *in vitro* and bacterial translocation after burn injury in mice. *Nutrition*, 10, 32-36.

Pan, M-H., Ghai, G., and Ho, C-T. (2008). Food bioactives, apoptosis, and cancer. *Molecular Nutrition and Food Research*, 52, 43-52.

Phillip, L. E., E. S. Idziak, and S. Kubow. 2000. The potential use of lignin in animal nutrition, and in modifying microbial ecology of the gut. In Eastern Nutrition Conference (ed. by. Animal Nutrition Association of Canada) Pages 1–9.

Pongnaravane, B., Goto, M., Sasaki, M., Anekpankul, T., Pavasant, P. and Shotipruk, A. 2006. Extraction of anthraquinones from roots of *Morinda citrifolia* by pressurized hot water: Antioxidant activity of extracts. Journal of Supercritical Fluids, 37, 390-396.

Pourali, O., Asghari, F. S., and Yoshida, H. (2010). Production of phenolic compounds from rice bran biomass under subcritical water conditions. *Chemical Engineering Journal*, 160, 259-266.

Pouteau, C., Dole, P., Cathala, B., Averous, L., and Boquillon, N. (2003). Antioxidant properties of lignin in polypropylene. *Polymer Degradation and Stability*, 81, 9-18.

Prasad, N. R., Ramachandran, S., Pugalendi, K. V., and Menon, V. P. (2007). Ferulic acid inhibits UV-B–induced oxidative stress in human lymphocytes. *Nutrition Research*, 27, 559-564.

Priefert, H., Rabenhorst, J., and Steinbüchel, A. (2001). Biotechnological production of vanillin. *Applied Microbiology and Biotechnology*, 56, 296-314.

Prior, R. L., Wu, X., and Schaich, K. (2005). Standardized methods for the determination of antioxidant capacity and phenolics in foods and dietary supplements. *Journal of Agricultural and Food Chemistry*, 53, 4290-4302.

Pronyk, C., and Mazza, G. (2009). Design and scale-up of pressurized fluid extractors for food and bioproducts. *Journal of Food Engineering*, 95, 215-226.

Rangsriwong, P., Rangkadilok, N., Satayavivad, J., Goto, M., and Shotipruk, A. (2009). Subcritical water extraction of polyphenolic compounds from *Terminalia chebula* Retz. fruits. *Separation and Purification Technology*, 66, 51-56.

Ren, W., Qiao, Z., Wang, H., Zhu, L., and Zhang, L. (2003). Flavonoids: Promising anticancer agents. *Medicinal Research Reviews*, 23, 19-534.

Ricke, S. C., P. J. van der Aar, G. C. Fahey, and L. Berger. (1982). Influence of dietary fibres on performance and fermentation characteristics of gut contents from growing chicks. *Poultry Science*, 61, 1335–1343.

Robert, R. S., Muzzy, J. D., and Faasa, G. S. (1988). Process for extracting lignin from lignocellulosic material using an aqueous organic solvent and an acid neutralizing agent. Patent No US 4,746,401.

Rodríguez, G., Lama, A., Rodríguez, R., Jiménez, A., Guillén, R., and Fernández-Bolaños, J. (2008). Olive stone an attractive source of bioactive and valuable compounds. *Bioresource Technology*, 99, 5261-5269

Roginsky, V., and Lissi, E. A. (2005). Review of methods to determine chain-breaking antioxidant activity in food. *Food Chemistry*, 92, 235-254.

Rosazza, J.P.N., Huang, Z., Dostal, L., Volm, T. and Rousseau, B. 1995. Biocatalytic transformations of ferulic acid: An abundant aromatic natural product. *Journal of Industial Microbiology*, 15, 472-479.

Rose, D. J., and Inglett, G. E. (2010). Two-stage hydrothermal processing of wheat (*Triticum aestivum*) bran for the production of feruloylated arabinoxylooligosaccharides. *Journal of Agricultural and Food Chemistry*, 58, 6427-6432.

Saija, A., Tomaino, A., Trombetta, D., Pasquale, A. D., Uccella, N., Barbuzzi, T., Paolino, D., and Bonina, F. (2000). *In vitro* and *in vivo* evaluation of caffeic and ferulic acids as topical photoprotective agents. *International Journal of Pharmaceutics*, 199, 39-47.

Sakagami, H., Hashimoto, K., Suzuki, F., Ogiwara, T., Satoh, K., Ito, H., Hatano, T., Takashi, Y., and Fujisawa, S.-I. (2005). Molecular requirements of lignin-carbohydrate complexes for expression of unique biological activities. *Phytochemistry*, 66, 2108-2120.

Sakagami, H., Kawazoe, Y., Komatsu, N., Simpson, A., Nonoyama, M., Konno, K., Yoshida, T., Kuroiwa, Y., and Tanuma, S.-I. (1991). Antitumor, antiviral and immunopotentiating activities of pine cone extracts: Potential medicinal efficacy of natural and synthetic lignin-related materials. *Anticancer Research*, 11, 881-888.

Sakagami, H., Oh-Hara, T., Kaiya, T., Kawazoe, Y., Nonoyama, M., and Konno, K. (1989). Molecular species of the antitumor and antiviral fraction from pine cone extract. *Anticancer Research*, 9, 1593-1598.

Sakagami, H., Satoh, K., Fukamachi, H., Ikarashi, T., Shimizu, A., Yano, K., Kanamoto, T., Terakubo, S., Nakashima, H., Hasegawa, H., Nomura, A., Utsumi, K., Yamamoto, M., Maeda, Y., and Osawa, K. (2008). Anti-HIV and vitamin C-synergized radical scavenging activity of cacao husk lignin fraction. *In Vivo*, 22, 327-332.

Sánchez-Moreno, C. (2002). Methods used to evaluate the free radical scavenging activity in foods and biological systems. *Food Science and Technology International*, 8, 121−137.

Satoh, K., Anzai, S., and Sakagami, H. (1998). Enhancement of radical intensity and cytotoxic activity of ascorbate by *Crataegus cuneata* Sieb et. Zucc. Extracts. *Anticancer Research*, 18, 2749-2753.

Sena-Martins, G., Almeida-Vara E. and Duarte, J. C. (2008). Eco-friendly new products from enzymatically modified industrial lignins. *Industrial Crops and Products*, 27, 189–195.

Serra, S., Fuganti, C., and Brenna, E. (2005). Biocatalytic preparation of natural flavours and fragrances. *Trends Biotechnology*, 23, 193-198.

Shao, S., Wen, G., and Jin, Z. (2008). Changes in chemical characteristics of bamboo (*Phyllostachys pubescens*) components during steam explosion. *Wood Science and Technology*, 42, 439-451.

Shimizu, N., Naoe, T., Kawazoe, Y., Sakagami, H., Nakashima, H., Murakami, T., and Yamamoto, N. (1993). Lignified materials as medicinal resources. VI. Anti-HIV activity of dehydrogenation polymer of p-coumaric acid, a synthetic lignin, in a quasi-*in-vivo* assay system as an intermediary step to clinical trials. *Biological and Pharmaceutical Bulletin*, 16, 434-436.

Slamenová, D., Horváthová, E., Bartková, M., Krajcovicová, Z., Lábaj, J., Košíková, B., and Mašterová, I. (2006). Reduction of DNA-damaging effects of anti-HIV drug 3'-azido-3'-dideoxythymidine on human cells by ursolic acid and lignin biopolymer. *Neoplasma*, 53, 485-491.

Slamenová, D., Kosíková, B., Lábaj J., and Ruzeková, L. (2000). Oxidative/antioxidative effects of different lignin preparations on DNA in hamster V79 cells. *Neoplasma*, 47, 349-353.

Sudheer, A. R., Muthukumaran, S., Devipriya, N., Devaraj, H., and Menon, V. P. (2008). Influence of ferulic acid on nicotine-induced lipid peroxidation, DNA damage and inflammation in experimental rats as compared to N-acetylcysteine. *Toxicology*, 243, 317-329.

Sun, J.-X., Xu, F., Sun, X.-F., Sun, R.-C., and Wu, S.-B. (2004). Comparative study of lignins from ultrasonic irradiated sugar-cane bagasse. *Polymer International*, 53, 1711-1721.

Suzuki, H., Iiyama, K., Yoshida, O., Yamazaki, S., Yamamoto, N., and Toda, S. (1990). Structural characterization of the immunoactive and antiviral water-solubilized lignin in an extract of the culture medium of *Lentinus edodes* mycelia (LEM). *Agricultural and Biological Chemistry*, 54, 479-48

Tabart, J., Kevers, C., Pincemail, J., Defraigne, J-O., and Dommes, J. (2009). Comparative antioxidant capacities of phenolic compounds measured by various tests. *Food Chemistry*, 113, 1226–1233.

Tapin, S., Sigoillot, J.-C., Asther, M., and Petit-Conil, M. (2006). Feruloyl esterase utilization for simultaneous processing of nonwood plants into phenolic compounds and pulp fibers. *Journal of Agricultural and Food Chemistry*, 54, 3697-3703.

Terpinc, P., and Abramovič, H. (2010). A kinetic approach for evaluation of the antioxidant activity of selected phenolic acids. *Food Chemistry*, 121, 366-371.

Thakkar, J. N., Tiwari, V., and Desai, U. R. (2010). Nonsulfated, cinnamic acid-based lignins are potent antagonists of hsv-1 entry into cells. *Biomacromolecules*, 11, 1412-1416.

Tsubaki, S., Iida, H., Sakamoto, M., and Azuma, J.-I. (2008). Microwave heating of tea residue yields polysaccharides, polyphenols, and plant biopolyester. *Journal of Agricultural and Food Chemistry*, 56, 11293-11299.

Tsujiyama, S-I., and Ueno, M. (2008). Formation of 4-vinyl guaiacol as an intermediate in bioconversion of ferulic acid by *Schizophyllum commune*. *Bioscience, Biotechnology and Biochemistry*, 72, 212-215.

Tung, Y. T., Wu, J. H., Hsieh, C. Y., Chen, P. S., and Chang, S. T. (2009). Free radical-scavenging phytochemicals of hot water extracts of *Acacia confusa* leaves detected by an on-line screening method. *Food Chemistry*, 115, 1019-1024.

Ugartondo, V., Mitjans, M., and Vinardell, M. P. (2008). Comparative antioxidant and cytotoxic effects of lignins from different sources. *Bioresource Technology*, 99, 6683-6687.

Ugartondo, V., Mitjans, M., and Vinardell, M. P. (2009). Applicability of lignins from different sources as antioxidants based on the protective effects on lipid peroxidation induced by oxygen radicals. *Industrial Crops and Products*, 30, 184-187.

Van der Aar, P.J., Fahey Jr., G.C. and Ricke, S.C. (1983). Effects of dietary fibers on mineral status of chicks. *Journal of Nutrition*, 113, 653-661.

Villaño, D., Fernández-Pachón, M. S., Troncoso, A. M., and García-Parrilla, M. C. (2005). Comparison of antioxidant activity of wine phenolic compounds and metabolites *in vitro*. *Analytica Chimica Acta*, 538, 391–398.

Vinardell, M. P., Ugartondo V. and Mitjans, M. (2008). Potential applications of antioxidant lignins from different sources. *Industrial Crops and Products*, 27, 220–223.

von Gadow, A., Joubert, E., and Hansmann, C. F. (1997). Comparison of the antioxidant activity of aspalathin with that of other plant phenols of Rooibos tea (*Aspalathus linearis*), α-Tocopherol, BHT, and BHA. *Journal of Agricultural and Food Chemistry*, 45, 632-638.

Walton, N. J., Narbad, A., Faulds, C. B., and Williamson, G. (2000). Novel approaches to the biosynthesis of vanillin. *Current Opinion in Biotechnology*, 11, 490-496.

Wang, K., Jiang, J.-X., Xu, F., and Sun, R.-C. (2010). Influence of steaming pressure on steam explosion pretreatment of lespedeza stalks (*Lespedeza cyrtobotrya*). II. Characteristics of degraded lignin. *Journal of Applied Polymer Science*, 116, 1617-1625.

Wang, Y., Marx, T., Lora, J., Phillip, L. E., and McAllister, T. A. (2009). Effects of purified lignin on *in vitro* ruminal fermentation and growth performance, carcass traits and fecal shedding of *Escherichia coli* by feedlot lambs. *Animal Feed Science and Technology*, 151, 21-31.

Wayman, M. and Lora, J. H. (1978). Aspen autohydrolysis. The effect of 2-naphtol and other aromatic compounds. *TAPPI*, 61, 55-57.

Wiboonsirikul, J., and Adachi, S. (2008). Extraction of functional substances from agricultural products or by-products by subcritical water treatment. *Food Science and Technology Research*, 14, 319-328.

Wu, G., Heitz, M., and Chornet, E. (1994). Improved alkaline oxidation process for the production of aldehydes (vanillin and syringaldehyde) from steam-explosion hardwood lignin. *Industrial and Engineering Chemistry Research*, 33, 718-723.

Yang, J., Wang, S., Lorrain, M.-J., Rho, D., Abokitse, K., and Lau, P. C. K. (2009). Bioproduction of lauryl lactone and 4-vinyl guaiacol as value-added chemicals in two-phase biotransformation systems. *Applied Microbiology and Biotechnology*, 84, 867–876.

Yin, M. C., Lin, C. C., Wu, H. C., Tsao, S. M., and Hsu, C. K. (2009). Apoptotic effects of protocatechuic acid in human breast, lung, liver, cervix, and prostate cancer cells: potential mechanisms of action. *Journal of Agricultural and Food Chemistry*, 57, 6468-6473.

Yoon, B. H., Jung, J. W., Lee, J-J., Cho, Y-W., Jang, C-G., Jin, C., Oh, T. H., and Ryu, J. H. (2007). Anxiolytic-like effects of sinapic acid in mice. *Life Sciences*, 81, 234-240.

Zemek, J., Valent, M., Pódová, M., Košíková, B., and Joniak, D. (1987). Antimicrobial properties of aromatic compounds of plant origin. *Folia Microbiologica*, 32, 421-425.

Zou, Y., Kim, A. R., Kim, J. E., Choi, J. S., and Chung, H. Y. (2002). Peroxynitrite scavenging activity of sinapic acid (3,5-dimethoxy-4-hydroxycinnamic acid) isolated from *Brassica juncea*. *Journal of Agricultural and Food Chemistry*, 50, 5884–5890.

In: Lignin
Editor: Ryan J. Paterson

ISBN: 978-1-61122-907-3
©2012 Nova Science Publishers, Inc.

Chapter 3

LIGNIN: PROPERTIES AND APPLICATIONS

Ufana Riaz and S. M. Ashraf*

Materials Research Laboratory, Department Of Chemistry
Jamia Millia Islamia, New Delhi, India

The development of commercially viable "green products" based on natural resources for both matrices and reinforcements for a wide range of applications is on the rise .Biodegradable plastics made from renewable resources have received increased attention. The driving force for this interest arises from sustainability gains and environmental amelioration provided through a reduced dependence on petroleum reserves, increased disposal options, and lower levels of greenhouse gases. This development has occured in spite of acknowledged challenges related with material properties, recycling, and cost. Most of the interest in biodegradable plastics is aimed at developing low cost composites that are economical in high-volume applications. Applications considered most relevant include packaging and consumer products. Lignin is one of the most abundant natural raw materials available on earth, after cellulose. Lignin research comes out of the tradition of organic chemistry. The applied R&D has primarily been driven by the needs for knowledge in relation to pulping and bleaching. Lignin recovery and utilization research has been highly empirical and raw material-based rather than consumer-oriented. The 300 million dollar lignin business today is dominated by products from sulphite-pulping liquors, originally based on the need to solve a pollution problem. It is now a chemical business associated to the pulp mill. The products are primarily directed to the process industry, but increasingly higher value functional products are produced. The lignin business is faced with a high cost of R&D and has to carefully select which markets to serve. There are presently several new fields of lignin application that are too complex and competitive and require synergistic effort to prove their viability for uses in fields like coatings and adhesives. The use of lignocellulosics as fillers and reinforcements in thermoplastics has been gaining acceptance in commodity plastics applications in recent years. Another major advancement in engineered lignin is in product and performance enhancement. Advanced engineered lignin based composites are currently being developed that will simultaneously meet the diverse needs of users for high-performance and economical commodity products. These engineered lignin biocomposites will provide advanced

---

* Corresponding author email: ufana2002@yahoo.co.in.

performance, durability, value, service-life, and utility while at the same time being a fully sustainable technology.

The focus of this review is to highlight the state of knowledge in processing, manufacturing, characterization, and the challenges in the integration lignin based coatings and adhesives for their potential applications in various fields.

1. LIGNIN: CHEMISTRY, OCCURRENCE AND STRUCTURE

Lignin is a complex 3-dimensional polymer that occurs predominantly in the xylem of most land plants, forming about 1/3 of the terrestrial woody biomass. As a major component of the cell wall of tracheids, vessels and fibers, lignin contributes to the compression strength of woody stems, and to the water proofing of conductive elements within the xylem.Lignification has allowed the evolution of large arborescent land plants capable of survival in relatively arid environments. As a major component of wood, lignin is of immense technological significance, particularly in the pulp and paper industry. Lignin consists of both aromatic and aliphatic portions, which has a heterogeneous branched network of phenyl propane units. The structure of lignin is amorphous and coexists, along with the hemicelluloses. The phenyl propane units are held together in many ways by ether and C-C bonds.Lignin molecules are derived mainly from three phenylpropane monomers: p-coumaryl alcohol, coniferyl alcohol, and synapyl alcohol, Figure.1.

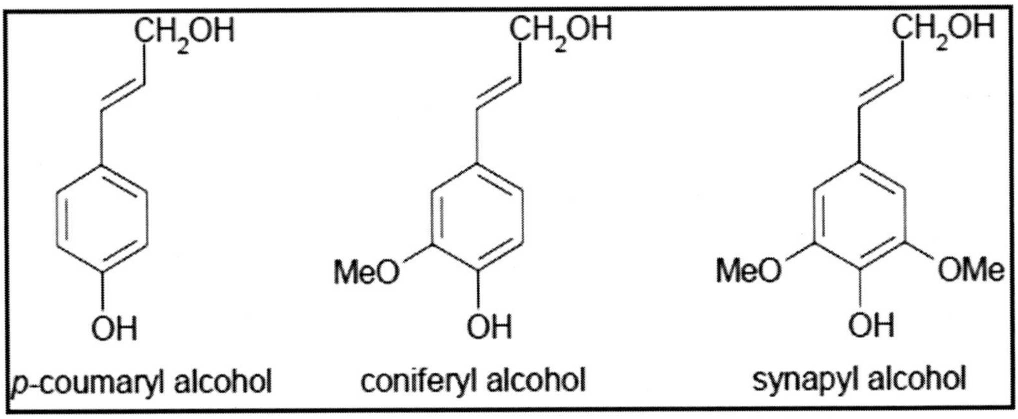

Figure 1. Lignin monomers.

These monolignols are polymerized by a radical coupling process that links them by carbon-carbon or ether bonds. A linkage may occur at any of several different locations on each phenolic unit, causing many different types of linkage to be possible. The most common linkage found in a lignin molecule are β-O-4, α-O-4, β-5, 5-5, 4-O-5, β-1, and β-β ,Figure.2 [1-2].

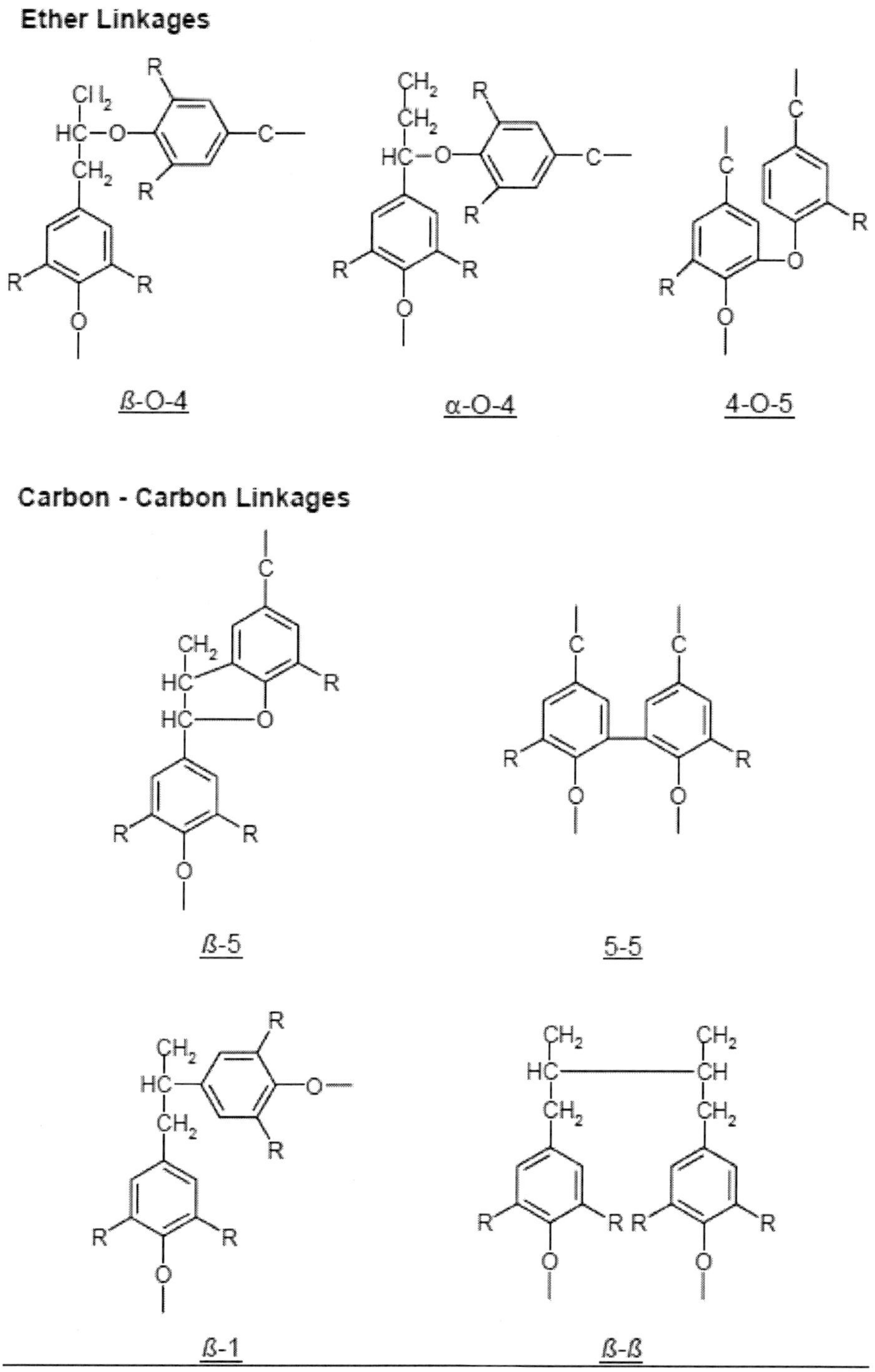

Ether Linkages

ß-O-4   α-O-4   4-O-5

Carbon - Carbon Linkages

ß-5   5-5

ß-1   ß-ß

Figure 2. Types of linkages present in Lignin.

Though these are the dominant linkages, at least 20 different linkage types have been identified [3-4]. The ether type linkages are known to dominate in native lignin, and are estimated to make up approximately one half to two thirds of the total number of native plant lignin linkages. Monolignols can be trifunctionally linked, forming branch points within the polymer and giving it a network-like structure. Lignin molecules cannot be depicted as a series of regular, defined repeating units, as traditional polymers are. In contrast, lignin is a highly irregular and complex polymer [1,3].Lignin is a thermoplastic polymer and has a glass transition temperature which may range between 127-193°C depending upon its molecular weight and chemical structure. Lignin exists in a wide range of molecular weights which for isolated lignins are in the range of 1000 to 12,000 depending upon the extent of degradation and condensation during isolation. Plant lignins can be divided into three broad classes, which are commonly called:

- ➢ softwood (gymnosperm),
- ➢ hardwood (dicotyledonous angiosperm), and
- ➢ Grass or annual plant (monocotyledonous angiosperm) lignins.

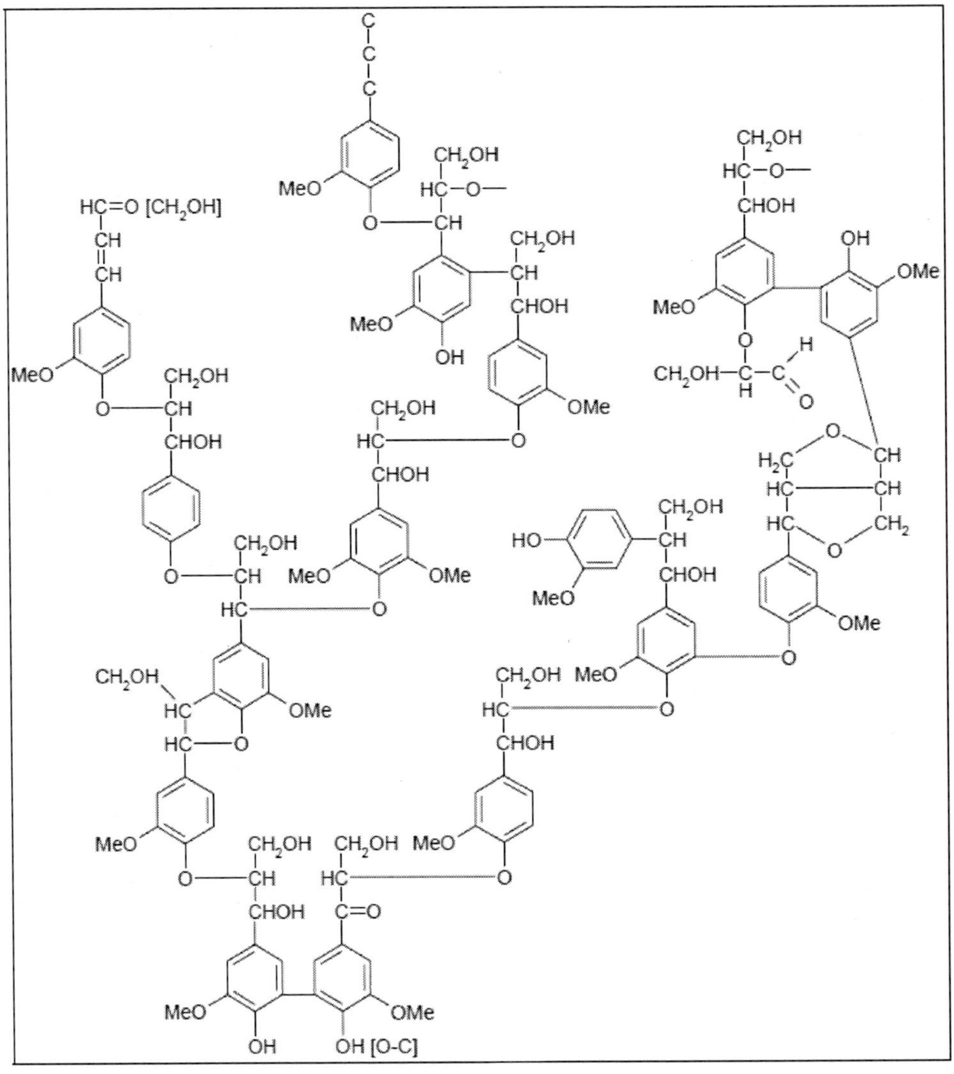

Figure 3. Structure of lignin.

Until now only about 2% of the lignins available in the pulp and paper industry is commercially used comprising of about 1,000,000 tons/year lignosulphonates originating from sulphite pulping [4] and less than 100,000 tons/year of kraft lignins produced in the kraft process. The existing markets are either very low value products (lignosulphonates mainly in dispersing and binding applications) or limited to very narrow market segments (high-quality dispersants from chemically modified kraft lignin). The possible development in these fields is limited to very low growth rates. On the other hand, the potential for lignin production in the existing pulp and paper industry is more than 50 million tons/year without counting other possible biomass. Based on already existing technical knowledge and on the technical characteristics that are already achieved with lignin products, this amount could be efficiently used to replace fossil resources.

Models have been proposed for lignin from several different sources, though due largely to lignin's complicated nature and the difficulties inherent in lignin analysis, no complete structure of a lignin molecule has ever been identified. The models that have been developed are only representations drawn from analyses of the relative proportions of each lignin unit type and each linkage type, Figure 3.

2. STATE OF THE ART

Lignin research has primarily been driven by the need for knowledge and has been highly empirical and raw material-based rather than consumer-oriented, Figure.4.The 300 million dollar lignin business today is dominated by products from sulphite-pulping liquors, originally based on the need to solve a pollution problem. The high value functional products are produced, particularly from kraft lignin. In general, the academic groups do not have the interdisciplinary resources to address complex applications or new markets. Technical service is poor and the research has to be multidisciplinary. The lignin business is faced with a high cost of R&D and has to carefully select which markets to serve. The fields of lignin application are too complex and competitive to be developed by a single partner and need synergistic effort to prove their viability for uses in fields involving for example bioactivity and biochemistry: anti-bacterial, anti-oxidant, animal feed and enzymes. To make the value added products of lignin scientifically, as well as economically viable, the new stakeholders need the basic detailed information and specific data about existing lignin know-how; state-of-the-art expertise and knowledge; specific and achievable aims. Apart from the novel application of lignin, it uses in the traditional fields like materials (polymers, adhesives) and specialty chemicals (oncrete additives, emulsifiers, binders) that can be diversified according to established needs for more natural products. Lignin has the image of a low-quality, low-price waste product and the needs of the users are very specific. Moreover the available sources are not sufficient to make industrialists feel comfortable to choose this raw material in respect of the quality, quantity or supply sources.

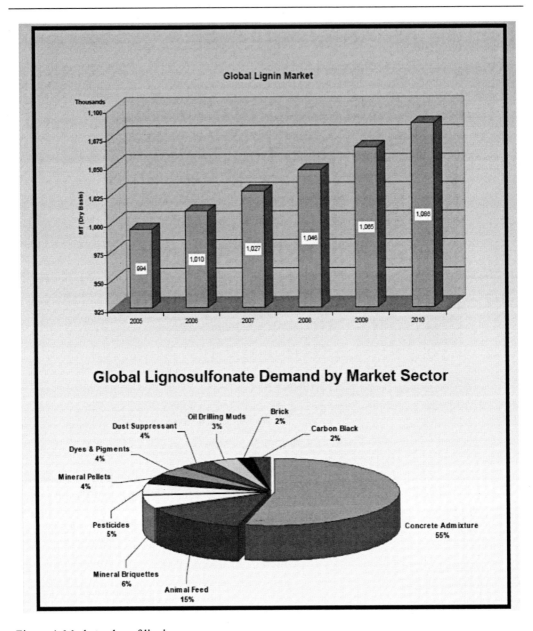

Figure 4. Market value of lignin.

3. ISOLATION OF LIGNIN FROM BIOMASS

Lignin is isolated from plant biomass by procedures that rely on depolymerization or derivatization of lignin, or both. Therefore, since its chemistry is inevitably changed by isolation, the polymeric properties of native lignin like molecular weight and glass transition temperature (Tg) are difficult to gauge. The weight-average molecular weight (Mw) of isolated lignins most closely resembling native lignins from softwood has been estimated at 20,000, whereas similar estimates for hardwood lignin have been somewhat lower. The polydispersity of lignin determined by these studies was between 2.5 and 3.0 [3].The Tg of native hardwood lignin is reported to be between 65 and 85°C, while that of native softwood lignin is between 90 and 105°C. Isolated lignin may display substantially different Tg's depending on the isolation procedure used [5].

There has been increasing interest in using agricultural biomass as a source of industrial fiber and chemical feedstock. Agricultural biomass conversion to produce ethanol for fuel in particular has been on the rise [6-8]. To use the plant polysaccharides, as in paper making, lignin must first be separated from the biomass. One major source of agricultural biomass that is being used to produce cellulosic ethanol is corn stover, the residue left after a corn harvest. Approximately 75 million dry tons of corn stover is produced annually. With its lignin content established at 11.4%, this represents a potential source of 8.5 million dry tons of lignin a year. Currently, this is a growing source of lignin that, without any established industrial uses, is largely considered waste. Like wood pulping, a variety of methods can be used to generate fermentable carbohydrates, and each method will produce a lignin byproduct with unique properties. Typically, the biomass is pretreated with weak mineral acid. Then the carbohydrate may be depolymerized and extracted by either continuing acid hydrolysis or by enzymatic treatment [8]. A second treatment process that has been investigated for use in biomass conversion is steam explosion [9]. In this process, biomass is briefly subjected to high pressure and temperature (200-250°C), and is then explosively released to atmospheric conditions [3]. The plant polysaccharides can then be degraded by enzymes, and the lignin can be extracted by aqueous alkaline solvents.

4. EXTRACTION PROCEDURES

There are numerous methods for isolating lignin from biomass. Large scale industrial processes exist in which lignin is separated from plant materials in order to obtain the polysaccharide component. These processes are optimized to produce plant fiber with uniform characteristics, while lignin is generally considered byproduct waste. Consequently, these industrial isolation procedures often significantly alter the chemical structure of lignin. Smaller laboratory scale procedures that result in less change to the native lignin structure are used in scientific studies on lignin.

(a) Pretreatment Methods

Pretreatments are designed to open the structure of lignocellulosic biomass prior to enzyme hydrolysis, to allow efficient production of C5 and C6 sugars. Hydrolysis of the major component, cellulose has received the most attention, as it can be used to produce ethanol by fermentation. Cellulose exists in nature as a compact and complex matrix with lignin and hemicellulose [10]. The cellulose is highly ordered and crystalline with amorphous regions. It is surrounded by lignin which acts as a physical barrier [11] and is associated with the hemicellulose. Obviously, reduction of crystallinity of cellulose and removal of lignin and hemicellulose are important goals for any pretreatment process [12]. A number of pretreatment methods have been developed to improve cellulose hydrolysis from lignocellulose. They include mechanical pulverization, pyrolysis, concentrated acid, dilute acid, alkali, hydrogen peroxide, auto hydrolysis, ammonia fiber explosion (AFEX), wet-oxidation, lime, CO2 explosion, organic solvent treatment, etc. Each method, in some way, decreases the size of the biomass and opens its physical structure. A summary of pretreatment methods is given in Figure.5.

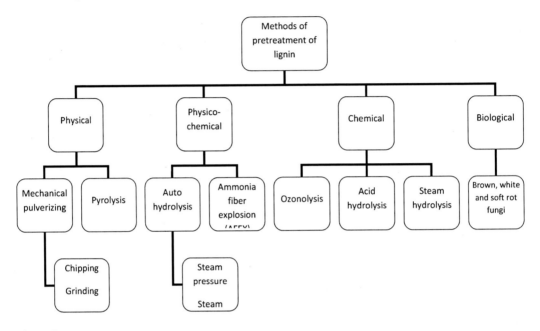

Figure 5.

(b) Various Techniques Used For Extraction

In plants, cellulose fibers are bonded to lignin. There are two main technologies for making chemical pulp i.e. isolating cellulosic fibers by chemical means which consists mainly in removing the lignin from wood or other plant material to release the fibers:

(i) The Acid Sulphite Procedure

In the acid sulphite technologies, wood is cooked with salts of sulphurous acid. The dark solution of the degraded lignin dissolved out of the wood is called spent sulphite liquor in the sulphite process and black liquor in the kraft process. In sulfite process, the reacts with sulfur dioxide and hydrogen sulfite, causing sulfonic acid groups to be attached to the lignin backbone, thereby making the lignin water soluble. Specific reactions in this process are controlled by modifying the pH between neutral and acidic conditions. Lignins produced by the sulfite pulping process are known as lignosulfonates.

(ii) Alkaline Kraft Process

In the alkaline kraft process, wood is cooked with a solution (white liquor) containing sodium hydroxide and sodium sulphide and is turned into black liquor which is concentrated and used as a fuel. This approach also recovers the alkali needed to prepare the cooking liquor and furnishes the heat, steam and electric power necessary for the industrial process. The most significant and problem is the profitable utilization of the vast quantities of lignin available as waste products or byproducts of the forest-using industries and, especially, the pulp and paper industry. For more than three quarters of a century the chemical pulping industry has been utilizing essentially only half the tree as pulp, and has been discarding the other half as a waste or as a source of heat. Pressures for the abatement of pollution effluents, the need for conservation to meet expanding requirements of our raw materials, and the urge for more economical production in the face of higher operating costs have been responsible for initiating research on the utilization of lignin, Figure.5

The main objective of wood pulping is to remove lignin from wood fiber in order to use the fiber in paper production. Approximately 52 million dry tons of lignin were isolated by the pulp and paper industry in 2002 [13]. There are several methods used for removing lignin, but the most common is kraft pulping which produces strong pulps for use corrugated board, liner board and paper-bags. The kraft process utilizes a solution of sodium hydroxide and sodium sulfide to treat wood chips at an elevated temperature around 170 °C. This treatment fragments the lignin molecules by attacking the ether linkages. As a result, the molecular weight of the lignin decreases and the number of phenolic hydroxyl groups increases. The phenolic groups are susceptible to ionization in alkaline conditions by extraction of the phenolic proton, causing the lignin chains to become soluble. In addition, aliphatic thiol groups introduced to the lignin give kraft lignin a distinct odor. "Organosolv" lignin is produced by pulping wood in organic solvents with catalysts in both acidic and alkaline conditions. This process is arguably more ecologically sound than kraft or sulfite pulping due to the absence of sulfur compounds.

(iii) Soda Pulping

The fact that neither liquid fuel production from biomass nor solvent pulping have reached commercial maturity has severely limited the market penetration of sulfur-free lignins. Sulfur-free soda pulping of non-wood feedstocks such as straw, sugar cane bagasse, flax, etc., is practiced widely around the world and offers a potentially more readily available source of such lignins. Because of their small size and economic constraints, many of the mills processing nonwood resources cannot afford the capital-intensive system developed for processing much larger quantities of spent pulping liquors, such as those used in wood pulping operations for recovery of energy and recycling of cooking chemicals. Consequently, many small nonwood mills are forced to dispulp their pulping effluents with little or no treatment into the environment. Lignin recovery from these effluents can significantly reduce the environmental impact of these operations. Furthermore, the remaining effluent (after lignin removal) can be more easily purified by biological treatment. Another source of nonwood lignins may come from larger nonwood mills that may have a conventional recovery process but want to incrementally expand their pulping capacity and need an economically and environmentally acceptable way to handle the excess liquor generated. Recovery of these lignins is based on precipitation followed by liquid/solid separation and drying. One of the difficulties of this type of process when applied to nonwood fiber sources is that the soda spent pulping liquors often contain silica, which may co-precipitate with the lignin, rendering it of lower quality. Hence several modification (derivatization) options have been adopted [15-16]. Soda lignins have been available in Asia are normally of low and variable quality These soda lignins all have in common that they are of (1) low molecular weight, (2) high phenolic hydroxy content, and (3) relatively low (but variable) glass transition temperature However, their properties sometimes vary significantly For instance, the thermal behavior of these lignins seems to depend on type of process and feedstock , and on the presence of contaminants that may act as plasticizers or antiplasticizers. Some lignins do not exhibit flow with temperature, whereas others readily undergo fluid flow when heated. In general, nonwood soda lignins may have a wider range of properties than have been observed with wood lignins

(iv) Laboratory Extraction of Lignin

Although there is no process that will separate lignin from plant polysaccharides without some degree of modification to the lignin chemistry, there are methods that can separate lignin with its native structure mostly intact. These methods are commonly used in laboratories attempting to better understand the structure or biochemistry of native lignin. The method that isolates lignin with the least alteration uses cellulose enzymes to break down the surrounding plant polysaccharides. This method is tedious, however, and a more common laboratory method uses ball milling to mechanically break down the plant fibers, followed by solvent extraction of the lignin [3]. Ball milling is known to affect lignin structure, but the extracted lignin is considered to be a relatively good indicator of native lignin chemistry. This method only recovers approximately 20 – 30% of the lignin present in a plant sample. The yield can be increased substantially by treating the ball milled plant fiber with polysaccharide enzymes [14]. Other lignin isolation procedures exist on a laboratory scale, but they are not commonly used outside of specific applications and alter the lignin structure so that they are not useful for native lignin chemistry studies [15-16]. Table.1 presents molecular weight properties and glass transition temperatures of lignin from several different sources and isolation procedures. As the table illustrates, these properties vary widely between the different lignins. There are discrepancies in these numbers in the literature between materials that should be nearly identical. For example, the Glasser et al. [9] reports a weight average molecular weight (Mw) value for kraft lignin derived from pine as 19,800 g/mol. This is clearly different from the value included in Table 1, which was obtained by a similar procedure.

Table 1. Isolation techniques of lignin

Isolation method	Species	Mw (g/mol)	Polydispersity	Tg (°C)
Milled wood	Pine	11,400	8.8	160
Milled wood	Alder	7000	6.4	110-130
Kraft	Pine	4300	3.3	169
Kraft	Hardwood mix	3000	2.9	-
Lignosulphonate	Mix	1000-10,000	6.8	-
Organosolv	Pine	1400	2.8	91
Acid hydrolysis	Pine	40000	50	96
Steam explosion	Aspen	2300	2.9	139

(v) Membrane Separation

The membrane separation processes have been widely studied in the past years because its implementation is of great interest in several fields such as food, chemical, biological and pharmaceutical industries. Membrane technology works out in a simple and effective separation, concentration and purification of products in batch and small–medium scale processes, as it is only necessary to use of a semi-permeable medium and to generate hydrostatic pressure difference as to driving force. Although there is the disadvantage associated with fouling and cleaning cycle or in-service life of the membrane. The extent or effectiveness of membrane technology depends on the type of membrane used and the particle size that it can retain . Thus four processes can be distinguished: micro filtration (MF), ultra filtration (UF), nano filtration (NF) and reverse osmosis (RO). All these methods differ in the membrane type, operating conditions required and its application field, Table.2

Table.2. Different membrane separation processes

Process and pore size	Membrane type	Membrane material	Applications
Membrane filtration	Symmetric microporous (0.1–10μm)	Ceramics, metal oxides (aluminium-, titanium-, zirconium-), graphite, polymers (cellulose nitrate or acetate	Sterile filtration, clarification
Ultra filtration	Asymmetric microporous (1–10 nm)	Ceramics, polysulfone, polypropylene, nylon 6, PTFE, PVC, acrylic copolymer	Separation of macromolecular solutions
Nano filtration	Thin-film membranes	Cellulosic acetate and aromatic polyamide	Removal of hardness and desalting
Reverse osmosis	Asymmetric skin-type (0.5–1.5 nm)	Polymers, cellulosic acetate, aromatic polyamide	Separation of salts and microsolute from solutions

5. FFECT OF EXTRACTION METHODS ON THE LIGNIN YIELD

The maximum yield of lignin in obtained in alkaline media, Figure.6 .The difference on the yield can be explained on the basis of mechanism of extraction of isolation by different methods

Lignin is known to posses physiochemical linkage and chemical bonds with cellulose and hemi cellulose .Alkaline delignification takes place at the solid liquid interface of lignin and alkali solution .The swelling of cellulose onto which lignin is grafted proceeds via dissolution of lignin from polysaccharide matrix during extraction. Better swelling of cellulose fiber in alkali causes several fold increase in the reactive surface area which enhances the delignification process

The spectral analysis of bagasse lignin, Figure.7 extracted by various procedures shows OH stretching band in solvent lignin at 3340 cm^{-1}.Slight spectral differences are observed in the finger print region .The band at 1236 cm^{-1} for solvent lignin is observed at 1275 cm^{-1} for solvent lignin. The C=O stretching band observed at 1700 cm^{-1} incase of acid and solvent lignin appear to be absent in alkaline lignin.

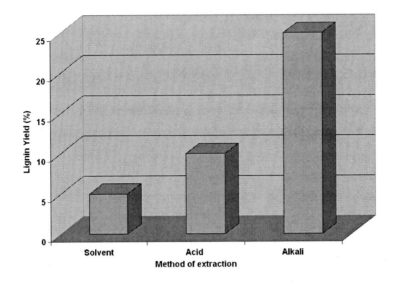

Figure 6. Percentage yield of lignin obtained by different extraction procedures.

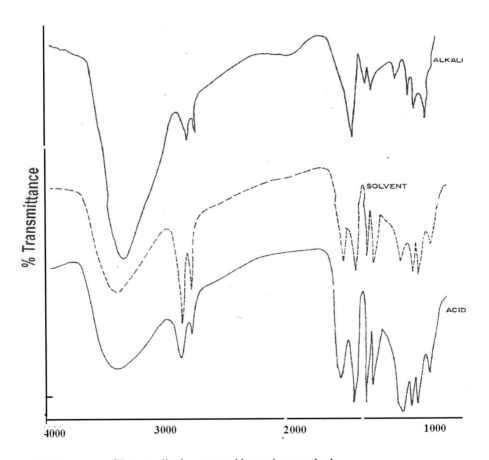

Figure 7. FT-IR spectra of Bagasse lignin extracted by various methods.

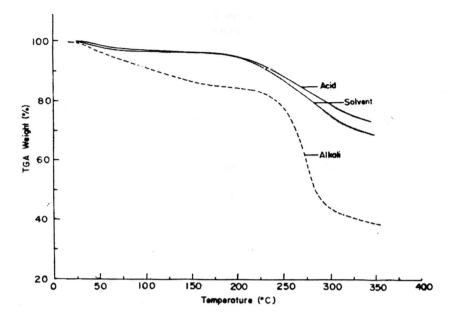

Figure 8. TGA thermograms of Lignin extracted by different methods.

The TGA thermograms of Lignin extracted by different methods, Figure.8, indicate that acid and solvent lignin have higher molar mass than alkali extracted lignin .In the former cases the decomposition is extremely slow indicating fragmentation of long molecular chains causing low loss in weight .Upon comparing the three lignins it is observed that alkali lignin is least stable and shows sharp and steep decomposition. This is attributed to the structural changes occurring during isolation. Acid isolation undergoes condensation reaction and forms a stable structure whereas alkaline isolation results in the cleavage of the β-O-4 ether linkages and hydrolysis of the ether linkages resulting in low molar mass.

6. ANALYTICAL METHODS ADOPTED FOR THE DETECTION OF FUNCTIONAL GROUPS PRESENT IN LIGNIN

(i) Elemental Analysis

Carbon, hydrogen, sulfur and nitrogen contents were determined using an elemental analyzer. After correction for ash content, the percentage of oxygen is calculated by difference.

(ii) Methoxyl Groups

For the determination of methoxyl groups , lignin (0.15 g) is treated for 10 min with refluxing concentrated sulfuric acid (10 ml). The reaction mixture was cooled, 70 ml of distilled water is added, and the methanol produced in the reaction was distilled off under vacuum and quantified by gas chromatography

(iii) Acetylation

A weighted amount of each lignin except lignosulfonate is acetylated for 48 h with a mixture of purified pyridine-acetic anhydride (1:1, v/v). Methanol is used to quench the remaining acetic anhydride. Finally, a flow of nitrogen was applied to evaporate the solvents and the samples were dried under vacuum

(iv) Hydroxyl Groups

Aliphatic and phenolic hydroxyl groups are determined by three wet chemical methods (aminolysis, ultraviolet spectroscopy and non-aqueous potentiometry) and two spectroscopic methods (^{1}H NMR and ^{13}C NMR). The two spectroscopy methods enabled aliphatic hydroxyl quantification.

(v) Aminolysis

The procedure is used to determine free phenolic hydroxyl groups in lignin . The acetylated lignin, dissolved in 1.0 ml of dioxane containing 5mg of 1-methylnaphtalene, was treated with 1.0 ml of dioxanepyrrolidine (1:1, v/v) solution, which initiated the aminolysis reaction. After the addition of pyrrolidine, samples were taken from the reaction mixture at different times (total reaction time was approximately 120 min) and analyzed by gas chromatography. The amount of 1- acetylpyrrolidine formed (equivalent to the amount of hydroxyl groups) is recorded as a function of time. The content of phenolic hydroxyl groups was calculated by extrapolation of the curve at zero time.

(vi) Phenolic Hydroxyl Groups by Ultraviolet Spectroscopy (Δε Method).

The content of various phenolic units in lignin samples was determined by UV spectroscopy . This method is based on the difference in absorption at 300 and 360 nm between phenolic units in neutral and alkaline solutions. The content of ionizing phenol hydroxyl groups can be quantitatively evaluated by comparing the Δε values of substances studied at certain wavelengths to the values of Δε of the products..

(vii)Carbonyl Groups

Carbonyl groups for all lignins were determined by two wet chemical methods: the Modified Oximating method and differential UV-spectroscopy. The Modified Oximating method present a correction technique, which is necessary for lignins containing carboxyl groups. Differential UV.It involves differential absorption measurements that take place when carbonyl groups are reduced at the benzylic alcohol corresponding with sodium borohydride.

This method determines some carbonyl lignin structures such as aldehydes and ketones structures

(viii) Carboxyl Groups

Carboxyl groups are determined by using three methods: acid number determination and aqueous and nonaqueous titration methods. These methods are described below.

(ix) Acid number determination. Carboxylic groups are determined by the pH of 100 ml of 95% ethanol in water adjusted to 9.0 using 0.1 mol/l sodium hydroxide in water. After adding 1 g of dried lignin, the mixture is stirred for 4 h and subsequently titrated back to pH 9.0 with 0.1 mol/l sodium hydroxide solution.

(x) Aqueous Titration Method

Llignin sample (1 g) is suspended in 100 ml of alkaline aqueous solution. After stirring for 3 h, the pH was adjusted to 12 with sodium hydroxide. After stirring again, the solution is potentiometrically titrated with 0.1 mol/l aqueous hydrochloride acid.

(xi) Non-Aqueous Potentiometry Method

This procedure involves a non-aqueous potentiometric titration of lignin with tetra-n-butyl ammonium hydroxide in the presence of an internal standard, which is p-hydroxybenzoic acid . The advantage of this method is that it determines not only the carboxyl groups in lignin but it concurrently determines the weakly acidic phenolic hydroxyl groups. When combined with an ion-exchange treatment, the aforementioned titrimetric procedure was also used to determine the strongly acidic groups (sulfonates groups) in lignosulfonate.

(xii) Expanded C9 Formulae

The expanded formulae C9 contain complete information about the lignin structure. They are obtained by combining the results from elementary analysis and functional groups analysis.

(xiii) Statistical Analysis

The methods for determining the functional groups in lignins are compared by applying paired two-sided t-tests at a 95% confidence level for mean values and combining the two methods. The results are presented as averages and their standard deviation.

7. EXTRACTION OF LIGNIN FROM VARIOUS RESOURCES AND THEIR CHARACTERIZATION

(a) Sugarcane Bagasse

Sugarcane (*Saccharum officinarum*) is a grass that is harvested for its sucrose content. After extraction of sugar from the sugarcane, the plant material that remains is termed bagasse. Sugarcane bagasse found at sugar mills contains both relatively easy and hard to degrade materials. The easily degraded materials appear to be from the leaf matter and the hard to degrade from the rind [18]. Bagasse has several advantages for use in ethanol production. Most importantly, unlike corn stover, bagasse is collected as part of the sugar production process, so it does not require a separate harvest. It is also physically ground as part of the extraction process [18]. Furthermore, bagasse is cheap, readily available, and has high carbon content. Its disadvantage is that it already has value as fuel to the producer, so any product values must exceed that of the fuel value.

(b) Eucalyptus Bark

Eucalyptus is important commercially and it is planted in many countries as a pulp resource because of its rapid growth and ease of adaptation to various environmental conditions. *Eucalyptus* woods generally contain high contents of polyphenols The absorption of ultraviolet light by polyphenols interferes with the visualization of lignins in cell walls in thin sections of some species of *Eucalyptus* by micro spectrophotometry. Similar effects of polyphenols on the analysis of lignins by ultraviolet microscpectrophotometry have been observed in several species.Therefore, in *Eucalyptus*, we must consider the effects of the polyphenols on lignin analysis. However, little information has been reported about the effects of alkali-soluble extractives, namely polyphenols, on the Mäule color reaction [19]. Thin sections and wood meals are prepared from the blocks. For recordings of visible- light absorption spectra after the Mäule color reaction and of ultraviolet absorption spectra, 10 μm transverse sections and 50 μm radial sections are cut. For measurements of Klason lignin content, about 3 g of wood meal, from 40 to 80 mesh, was prepared. Wood meal and thin sections are extracted with a mixture of ethyl alcohol and benzene(1: 2, v/v) for 6 h. Thin sections and wood meal that had been extracted with a mixture of ethyl alcohol and benzene are extracted with solutions of NaOH in an autoclave. Alkali extraction is performed under the following conditions: concentration of NaOH, 0.1, 0.4, 1.0, 4.0 or 10.0%; temperature, 105 °C; and duration, 60 min. Samples for alkali extraction are weighed approximately 0.05 g in the case of thin sections and 0.5 g in the case of wood meals. After alkali extraction, samples are cooled and liquid is decanted through a ground-glass filter (pore size: 40–100 μm). Then samples are washed successively with distilled water, 10% acetic acid and distilled water. Contents of alkali-soluble extractives are calculated from weights of residues of wood meal that is extracted with a mixture of alcohol and benzene (1: 2, v/v).

(c) Coconut Coir

Coir is a fibrous material found between the leathery covering and the shell of a coconut. The natural color of coir varies from light brown to very dark brown, depending on the variety and maturity of the nut from which it was extracted, and the processing conditions. The fibers are stiff coarse, resilient, pliable and quite resistant to bacterial attack[20]. The fibers consist mainly of lignin and cellulose. Cellulose and hemicelluloses make up the bulk in the ground tissue of the husk. Lignin, the other main fiber constituent, is responsible for the stiffness of the coir. It is also responsible partly for the natural color of the fiber.

Table 3. Composition of coconut coir

Constituent	Percent
Moisture	15
Lignin	43
Ash	8.26
Alkalinity of ash	37.5

Raw coconut coir pit from retted husks is taken and digested using 2% of sodium sulphite solution. The material to liquor ratio is maintained as 1:10 and the pH is maintained in the acidic range using dilute hydrochloric acid solution. The digestion is carried out in autoclave at 115 ^{0}C for 30 minutes and then the temperature is raised to 135 ^{0}C for 90 minutes .The lignin is extracted in two stages –in the first stage lignin reacts to form lignosulphonic acid which in the second stage is made soluble by hydrolysis .This soluble compound is acidic and combines with a base to form salt.Since sodium is the base present during cooking, the soluble sodium lignosulphonate is formed as a black liquor

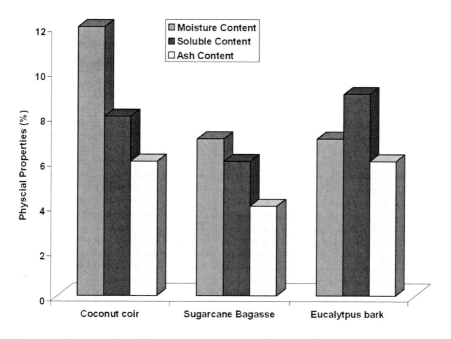

Figure 9.Physical characteristics of Sugarcane bagasse, eucalyptus bark and coconut coir.

7.1. Physical Characteristics of Sugarcane Bagasse, Eucalyptus Bark and Coconut Coir

The moisture content is highest incase of coconut coir while incase of other agro-wastes it is comparable. The soluble content is slightly higher for eucalyptus bark while the ash content is lowest for sugarcane bagasse.The % yield of lignin from the above sources is shown in Figure 9.

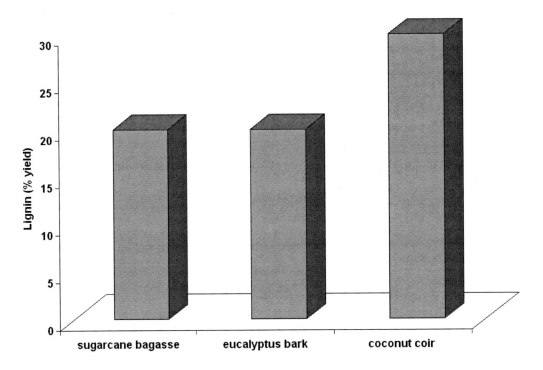

Figure 10. Percentage yield of lignin from different agro waste.

As observed lignin yield was found to be highest for coconut coir ≈ 30% while for bagasse and eucalyptus bark the yields were comparable ≈ 20%.

Table.4. Chemical characteristics of Sugarcane bagasse, eucalyptus bark and coconut coir

Lignin source	Hydroxyl value	Elemental Analysis % C % H %O		
Sugracane Bagasse	296	44.39	5.46	50.15
Eucalyptus bark	236	50.07	4.68	45.25
Coconut coir	212	45.21	5.62	49.17

The highest hydroxyl value was observed for bagasse lignin and lowest for coconut coir,Table.4 , which confirms that baggasse has highest number of OH groups [20] .However, Bagasse and coconut coir have the same % of elemental constituents. It appears that some oxygen of coconut coir is present as aldehyde or ketone group while the other possibility is the presence of methoxy groups on the aromatic ring of coconut coir.

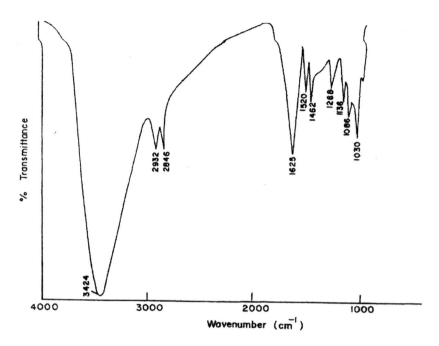

Figure 11. FT-IR spectrum of Bagasse lignin.

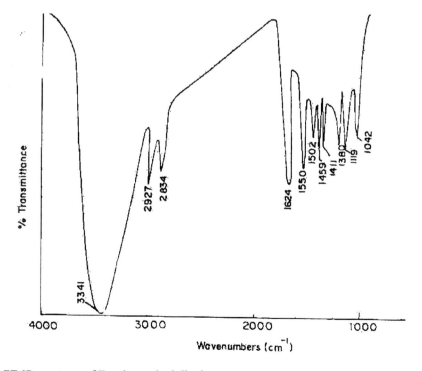

Figure 12. FT-IR spectrum of Eucalyptus bark lignin.

7.2. Spectral Analysis of Sugarcane Bagasse, Eucalyptus Bark and Coconut Coir Pith Lignins

The FT-IR spectrum of bagasse lignin, Figure11, shows a peak at 1625 cm⁻¹ assigned to aromatic benzene ring vibrations .The bands in the region of 1500 cm⁻¹ are also attributed to the presence of aromatic ring. The band at 1462 cm⁻¹ is due to the methyl group of methoxy while the band at 1268 cm⁻¹ is correlated to the C-O of guaiacyl ring the bands at 1136 and 1030 cm⁻¹ are assigned to the guaiacyl and syringyl CH.[20-21]. The band at 1086 cm⁻¹ is assigned to C-O of secondary alcohols that is typical of soft wood lignins.High intensity of OH peaks indicates the presence of higher number of hydroxyl groups .

The FTIR spectra of eucalyptus bark lignin ,Figure.12, shows a band at 3341 cm⁻¹ characteristic of OH peak while the bands at 2927 and 2834 cm⁻¹ are assigned to asymmetric and symmetric vibrations of CH .The bands present at 1550 cm⁻¹ and 1502 cm⁻¹are due to the aromatic skeletal vibrations.The band at 1459 cm⁻¹ is assigned to CH deformation of the methyl group of epoxy .The band at 1411 cm⁻¹ is due to CH in plane deformation of aromatic ring stretching. One additional band is noticed at 1380 cm⁻¹ which is attributed to the C-O syringly unit and is absent in bagasse lignin .The band at 1119 cm⁻¹ is assigned to CH stretching in syringyl while the band t 1042 cm⁻¹ is assigned to CO of primary alcohol.

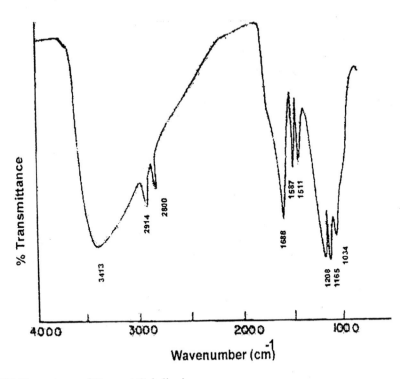

Figure 13. FT-IR spectrum of Coconut Coir lignin.

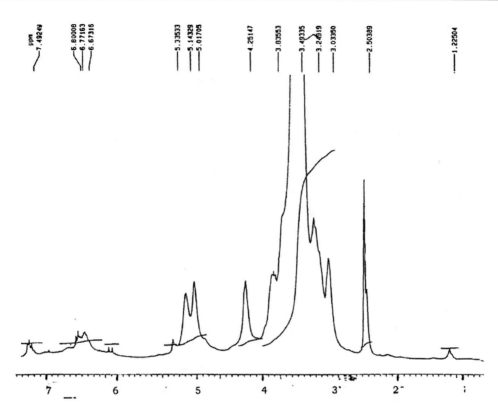

Figure 14. NMR spectrum of Bagasse lignin.

The spectra of coconut coir lignin Figure.13, shows a broad band at 3413 cm^{-1} and bands at 2914 and 2800 cm^{-1} due to OH and CH stretching vibrations respectively. The band at 1688 cm^{-1} is assigned to C=O stretching vibration in aldehydes and ketones while bands at 1587 cm^{-1} and 1511 cm^{-1} are due to aromatic skeleton vibrations. The bands at 1208 cm^{-1} and 1165 cm^{-1} are due to C-O of secondary alcohol .The band at 1034 cm^{-1} is due to guaiacyl CH while the band of guaiacyl ring is observed at 1208 cm^{-1} towards lower side as compared to bagasse lignin. The carbonyl peak appears to be more pronounced indicating the presence of relatively larger number of CO and CHO groups[20-21].

The NMR spectrum of Bagasse lignin, Figure.14, shows a peak at δ 1.22 ppm due to the proton attached to the β-carbon. Very broad and intense peaks are observed in the range of δ 3.03-3.50 ppm.These are attributed to benzylic proton with substituents H or CH$_2$OH at α-carbon. The shoulder peak at 3.8 ppm is attributed to methoxy protons .Intense peaks in the range of 4.2-5.3 ppm are due to aliphatic protons present in the side chain. The peak at δ 4.25 ppm is due to proton of saturated β carbon having OC4 (β-O-4 linkage), while the peaks in the range of 5.01-5.33 with center at 5.14 ppm are attributed to β proton with OH and CH substituents .The peaks in the range of 6.6 -6.8 are typical of aromatic protons .A small peak at 8.3 ppm is attributed to phenolic OH proton.

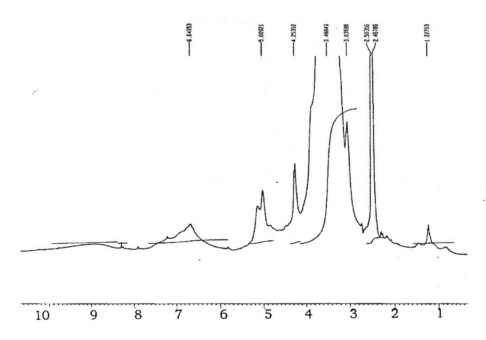

Figure 15. NMR spectra of bark lignin.

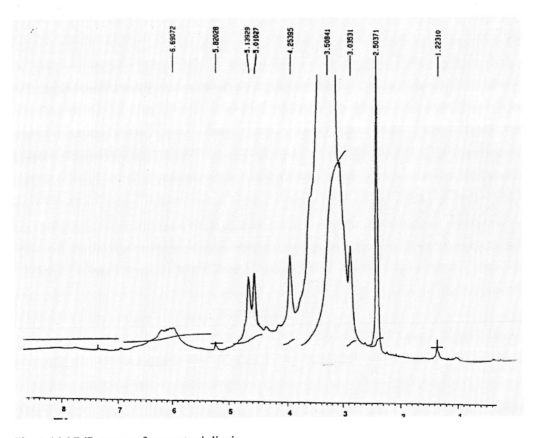

Figure 16. NMR spectra of coconut coir lignin.

The NMR spectra of bark lignin ,Figure.15, shows β-carbon peak at 1,22ppm ,aromatic proton peak at 6.64 ppm and methoxy peak at 3.8 ppm .The peaks in the range of 3-3.8 are more pronounced .The NMR spectra of coconut coir lignin, Figure.16, also reveals all the characteristics peaks but the peaks in this case are. some what broad.The peak at δ 1.22 ppm is due to β-carbon protons while the peak at 2.5 ppm is due to α-carbon protons .It is observed that the peaks in the range of 3-3.8 show fair spreading which shows higher number of protons of this type .It can be inferred that all lignins are composed of aromatic rings and the shifting of peaks is attributed to the presence of different types of substituents of aromatic rings giving rise to different proton environments.

7.3. Thermal Analysis of Sugarcane Bagasse, Eucalyptus Bark and Coconut Coir Pith Lignins

The DSC thermograms of different lignins are shown in Figure.17. The glass transition temperature is all lignins appear to be comparable and shows slight variation. The thermogravimetric analysis ,Figure.18, of bagasse lignin shows 8% weight loss at 100^0C.while in the temperature range of 200-300^{0}C shows a major thermal event resulting in 42% loss .Beyond 300^{0}C slow decomposition and steady weight loss is observed[22] . The bark lignin also shows similar thermal event .The temperature range of 100^{0}C-200^{0}C exhibits 44% weight loss while the second major event is observed in the range of 200^{0}C -250^{0}C.In this region 13% weight loss is observed. Beyond 250^{0}C very slow and steady weight loss takes place upto350^{0}C.Incase of coconut coir lignin only 1% weight loss is observed till 100^{0}C.In the temperature range of 200-300^{0}C major weight loss event is observed with 52% weight loss. All the samples loose maximum weight in the temperature range of 250-300^{0}C .Pyrolytic degradation of lignin in this temperature range is due to the fragmentation of inter unit linkages releasing monomeric phenols into vapour phase [22].

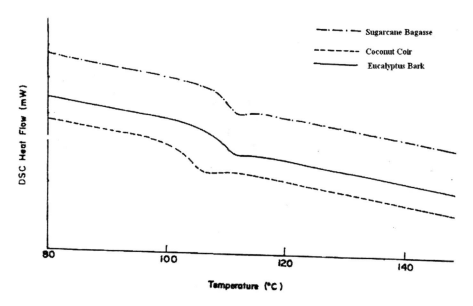

Figure 17. DSC thermograms of different lignins.

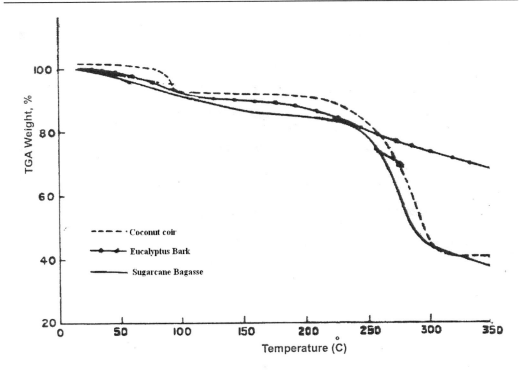

Figure 18. TGA thermograms of different lignins.

8. APPLICATIONS OF LIGNIN

The potential use of lignin for the production of plastics has been established by a number of researchers in past years [23-30]. Kratzl et al.[31] have investigated the reaction of lignin with phenol and isocyanates. These preliminary studies were designed to determine the potential utilization of lignin in the plastics industry. Glasser et al [33-35] have focused particularly on the possible use of lignin as a component in engineering plastics. Lignin is considered as a high-impact strength, thermally resistant thermoset polymer. It thus inherently has properties similar to those sought in some uses of engineering plastics. Central to this research effort was the chemical modification of lignin by hydroxyalkylation - forming hydroxy propyl lignin derivatives - to improve their viscoelastic properties as prepolymers for thermoset engineering plastics. A considerable amount of data has been accumulated regarding the modification of lignins to engineering plastics.[32].Unfortunately, the incorporation of various monomers and polymers, such as di- and polyvalent epoxy phenols, esters and isocyanates, in the lignin structure in most cases resulted in brittle or tarry materials whose properties designated them as potential adhesives, lacquers, dispersants and films, but not as structural materials. [33-35]. Recent systematic study on the relation between network structure and substituent in kraft lignin or steam exploded lignin have shown that the lignin containing networks can be modified in new ways. [36]. The toughening of glassy, structural thermoset can be achieved by incorporating a variety of polyether and rubber-type soft segment components in the polymer network structure.

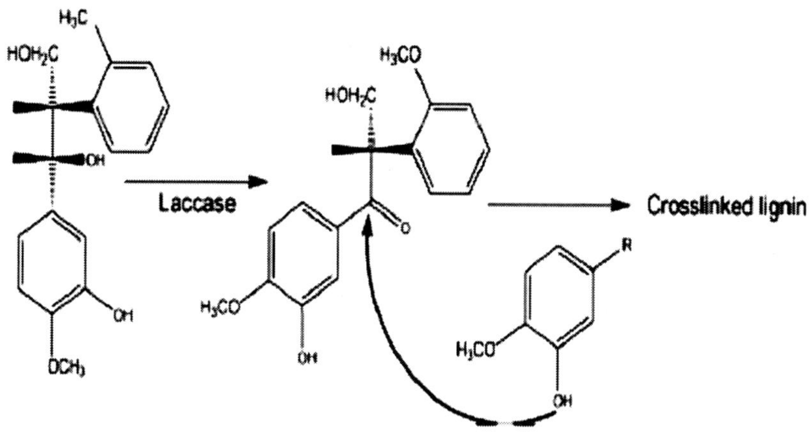

Figure 19. Lignin moieties generated.

(a) Lignin Based Wood Adhesives

The principle point of entry for lignin in the adhesives sub sector is likely to be in fiberboard production. Technologies have been trialed by Lund et al. [37] and Felby et al. [38-39] wherein the in situ cross-reaction of lignin via laccases (oxidoreductases) were used at the pilot plant level to provide boards comparable in strength to those produced using urea–formaldehyde adhesives. The inclusion of hot pressing to this process further increased the crosslinking. Moreover, solubilization of lignin in PF glue (30%, w/w) led to increases in plywood shear strength. Gel permeation experiments following the in situ treatment of lignin in beech fibres with laccase showed distinct increases in molecular weight of the lignin directly supporting the suggestion that lignin covalent inter-bonding was occurring [34-35]. In addition, model compound studies [36] have shown that the laccase treatment of lignin model compounds (and by inference lignin) generates moieties of increasing molecular weight that can react further to increase crosslinking, Figure.19.

The fiberboard market has a large turnover but operates on a low-profit margin and this makes the introduction of new technologies notoriously difficult [37]. However, fiberboard production is predicted to increase from 15.4 to $16.8 Mm^3$ over the period 2002–2006, an 8.3% increase. Given the increasing legislation surrounding the restriction of chemical usage in fiberboard formation it is likely that alternative, safer sources will be sought and the laccase system should benefit from this

It has been suggested that the most promising application of lignin-polymer systems in recent years has been the development of thermosetting resins used as wood adhesives. Presently, the adhesives used in these applications have been formaldehyde based, manufactured from gas and petroleum such as urea-formaldehyde, melamine formaldehyde and phenol-formaldehyde.[38] More than half of all formaldehyde produced in North America is used for the manufacture of these polymers. [39-41]. Their dependence on gas and petroleum prices was of particular concern during the energy crisis of the early 1970's, which saw a levelling off of wood board production. Furthermore, the detrimental effects of formaldehyde gas emitted from formaldehyde foam insulation prompted official banning of these products used in construction, and consequently has focused attention on the use of formaldehyde in the manufacture of building products. Hence, despite continued supply at low cost, there is an ever increasing demand for new adhesive products which are independent of the cost of gas and oil, and are nontoxic. For this reason, research on lignin-polymer systems is of continued importance in the area of lignin-phenol formaldehyde (L-PF), lignin-urea formaldehyde (L-UF), and lignin-polyisocyanate (L-PU) adhesives.

(i) Lignin-Phenol Formaldehyde (L-PF) Systems

Lignin-phenol formaldehyde (L-PF) formulations have been used in the manufacture of particleboard, fiber board, and plywood, as well as other adhesive and board products. The reason for the development of these new adhesives must also be seen from the point of view of lowering the overall cost of production, since lignin, in the form of spent sulfite liquor (SSL) or kraft black liquor (KL), is less costly than PF [42]. Roffael and Rauch [43-46] demonstrated the applicability of using sodium and ammonium-based lignin SSL as a substitute for conventional phenolic adhesives. Approximately 25 and 35% of phenolic resin may be substituted in 20-mm thick one-layer, and three-layer particleboard respectively, without significant decrease in the mechanical properties of the board. Particleboard properties were also reported to be a function of the alkali content of the resin and on pressing conditions. The crosslinking of the lignin macromolecular units occurred via a Lederer-Manasse type reaction, with side chains being linked by a Tollens reaction. The copolymerization process of lignin with PF is shown in Figure.20.

Dolenko and Clarke [47-48] prepared a similar adhesive compound for plywood and wafer board production using methylolated kraft lignin and a phenol formaldehyde resol, Figure 21. The introduction of methyl groups into the lignin molecule by reaction with formaldehyde increased the reactivity of the lignin such that crosslinking could occur more readily.

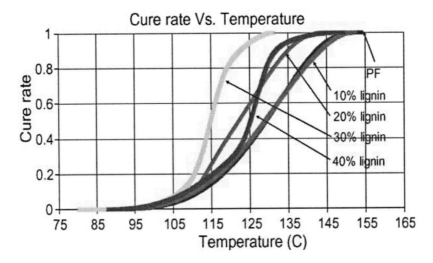

Figure 20. Cure conversion peaks for lignin-PF resin.

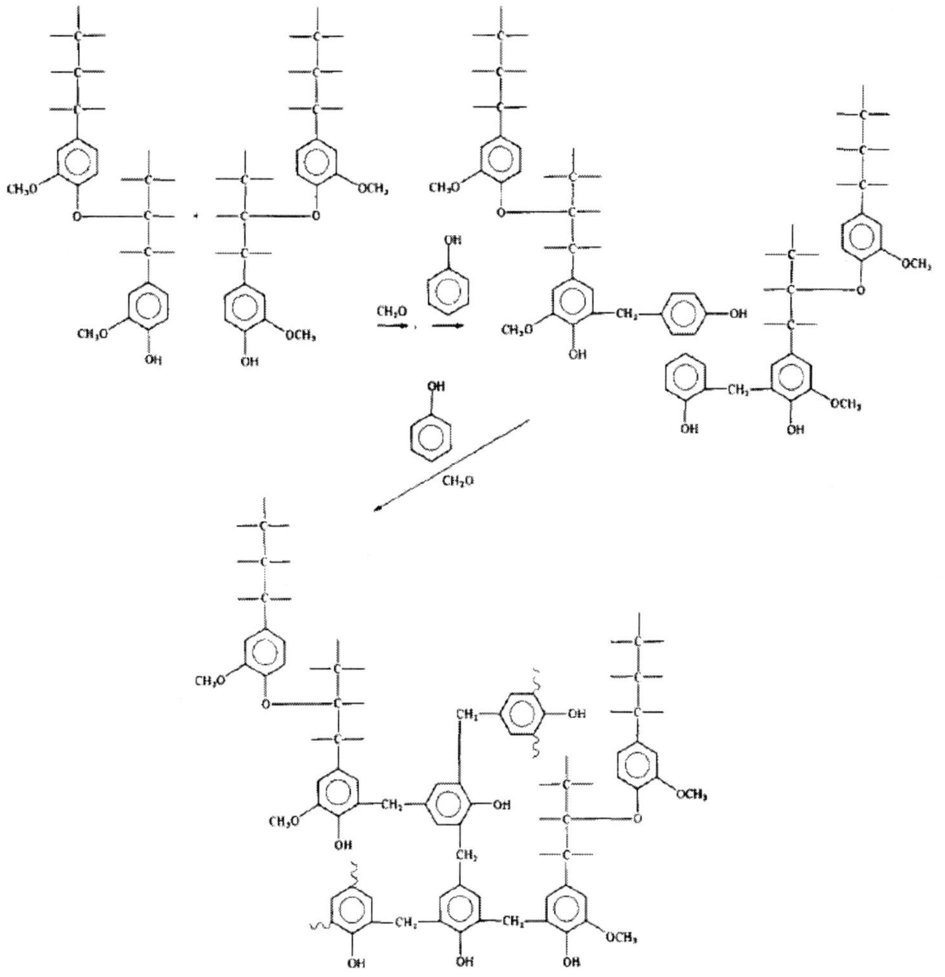

Figure 21. Structure of phenol/formaldehyde lignin.

Studies conducted to evaluate adhesive characteristics for use in plywood products indicated that board strengths increased with a corresponding increase in the lignin-PF ratio and up to 50% of the PF adhesive could be substituted without adversely affecting board properties. Furthermore, the curing properties of the adhesives were influenced by the pH of the adhesive, with board strengths increasing with decreasing pH. It is surmised that curing in acidic conditions causes cross-linking to occur via ether linkages accounting for the superior resin properties. It has been suggested that the low viscosity of lignin in acidic solutions is a problem with this approach since low solubility makes it difficult to produce low viscosity solutions with a high solids content. [48]. Adams and Schoenherr [49] produced a high solids (40%) low viscosity (10,000 cps) lignin solution by dissolving kraft lignin in phenol-formaldehyde alkali solutions. This copolymer, when compounded into a plywood adhesive used in the manufacture of three ply panels of Douglas fir, produced wood failures of 92%. Hollis and Schoenherr [50] used the same solution system in a two-step process to yield a resin having 37% solids content and 460 cps viscosity. The resin was formulated into an adhesive for use in plywood manufacture. Tests conducted on three-ply panels pressed at

140°C and 1.2 MPa for 6 min showed 94% wood failure.Forss and Fuhrmann [51] used ultra filtration to extract the high molecular weight (MW) fraction of alkali lignin to formulate a lignin-PF adhesive (Karatex) suitable for use in the production of fiberboard, plywood and particle board. The authors suggest that in order to obtain a three-dimensional molecular network, less PF is required to crosslink the high MW fraction in comparison to the low MW fraction of alkali lignin. Tests on three-layer plywood, using 40-70% lignin content as adhesive, show that board properties are similar to those adhered with conventional PF adhesives. In contrast, Shen and Calv [52] were able to demonstrate that the low molecular weight fraction of lignin sulfonates yielded better bonding qualities and shorter curing times compared to crude lignin sulfonates. Their patented process is presently being exploited commercially in Canada, where the low molecular weight fraction is obtained by ultra filtration..

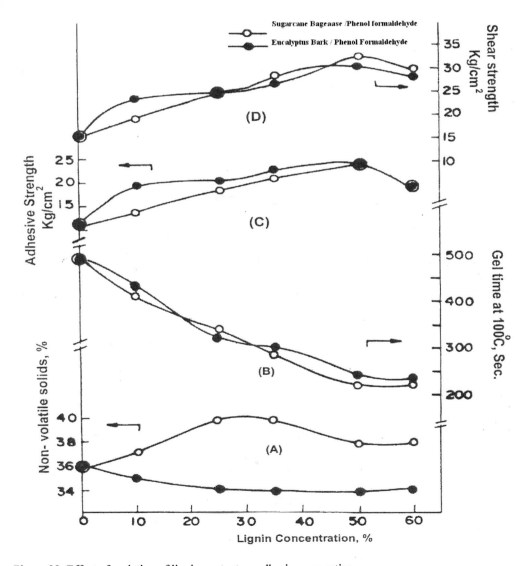

Figure 22. Effect of variation of lignin content on adhesive properties.

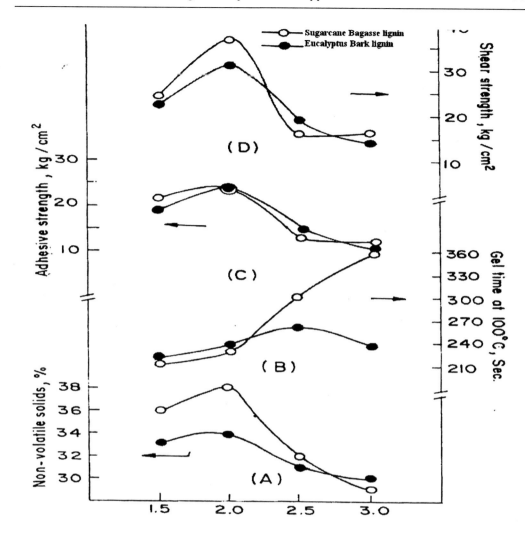

Figure 23. Effect of formaldehyde/Phenol concentration on 50% lignin substitution.

Ammonium-SSL-PF is of interest because it yields particleboard of exterior quality, practically free of formaldehyde emission. This adhesive, however, is more demanding compared to UF resin because it requires longer pressing times at higher temperatures. Studies have been conducted with a view to further increasing the lignin content of L-PF products, to 60% levels. Results, based on activation energies and cure rates obtained from DSC thermal analysis, indicated that prepolymers of kraft lignin (KL) and steam explosion lignin (SEL), when chemically modified with formaldehyde and phenol, have no inhibitory influence on the resin cure at the 50-60% phenol substitution level. The ability of hydroxymethylated lignin prepolymers to condense with phenol was found to be dependent on the particular lignin structure, and KL from pine proved the most amenable to chemical modification with formaldehyde and phenol, in comparison to SEL, bagasse and eucalyptus bark substituted lignin formaldehyde adhesives.The adhesive strength and gel yime has been studied, Figure 22.

A gradual increase in the adhesive strength is observed with increase in lignin conent.The adhesive strength increase to 11 kg/cm² upon 10% addition and continuously increases upto

50% to 24 kg/cm^2 ,on further increasing the lignin content the adhesive strength decreases to 20 kg/cm^2 .The shear strength is also found to increase with the lignin content which increases to 18 kg/cm^2 at 10% loading of bagasse lignin and 23 kg/cm^2 for eucalyptus bark lignin. At 50% loading the values of shear strengths for bagasse lignin are found to be 32 kg/cm^2 and for eucalyptus bark lignin it is observed to be 31 kg/cm^2 .upon further increasing the lignin content a decrease in the shear strength is observed which is noted to be 30 kg/cm^2 and 27 kg/cm^2 for bagasse lignin and eucalyptus bark lignin respectively. The adhesive strength and shear strength depends on the crosslink density and incorporation of lignin in phenol formaldehyde resin provides a better reinforcement structurally [53].

The effect of formaldehyde concentration has also been studies by varying the formaldehyde to phenol ratio form 1.5 -3.0 at 50% lignin substitution,Figure.23. At formaldehyde /phenol ratio of 1.5, the non volatile content is found to be 36% and 33 % for bagasse lignin and eucalyptus bark lignin respectively. A sharp increase in the nonvolatile content is observed upon increasing the formaldehyde /phenol ratio to 2.0 which is found to be 35% for bagasse lignin and 34% for eucalyptus bark lignin respectively. On further increasing the formaldehyde/phenol ratio to 3% a decrease in the non volatile content is observed. The values at this ratio are found to be 26% and 30% for bagasse lignin and eucalyptus bark lignin. The higher values of non volatile content at lower values of formaldehyde /phenol content are attributed to the reaction of formaldehyde in crosslink and network formation. The non volatile content is higher in eucalyptus bark lignin than in bagasse lignin due to the structural differences in the two lignins at molecular levels.

(ii) Lignin-Urea Formaldehyde (L-UF) Systems

The main advantages of UF are its low cost and excellent bonding qualities. As mentioned previously, UF slowly hydrolyses and emits formaldehyde during the life of the board. Generally, between 10-50% of UF binder may be replaced by SSL, depending on the SSL base. Studies undertaken in Germany to evaluate the effect of producing particleboards with 15% of the UF adhesive substituted by SSL indicated that the board properties were not imparied by the addition of various SSL. Furthermore, it was shown that formaldehyde release was reduced between 10 and 18%, corresponding to the amount of SSL in the UF-SSL formulation. In the case of NH4-SSL, the release of formaldehyde is decreased to a greater extent than the percentage of NH4-SSL in the formulation. Edler [55] formulated a high solids SSL-UF adhesive consisting of urea-formaldehyde copolymer, waste sulfite liquor, wheat flour filler and ammonium ion. TM Tests to evaluate the dry and wet shear strength of the adhesive indicated that it would be suitable for use in the manufacture of plywood Studies carried out at the Composite Wood Products Laboratory of Forintek in Canada, show similar results with L-UF resin, where the NH4-based SSL replaced up to 50% of the UF resin. [56] .The largest saving in resin cost was obtained with SSL as UF extender, replacing 45% of the UF resin with NH4-SSL for a saving of 28% resin and a reduction in formaldehyde emission of 50%. The results of these studies are shown in Table.5, indicating that despite a lowering of mechanical properties in comparison to neat UF polymer, the board still conformed to Canadian standard requirements for interior grade particle boards.

Table 5. Properties of UF-SSL particle board

Adhesive percent UF: SSL	Modulus of rupture (mpa)	Modulus of elasticity (GPa)	Internal bond (MPa)
8.5: 0	22	3.42	1.07
6.8 :1.7	20.2	3.71	0.70
5.78: 2.72	19.6	3.58	0.65
5.29 :3.32	20.1	3.42	0.62
3.74 4:.76	18.1	3.68	0.58

(iii) Lignin-Isocyanate (L-PU) Systems

The polyester-polyether polyol was reacted with various isocyanates (e.g. hexamethylene diisocyanate (HDI), methylene polydiisocyanate (MDI), toluene diisocyanate (TDI)) to produce a series of wood resin adhesives [56-57].These results indicate that L-PU polymers compare favourably with resorcinol-formaldehyde and epoxy polymers both in terms of shear strength and percentage of wood failure.Lambuth [58] patented a process for the formulation of an aqueous polyisocyanate lignin adhesive. It was noted that although polyisocyanates are strongly hydrophobic, those of low molecular weight react only very slowly with water at room temperature. Hence an aqueous solution of lignin, provided by waste liquor derived from chemical pulping, could be used without any modification. It was reported that gas which does form, due to the reaction of water and isocyanate, would not be a problem since the wood matrix, on which the adhesive is applied, was sufficiently porous to absorb the gas without affecting the quality of the bond. Thus, provided a low molecular weight fraction of polyisocyanate is used (MW = 300-2000) to ensure optimum viscosity for bonding, any aqueous solution of either KL or SSL may be used

Fiberboards which were manufactured containing 33% waste liquor in combination with isocyanate polyol met particle board standards and compared favourably with the performance of UF and PF binders. Studies initiated by Gamo [59-60] on the utilization of lignin, tannin, and other materials from natural sources for use as wood binding adhesives, formed basis from which kraft lignin formaldehyde (KLF) polymers modified with isocyanates were developed. . The KL reaction with formaldehyde and the addition of isocyanate were found to be necessary in order to obtain sufficient bonding strength and water resistance. It was shown that the concentration of hydroxyl groups in the KL alone was not sufficient to permit adequate copolymerization with the isocyanate. Additional hydroxyl groups from the reaction with formaldehyde were needed to yield adequate crosslinking reactions with the isocyanate. Bond shear strength and water resistance of wood adhesive bonds were tested on three-ply plywood test pieces. It was found that both the shear strength and the water resistance of the adhesive increased as the amount of isocyanate in the solution increased.

(iv) Lignin-Elastomer Adhesives

The literature contains almost as many references to the use of kraft lignin as an extender, modifier and reinforcing pigment in rubber compounding as it doesto resin formation) [61]. Many types of lattices such as natural, nitrile and neoprene are compatible with lignin either in the acid or sodium salt form, and will tolerate the addition of appreciable quantities without coagulation. The lignin reportedly improves the quality and strength of fibres made from the latex. Mixtures of rubber with kraft lignin are also suitable for use as rubber to glass adhesives.The lignin after the addition of mineral acid reduces spurting and evaporation) [62]. Kraft lignin also compares very favorably with other reinforcing agents such as carbon black because of its lower density. At high loadings, high tensile is obtained. Various treatments of kraft lignin have been suggested for improving its rubber reinforcing properties. By passing oxygen through an alkaline solution of the lignin, a superior reinforcing pigment for butadiene-styrene rubbers can be produced. An alkali hydrolyzed lignin which is at least 90% soluble in acetone has been claimed to function as an improved filler for butadiene acrylonitrile copolymers.In addition to the use of straight lignin-rubber master batches as such, Mills and Haxo [63] have patented the incorporation of an organic polyisocyanate as a co reactant. Papst [64] in Germany has claimed that almost any lignin can be satisfactory reinforcing filler if the rubber is a graft copolymer or block copolymer obtained by polymerization of natural rubber with vinyl compounds containing tertiary amine groups. Lyubeshkina [65] indicated that a distinctive feature of the preparation of lignin-filled rubber is that lignin is introduced at the latex stage, since the introduction of powdered lignin into dry rubber does not lead to the reinforcement. It is believed that the reason for this is the binding of lignin particles to one another by hydrogen bonds, which leads to condensation and growth of its particles and prevents the dispersion of the lignin particles during the rolling of rubber.The introduction of lignin into rubber at the latex stage greatly complicates and increases the cost of the technology of its manufacture, since the usual method for the preparation of the parent lignin-rubber mixture consists in the dissolution of lignin in an aqueous solution of alkali, mixing it with the rubber latex, heating the mixture, and the joint coagulation of the rubber and lignin by pouring the mixture into an acidic solution while stirring. The reinforcing effect of lignin is determined by its particle size, which depends on the precipitation conditions and certain other factors. In the parent mixture lignin is not bound to rubber by chemical bonds. The reinforcing effect of lignin increases appreciably when lignin is precipitated in the presence of sodium silicate. The mechanism of the reinforcement of raw rubbers by sulphate lignin can be accounted for mainly by its surface active properties and the coagulate is therefore precipitated in the form of fine particles.This leads to the formation of a material with low stress parameters for 300% elongation and a high hardness, so that the lignin-filled vulcanisates exhibit a lower resistance to wear than carbon-black filled rubbers. In order to eliminate the above defects, the lignin-rubber mixture is treated with polyisocyanates and is used in the manufacture of rubbers for car tyres. The polyisocyanates employed are di- and tri-isocyanates of different chemical structure in amounts ranging from 1 to 10 parts per 100 parts of rubber (dry weight)).The articles obtained from such a composition have improved operating characteristics, in particular the stress corresponding to 300% elongation The mechanism of the reinforcement of rubbers by lignin is fairly complex and has not been completely investigated. Apart from the physical explanation, associated with the surface active properties of lignin, there is a chemical explanation of this phenomenon.The reinforcement of rubbers by alkali lignin can be accounted for by the interaction of the active groups of lignin with the double bonds or active groups of the rubbers - the formation of hydrogen bonds between the hydroxyl groups of lignin and the π electrons of the double bonds of the rubbers When isocyanates are introduced, the high reactivity of the isocyanate groups promotes the rupture of the hydrogen bonds, which are unable to reinforce the dry lignin and the latter is thus converted into an active series.

Lignin filled rubber can be obtained in two ways:

(1) the coprecipitation of lignin and rubber from a lignin-latex mixture by the method indicated above, and

(2) by introducing hydrated lignin into rubber with the aid of rubber mixing equipment. The vulcanisates filled with dry lignin are characterized by a tensile stress not exceeding $6.0 mN/m^2$. Lignin-oil pastes can be used as lignin products isolated from black liquors for the reinforcement of raw and cured rubbers. They are obtained by mixing hydrated lignin, containing approximately 60% of moisture, with a plasticizer (tall oil, pine resin), after which moisture is evaporated from the resulting mixture by continuous stirring at 60-70°C. Simultaneously with drying, lignin is dispersed in the plasticizer, which prevents the agglomeration of the lignin particles and tends to increase the degree of their dispersion. The lignin-oil paste consists of a powder or solid lumps of dark-brown colour, depending on the type of plasticizer. Simultaneously with drying lignin is dispersed in the plasticizer, which prevents the agglomeration of the lignin particles and tends to increase the degree of their dispersion. The lignin-oil paste consists of a power or solid lumps of dark-brown colour, depending on the typeof plasticizer. Such a paste can be introduced into rubbers using the normal rubber mixing equipment. [66].

Another method for the isolation of lignin from black liquors involves its precipitation in the presence of water-soluble additives (low molecular weight alcohols, aldehydes, ketones, or emulsifiers) which promote the formation of a highly disperse product. The vulcanisates obtained using these lignins have high values of tensile stress. TM Dry sulphate lignin is of considerable interest for filling and modification of thermoplastics in order to obtain frost-resistant polypropylene compositions and also in the reprocessing of wastes from polymeric materials. Studies have been carried out on the creation of composite materials using sulphate alkali lignin based on polyolefins-materials ,including preparations based on secondary polyethylene obtained from used articles.[67-73].The nature of the interaction of the ingredients of the polymer composition under processing conditions is very important for the specific modification of the basic polymer by sulphate lignin (when the content of the latter is up to 5%) and for the determination of the technological regime for the conversion of poprolin materials into articles.It has been established that a radical interaction took place between polypropylene and lignin. The results agree with data obtained in the study of free radicals produced during the formation of the carbon black-raw rubber gel. It has been observed that during the rolling of the rubber in the course of its processing, free radicals are formed and interact with the polymer via a radical mechanism. [68-69].An interesting feature of this process is the approximately linear relation between the ratio of the EPR line widths in air and *in vacuo* and the specific surface of the carbon black, which indicates the surface character of the interaction of some of the spins, capable of participating not only in the formation of the carbon black-raw rubber gel but also in the agglomeration of the carbon black particles.In order to improve the operating characteristic of poprolin (to reduce the embrittlement temperature and increase the frost resistance), a low molecular weight plasticizer, dioctyl sebacate (DOS), was introduced into the polypropylene- lignin system, which resulted in the formation of a frost-resistant material - poprolin-1

. It follows from the EPR data that the introduction of DOS into poprolin disrupts the interaction between lignin and polypropylene, which entails a decrease of *ca/Co.* The attempts to intensify the disrupted interaction by introducing an additional amount of lignin (in excess of the optimum value for the given composition) did not lead to the desired results.

Tests showed that the disrupted interaction cannot be restored by introducing fillers of the type of titanium dioxide, which has been stated to be capable of adsorbing lignin, wetting its surface and leading to a more effective dispersion of the plasticizer and lignin in the polymer matrix. Furthermore, the introduction of titanium dioxide prevents the interaction of lignin with DOS, stimulating the interaction of lignin with polypropylene and leading to an enhanced stability of the system. These data indicate the possibility of regulating by the above procedures the properties of complex materials important for practical applications. The study of rheological characteristics of poprolin-1 showed that the introduction of the optimum amount of lignin (three parts) reduces its fluidity [74-75].This greatly facilitates the processing of poprolin-1.

The technology of the processing of plastics designed to increase the quality of the polymeric materials and to increase the durability of the articles obtained from them is known to involve the introduction of antioxidants in order to suppress oxidation processes occurring under the influence of various factors (high temperature, solar radiation, etc.).[65]. Phenols, as inhibitors absorbing in the ultraviolet range, have come to be widely used among antioxidants. The photoreaction of phenols leads to the formation of phenoxy-radicals, which in their turn undergo a photoreaction in the polymer matrix and their subsequent fate depends on their reactivity, temperature, concentration, and the type of polymer matrix. In terms of their chemical structure, lignins can be assigned to the class of sterically hindered phenols. On the other hand, the involvement of these phenols in radical reactions reduces to the initial abstraction of the hydrogen atom from the hydroxyl groups of phenols with formation of a phenoxy radical, i.e. the behaviour of the sterically hindered phenol in radical reaction is determined by the properties of the phenoxy radicals formed. It is known that the O-H bond energy in sterically hindered phenols is somewhat lower than in unsubstituted phenols, which facilitates, on the one hand, the homolytic dissociation, and on the other hand, leads to the formation of thermodynamically more stable species - stable free radicals.[76].The stability of the radicals formed in the presence of the antioxidant has a significant influence on the kinetics of the ageing of polymers. The experimental results, showing that lignin or its degradation products can be used as thermo stabilizers for polypropylene at its processing temperature (in excess of 200°C) and as inhibitors in its photo ageing process, may serve as confirmation of the above considerations. . However, data have been obtained showing that sulphate lignin can be used as a photo modifying additive to polypropylene and that it does not affect its degradation on prolonged irradiation. . At the same time it can greatly reduce the efficiency of light stabilizers operating in accordance with an inhibition mechanism, so that, in order to improve the light stability of lignin containing compositions, it is desirable to introduce additional substances belonging to the class of UV absorbers or dyes. Since the ligno-tall product contains simultaneously structure- forming (lignin), reinforcing (aliphatic acid salts), and plasticizing (tall oil) substances, it is considered a complex modifying agent. The use of lignin derivatives as stabilizers for synthetic polymers like polyethylene, polypropylene and poly(vinyl chloride) has been investigated by several authors. Gul et al. [77] have shown that lignin derivatives increase the cold and UV light resistance of polypropylene.Levon et al. [78] found that blending the additive lignosulphonate with polyethylene in a plastic order increased the stability, as indicated by rheology and changes in spin density, if the blending temperature was higher than 463 K. A high carbohydrate content in the lignin gave poorer results. Bronovitskij [79] found that the antioxidant effect of hydroxylated lignin on polyethylene, measured as oxygen uptake, was about 10 times less than for commercial antioxidants.

The synthesis of poly(phenylene sulphide)-like polymers from lignin, socalled sulphur lignin, also enables electrically conducting polymers to be synthesized from these materials. [80].Thus the conductivity of sulphur lignin doped with ferric chloride, a Lewis acid, has been found to be in the range 10 -3 to 10 -2 S/cm, corresponding to a good semiconductor. Similar results have also been obtained with other dopants. The long-term stability of the materials also seems to be quite good. Sodium lignosulphonate and sulphur lignin have been investigated as components in electrically conducting polymers) .A thermostable and probably nontoxic polymer, sulphur lignin, has been obtained by modifying lignin-based sodium lignosulphonate through reactions with elemental sulphur in an autoclave at 473-513 K. Sodium lignosulphonate and sulphur lignin are normally electrical insulators, but their conductivity can be increased for instance by compounding with graphite or doping with various electron acceptors or donors. The combination of compounding with graphite followed by doping with bromine has been investigated: The conducting properties were measured on samples ground and pressed to the size of IR-pellets. The measurements were made with standard four-point probe or two-point probe techniques. The doping was followed by infrared and electron spin resonance spectroscopy study. The percolation phenomenon in the combination of sulphur lignin and graphite was determined by measuring conductivities. The combination of sulphur lignin, graphite, and bromine was studied through the same way.

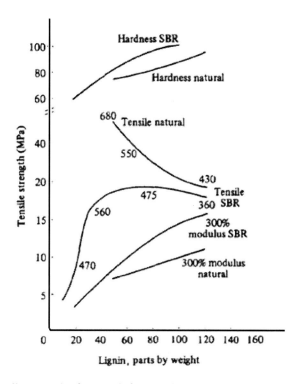

Figure 24. Effect of tensile strength of SBR reinforced with purified kraft lignin on precipitation temperature.

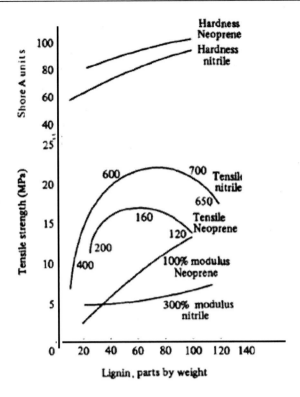

Figure 25. Variation of tensile strength properties reinforced with purified kraft lignin.

The use of lignins in conjunction with elastomers is hampered by their performance deficiencies, which are related to both physical and chemical properties.Factors which contribute to these deficiencies, include:

[1] The structural complexity of lignin;
[2] The tenacious resistance of lignin to any kind of degradation;
[3] The inherent variability of lignin, which is aggravated in isolation.

These significant factors have not served as a deterrant to continuing research, however, which indicates the optimism of researchers with respect to overcoming these difficulties. In its unaltered natural state, as was already mentioned, lignin acts as a binder, the matrix for cellulose fibres within the structure of wood. This indicates that lignin could be suitable for improving the mechanical properties of elastomers.While many reinforcing agents can be dry milled into elastomers, this method cannot be utilized with lignins. Attempts to dry mill lignin into styrene-butadiene rubber (SBR) showed almost no reinforcing effect. This can be attributed to the fact that lignin particles, under the influence of heat generation, coalesce by hydrogen bonding and thus cannot be finely dispersed into the rubber. Lignin must be incorporated into rubber by latex master batching in order that its reinforcing properties can be realized. To prepare a lignin-rubber master batch, it is usual to dissolve the lignin in an aqueous alkali, then add the alkali-lignin solution to the rubber latex. This mixture is heated and coagulated by running the mixture into an acid solution with adequate agitation. The reinforcing characteristics of a lignin in a master batch can be affected by several variables:

[1] Concentration of the alkaline-lignin solution and the rubber latex;
[2] Concentration of the coagulating acid solution;
[3] Temperature of the ingredients;
[4] Order of addition of the ingredients;
[5] Rate of agitation during co-coagulation,

Lignin precipitation temperature and particle size also affect its reinforcing characteristics. The tensile strength of SBR reinforced with purified kraft lignin is very sensitive to the precipitation temperature at which the lignin is isolated, Figure.24.

When the particle size of lignin is reduced below a critical limit(which gives a surface area of 45 mE/g or more), the lignin undergoes excessive coalescence and thus loses its reinforcing ability within elastomers. The type of lignin used can affect results, as hardwood technical lignins are inferior to their softwood counterparts as dispersants, emulsion stabilizers and elastomer reinforcing agents. Alkali lignin is an excellent reinforcing agent for SBR when co precipitated with rubber from latex. Problems can still occur because low molecular weight lignin or non-lignin constituents of kraft lignin prevent lignin from dispersing in SBR. These constituents act as agglomerating or coupling agents which promote the clustering of lignin particles. The removal of these detrimental materials can raise the softening temperature and reinforcing ability of lignin. Keilen et al.[81] have demonstrated that lignin (sulfate process) co-precipitated with natural, styrene-butadiene , nitrile, or Neoprene rubbers can yield tensile strengths comparable (at the same volume loadings) to carbon blacks,Figure 25. This could potentially consume large quantities of polymeric lignin for commercial markets. The polyblending of silicone rubber sealants with lignin does not require the involved process of latex master batching. Recently investigations by Beznaczuk [82] utilized the uncured silicone sealant (building construction grade), based on polydimethylsiloxane (PDS), and kraft lignin, polyblended by mechanical mixing only. Some 12.5-mm butt joint specimens (5, 10 and 15% lignin by weight) were cast between substrates of wood, aluminium and mortar. After a 14-day cure, specimens were exposed to laboratory control conditions (24°C, ~ 35% R.H.), accelerated weathering (4 cycles daily between - 30°C and + 30°C for a total of 400 cycles), and natural weathering conditions in a relatively polluted location of downtown Montreal. In comparison with the PDS sealants (no lignin), there were two cases where the strength and/or toughness of the polyblends was higher after tensile testing.The first case was that of the 5% lignin polyblend on a mortar substrate in control conditions, revealed that the reinforcing ability of alkaline lignin, for GR-S rubber, was greatly enhanced with increasing oxidation of the lignin prior to coprecipitation with the latex [83]. The oxidation of lignin is performed by bubbling air through an aqueous solution of lignin acid salt or lignin sodium salt; by purposely oxidizing the black liquor prior to or during its precipitation treatment; or by oxidizing the lignin acid salt in dry powdered form in a current of hot air or oxygen with or without the aid of a catalyst

To improve the reinforcing characteristics of lignin, various modifications have been proposed, such as treatment with urea-formaldehyde, heat treatment in the presence of an aldehyde, combination with phenol-formaldehyde resin, treatment with organic polyisocyanate, and combination with water. Lignin-rubber master batches that are at least partially treated with a diisocyanate can exhibit improved hot tensile strength, abrasion, and reduced torsional hysteresis. Heating of lignin synthetic rubber master batches for 5-40 min at 150-175°C tends to improve abrasion resistance and lower torsional hysteresis. [84].Lignin has the potential for providing various other properties to elastomers besides mechanical strength reinforcement. Murray and Watson [85] have observed that lignin acts as a stabilizing agent, against oxidation, for SBR. De Paoli *et al.* [86] investigated the use of lignin from sugar cane bagasse as a photo stabilizer for butadiene rubber. The photo stabilizing effect of lignin was further enhanced by the addition of dioctyl-p-phenylenediamine, and can be compared to commercially used stabilizers. Thus the potential for widespread commercial use of lignins in conjunction with elastomers exists if researchers can demonstrate that all performance requirements can be satisfied. Lignin, for example, can compete actively with carbon black as reinforcement for rubber if hardness can be reduced and modulus improved. This and other industrial uses of lignin--elastomer systems could prove to be beneficial worldwide, giving a useful function to a by-product of the wood pulping process.

(b) Lignin Based Foams and Films

Moorer et al. [87] employed lignin in the formation of polyurethane foams, by dissolving lignin in glycol and then reacting it with diisocyanate. The reaction was described in terms of the isocyanate acting as a crosslinking agent, linking the two kinds of polyol. The hydroxyl groups in lignin were assumed to be the key to this reaction. Kratzl and co-workers [88] sought to increase the number of reactive positions in lignin by condensation with ethylene oxide, propylene oxide, and an alkyl sulphide. The reactions were shown to produce hydroxyl groups suitable for mixing and reacting with diisocyanates in the formation of rigid polyurethane foam. Low density polyurethane foams with acceptable strength and excellent flammability properties were formulated by Glasser [89] with a commercial furan polyol containing 20% hydroxypropyl lignin. Propylene oxide-modified lignin from two sources, kraft and organosolv lignin, were employed. The organosolv lignin derivative exhibited better foaming characteristics than the corresponding kraft lignin, which collapsed when the rising foam was touched for testing. The weight contribution of lignin derivative was limited to 20% by compatibility with the fluorocarbon blowing agent, and solubility in polyol. Preliminary tests encourage further research on structural materials containing hydroxypropyl lignin derivatives.

Glasser [89] has undertaken a series of studies related to the formation of polyurethane films prepared from chain-extended hydroxypropyl lignin (CEHPL). The thermal, mechanical, and network properties of these PUs were investigated by DMA and DSC analysis. All films exhibited a single glass transition (Tg) which varied between -53 ° and 101°C, depending on lignin content. From swelling experiments, molecular weight between crosslinks was determined and found to vary over 2.5 orders of magnitude. These values were related to the change in Tg that accompanied network formation. Stress-strain experiments showed a variation in Young's modulus between 7 and 1300 MPa. Most of the variations in material properties was related to lignin content and to a lesser extent to diisocyanate type (i.e. hexamethylene diisocyanate or toluene diisocyanate). The source of the CEHPL had no effect on the properties of the polyurethane and these products can be controlled and engineered for a wide variety of practical uses. Lignins from various sources are employed as in the formulation of polyurethane network films. [90] The lignins included a milled wood lignin (red oak), a kraft lignin, an organosolv lignin, a steam explosion lignin and an acid hydrolysis lignin. The network like materials is evaluated by DSC, DMA and swelling. On the basis of the results a classification method of lignin prepolymers for use in (polyurethane) network-like materials was advanced, which is based on the increase of the Tg, as a result of crosslinking. This increase in Tg can be expressed as average molecular weight between crosslinks, using relationships proposed previously. Agreement between experimental (based on rise of Tg) and theoretical (based on synthesis parameters and stoichiometry) data suggests good network forming characteristics of a prepolymer, and disagreement indicates the formation of a non-uniform network with a large fraction of soluble material. The hydroxypropyl derivatives were comparable to the corresponding hydroxypropyl lignins, which in turn outperformed the parent lignins. Lignins ranked in the order of increasing qualifications for network formation as follows: acid hydrolysis, milled wood, kraft, organosolv and steam exploded lignin. The results suggest that solubility is the key parameter determining uniformity and structural material properties of the lignin prepolymers in thermosetting networks. Polyurethane films were produced by solution casting from a three-component polyurethane system consisting of a non-derivatized, low molecular weight fraction of kraft lignin, polyethylene glycol of various molecular weights and toluene diisocyanate.

At low contents of kraft lignin (< 10%) and low molecular weights of polyethylene glycol (300-600), weak but very flexible (ultimate strain 100%) polyurethane films were obtained whereas some of the films obtained with intermediate kraft lignin contents (15-25%) and low molecular weight polyethylene glycol showed considerable toughness. Use of higher molecular weight polyethylene glycol (4000) resulted in high values of ultimate strain (500-600%) also at high (30%) kraft lignin content. The crosslink densities of the films were generally very low.

(c) Lignin-Epoxy Systems

Epoxy resins based on lignin were reported by Mihailo et al. [91]in 1962. The lignin phenol resin was obtained by heating the 100 parts lignin with 140 parts phenol and 3 parts H_2SO_4 at 140-180°C, at which temperature phenol and H_2O distilled and the rest of phenol was removed in vacuum to obtain the lignin phenol resin.Epichlorohydrin, methyl-ethyl-ketone and lignin phenol resin were mixed and 40% NaOH added during stirring and boiling. The mixture was filtered and distilled, separating the epoxy resin. Ball et al. [92] reported epoxy-lignin resins in 1962. A resin casing was produced by heating 50 grams of an epoxy intermediate of the diglycidly ether of bisphenol to 190°C, and 17.5 grams of lignin was added and dissolved. The resulting solution was cooled to 120°C, and 33 grams of phthalic anhydride were added. The temperature was then raised to 130°C to assure dissolving of the phthalic anhydride in the lignin-epoxy solution. The solution was poured into molds which were heated to 160°C and kept at that temperature for 45 min, by which time an initial curing had occurred. The initially cured resin was removed from the molds after cooling and then subjected to a temperature of 136°C for 60 hr to attain a final curing. Nabuo Shiraishi [93] has also made an attempt to prepare lignin--epoxy resin adhesives through a chemical way. However, in order to improve its reactivity, kraft lignin was first phenolated with bisphenol-A.For phenolation, a small amount of aqueous hydrochloric acid or BF_3-ethyl etherate was used as catalyst. The phenolation with bisphenol-A was found to enhance the solubility of the lignin derivative. In fact, the lignin-epoxy resins obtained were found to be completely soluble in certain organic solvents, including acetone. The adhesive qualities of two types of epoxy-resins, prepared from lignins with different degrees of purity (Tokai Pulp kraft lignin F with < 4.5% sugars and Oji kraft lignin with 11.7% sugars) were examined. The enhanced adhesion was noted for the less purified lignin, suggesting active participation of the sugar components. Both epoxy resins gave satisfactory dry- and wet-bond strengths after 5 min of hot-pressing at 140°C. The adhesive qualities of the lignin-epoxy resin adhesives was found to be improved by the addition of calcium carbonate (50% by weight) to the liquid resin. This must be attributed to the nature of the weak alkali in calcium carbonate as a cure accelerator, and to the reinforcement effect of fillers.Besides hydrochloric acid, BF_3 is known to be an effective catalyst for the introduction of phenol groups into lignin. BF_3 is also known as a catalyst which can promote glycidylation not only of phenolic but also of alcoholic hydroxyl groups. Thus, the preparation of lignin-epoxy resins with BF_3 as catalyst was also studied. The epoxy value found for the standard lignin-epoxy prepared in the presence of BF_3 was 0.48, whereas that of the epoxy with HCl as catalyst was 0.38. The lignin-epoxy resins from the reaction in the presence of BF3 were also tested as adhesives for plywood, using triethylenetetramine as curing agent with hot-pressing at 140°C. The results of adhesion tests showed that waterproof adhesive strengths were improved by use of BF_3 as catalyst. This permitted the adoption of hot-pressing times as short as 3 min for preparing three-ply plywood panels with 6 mm thickness and resulted in satisfactory waterproof adhesive performance. The addition of calcium carbonate (50% by weight) to the liquid adhesive was again found to enhance the waterproof adhesion.

An epoxy resin was synthesized on the basis of lignin by reaction of epichlorhydrin with hydroxypropyl lignin which has been reported by Nieh and Glasser [94]. A mixture of a quaternary ammonium salt and potassium hydroxide was used as catalyst. Additional KOH was added stepwise at a rate which compensated for KOH consumption during dechlorohydrogenation. The epoxidation reaction was first studied using hydroxypropyl quaiacol as a lignin-like model compound. At room temperature, and when epichlorohydrin was in excess, the reaction was completed in five days. The reaction was found to be highly dependent on the stepwise addition of KOH, and it was independent of epichlorohydrin concentration. The maximum conversion of hydroxy to epoxy functionality was found to be 100% to 50% for model compound and lignin derivative, respectively. The lignin-epoxy resin was crosslinked with diethylenetriamine, amine terminated poly(butadiene-co-acrylonitrile) and phthalic anhydride. Sol fraction, swelling behaviour, and dynamic mechanical characteristics of the cured lignin-epoxy were studied in relation to cure conditions. Tomita and his coworkers [95] reported the preparation of pre-reacted ozonized lignin/epoxy resins. Ozone has recently become industrially available at low cost. In the experiment, the lignin (50 g) was dissolved in a mixture of dioxane (500 ml) and methanol (1000 ml), and ozonized at 0°C with an oxygen flow rate of 0.5 ml/min and ozone concentration of 30. After ozonization, the solution was treated with an excess amount of ether, and the insoluble fraction was filtered off, followed by drying under vacuum. The ozonized lignin was mixed by stirring with DGEBA and heated at 120°C. After heating for 30 min, the mixture was cooled to room temperature, the solidified reactants were dissolved in acetone (2 ml) and the curing reagents, diethylenetriamine or hexamethylenediamine, were added at 900 of the stoichiometric amount to epoxy equivalent.Curing was generally done by heating at 130°C for 2 hr and allowing the product to stand at room temperature for one day. In order to demonstrate the effect of various lignin functional groups , kraft lignin and several of its derivatives have been investigated as potential additives. The results show that polymerization of olefins is influenced mainly by hydroxyl (OH) groups present in the lignin.

(d) Polymer-Lignin Systems

Chodak *et al.[96]* investigated the influence of lignin addition on the crosslinking characteristics of polypropylene (PP). Unmodified kraft lignin has the greatest effect on the radical transformation reaction, since it contains many more OH groups than the lignin derived additives. It was also observed that the reactivity of aromatic OH groups (i.e. those attached to the aromatic ring in lignin) was greater than those of the aliphatic OH groups. Klason [97] has used lignin as a filler in high density polyethylene (HDPE) and PP. Lignin appears to behave as a reinforcing filler which, when combined with either HDPE or PP, serves to moderately increase the modulus of these polyolefins.These results were satisfactorily predicted using the Halpin-Tsai equation as modified by Niesen. Glasser *et* aL [23] have conducted a study to elucidate the mode of formation of new polyurethanes synthesized from lignin isocyanate combinations. They demonstrated that lignin may be carboxylated by reacting it with maleic anhydride to form a nonhydrolyzable copolymer. The carboxyl and phenolic groups of this copolymer offer reactive sites for oxyalkylation to a polyol. This may be achieved by reacting the saponified copolymer with propylene oxide in the presence of an alkali catalyst. The carboxyl groups are esterified, and the phenolic and aliphatic hydroxyl groups are etherified to yield a highly viscous, homogeneous active and polyfunctional polyester polyether polyol suitable for mixing and for reaction with diisocyanates. To further demonstrate the versatility of lignin as a co reactant in polyurethane formation, Glasser *et al.* [57] undertook the production of polyurethane foams from carboxylated lignin. The structure of any thermosetting resin can be described in terms of a relatively high molecular weight backbone polymer that is tied together by crosslinking segments. The properties of the network are then determined by the nature of the two components (main polymer and crosslinking agent) and the number of crosslinking sites introduced into the system (crosslinking density). By changing any one, or any combination of these structural features, the properties of the network are changed, allowing for the formulation of products with a wide range of properties. Based on this premise, Glasser *et* a/. [23] conducted a series of tests on hydroxypropyl lignin polyol-isocyanate derivatives. They established the effect of the crosslink density on polyurethane film properties by varying both the NCO/OH stoichiometry and the hydroxyl content of hydroxypropyl-lignin polyol. The maximum effective crosslinking was found to occur at a NCO/OH ratio of 3, after which the density levelled off. It is suggested that the higher concentration of diisocyanate, in comparison to conventional polyurethane systems (i.e. 3 : 1 vs 2 : 1), is due to the high polydispersity of the lignin polyol, which excludes incorporation of lower molecular weight fractions into the network. Thus the influence of the sol fraction on the physical properties of the polymer network was significant, but the manner and the degree to which it affected the overall behaviour of the network could not yet be ascertained. The crosslink density was found to vary directly with the hydroxyl content of the polyol. The Tg of the polyurethane films varied linearly with the crosslink density of the network. The dynamic mechanical properties of the films were found to be sensitive to the weight fraction of extractable in the network.

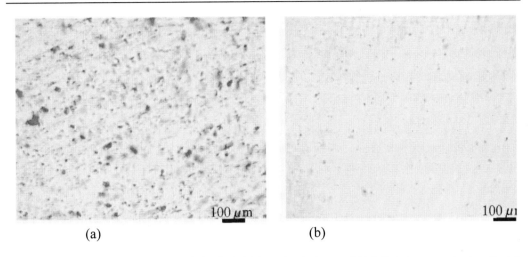

(a)   (b)

Figure 26. Blend morphology of Kraft lignin (1%) in polyethylene and Poly(butylene succinate-adipate) (visible microscopy).

Saraf *et al. [21]* studied polyurethane synthesized from blends of hydroxypropyllignin (HPL) derivatives with polyethylene glycol (PEG). The effect of soft segment incorporation in relation to content and molecular weight was examined on the basis of thermal and mechanical properties. A steady drop in Tg was observed, with increasing PEG (soft segment) content and a significant increase in Tg with rising molecular weights. The tensile behaviour was considerably modified by the addition of even minor percentages of PEG. Both Young's modulus and tensile strength decreased with increased PEG content, whereas the ultimate strain at break increased for similar percentages of PEG in the blend. Hence the mechanical properties were particularly sensitive to glycol content. Uniformity in structure, reduction in brittleness, and considerable improvement in mechanical properties with the inclusion of minor PEG constitutents, indicated that the lignin-based network polyurethane could be synthesized with a wide range of performance characteristics. In a subsequent experimental program, the effect of concentration of polybutadiene glycol(PBDG) in a mixture of HPL was examined in relation to the thermal and mechanical properties. A phase separation between the two polymer components in the polyurethanes was detected by thermal and mechanical analysis. This effect was evident for nearly all degrees of mixing. It was concluded that polyurethane films behaved like rubber-toughened lignin networks when PBDG was the discrete phase, and like lignin-reinforced rubber when the lignin derivative phase was discrete.Yoshida [98] found that at low NCO/OH ratios, kraft lignin contributed effectively to the formation of three-dimensional polymer network. When an optimum quantity of kraft lignin (5-20% depending on the NCO/OH ratio) is used, the resulting polyurethanes show improved mechanical properties compared to the polyurethanes synthesized with the polyether polyol as the only polyol component. At high contents of kraft lignin (> 30%) the polyurethanes were hard and brittle regardless of the NCO/OH ratio used. This is attributed to the combined effect of increased crosslink density and of an increase in chain stiffness.

Yoshihiro Sano and his coworkers [99] have investigated the utilization of lignin in resorcinol resins (RF). High molecular weight solvolysis lignin (B-CWL-1) and low molecular weight solvolysis lignin (B-CWL-2) were isolated in yields of 14.8 and 12.8%, respectively, from spent liquor which was prepared in the cooking of beech wood with cresols and water (7 : 3) including a small amount of acetic acid for 2 hr at 170°C. The pulp was obtained with 7.3% Klason lignin.B-CWL-1 with MW 3500 (by GPC) and B-CWL-2 with less than 700MW. Three different approaches were tried in preparing lignin-based resorcinol resin adhesives. Adhesive-1 was a mixture of lignin-based phenol resin (LPE) and resorcinol. LPF was made by condensing B-CWL-1 and phenol (9:1) with HCHO. Adhesive-2 was a mixture of RF and an aqueous NaOH solution of B-CWL-1 or 2. RF was prepared by the usual method using resorcinol and HCHO. Adhesive-3 was a mixture of RF and lignin-based resin (LF). LF was prepared by reacting B-CWL- 1 or 2 with HCHO. Paraformaldehyde and a filler were added to each of the adhesives for use in making 3-ply plywood panels. Adhesive-3 was the best of the lignin-based resorcinol resins. Tensile shear strength tests showed that as much as 80% of the resorcinol resins could be replaced by B-CWL- 1, and also as much as 60% could be replaced by B-CWL-2. With adhesive-2 being more convenient, more than 60% of the resorcinol resins could be replaced by B-CWL-1. Utilization in phenolic resins was also studied in terms of different solvolysis lignins. Hence solvolysis lignins derived from oak wood and beech wood and isolated by solvolysis pulping using a cresol-water system were studied in the context of their potential use as an extender or modifier of a phenolic based adhesive system. It was found that at least 90% of the phenolic resins can be replaced by each lignin. If necessary, each of the lignins can be condensed with HCHO in order to produce phenolic-type resins, and also the temperature of the hydraulic press for the preparation of the tested plywood can be lowered from 140 to 120°C.

L. P. Kovrizhnykh et al. [100] have studied a system based on urea resins with industrial lignosulphonates .Replacement of 20-30% HCHO-urea resin with Na or Ca lignosulphonate gave good-quality binder for particleboards. The binder cured rapidly in the presence of 2% $(NH_4)_2S_2O_s$. The obtained boards had good physico-chemical properties and a lower emission of HCHO than boards bonded with pure HCHO-urea resin. Lignosulphonates treated with $(NH_4)_2S_2O_s$ were used in combination with urea resin as binders in the manufacture of particleboards.The curing of binder containing 30% lignosulphonates and 2% $(NH_4)_2S_2O_s$ was accompanied by a sharp decrease in free HCHO. The curing was a 2nd order reaction with reaction constant 0.083 L/mol-s for lignosulphonate-containing urea vs 0.022 L/ mol-s for unmodified urea resin. The particleboards based on urea resin lignosulphonate binders exhibited high strength, good water resistance and low toxicity. The cost of particleboards was significantly lower by replacing 20-30% of urea with lignosulphonates. It is also possible to modify lignin by grafting using, e.g., styrene or acrylic monomers. The grafting process is a free radical reaction which can be initiated either by radiation or by peroxide, as used in ordinary polymerization processes. Thus, acrylic monomers can be grafted onto lignosulphonates in aqueous solution using hydrogen peroxide/iron catalyst, Experimental evidence has indicated that radiation can also be used to graft styrene onto kraft lignin to give a product with similar solubility properties as polystyrene. The reaction seems to be promoted by methanol. [101]. Also, acrylonitrile may be grafted by radiation to lignosulphonates with grafting yields of about 20%. However, the compatibility of the grafted product as a filler in SBR seems to be inferior to that of the original lignosulphonate.[102]. Meister and Pati[103] have developed a method to chemically initiate free radical polymerization of 2-propenamide on kraft pine lignin with a molecular weight of 22,000. Proof that lignin and poly(1-amindoethylene) are grafted together has been obtained from size exclusion chromatography, solubility, dialysis, and fractionation. The grafted lignin is an amorphous brown solid which can be crushed, and is insoluble in most organic solvents but does dissolve in water.

Poly(lignin-g-(1-amindoethylene)) and its hydrolyzed derivative, poly(lignin-((1-amidoethylene)-co-(sodium 1-carboxylatoethylene))), exhibit a lower yield point, lower gel strength, and lower API (American Petroleum Institute) filtrate volume. [103].

High-performance polymers having 4-hydroxyphenyl, guaiacyl and syringyl groups were synthesized from lignin degradation products} [104]. Physical properties of the polymers were investigated by differential scanning calorimetry, thermogravimetry, gel permeation chromatography, viscosity measurement, etc. The relationship between the chemical structure and physical properties of polymers was analyzed from the viewpoint of molecular design. It was found that physical properties such as molecular weight, solubility for solvents, crystallinity, relaxation in glassy state, thermal decomposition temperature, etc., could be controlled by the appropriate arrangements of chemical bonds and functional groups such as phenylene group, methoxyl group, alkylene group, etc. Acrylation of lignin has been reported by Naveau [105] who employed both methacrylic anhydride and methacrylyl chloride with kraft lignin to achieve substitution levels of between 17 and 72% of the monomeric units in lignin. Another avenue to acrylation presents itself through the availability of difunctional monomers having both acrylate (or vinyl) and isocyanate functionality.Isocyanatoethyl methacrylate is one such monomer that may be used for converting terminal hydroxyl functionality into acrylate functionality} Glasser *et a) [106]* examined the synthesis of acrylated lignin derivatives using isocyanatoethyl methacrylate, and the copolymerization characteristics of these derivatives with methylmethacrylate and styrene. Hydroxy butyl lignin, as well as lignin and lignin derivative-like model compounds, were chemically modified by reaction with isocyanatoethyl methacrylate. The acrylated derivatives were studied with regard to copolymerization characteristics with styrene and methyl methacrylate. The reactivity ratios obtained suggest that lignin (model compound) acrylates form preferentially alternating copolymers with methyl methacrylate and styrene, and that the degree of deviation from azeotropic composition varies with chemistry. Azeotropic points were between 17 and 30mole % lignin (model compound) acrylate. Reaction of acrylated lignin derivatives with vinyl monomers produces network polymers with gel structure that shows the expected rise in sol fraction as vinyl equivalent weight increases.

Researchers attempted to obtain many kinds of three-dimensional polymers from lignin, utilizing the characteristics of lignin as a polymer. It is generally recognized that polyurethane is one of the most useful three-dimensional polymers, because polyurethane has unique features, various forms of materials can be obtained, and their properties can easily be controlled. Therefore, many attempts to obtain polyurethane using lignin as a raw material have been made.Hirose *et al) [106]* reported that solvolysis lignin was obtained by cooking using cresol-water. Polyurethanes were prepared from solvolysis lignin, polyethylene glycol and 4,4'-diphenylmethane diisocyanate. Thermal degradation was investigated by thermogravimetry. It was found that lignin in polyurethane retarded the thermal degradation of polyurethane in air. The retardation was not observed in the degradation of polyurethane in nitrogen. These results indicated that the retardation was caused by the oxidative condensation of lignin. Polyurethane containing polyol with phosphorus were also prepared. It was found that polyurethane became non flammable in air with the addition of polyol with phosphorus.

9. VALUE-ADDED PRODUCTS FROM LIGNOCELLULOSIC WASTES

(i) Reducing Sugars

Fermentable sugars comes first in the value chain of processed lignicellulosic wastes (LCW) with glucose, xylose, xylitol, cellobiose, arabinose, pentose and galactose being the main reduced sugars produced [107-116]. In these , hydrolysable sugars yield of up to 83.3% has been achieved at the reaction temperatures of 37 - 50°C for 6 – 179 h at pH 5 - 6. The size of substrate added determines the amount of the saccharification products [117-118]. Some transgenic plant residues have been reported to yield nearly twice as much sugar from cell walls compared to wild-types [119] .Glucose seems to be the major monosaccharide product from LCW. The challenge facing depolymerization of hemicellulose into fermentable sugars is the requirement for a consortium of enzymes to complete the hemicellulose hydrolysis, leading to high enzyme costs. Efforts to overcome the problem include process improvement and the use of modified microorganisms that produce the required hemicellulose enzymes [120-121].

(ii) Enzymes

Lignocellulosic enzymes, mainly from fungi and bacteria, are important commercial products of LCW bioprocessing used in many industrial applications including chemicals, fuel, food, brewery and wine, animal feed, textile and laundry, pulp and paper and agriculture [122]. Overall, extra cellular enzymes are secondary metabolic products released in the presence of inducers at N-limited media [123]. They include hydrolytic enzymes such as cellulases; hemicellulases and pectinases; degradative enzymes like amylases, proteases; and ligninolytic enzymes like laccases, peroxidases and oxidases. Cellulases production from LCW has been extensively studied [124-134]. Phytases, mannanases and amylases are also produced by microorganisms using LCW as the main feedstock [135-136] .On the other hand, hemicellulolytic enzymes, mainly xylanases, are produced from a wide range of LCW biomass [137-143].Pectinases such as endo polygalacturonase(endo-PG), exo-polygalacturonase(exo-PG) and pectinliase are mainly produced from solid-state fermentation processes utilizing agricultural residues [144-145] ,while protease has been produced by *Penicillium janthinellum* in submerged cultures [146].Among the ligninases produced from LCW, laccases are the mostly studied [147-148], followed by Manganese peroxidase and lignin peroxidase [149-150].Very high enzyme activities (31,786 U/L) have been reported when the experiments are carried out under optimal conditions (pH 5.5 - 6: temperature 30 - 45°C)(Rosales et al., 2007). Recovery of pure enzymes is achieved through 50 - 80% (NH4)2SO4 saturation followed by chromatographical purification techniques [151-152].Several efforts have been made to increase the production of enzymes through strain improvement by mutagenesis and recombinant DNA technology. Cloning and sequencing of the various genes of interest could economize the enzymes production processes [153].

(iii) Biofuels

Conversion of LCW to biofuels provides the best economically feasible and conflict-free second generation renewable alternatives [154-155].Significant advances have been made towards bioconversion of plant biomass wastes into bioethanol, biodiesel, biohydrogen, biogas (methane).Production of ethanol from sugars or starch from sugarcane and cereals, respectively, impacts negatively on the economics of the process, thus making ethanol more expensive compared with fossil fuels. Hence, the technology development focus for the production of ethanol has shifted towards the utilization of residual lignocellulosic materials to lower production costs [156].Currently, research and development of saccharification and fermentation technologies that convert LCW to reducing sugars and ethanol, respectively, in eco-friendly and profitable manner have picked tempo with breakthrough results being reported [157-158].Ethanol yield of 6 - 21%has been obtained through fermentation of agricultural and municipal residues [159-160]. While micro aeration enhances productivity of bioethanol from LCW using ethanologenic *E. coli [161]* (Okuda et al., 2007), simultaneous saccharification and fermentation (SSF) using recombinant *Saccharomyces cereviasiae* result to as high as 62% of the theoretical value [162]. The principal benefits of performing the enzymatic hydrolysis together with the fermentation, instead of in a separate step after the hydrolysis, are the co fermentation of both hexoses and pentoses during SSF,reduced end-product inhibition of the enzymatic hydrolysis and the reduced investment costs [163-164] . Life cycle assessment (LCA) shows that bio-ethanol from LCW results to reductions in resource use and global warming [165]. The long-term benefits of using waste residues as lignocellulosic feedstocks will be to introduce a sustainable solid waste management strategy for a number of lignocellulosic waste materials; contribute to the mitigation in greenhouse gases through sustained carbon and nutrient recycling; reduce the potential for water, air, and soil contamination associated with the land application of organic waste materials; and to broaden the feedstock source of raw materials for the bio-ethanol production industry [166].Biodiesel is a renewable fuel conventionally prepared by transesterification of pre-extracted vegetable oils and animal fats of all resources with methanol, catalyzed by strong acids or bases [167]. They are fatty acid methyl or ethyl esters used as fuel in diesel engines and heating systems [168].Production of biodiesel from lignocellulosic residues such as olive oil wastes has been a subject of research towards improving the thermal waste treatment systems and cleaner energy production [169]. Since the current supplies from LCW based oil crops and animal fats account for only approximately 0.3%, biodiesel from algae is widely regarded as one of the most efficient ways of generating biofuels and also appears to represent the only current renewable source of oil that could meet the global demand for transport fuels [170] . Hydrogen has been considered a potential fuel for the future since it is carbon-free and oxidized to water as a combustion product [171]. While conventional burning or composting seem to be the most cost-effective hydrogen production methods, bacteria such as *Enterobacter aerogenes* and *Clostridium* isolates can convert saccharified LCW biomass into biohydrogen [168]. Biohydrogen production from agricultural residues such as olive husk pyrolysis [172]; conversion of wheat straw wastes into biohydrogen gas by cow dung compost [173]; bagasse fermentation for hydrogen production generate up to 70.6% gas yields. System optimization for accessibility of polysaccharides in LCW and the use of genetically efficient bacterial strains for agro waste-based hydrogen production seem to be the ideal option for clean energy generation. Hydrogen generation from inexpensive abundant renewable biomass can produce cheaper hydrogen and achieve zero net greenhouse emissions [174] .Biogas production from lignocellulosic materials is a steady anaerobic process where methane rich biogas comes mostly from hemicellulose and cellulose. Anaerobic biomethane production is an effective process for conversion of a broad variety of agricultural residues to methane to substitute natural gas and medium calorific value gases. Biogas containing 55 - 65% methane can be produced from jute caddis - a lignocellulosic waste of jute mills by anaerobic fermentation, using cattle dung as sole source of inoculum [175]. Anaerobic digestion of poultrydroppings, cow dung and corn stalk can give up to137.16 L of biogas from 0.28 m^{3}digester. Mesophilic aerobic pretreatment to delignify sisal pulp waste prior to its anaerobic digestion has been shown to improve methane yields. The overall, the success of biofuels production from LCW is dependent on the optimal performance and cost effectiveness of pretreatment and product generation processes.

(iv) Organic Acids

Organic acids are some of the products of ligninolytic residues fermentations via environmentally friendly integrated processes. Volatile fatty acids including acetic acid, propionic acids and butyric acid are produced from a wide range of LSW such as cereal hulls [176]; bagasse residues [177]; food wastes [178] and sisal leaf decortications residues [179]. In addition, lactic acid is produced from waste sisal stems [180], sugarcane bagasse [181] and kitchen waste [182] by using *Lactobacillus* isolates. Furthermore, formic acid, levulinic acid, citric acid, valeric acid, caproic acid and vanillinic acid are obtainable from bioprocessing of LCW [183]. Overall, organic acids production requires batch or continuous incubation conditions, the average reaction parameters being 35°C,pH 6.0, hydraulic retention time (HRT) of up to 8 days and organic loading rates of 9 g/l d. Product yields of upto 39.5 g/l have been reported [178].

(v) Compost

Compost, a nutrient-rich, organic fertilizer and soil conditioner, is a product of humification of organic matter. This process is aided by a combination of living organisms including bacteria, fungi and worms which transform and enhance lignocellulosic waste into humic-like substances [184]. Vermicomposting is the bio-oxidation and stabilization of organic matter involving the joint action of earthworms and microorganisms, thereby turning wastes into a valuable soil amendment called vermicompost [185]. Substrates suitable for making humus–rich compost include cereal straw and bran [186]; urban wastes [187]; water hyacinth [188]; lemon tree prunings,cotton waste and brewery waste [189]; horticultural wastes [190]; olive, palm and grape wastes [191]. While bacteria inoculants such as *Bacillus shackletonni*, *Streptomyces* are used to improve the composting process [192], ligno-cellulolytic fungi may also be used in a pretreatment process before composting in order to reduce the resistance of the substrate to biodegradation [192-193].

(vi) Biocomposites

Biodegradable polymers constitute a loosely defined family of polymers that are designed to degrade through the action of living organisms. Such commercially available biodegradable polymers are polycaprolactone,poly (lactic acid), polyhydroxyalkanoates, poly (ethylene glycol), and aliphatic polyesters like poly (butylenesuccinate) (PBS) and poly (butylene succinate -co- butyleneadipate) [194].Lignocellulosic material-thermoplastic polymer composites are among the emerging products of LCW. In most cases, lignocellulosic biomass flour is used as the reinforcing filler and polypropylene as the thermoplastic matrix polymer to manufacture particle-reinforced composites.Natural fibres from LCW are considered to be of low-cost by-products, environmentally friendly and practically sustainable raw materials [195]. Evaluations of LCW fiber plastic composites utilizing wood fiber wastes [196] ; wheat and rice straw [197] ; jute/cotton, sisal/cotton andramie/cotton hybrid fabrics [198]; non-wood plant fibres [199]; waste newsprint paper [200]; flax and hemp [194]; oil palm wastes [201] cotton gin waste [202]; banana fibres [203]; cereal husks [190]; tissue paper wastes and corn peels [204]; bagasse [205] and nanofibers from the agricultural residues [206] have shown that such composites are suitable for making products that have improved biodegradability, mechanical strength, thermal stability, electrical conductivity and recyclability.Treated LCW wastes are also used in the construction industry for manufacturing of light-weight agro-gypsum panels [207] and lightweight sand concretes [208] with improved structural and thermal properties.Biocompositesare very promising in producing sustainable current and future green materials to achieve durability without using toxic chemicals. The challenge facing the biocomposite industry is to make materials that have better rubber/fiber interface, improved wettability and compatibility.

(vii) Food and Feed

Bioconversion of lignocellulosic agro-residues through mushroom cultivation and single cell protein (SCP) production offer the potential for converting these residues into protein-rich palatable food and reduction of the environmental impact of the wastes. Mushrooom cultivation provides an economically acceptable alternative for the production of food of superior taste and quality which does not need isolation and purification [209]. Cultivation of edible mushrooms such as *Lentinus* spp and *Grifola*spp is achievable on a wide range of LCW substrates such as wood waste, corncob meal, wheat straw, barley straw, soybean straw, cereal bran, cotton waste, sorghum stalk, banana pseudo stem, hazelnut husks, waste tea leaves, dry weed plants, peanut shells, wastepaper and olive mill wastewater. Mushrooms with increased number of fruit bodies and high contents of protein and total carbohydrates are obtained when LCW substrates are used in combination. On the other hand, SCP production from LCW offers a potential substrate for conversion of low-quality biomass into an improved animal feed and human food. SCP is the protein extracted from cultivated microbial biomass. It can be used for protein supplementation of a staple diet by replacing costly conventional sources like soy meal and fishmeal to alleviate the problem of protein scarcity. Moreover, bioconversion of agricultural and industrial wastes to protein-rich food and fodder stocks has an additional benefit of making the final product cheaper [210]. Removal of nucleic acids and toxins from SCP is key to ensure the safety of food and feed. Among the

SCP obtained from LCW using agricultural wastes as the main growth media, *Saccharomyces cerevisiae*, and *Kluyveromyces marxianus* top the list [211] . SCP yield of 51 and 39.4% efficiency of conversion of beet-pulp into protein has been reported from the above strains. Solid-state fermentation of LCW seems to be the most preferred culturing method, while cloning is being considered as a suitable technique for improvement of SCP production [210].

(viii) Medicines and Adsorbents

LCW provides a suitable growth environment for mushrooms that comprise a vast source of powerful new pharmaceutical products. In particular, *Lentinula edodes*, *Tremella fuciformis* and *Ganoderma lucidum* contain bioactive compounds such as anti-tumor, anti-inflammatory, anti-virus and anti-bacterial polysaccharides.Moroever, they contain substances with immuno modulating properties, as well as active substances that lower cholesterol [209]. Future prospects for research on bioactive compounds from fungi grown on such cheap and ubiquitous substrates look bright and could lead to breakthroughs in the search for antibacterial, antiviral and anticancer chemotherapies. Adsorbents obtained from plant wastes are feasible replacements for costly conventional methods of removing pollutants such as heavy metals ions, dyes, ammonia and nitrates from the environment. The use of lignocellulosic agro wastes is a very useful approach because of their high adsorption properties, which resultsfrom their ion-exchange capabilities. Agricultural wastes can be made into good sorbents for the removal of many metals, which would add to their value, help reduce the cost of waste disposal, and provide a potentially cheap alternative to existing commercial carbons [212]. Chemically modified plant wastes such as rice husks/rice hulls, spent grain, sugarcane bagasse/fly ash, sawdust, wheat bran, corncobs, wheat and soybean straws, corn stalks, weeds, fruit/vegetable wastes, cassava waste fibres, tree barks, azolla (waterfern), alfalfa biomass, coir pith carbon, cotton seed hulls, citrus waste and soybean hulls show good adsorption capacities for Cd, Cu, Pb, Zn and Ni [213]. They are usually modified with formaldehyde in acidic medium, NaOH, KOH and CO_2, or acid solution or just washed with warm water [214]. Scanning electron micrographs with energy spectra shows that heavy metals are immobilized via two possible routes: adsorption and cation exchange on hypha, and the chelation by fungal metabolite [215].LCW have also been shown to be able to adsorb dyes from aqueous solutions. Adsorption of reactive dyes by sawdust char and activated carbon [216];ethylene blue by waste *Rosa canina* sp. seeds [217]; anionic dyes by hexadecyltrimethylammonium modified coir pith [218]; and methylene red by acid-hydrolyzed beech sawdust [219] have been reported. Ammonia and nitrate removal by using agricultural waste materials as adsorbents or ion exchangers have also been studied [220].Prehydrolysis enhances the adsorption properties of the original LCW material due to the removal of the hemicelluloses during sulphuric acid treatment, resulting in the 'opening' of the lignocellulosic matrix's structure, the increasing of the surface area and the activation of the material's surface owing to an increase in the number of dye binding sites [219-220].

10. COMPUTER SIMULATION REACTIONS OF LIGNIN

As a result of the complexity of lignin, computer modeling has played an important role in the development of the field. There are two complementary approaches to lignin modeling: quantum chemical calculations on small polymers and simulation of the formation of a large polymer. Lignols are added one at a time to the growing polymer; reactions are accepted or rejected both on the basis of input parameters (monolignol radical reactivities, bond distribution, etc.) and according to the evolving statistical properties of the simulated polymer.Essentially,the intrinsic chemistry of the monolignols is modified during the simulation in order to reconcile the model to experimental data. In the model of Lange and coworkers [221] based on the Flory-Stockmayer treatment of lignin by Bolker and Brenner,[222], lignification is viewed as a stepwise process: First, dilignols are formed which then oligomerize into chains, and finally the chains form cross-links. There is some experimental support for this view of lignification, but it is unclear whether occasional random deviations from this scenario (e.g., early formation of polymer branch points)play a role in the overall structure of the polymer. An effort is made to construct realistic lignin structures in space by taking careful account of the size and geometry of the monolignols as they are added to the polymer. Each of these models proceeds from a particular view of the lignification process to an implementation in the form of a sequentially executed program. Although this class of models seems to be particularly effective in detecting stoichiometric anomalies in experimental data, it is not clear that they necessarily produce realistic spatial structures for the lignin polymer. In contrast to the studies outlined above, there have been attempts to capture the dynamics of lignification by establishing some simple transport and bonding rules and allowing these to run their course. The 'physics" of the model is formed at the outset and mimics in vivo physics, including the "parallelism" resulting from the simultaneous presence of many molecules in the reaction space. In other words, attempt to recreate the conditions of the reaction care carried rather than attempt to generate the products directly. Discrete simulation methods for efficiency and simplicity are applied where both time and space are discretized; in particular, space is represented by a square grid of 256 x 256 cells with periodic boundary conditions. Planar models are best suited to modeling lignin growing in restricted spaces such as the secondary wall of a wood fiber. Also, while it might be tempting to associate planar models with lignin monolayers ,it would be wrong since monolayers are produced by spreading *bulk* lignin on an interface rather than by growing a polymer there.Furthermore, authors have opted to use purely local rules to control the polymerization process; the reactivity of the monolignol radicals makes this a reasonable model feature.At every time step, a coordinate origin is randomly selected and the plane is cut up into 2 x 2 blocks of cells. There are three types of "moves" in these blocks: influx, diffusion, and bonding. Influx is the addition of monomer radicals to the reaction space; in vivo, this would correspond to a combination of the synthesis, release, and catalytic dehydrogenation of monolignols by plant cells. In this simple model, only the monomers diffuse; once a dilignol (or higher oligomer) has formed, it is fixed in space.It is known that diffusion-limited aggregation models are insensitive to the oligomer diffusion constant, provided that this is lower than the monomer diffusion constant. Some rules must be provided for bonding monolignols to monolignols and monolignols to oligomers. Oligomers do not, in this model, directly bond to one another because they cannot diffuse into proximity. Rather, they are bridged by monolignols when they grow sufficiently close to one another. Bonds are formed combinatorially by joining loci on adjacent molecules. This maximizes the number of bonds considered, while minimizing the complexity of the model. All of the rules are executed in parallel. The implementation architecture for this model is a cellular automata machine (CAM) [222].The principal advantages of this modeling environment are speed and ease of programming. Since it is a Monte Carlo simulation of the sort authors have proposed one bit-plane for random number generation, this effectively limits models to three bits of state information; i.e., only eight states are available, of which one must be set aside to represent empty space.Given the limited number of states and the desire to incorporate hemicellulose and branching, each monolignols is represented by a point. Three bonding sites are considered, namely, 0-4, C-β, and C-5 (numbering scheme as in Figure.27); roughly 75% of the bonds in native lignin can thus be formed combinatorially.

To keep the state space small, in addition to permitting only bonds for which there is experimental evidence, the assumption is made that C-5 cannot be bound. (This allows the state of any cell in the simulation to be held in three bits, given the desire to represent hemicellulose.) A branch point is formed whenever all three sites on a single monolignol are bound. The slightly artificial bonding constraint described above favors the formation of a polymer backbone composed primarily of β-0-4 bonds with branching due mostly to other bond types, in accord with most common descriptions of the lignin polymer.Furthermore, one cell state is reserved for hemicellulose.C-β to hemicellulose bonds are formed, i.e., a hemicellulose site is assumed to be capable of as many bonds as it makes contacts with lignin.However, C-α to sugar bonds have been demonstrated. Given the similarity in chemical properties of the C-α and C-β sites, assuming that the latter can also form bonds to hemicellulose is reasonable.Briefly, the influx, diffusion, and bonding rules are as follows:

Influx. Every nth step, one quarter of the blocks are randomly selected. For every selected block, a single position (of the four available) is designated for influx. If this position is empty, a monolignol is added. Influx is turned off once the target surface density D* is reached.

Diffusion. At every step, a random shuffle of each block is formed. If this shufle leaves bonded monolignols undisturbed and conserves the number of monomers, it is executed.

Bonding. In each block, attempt to form bonds inthe following order:

With probability 1/16, attempt to form a bond between C-β and hemicellulose.
Attempt to form a 5-5 bond.
Attempt to form a β-5 bond.
Attempt to form a 4-0-5 bond.
Attempt to form a β −0-4 bond.

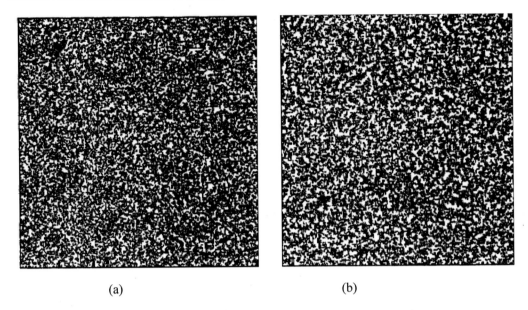

(a)   (b)

Figure 28. Model lignin polymer grown at (a) n = 2 and (b) 2000; in both cases **t** = 100 and D* x 0.476. Each pixel represents a monolignol, so each cell (occupied or not) represents a square of side about 4 ^{0}**A;** the simulation space is composed of 256 x 256 cells, corresponding therefore to a square physical space whose side is just over 1000 ^{0}**A** long. At larger values of n, the polymer is more highly coagulated. This is related to the branching: The fraction **of** trifunctional monomers rises from **6%** to just under 20% as n rises from 2 to 2000.

The first type of bond cannot be formed because the probabilistic test failed or because either or both of the required bonding partners is unavailable, try the next and so on until all possibilities have been exhausted. Since position C-5 cannot be bound before 0-4,this ordering of the bonds results in roughly the right proportion of each bond type, i.e., , β -0-4 > 5-5 > β -5> 4-0-5 > β- β this sequence varies somewhat from one lignin source to another but can in any event easilybe altered by reordering the bond attempts in the above list. The hemicellulose bonding probability was chosen capriciously, there being a dearth of experimental data on this subject. The model incorporates a relaxation time τ; when τ time steps pass without any new bonds being formed, the simulation is halted. The model has three parameters: the relaxation time τ, the target surface density D*, and the influx period n. The effects of these parameters on the model properties have been examined. As expected, since τ only serves to recognize a completed simulation, it has very little effect on the outcomes. On the other hand, both the density and the influx period have measurable effects on the model behavior. The dependence of some of the model's properties on D* and n is not unexpected: The fractional degree of polymerization (DP; the number of bonds divided by the number of monomers) rises and the phenyl-OH content decreases with both D* and n, the influx period, the latter result being in agreement with the outcomes of in vitro lignification experiments in which the rate of influx was varied. The proportion of trifunctional (branch point) monolignols also varies with these parameters, rising with both D* and n. The influx period n has a greater effect on the polymer morphology than might at first be suspected. This is due to the branching. At low n, the reaction space quickly becomes so crowded that almost every monolignol has several bonding opportunities at every time step, resulting in a very rapidly formed, uniform polymer with very few branch points. For larger n the polymer coagulates, forming a much larger number of branch points. Figure 28 illustrates these observations.

In addition, the percolation behavior of the Percolation theory has previously been used to understand the delignification process. The percolation threshold can be identified by plotting f, the ratio of the mass of the largest polymer formed to the total simulation mass, against D*; at the threshold, this fraction rises sharply. Alternatively, we can frame the percolation behavior in terms of polymer gelation theory; in this theory, the percolation threshold is called the gel point. The gel point separates an initial period of increase in the mean molecular weight of the sol phase from a period of decrease. We can understand the percolation gelation behavior of the model as that of a multimember growth process. Since, only monomers are subject to diffusion, after the establishment of a few small oligomers early in the simulation, most of the polymerization consists of addition of monomers to existing polymers. The mean radii of these polymers grow until, near the percolation threshold, they begin to merge. This causes rapid growth of the "gel" phase at the expense of the larger polymers in the "sol" phase as the latter are joined to the former, at first by single-monomer bridges but eventually by a complex network of cross-links. The fractional DP and the degree of branching should be lower and the phenyl- OH content higher in the two-dimensional system.On the other hand, if the monolignol bonds to hemicellulose, because of the flatness of the interface and because only one lignin-hemicellulose bond is permitted per monomer, two of the hemicellulosic neighbors can only block the approach of other monolignols to the one anchored to the surface; worse, if a monolignol in contact with the hemicellulose reacts with another monolignol through its $C\text{-}\beta$ functionality, three neighbor sites are blocked. Thus, we expect the DP and degree of branching to drop and the phenyl-OH content to rise as the number of bonding opportunities at the hemicellulose interface is depressed. Accordingly, a rough lignin hemicellulose interface with an average contact number of less than 3 should increase the DP of the model. There are various ways to measure the length of an interface; the number of sites which touch the hemicellulose. Using any reasonable measure, however, the same result is obtained: Crenellation doubles the length of an interface. Thus, a space with a crenellated interface of width w and length 1 is equivalent to two spaces with smooth walls of width w/2 and identical length. In other words, we should compare the statistical properties of lignin grown in contact with crenellated hemicellulose in a space of width w with lignin grown between flat walls in a space of width w/2. Although site exclusion effects are obviously important in two dimensions, they might also be significant in three dimensions, depending on the geometry of the lignin-hemicellulose complex. If there are relatively many lignin-hemicellulose bonds and if typical bonding geometries result in the steric unavailability of many monolignol binding sites, effects similar to those observed above might be found. At least some of the differences between lignins obtained from different parts of the tree might have their origins in hemicellulose interactions [222].

11. RECENT DEVELOPMENTS IN LIGNIN BASED NANOCOMPOSITES

Polymer nanocomposites are produced by incorporating materials that have one or more dimensions on the nanometer scale (<100 nm) into a polymer matrix. These nanomaterials are in the literature referred to as for example nanofillers, nanoparticles, nanoscale building blocks or nanoreinforcements.Nanocomposites have improved stiffness, strength, toughness, thermal stability, barrier properties and flame retardancy compared to the pure polymer matrix. Nanoreinforcements are also unique in that they will not affect the clarity of the polymer matrix. Only a few percentages of these nanomaterials are normally incorporated (1–5%) into the polymer and the improvement is vast due to their large degree of surface area [223]. Because of the nanometric size effect, these composites have some unique outstanding properties with respect to their conventional micro composite counterparts.

The properties of nanocomposite materials depend not only on the properties of their individual parents, but also on their morphology and interfacial characteristics. In the particular case of polymer reinforced with rigid nanofillers, various parameters seem to be of importance in characterizing the fillers: geometrical factors such as the shape, the size, and the aspect ratio; intrinsic mechanical characteristics such as the modulus or the flexibility; surface properties such as specific surface area and surface treatment [224]. The type of polymer matrix used and the possible effects of nanofillers on its microstructure and its intrinsic properties are also essential parameters determining the composite properties. In addition, the processing conditions can affect on composite properties.

Whereas the general class of inorganic nanocomposites as enjoyed much discussion and is still a fast-growing area of research, exciting new research on bio-based nanocomposites have a greater potential because the bio-resource can be both sustainable and genetically manipulated. Wood cellulose nanofibrils have about 25% of the strength of carbon nanotubes, which are expected to be the strongest fibers that can be produced. Their potential cost, however, might be 10 to 100times less, giving cellulose nanofibrils a unique economic advantage [225].

Natural fibers are pervasive throughout the world in plants such as grasses, reeds, stalks, and woody vegetation. They are also referred to as cellulosic fibers, related to the main chemical component cellulose, or as lignocellulosic fibers, since the fibers usually often contain a natural polyphenolic polymer, lignin, in their structure. The use of lignocellulosic fibers derived from annually renewable resources as a reinforcing phase in polymeric matrix composites provides positive environmental benefits with respect to ultimate disposability and raw material use [226]. By comparing with inorganic fillers, the main advantages of lignocellulosics are;

1. Renewable nature
2. Nonfood agricultural based economy
3. Low energy consumption and cost
4. Low density
5. High specific strength and modulus
6. High sound attenuation of lignocellulosic based composites
7. Relatively reactive surface, which can be used for grafting specific groups.

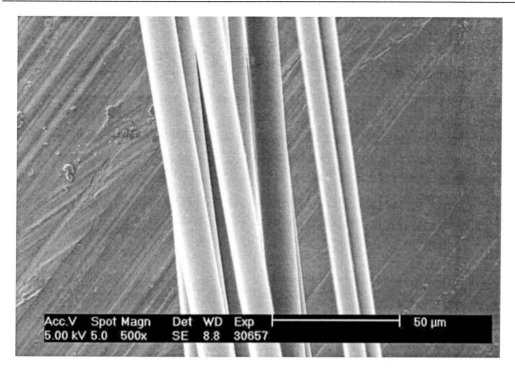

Figure 29. Scanning electron micrograph of carbonized lignin blend fibers produced during multifilament spinning.

Wood fibers, the most abundant biomass resource on earth, are hollow tubes made up of cellulose embedded in a matrix of hemicellulose and lignin. Most of the cell-wall materials are located in the second layer, which consists of a helically wound framework of microfibrils, Figure 29 [227]. The most important attribute of wood is its mechanical properties, in particular its unusual ability to provide high mechanical strength and high strength-to-weight ratio while allowing for flexibility to counter large dimensional changes due to swelling and shrinking. These unique properties of wood are a direct result of its hierarchical internal structure. Nanotechnology has resulted in a unique next generation of wood based products that have hyper-performance and superior serviceability when used in severe environments. They will have strength properties now only seen with carbon-based composites materials, and they will be both durable while in service and biodegradable after their useable service-life. Nanotechnology will also promote the development of intelligent wood- and biocomposite products with an array of nanosensors built in. Building functionality onto lignocellulosic surfaces at the nanoscale could open new opportunities for such things as self- sterilizing surfaces, internal self-repair and electronic lignocellulosic devices. The high strength of nanofibrillar cellulose together with its potential economic advantages will offer the opportunity to make lighter weight, strong materials with greater durability [225].However, as in all markets, technology and shifting demographics give rise to hanging market demands. Materials and products used in housing construction are not immune to such changes. Because a home or a commercial building is typically the largest purchase a family will make and one of the larger investments a corporation will make, consumers want structures that maintain their value over time and are safe and secure, healthy, comfortable, long-lasting (durable), low maintenance, affordable (lower in cost and providing more value for the dollar), easily adaptable to new and modified architectural designs, and allow for personalized customization, have smart system. Capabilities, and reduce costs for heating and air conditioning. Wood-based construction materials function extremely well under a variety of end-use conditions. Under wet conditions, however, they can be prone to decay, mold, mildew, and insect attack. Wood can be protected from biodeterioration by treatments using toxic chemicals or by maintaining low moisture content in wood. Achieving control of moisture is a major opportunity for nanotechnology to aid in preventing biodeterioration of wood and wood-based materials. New non- or low-toxicity nanomaterials such as nanodimensional zinc oxide, silver, titanium dioxide, and even possibly clays might be used as either preservative treatments or moisture barriers. In addition, resistance to fire might be enhanced by use of nanodimensional materials like titanium dioxide and clays. Composites allow an array of disparate materials with greatly differing properties to be engineered into products matched to end-user needs and performance requirements. For example, future nanocomposite construction materials may use combinations of wood, wood fiber, plastics, steel, and concrete. To achieve this, it will be necessary to be able to make hydrophilic materials compatible with hydrophobic materials such as wood and plastics [228].

CONCLUSION

Engineered wood- and lignocellulosic-composite technologies allow users to add considerable value to a diverse number of wood- and lignocellulosic feedstocks including small-diameter timber, fast plantation-grown timber, agricultural fiber and lignocellulosic residues, exotic-invasive species, recycled lumber, and timber removals of hazardous forest-fuels. Another potential advantage of this type of economic- and materials-development scenario is that developing industrial composite processing technologies will provide producers an ability to use, and to adapt with, an ever-changing quality level of wood and/or other natural lignocellulosic feedstocks. However, the current level of performance of our state-of-the-art engineered composite products sometimes limit broader application into commercial, non-residential and industrial construction markets because of both real and perceived issues related to fire, structural-performance, and service-life. The worldwide research community has recognized this and is currently addressing each of these issues. From a performance standpoint, this developing knowledge has already and will continue to provide the fundamental understanding required to manufacture advanced engineered composites. From a manufacturing and a resource sustainability standpoint, with this evolving fundamental understanding of the relationships between materials, processes, and composite performance properties we now can in some cases, or may soon be able to, recognize the attributes and quality of an array of bio-based materials then adjust the composite manufacturing process to produce high-performance composite products. As this fundamental understanding is developed, we will increasingly be able to produce advanced, high-performance wood- and bio-composites. Then we must use those technologies as tools to help forest and land managers fund efforts to restore damaged eco-systems and which in turn may further promote sustainable forest management practices.

In summary, the utilization of engineered lignin and its biocomposite materials that meet user needs and maximize the environmental sustainability is fast becoming a reality. We must commit ourselves to developing the fundamental and applied science and technology necessary to provide improved value, service-life, and utility so the world can use sustainable bio-based materials. Then as forest resource options change, as excess bio-based waste-streams become available, as alternative non-wood and non-lignocellulosic materials become more economical and available, and/or as air and water quality regulations become more stringent, we will have the tools to address the problem of achieving resource sustainability and enhancing economic development through lignin utilization

REFERENCES

[1]   Dence C.W., Lin, S.Y. (1992). General structural features of lignin.In, *Methods in Lignin Chemistry*. Springer, Berlin, Heidelberg, pp. 3–6 and 458–463.

[2]   Whetten, R. W., MacKay, J. J., and Sederoff, R. R. (1998). Recent advances in understanding lignin biosynthesis. *Annual Review of Plant Physiology and Plant Molecular Biology* 49(1), 585-609.

[3]   Sjöström, E. (1993). *Wood Chemistry Fundamentals and Applications* (2nd edition). SanDiego, CA, Academic Press, Inc.

[4]   Gargulak, J.D., Lebo, S.E. (2000). Commercial use of lignin-basedmaterials. In, Glasser, W.G., Northey, R.A., Schultz, T.P. (Eds.),Lignin, Historical, Biological, and Materials Perspectives. *ACS Symposium Series*. American Chemical Society,p. 307.

[5]   Glasser, W. G. (2000). Classification of lignin according to chemical and molecular structure. *In "Lignin, Historical, Biological, and Materials Perspectives"* (W. G.Glasser, R.A. Northey, and T. P. Schultz, eds.), pp 216-238. ACS Symposium Series No. 742. American Chemical Society, Washington, D.C.

[6]   Lynd, L. R. (1996). Overview and evaluation of fuel ethanol from cellulosic biomass, technology, economics, the environment, and policy. *Annual Review of Energy and the Environment* 21, 403-465.

[7]   D. O. E. (2000). *Annual Energy Outlook 2001 with Projections to 2020.* Report DOE/EIA-03833. Energy Information Administration, U.S. Dept. of Energy. Table 33. Dec 22.

[8]   Schell, D. J., Riley, C. J., Dowe, N., Farmer, J., Ibsen, K. N., Ruth, M. F., Toon, S. T., and Lumpkin, R. E. (2004). A bioethanol process development unit, initial operating experiences and results with a corn fiber feedstock. *Bioresource Technology* 91(2),179-188

[9]   Glasser, W. G., Barnett, C. A., Muller, P. C., and Sarkanen, K. V. (1983). The chemistry of several novel bioconversion lignins. *J.Agr. Food. Chem.*31(5), 921-930.

[10]   Côté, W. (1982). *Biomass utilization*. Serier A, Life Sciences Vol. 67. Plenum press. NewYork.

[11]   Fox, D., Gray, P., Dunn, N., and Marsden, W. (1987). Factors affecting the enzymic susceptibility of alkali and acid pretreated sugarcane bagasse. *J. Chem. Tech. Biotechnol.* 40,117-132.

[12] Wu, Z., and Lee, Y.Y. (1997). Ammonia recycled percolation as a complementary pretreatment to the dilute-acid process. *Appl. Biochem. Biotech.* 63-65,21-34.

[13] Anonymous (2005). Biomass as a feedstock for a bioenergy and bioproducts industry, the technical feasibility of a billion ton annual supply,*http,//www1.eere.energy.gov/biomass/pdfs/final_billionton_vision_report2.pdf*

[14] Sun, R., Xiao, B., and Lawther, J. M. (1998). Fractional and structural characterization of 20 ball-milled and enzyme lignins from wheat straw. *Journal of Applied Polymer Science* 68(10), 1633–1641.

[15] Obst, J. R. and Kirk, T. K. (1988) Isolation of lignin. *In "Methods in Enzymology;* Volume161; Biomass; Part B; Lignin, Pectin, and Chitin." (W. A. Wood an S. T. Kellog, eds.), pp 3-12. Academic Press, Inc., San Diego.

[16] Glasser W.G (2001) in F. C. Beall (ed.) *The Encyclopedia of. Materials, Science and Technology*, Elsevier Science Oxford, UK.

[17] Glasser W.G. and Sarkanen S.(1989) Crosslinking options for lignin , *ACS Symp. Ser. No.* 397, **546**.

[18] Fox, D., Gray, P., Dunn, N., and Marsden, W. (1987). Factors affecting the enzymic susceptibility of alkali and acid pretreated sugarcane bagasse. *J. Chem. Tech. Biotechnol.* 40,117-132.

[19] Khan M. A., Ashraf S.M., Malhotra V.P. (2004) . Eucalyptus bark lignin substituted phenol formaldehyde adhesives: A study on optimization of reaction parameters and characterization, *J. App. Polym.Sci.,* 92(6) ,3514-3523.

[20] http://jmi.nic.in/Research/ab2003_chemistry_mozaffaralamkhan.htm

[21] Khan M. A., Ashraf S.M., Malhotra V.P. (2004) . Development and characterization of a wood adhesive using bagasse lignin „*J.Adh.Sci.Tech.* 24(6), 485-493

[22] Khan M.A and Ashraf S.M (2007) Studies on the thermal characterization of Lignin „*J.Therm.Anal,.Cal.,* 89(3),993-1000

[23] Feldman D. (1963). Pheonplasts from Lignin, *Celluloza Hirtie (Bucharest),* 12(8-9), 275-280

[24] Wu L.C.F and Glasser W.G. (1984) . Engineering plastics from lignin. I. Synthesis of hydroxypropyl lignin from lignin II.[Derivatives by batch reaction at a temperature of 180 degrees Celsius], *J. Appl. Polym. Sci.* 29, 1111-1123

[25] Glasser W.G., Barnett C.A., Rials T.G and Saraf V.P.(1984) Engineering plasticslignin derivatives J. *Appl. Polym. Sci.* 29,**1815**-1830.

[26] . Saraf V.P. and Glasser W.G.(1984) Engineering plastics from lignin. III. Structure property relationships in solution cast polyurethane films J. *Appl. Polym. Sci.* 29, 1831-1841 .

[27] Rials T.G. and Glasser W.G.(1984). Engineering plastics *Holzforschung ,* 38(4), 191-199.

[28] Muller P.C and Glasser W.G.(1984) . Engineering Plastics from Lignin. VIII. Phenolic Resin Prepolymer Synthesis and Analysis J. *Adhes.* 17, 157-174.

[29] Rials T.G.,Glasser W.G (1984). Engineering Plastics from Lignin - V. Effect of Crosslink Density on Polyurethane Film Properties — Variation in Polyol Hydroxy Content, *Holzforschung,* 38(5), 263-267 .

[30] Muller P.C ,Kelley S.S and Glasser W.G. (1984) Engineering Plastics from Lignin. IX. Phenolic Resin Synthesis and Characterization J. *Adhes.* 17, 185-206.

[31] Kratzl K., Buchtela K., Gratzl J., Zauner J. and Ettinghausen O (1962) . Lignin and plastics, the reaction of lignin with phenol and isocyanates, *TAPPI 45* No. 2,113-119

[32] Rials T.G ,Glasser W.G . (1986). Engineering Plastics from Lignin - XIII. Effect of Lignin Structure on Polyurethane Network Formation, *Holzforschung* 40(6), 353-360

[33] Rials T.G and Glasser W.G (1984) Engineering Plastics from Lignin. X. Enthalpy Relaxation of Prepolymers , *J. Wood Chem. Technol.* 4(3), 331-345.

[34] Glasser W.G. (1985), Engineering Plastics from Lignin - XII. Synthesis and Performance of Lignin Adhesives with Isocyanate and Melamine *Holzforschung* 39(6) 345-353.

[35] Kelley S.S and Glasser W.G.(1988) Engineering plastics from lignin. XV. Polyurethane films from chain-extended hydroxypropyl lignin *J. Appl. Polym. Sci* 36(4), 759-772

[36] Falkenhagen S.I (1972) *U.S. Pat.* 3,697,497759-772 0988).

[37] Lund M., Hassingboe J., Felby C.(2000).Oxidoreductase catalyzed bonding of wood fibers. In, Proceedings of the 6th European Workshop on Lignocellulosics and Pulp, Bordeaux, France,113–116.

[38] Felby C., Hassingboe J., Lund M.(2002). Pilot-scale production offibreboards made by laccase oxidized wood fibers, boardproperties and evidence for cross-linking of lignin. Enzyme, *Microb. Technol.* 31,736–741

[39] Felby C., Thygesen L.G., Sanadi S., Barsberg S.(2004). Nativelignin for bonding of fibre boards—evaluation of bonding mechanisms in boards made from laccase-treated fibres of beech (*Fagus sylvatica*). *Ind. Crops Prod.* 20, 181–189.

[40] Bohlin C., Persson P., Gorton L., Lundquist K., Johnson, L.J.(2005).Product profiles in enzymic and non-enzymic oxidations of the lignin model compound erythro-1-(3,4-dimethoxyphenyl)-2-(2-methoxyphenoxy)-1,3-propanediol. *J. Mol. Catal. B, Enzym.* 35, 100–107.

[41] Anon. (2004). Global commodity price reporting and market intelligence, Phenol (Europe). *http,//www.icislor.com*

[42] Nimz H.H.(1983) Lignin-based wood adhesives, *Wood Adhesives* (A. Plzz! Ed.), Marcel DekkerInc., N.Y..

[43] Roffael E. And Rauch W. (1971). Über die Herstellung von Holzspanplatten auf Basis von Sulfitablauge. I. Stand der Technik und eigene Untersuchungen *Holzforschung* 25, 4, 112-116 .

[44] Roffael E. And Rauch W. (1971). Über die Herstellung von Holzspanplatten auf Basis von Sulfitablauge. II. Über ein neues und schnelles Verfahren zur Herstellung Sulfitablauge-gebundener Spanplatten , *Holzforschung* 25, 5, 149-155 .

[45] Roffael E. And Rauch W. (1972). Über die Herstellung von Holzspanplatten auf Basis von Sulfitablauge. III. Möglichkeiten zur Verkürzung der thermischen Nachbehandlung von Sulfitablauge-gebundenen 9 mm dicken Spanplatten *Holzforschung* 26, 6, 197-202 (1972).

[46] Roffael E. And Rauch W. (1973). Über die Herstellung von Holzspanplatten auf Basis von Sulfitablauge. Iv. Möglichkeiten zur Verkürzung der thermischen Nachbehandlung von Sulfitablauge-gebundenen 9 mm dicken Spanplatten *Holzforschung* 27, 6, 214-2

[47] Dolenko A.J. and Clarke M.R.(1978) Resin binders from kraft lignin *Forest Prod.* J. 28, 8, 41-46 .

[48] Dolenko A.J. and Clarke M.R. (1973) *U.S. Pat.* 4,112,675 (12 Sept.).

[49] Adams J.W. and Scoenhepr M.W. (1981), *U.S. Pat.* 4,306,999 (22 Dec.).

[50] Hollis J.W. and Schoenhepr M.W.(1981) *U.S. Pat.* 4,303,562 (1 Dec)

[51] Forss K.G. and Fuhrmann A. (1979). Finnish plywood, particleboard, and fireboard made with a lignin-base adhesive, *Forest Prod. J.* 29, No. 7, 39-43 (1979).

[52] Shen KC and Calvin L.(1979)In, *Proceedings of the 13th International Particleboard Symposium,*pp. 369-379.

[53] Khan M.A and Ashraf S.M (2005) Development and characterization of a lignin–phenol–formaldehyde wood adhesive using coffee bean shell, *J.Adh.Sci.Tech.* 19(6), 493-507.

[54] Khan M.A and Ashraf S.M (2006) Development and characterization of groundnut shell lignin modified phenol formaldehyde wood adhesive, Int.J.Chem.Tech. 347-352.

[55] Edler F.J. (1981), *U.S. Pat.* 4,194,997 (13 Jan.)

[56] Glasser W.G and Hsu O.H.H (1977), *U.S. Pat.* 4,017,474.

[57] Glasser W.G.,Hsu O.H.H. ,Reed D.L., Forte R.C. and Wu L.C.F. (1982), *Urethane Chemistry and Applications* (K. N. EDWARDS Eds), ACS Syrup. Scr. No. 172, pp. 311-338.

[58] Lamauth AL.(1981), *U.S. Pat.* 4,279,788 (July).

[59] Gamo M. (1978). J. *Adhes. Soc. Jap.* 14, No. 2, 56-66.

[60] Gamo M. (1979). *J. Adhes. Soc. Jap.* 15, No. 10, 435-439.

[61] Bardet M. , Foray M.F. and Robert D.(1985). Use of the DEPT pulse sequence to facilitate the 13C NMR. *Makromol. Chem.* 186, 1495-1504 . C.H

[62] Hoyt and Goheen D.W.(1971) Lignins, occurrence, formation, structure and reactions, p. 855,Wiley-Interscience, New York .

[63] Mills G.S and Haxo H.E (1959), *U.S. Pat.* 2,906,718.

[64] Lyubeshkina E.G (1983) Lignins as Components of Polymeric Composite Materials, *Russ. Chem. Rev.* 52(7), 675-692.

[65] Kopylov G., Lazaryants E.G.,Bogomolov B.D and Gel E.D.(1996) *Syrup. on the Chemistry of Wood,* Vol. 1, p. 375, Izd. Zinatne, Riga (1966).

[66] Gul V.E ,Lyubeshkina E.G. and Shargorodskii A.N.(1965) Mechanical Properties of polypropylene modified by decomposition products of alkali sulphate lignin *Mekhanika Polimerov* 6, 3-10.

[67] Lyubeshkina E.G.,Torner R.V and Gul V.E. (1967) Viscosity of lignin-modified polypropylene determined by capillary viscometry , *Mekhanika Polimerov* 2, 200-210.

[68] Lyubeshkina E.G. ,Belova L.T (1975) *All-Union Conf. on the Applications of Polym. Mater. in the Food Ind.,* p. 19, Kalingrad .

[69] Belova L.T., Lyubeshkina E.G,Gorodetskaya N.N. and Gul V.E. ,\(1976) *IVth All-UnionConf. on the Ageing and Stabilisation of Polym.,* p. 69, Tashkent.

[70] Seymour R.B (1969) Plastic Chemistry and Engineering *Ind.Engg.Chem.,* 61(8) 27-41.

[71] BerezkanV.I., Lyubeshkina E.G., Vilnits S.A and Gul V.E, (1977).*All-Union Sci. and Engng Conf. on the Processes,* No. 2, p. 26, Izd.MIKhM, Moscow

[72] Zavgoro V.K. and Lyubeshkina E.G , (1980).*IVth All-Union Sci. and Engng Conf. on Problems of the Creation and Applications of Polym. Mater. in Branches of Ind. Manufacturing Food Products,* p. 118, Minsk

[73] Stewart D. (2008). Lignin as a base material for materials applications:Chemistry, application and economics, *Ind.Crop Prod.* 27,202-207

[74] Lyubeshkina E.G, Belova L.T and Gul V. E. (1976) *Vlth All-Union Conf. on the Chem. And Utilization of Lignin,* p. 146, Izd. Zinatne, Riga.

[75] Pokhodenko V.D. (1969). *Phenoxy-Radicals,* p. 196, Izd. Naukova Dumka, Kiev

[76] Gosselink R.J.A , Jong E. de, Guran B., Abächerli A. (2004). Co-ordination network for lignin—standardisation,production and applications adapted to market requirements (EUROLIGNIN),*Ind.Crop.Prod.* 20, 121-129

[77] Levon K.,Huhtala J. Malm B.,Lindberg J.J .(1987). Improvement of the thermal stabilization of polyethylene with lignosulphonate , *Polymer 28,* 745-750.

[78] Pouteau C., Baumberger S., Cathala B., Dole P.(2004) Lignin–polymer blends: evaluation of compatibilityby image analysis *C. R. Biologies* 327 935-943.

[79] Levon K. Kinanen A. and Lindberg J.J. (1986). Electrically conducting compound synthesized from lignin *Polym. Bull. 16,* 433 .

[80] Keilen J.J and Pollak A. (1947) Lignin-Reinforced Nitrile, Neoprene, and Natural Rubbers, *Ind. Engng Chem.* 39, 480 *Rubb. Chem. Technol.* 20, 1099-1120 .

[81] Beznaczuk L.M. (1985) *Improving the Durability of Silicone Sealants through Polyblending,* Master's thesis, Concordia University, pp. 105-109 .

[82] Raff R.A.V. and Txomlinson G.H.(1949), *Can. J. Res.* F 27, 399-418.

[83] Haxo H.E Jr and Mills G.S. (1960) *Can. Pat.* 591,081, .

[84] Murray and Watsot W. H. (1948) *India Rubb. WId* 118, 677 .

[85] De Paoli M.A ,Furlan L.T and Rodrigues M.A (1983) *Quim. nova,* 121-122

[86] Moorer H.H., Dougirtv W. K.and Ball F.Y. (1970) *U.S. Pat.* 3,519,518

[87] Kratzl K, Buchtela K. ,Gratzl J. ,Zauner J. and Ettinghausen O. (1962), *TappiI 45* No. 2,113-119 .

[88] Kelley S.S. and Glasser W.G (1988) J. *Wood Chem. Technol.* 8(3), 341-359.

[89] Rials T.G (1986), *Holzforschung* 40(6), 353-360 (1986).

[90] Mihailo M. and Budevsce CH.(1962), *Comp. Rend. Bulgare Sei. 15* No. 2, 155-158 .

[91] Ball F.T , Doughty W. K. and Moorer H.H (1962), *Can. Pat.* 654,728 .

[92] Shiraishi N. (1989), *Lignin Properties and Materials*), ACS Symposium Series 397, ACS, Washington, DC, pp. 488-495.

[93] Nieh W.L.S and Glasser W.G (1989), *Lignin Properties and Materials* ACS Symposium Series 397, ACS, Washington, DC , pp. 506-513.

[94] Tomita B, Kurozumi A. and Hosoya S. (1989), *Lignin Properties and Materials* ACS Symposium Series 397, ACS, Washington, DC, pp. 496-505.

[95] Chodak I (1986), *Chem. Papers* 40(4), 461-470.

[96] Hsu O.H.-H. and Glasser W.G. (1975), *Appl. Polym. Syrup. 28,* 297-307 .

[97] Yoshida H.(1987) J. *Appl. Polym. Sci.* 34, 1187-1198 .

[98] Sono Y (1987), *Makuzai Gakkaishi,* 33(1), 47-52 .

[99] Kovrizhnvkh L.P. (1988), *Khim. Drev.* 1, 90-94.

[100] Hatakeyama (1975), *Jap. Pat.* 119,091 .

[101] Philips R.B (1975) J. *Appl. Polym. Sci.* 17, 443.

[102] Meister J.J. and Patil D.R (1985), *Polym. Mater. Sci. Eng.* 52, 235-239 (1985).

[103] Hatakevama H. ,Hirose S. and Hatakevama T. (1989) *Lignin Properties and Materials,* ACS Symposium Ser. 397, ACS, Washington, DC p. 205.

[104] Navaeau (1975), *Cell. Chem. Technol.* 9, 71 .

[105] Hirose S., Yano S., Hatakevama T. and Hatakevama H. (1989) *Lignin Properties and Materials,* ACS Symposium Ser. 397, ACS, Washington,DC ,p382.

[106] Akmar P.F., Kennedy J.F. (2001). The potential of oil and sago palm trunk wastes as carbohydrate resources. *J. Wood Sci. Technol.* 35(5), 467- 473.

[107] Saha B.C. (2003). Hemicellulose bioconversion. *J. Ind. Microbiol.Biotechnol.* 30, 279-29.

[108] Rodríguez-Chonga A., Ramírez J.A., Garrote G., Vázquez M. (2004).Hydrolysis of sugar cane bagasse using nitric acid, a kinetic assessment. *J. Food Eng.* 61(2), 143-152.

[109] Yáñez R., Alonso J.L., Parajó, J.C. (2004). Production of hemicellulosic sugars and glucose from residual corrugated cardboard. *Proc.Biochem.* 39(11), 1543-1551.

[110] Sepúlveda-Huerta E., Tellez-Luis S.J., Bocanegra-García V., Ramírez J.A. ,Vázquez M. (2006). Production of detoxified sorghum straw hydrolysates for fermentative purposes. *J. Sci. Food Agr.* 86(15),2579-2586.

[111] Tabka M.G., Herpoël-Gimbert I., Monod F., Asther M., Sigoillot J.C. (2006).Enzymatic saccharification of wheat straw for bioethanol productionby a combined cellulase xylanase and feruloyl esterase treatment.*Enz. Microbial. Technol.* 39(4,2), 897-902.

[112] Hanchar R.J., Farzaneh T.F, Nielson C.D., McCalla D., Stowers M.D. (2007). Separation of glucose and pentose sugars by selectiveenzyme hydrolysis of AFEX-treated corn fiber. *J. Appl. Biochem.Biotechnol.*, 137-140(1-12), 313-325

[113] Li W., Xu J., Wang J., Yan Y.J., Zhu X.F., Chen M.Q., Tan Z.C. (2008). Studies of monosaccharide production through lignocellulosic waste hydrolysis using double acids. *Energy Fuels*, 22(3), 2015–2021.

[114] Singh P., Suman A., Tiwari P., Arya N., Gaur A., Shrivastava A.K. (2008).Biological pretreatment of sugarcane trash for its conversion tofermentable sugars. *World J. Microbiol. Biotechnol.* 24(5), 667-673.

[115] Kim Y., Hendrickson R., Mosier N.S., Ladisch M.R., Bals B., Balan V., Dale B.E. (2008).Enzyme hydrolysis and ethanol fermentation of liquid hotwater and AFEX pretreated distillers' grains at high-solids loadings.*Bioresour. Technol.* 99(12), 5206-5215.

[116] Baig M.M.V, Baig M.L.B, Baig M.I.A, Yasmeen M. (2004). Saccharification of banana agro-waste by cellulolytic enzymes. *Afric. J. Biotechnol.* 3(9),447-450.

[117] Marques S., Alves L., Roseiro J.C., Gírio F.M. (2008). Conversion ofrecycled paper sludge to ethanol by SHF and SSF using *Pichiastipitis*. *Biomass Bioenergy*, 32(5), 400-406.

[118] Chen F., Dixon R.A. (2007). Lignin modification improves fermentable sugar yields for biofuel production. *Nat. Biotechnol.* 25, 759-761.

[119] Haan R.D., Rose S.H., Lynd L.R., vanzyl W.H (2007). Hydrolysis and fermentation of amorphous cellulose by recombinant *Saccharomycescerevisiae*. *Metabolic Eng.* 9(1), 87-94.

[120] Lu Y., Mosier N.S. (2007). Biomimetic catalysis for hemicellulosehydrolysis in corn stover. *Biotechnol. Prog.*, 23(1), 116-123.

[121] Howard R.L., Abotsi E., Jansen van Rensburg E.L., Howard S. (2003). Lignocellulose biotechnology, Issues of bioconversion and enzyme production. Afr. J. *Biotechnol.* 2(12), 602-619.

[122] Mtui G., Nakamura Y. (2007). Characterization of lignocellulosic enzymes from white-rot fungus *Phlebia crysocreas* isolated from a marine habitat. *J. Eng. Appl. Sci.* , 2, 1501-1508

[123] Jecu L. (2000). Solid state fermentation of agricultural wastes for endoglucanase production. *Ind. Crops and Products.* 11(1), 1-5.

[124] Emtiazi G., Nahvi I. (2000). Multi-enzyme production by *Cellulomonas* sp. grown on wheat straw. *Biomass Bioenergy*, 19 (1), 31-37.

[125] El-hawary F.I., Mostafa Y.S. (2001). Factors affecting cellulase production by *Trichoderma koningii J. Acta Alimentaria*, 30(1), 3-13.

[126] Ögel Z.B., Yarangümeli K., Dündar H., Ifrij I. (2001). Submerged cultivation of *Scytalidium thermophilum* on complex lignocellulosic biomass for endoglucanase production. Enzyme Microbial. Technol. 28(7-8), 689-695.

[127] Ojumu T.V., Solomon B.O., Betiku E., Layokun S.K., Amigun B. (2003). Cellulase Production by *Aspergillus flavus* Linn Isolate NSPR 101fermented in sawdust, bagasse and corncob. *Afr. J. Biotechnol.* 2(6), 150-152.

[128] Raj K., Singh R. (2001). Semi-solid-state fermentation of *Eicchorniacrassipes* biomass as lignocellulosic biopolymer for cellulase and glucosidase production by co-cultivation of Aspergillus niger RK3 and *Trichoderma reesei* MTCC164. *Appl. Biochem. Biotechnol.* 96(1-3),71-82

[129] Wen Z., Liao W., Chen S. (2005). Production of cellulase by *Trichodermareesei* from dairy manure. *Bioresour. Technol.* 96(4), 491-499.

[130] Muthuvelayudham R., Viruthagiri T. (2006). Fermentative production and kinetics of cellulase protein on *Trichoderma reesei* using sugarcane bagasse and rice straw. *Afr. J. Biotechnol.* 5 (20), 1873-1881.

[131] Pothiraj C., Balaji P., Eyini M. (2006). Enhanced production of cellulases by various fungal cultures in solid state fermentation of cassavawaste. *Afr. J. Biotechnol.* 5(20), 1882-1885.

[132] Gao J., Weng H., Zhu D., Yuan M., Guan F., Xi Y. (2008). Production and characterization of cellulolytic enzymes from the thermoacidophilic fungal *Aspergillus terreus* M11 under solid-state cultivation of cornstover. *Bioresour. Technol.* , 99(16), 7623-7629

[133] Daroit D.J., Silveir S.T., Hertz P.F., Brandelli A. (2007). Production ofextracellular _- glucosidase by *Monascus purpureus* on different growth substrates. *Process Biochem.* 42(5), 904-908.

[134] Mabrouk M.E.M., El-Ahwany A.M.D (2008). Production of mannanase by *Bacillus amylolequifaciens* cultured on potato peels. *Afr. J.Biotechnol.* 7(8), 1123-1128.

[135] Bhavsar K., Shah P., Soni S.K., Khire J.M. (2008). Influence of pretreatment of agriculture residues on phytase production by *Aspergillus niger* NCIM 563 under submerged fermentation conditions. *Afr. J. Biotechnol.* 7(8), 1101-1106.

[136] Abdel-Sater M.A., El-Said A.H.M. (2001). Xylan-decomposing fungi andxylanolytic activity in agricultural and industrial wastes. *Int.Biodeterioration Biodegrad.*, 47(1), 15-21.

[137] Pandey P., Pandey A.K. (2002). Production of cellulase-free thermostable xylanases by an isolated strain of *Aspergillus Niger* PPI, utilizing various lignocellulosic wastes *World J. Microbiol. Biotechnol.* 18(3),281-283.

[138] Rezende M.I., Barbosa A.M., Vasconcelos A.F.D., Endo A.S. (2002).Xylanase production by *Trichoderma harzianum* Rifai by solid state fermentation on sugarcane bagasse. Braz. J. Microbiol. 33(1), 67-72.

[139] Isil S., Nilufer A. (2005). Investigation of factors affecting xylanase activityfrom *Trichoderma harzianum* 1073 D3. *Braz. Arch. Biol. Technol.*48(2), 1516-8913.

[140] Elisashvili V., Penninckx M., Kachlishvili E., Asatiani M., Kvesitadze G.(2006). Use of *Pleurotus dryinus* for lignocellulolytic enzymes production in submerged fermentation of mandarin peels and tree leaves. *Enz. Microbial. Technol.* 38(7), 998-1004.

[141] Dobrev G.T., Pishtiyski I.G., Stanchev V.S., Mircheva R. (2007).Optimization of nutrient medium containing agricultural wastes for xylanase production by *Aspergillus niger* B03 using optimal composite experimental design. *Bioresour. Technol.* 98(14), 2671-2678.

[142] Mohana S., Shah A., Divecha J., Madamwar D. (2008). Xylanaseproduction by *Burkholderia* sp. DMAX strain under solid state fermentation using distillery spent wash. *Bioreasour. Technol.* 99(16),7553-7564.

[143] Botella C., de Ory I., Webb C., Cantero D., Blandino A. (2005). Hydrolytic enzyme production by *Aspergillus awamori* on grape pomace. *Biochem. Eng. J.* 26(2-3), 100-106 .

[144] Silva E.M., Machuca A., Milagres A.M.F.(2005). Effect of cereal on *Lentinula edodes* growth and enzyme activities during cultivation onforestry waste. Letters in *Appl. Microbiol.* 40(4), 283-288

[145] Oliveira L.A., Porto A.L.F., Tambourgi E.B.(2006). Production of xylanase and protease by *Penicillium janthinellum* CRC 87M-115 from different agricultural wastes. Bioreasour. Technol. 97(6), 862-867.

[146] Nazareth S.W., Sampy J.D. (2003). Production and characterisation of lignocellulases of *Panus tigrinus* and their application. Int. Biodeterioration *Biodegradation*, 52(4), 207-214..

[147] Couto S.R., López E., Sanromán M.A. (2006). Utilisation of grape seeds forlaccase production in solid-state fermentors. *J. Food Eng.* 74(2), 263-267 .

[148] Wuyep P.A., Khan A.U., Nok A.J. (2003). Production and regulation of lignin degrading enzymes from *Lentinus squarrosulus* (mont.). *Afr. J.Biotechnol.* 2(11), 444-447.

[149] Velázquez-Cedeño M.A., Farnet A.M., Ferré E., Savoie J.M. (2004).Variations of lignocellulosic activities in dual cultures of *Pleurotusostreatus* and *Trichoderma longibrachiatum* on unsterilized wheat straw. *Mycologia,* 96(4), 712-719.

[150] A-el-Gammal A., M-Ali A., Kansoh A.L. (2001). Xylanolytic activities of *Streptomyces* sp. 1--taxonomy, production, partial purification and utilization of agricultural wastes. *Acta Microbiol. Immunol. Hung.*48 (1), 39-52.

[151] Mtui G., Masalu R. (2008). Extracellular enzymes from Brown-rot fungus *Laetioporus sulphureus* isolated from mangrove forests of Coastal Tanzania. *Sci. Res. Essay* 3, 154-161

[152] Kumar R., Singh S., Singh O.V. (2008). Bioconversion of lignocellulosic biomass, Biochemical and molecular perspectives. *J. Ind. Microbiol.Biotechnol.* 35(5), 377-391.

[153] Hahn-Hägerdal B., Galbe M., Gorwa-Grauslunda M.F., Lidén G., Zacchi G. (2006). Bio-ethanol- the fuel of tomorrow from the residues of today. *Trends Biotechnol..* 24(1), 549-556.

[154] Rubin E.M. (2008). Genomics of cellulosic biofuels. *Nat.* 454(14), 841-845.

[155] Howard R.L., Abotsi E., Jansen van Rensburg E.L., Howard S. (2003).Lignocellulose biotechnology, Issues of bioconversion and enzymeproduction. *Afr. J. Biotechnol.* 2(12), 602-619.

[156] Lin Y., Tanaka S. (2006). Ethanol fermentation from biomass resources, Current state and prospects. *J. Appl. Microbiol. Biotechnol.* 69 (6),627-642.

[157] Prasad S., Singh A., Joshi H. (2007). Ethanol as an alternative fuel from agricultural, industrial and urban residues. *Reasourc. Conserv.Recycling*, 50(1), 1-39

[158] Akin-osanaiye B.C., Nzelibe H.C., Agbaji A.S. (2005). Production of ethanol from Carica papaya (pawpaw) agro waste, effect of saccharification and different treatments on ethanol yield. *Afric. J. Biotechnol.* 4(7),657-659

[159] Sørensen A., Teller P.J., Hilstrøm T., Ahring B.K. (2008). Hydrolysis of *Miscanthus* for bioethanol production using dilute acid presoaking combined with wet explosion pre-treatment and enzymatic treatment.*Bioresour. Technol.* 99(14), 6602-6607.

[160] Okuda N., Ninomiya K., Takao M., Katakura Y., Shioya S. (2007). Microaeration enhances productivity of bioethanol from hydrolysate ofwaste house wood using ethanologenic *Escherichia coli* KO11. *J.Biosci. Bioeng.* 103(4), 350-357.

[161] Itoh H., Wada M., Honda Y., Kuwahara M., Watanabe T. (2003).Bioorganosolve pretreatments for simultaneous saccharification andfermentation of beech wood by ethanolysis and white rot fungi. *J. Biotechnol.* 103(3), 273-280.

[162] Kádár Z, Réczey K (2004). Simultaneous saccharification and fermentation (SSF) of industrial wastes for the production of ethanol. *Ind.Crops Prod.* 20(1), 103-110.

[163] Olofsson K., Bertilsson M., Lidén G. (2008). A short review on SSF, aninteresting process option for ethanol production from lignocellulosic feedstocks. *Biotechnol Biofuels* 1(7), 1-14.

[164] von Blottnitz H., Curran M.A. (2007). A review of assessments conducted on bio-ethanol as a transportation fuel from a net energy, greenhouse gas, and environmental life cycle perspective. *J. Cleaner Prod.* 15(7),607-619 .

[165] Champagne P. (2007). Feasibility of producing bio-ethanol from waste residues. *Resourc. Conserv. Recycling* 50(3) 211-230.

[166] Liu B., Zhao Z.K. (2007). Biodiesel production by direct methanolysis ofoleaginous microbial biomass. *J. Chem. Technol. Biotechnol.* 82(8),775-780.

[167] Ito T., Nakashimada Y., Senba K., Matsui T., Nishio N. (2005). Hydrogen and ethanol production from glycerol-containing wastes discharged after biodiesel manufacturing process. *J. Biosci. Bioeng.* 100(3), 260-265.

[168] Arvanitoyannis I.S., Kassaveti A. (2007). Current and potential uses ofcomposted olive oil waste. *Int. J. Food Sci. Technol.* 42(3), 281 –295.

[169] Schenk P.M., Thomas-Hall S.R., Stephens E., Marx U.C., Mussgnug Posten J.H.C., Kruse O., Ben Hankamer B. (2008). Second Generation Biofuels, High-Efficiency Microalgae for Biodiesel Production. *J. BioEnergyRes.* 1(1), 20-43.

[170] Najafpour G., Ismail K.S.K., Younesi H., Mohamed A.R., Kamaruddin A.H. (2004). Hydrogen as clean fuel via continuous fermentation by anaerobic photosynthetic bacteria, *Rhodospirillum rubrum*. *Afr. J.Biotechnol.* (10), 503-507.

[171] Çalar A., Demirba A. (2002). Hydrogen rich gas mixture from olive huskvia pyrolysis. *Energy Conversion Manag.* 43(1), 109-117.

[172] Fan Y.T., Zhang Y.H., Zhang S.F., Hou H.W., Ren B.Z. (2006). Efficient conversion of wheat straw wastes into biohydrogen gas by cow dung compost. *Bioresour. Technol.* 97(3), 500-505.

[173] Zhang M., Cui S.W., Cheung, P.C.K., Wang Q. (2007). Antitumor polysaccharides from mushrooms, A review on their isolationprocess, structural characteristics and antitumor activity. *Trends Food Sci. Technol.* 18(1), 4-19.

[174] Banik S (2004). Jute caddis, A new substrate for biogas production. *J.Sci. Ind. Res.* 63(9), 747-751.

[175] Jin F., Zhou Z., Kishita A., Enomoto H. (2006). Hydrothermal conversion ofbiomass into acetic acid. *J. Mater. Sci.* 41(5), 1495-1500.

[176] Henrique M., Baudel H.M., Zaror C., de Abreu C.A.M .(2005). Improving thevalue of sugarcane bagasse wastes via integrated chemical production systems, An environmentally friendly approach. *Ind. Crops Products*, 21 (3), 309-315.

[177] Lim S.J., Kim B.J., Jeong C.M., Choi J., Ahn Y.H., Chang H.N. (2008).Anaerobic organic acid production of food waste in once-a-day feeding and drawing-off bioreactor. *Bioresour. Technol.* 99(16), 7866-7874.

[178] Mshandete A.M., Björnsson L., Kivaisi A.K., MST Mattiasson B. (2008).Effect of aerobic pre-treatment on production of hydrolases and volatile fatty acids during anaerobic digestion of solid sisal leaf decortications residues. *Afric. J. Biochem. Res.* 2(5), 111-119.

[179] Muruke M.H.S, Hosea K.M., Palangyo A., Heijthuijsen J.H.F.G. (2006). Production of lactic acid from waste sisal stems using a *Lactobacillus* isolate. *Discovery Innov.*, 18(1), 1-5.

[180] Adsul M.G., Varma A.J., Gokhale D.V. (2007). Lactic acid production from waste sugarcane bagasse derived cellulose. *Green Chem.* 9, 58-62

[181] Ohkouchi Y., Inoue Y. (2007). Impact of chemical components of organic wastes on l(+)-lactic acid production. *Bioresour. Technol.* 98(3), 546-553.

[182] Olson E.S. (2001). Conversion of lignocellulosic material to chemicals and fuels. National Energy Technology Lab., Pittsburgh, PA (US);National Energy Technology Lab., Morgantown, WV(US). Report No.FC26-98FT40320-19.

[183] Eyheraguibel B., Silvestre J., Morard P. (2008). Effects of humic substances derived from organic waste enhancement on the growth and mineral nutrition of maize. *Bioresour. Technol.* 99(10), 4206-4212.

[184] Benitez E., Sainzh H., Nogales R. (2005). Hydrolytic enzyme activities ofextracted humic substances during the vermin composting of alignocellulosic olive waste. *Bioresour. Technol.* 96(7), 785-790.

[185] Hart T.D., Lynch J.M.,De Leij F.A.A.M. (2003). Production of *Trichurusspiralis* to enhance the composting of cellulose-containing waste.*Enz. Microbial. Technol.* . 32(6),745-750.

[186] Taiwo L.B., Oso B.A. (2004). Influence of composting techniques onmicrobial succession, temperature and pH in a composting municipal solid waste. *Afr. J. Biotechnol.* 3(4), 239-243

[187] Chatterjee P., Metiya G., Saha N., Haldar M., Mukherjee D. (2005). Methods and substrate oriented composting of lignocellulosicmaterials for the production of humus rich composts. *Environ. Ecol.*23 (1), 55-60.

[188] García-Gómez A., Bernal M.P., Roig A. (2005). Organic matter fractions involved in degradation and humification processes during composting. *Compost Sci. Util.* 13(2) 127-135.

[189] Lopez M.J., Carmen M., Vargas-García C., Suárez-Estrella F., Moreno J.(2006). Biodelignification and humification of horticultural plant residues by fungi. *Int. Biodeterioration Biodegradation.* 57(1), 24-30.

[190] Salètes S., Siregar F.A., Caliman J.P., Liwang T. (2004). Lignocellulose composting, Case study on monitoring oilpalm residuals. *Compost. Sci. Util.* 12(4), 372-382

[191] Vargas-Garcı M.C., Suárez-Estrella F., López M.J., Moreno J. (2007).Effect of inoculation in composting processes, Modifications in lignocellulosic fraction. *Waste Manag.* 27(9), 1099-1107.

[192] Suthar S. (2007). Production of vermin fertilizer from guar gum industrial wastes by using composting earthworm *Perionyx sansibaricus*(Perrier). *J. Environmentalist* 27(3), 329-335.

[193] Tserki V., Matzinos P., Zafeiropoulos N.E., Panayiotou C. (2006).Development of biodegradable composites with treated and compatibilized lignocellulosic fibers. *J. Appl. Poly. Sci.* 100(6), 4703-4710.

[194] Georgopoulos S.T., Tarantili P.A., Avgerinos E., Andreopoulos A.G. ,Koukios E.G. (2005).Thermoplastic polymers reinforced with fibrous agricultural residues. *Poly. Degrad. Stability*, 90(2), 303-312.

[195] Bhattacharyya D., Jayaraman K. (2003). Manufacturing and evaluation ofwoodfibre-waste plastic composite sheets. *Polymers & polymer composites.* 11(6), 433-440

[196] Digabel F.L., Boquillon N., Dole P., Monties B., Averous (2004). Properties of thermoplastic composites based on wheat-straw lignocellulosic fillers. *J. Appl. Poly. Sci.* 93(1), 428 – 436.

[197] Mishra S., Mohanty A.K., Drzal L.T., Misra M., Hinrichsen G. (2004). A review on pineapple leaf fibers, sisal fibers and their biocomposites.*Macromolecular Mater. Eng.* 289(11), 955-974.

[198] Ndazi B., Tesha J.V., Bisanda T.N. (2006). Some opportunities andchallenges of producing bio-composites from non-wood residues. *J.Mater. Sci.* 41(21), 6984-6990.

[199] Madani M., Altaf H.B., Abdo A.E., Houssni E. (2004). Utilization of wastepaper in the manufacture of natural rubber composite for radiation shielding. *Progress in rubber, plastics Recycling Technol.* 20(4), 287-310.

[200] Shaji J., Kuruvilla J., Sabu T. (2006). Green composites from natural rubber and oil palm fiber, physical and mechanical properties. *Int. J.Poly. Mater.* 55(11), 925-945.

[201] Bourne P.J., Bajwa S.G., Bajwa D.S. (2007). Evaluation of cotton gin wasteas a lignocellulosic substitute in wood fiber plastic composites. *For.Products J.* 57,127-131

[202] Pothan L.A., George C.N., Jacob M., Thomas S .(2007). Effect of chemical modification on the mechanical and electrical properties of banana fiber polyester composites. *J. Comp. Materials*, 41(19), 2371-2386.

[203] Lertsutthiwong P., Khunthon S., Siralertmukul K., Noomun K.,Chandrkrachang S. (2008). New insulating particleboards prepared from mixture of solid wastes from tissue paper manufacturing and corn peel. *Bioresour. Technol.* 99(11), 4841-4845.

[204] Habibi Y., El-Zawawy W.K., Ibrahim M.M., Dufresne A .(2008). Processing and characterization of reinforced polyethylene composites made with lignocellulosic fibers from Egyptian agro-industrial residues. *Composites Sci. Technol.* 68(7-8), 1877-1885.

[205] Alemdar A., Sain M. (2008). Isolation and characterization of nanofibersfrom agricultural residues – Wheat straw and soy hulls. *Bioresour.Technol.* 99(6), 1664-1671.

[206] Basta A., Abd El-Sayed E.S., Fadl N.A. (2002). Lignocellulosic materials inbuilding elements. Part III. Recycled newsprint waste paper in manufacturing light-weight agrogypsum panels. *Pigment & Resin Technol.* 31(3), 160-170.

[207] Reis J.M.L. (2006). Fracture and flexural characterization of natural fiber reinforced polymer concrete. *Con. Build. Mat.* 20(9), 673-678.

[208] Israilides C., Philippoussis A. (2003). Bio-technologies of recycling agroindustrial wastes for the production of commercially important polysaccharides and mushrooms. *Biotechnol. Gen. Eng. Reviews*, 20(2003), pp. 247–259.

[209] Anupama A., Ravindra P. (2000). Value-added food, Single cell protein. *Biotechnol. Adv.* 18(6), 459-479.

[210] Robinson T., Nigam P. (2003). Bioreactor design for protein enrichment of agricultural residues by solid state fermentation. *Biochem. Eng. J.* 13(2-3), 197-203.

[211] Krishnani K.K., Ayyappan S. (2006). Heavy metals remediation of water using plants and lignocellulosic agro wastes. *Rev. Environ. Contam.Toxicol.* 188, 59-84.

[212] Zubair A., Bhatti H.N., Hanif M.A., Shafqat F. (2008). Kinetic and equilibrium modeling for Cr (iii) and Cr(vi) removal from aqueous solutions by *Citrus reticulata* waste Biomass. *J. Water Air Soil Poll.*, 191(1-4), 305-318 .

[213] Tsai W.T., Chang C.Y., Wang S.Y., Chang C.F., Chien S.F., Sun H.F. (2001).Utilization of agricultural waste corn cob for the preparation of carbon adsorbent. *J. Environ. Sci. Health Part B* 36(5), 677-686.

[214] Huang D.L., Zeng G.M., Feng C.L., Hu S., Jiang X.Y., Tang L., Su F.F., Zhang Y., Zeng W., Liu H.L. (2008). Degradation of lead-contaminated lignocellulosic waste *by Phanerochaete chrysosporium* and thereduction of lead toxicity. *Environ. Sci. Technol.* 42(13), 4946–4951.

[215] Gan Q., Allen S.J., Matthews R .(2004). Activation of waste MDF sawdust charcoal and its reactive dye adsorption characteristics. *Waste Management.* 24(8), 841-848.

[216] Gürses A., Doar Ç., Karaca S., Açikyildiz, M., Bayrak R. (2006).Production of granular activated carbon from waste *Rosa canina* sp.seeds and its adsorption characteristics for dye. *J. Hazard. Mater.* 131(1-3), 254-259.

[217] Namasivayam C., Sureshkumar M..V (2006). Anionic dye adsorption characteristics of surfactant-modified coir pith, a waste lignocellulosic polymer. *J. Appl. Poly. Sci.* 100(2), 1538-1546

[218] Batzias F.A., Sidiras D.H. (2007). Dye adsorption by prehydrolysed beech sawdust in batch and fixed-bed systems. *Bioresour. Technol.* 98(6),1208-1217.

[219] Kishore K.K., Parimala V., Gupta B.P., Azad I.S., Meng X., Abraham M.(2006). Bagasse-assisted bioremediation of ammonia from shrimp farm wastewater. *Water Environ. Res.* 78(9), 938-950(13).

[220] Lange, H.; Wagner, B.; Yan, J. F.; Kaler, E. W.; McCarthy,J. L. *Seventh International Symposium on Wood and Pulping Chemistry; Beijing*, 1993; Vol. 1, pp 111-122.

[221] Bolker, H. I.; Brenner, H. S. (1970). Polymeric Structure of Spruce Lignin , *Science* 170, 173-176.

[222] Rousselt M.R.(1995) Dynamic Model of Lignin Growing in Restricted Spaces, *Macromolecules*, 28,370-375.

[223] Alexandre M., Dubois P.: Polymer-layered silicate nanocomposites: preparation, properties and uses of a new class of materials. *Materials Science and Engineering, Reports*, 28, 1–63 (2000).

[224] Chazeau L., Gauthier C., Vigier G., Cavaillé J-Y. (2003).:Relashionships between microstructural aspects and mechanical properties of polymer-based nanocomposites.in *'Handbook of Organic-Inorganic Hybrid Materials and Nanocomposites'* (ed.: Nalwa H. S.),American Scientific Publishers, Los Angles

[225] Wegner T. H., Winandy J. E., Ritter M. A.: (2005). Nanotechnology opportunities in residential and non-residential construction. in '2nd International Symposium on Nanotechnology in Construction, Bilbao, Spain'

[226] Eichhorn S. J., Baillie C. A., Zafeiropoulos N.,Mwaikambo L. Y., Ansell M. P., Dufresne A.,Entwistle K. M., Herrera-Franco P. J., Escamilla G.C., Groom L., Hughes M., Hill C., Rials T. G., Wild P.M. (2001). Current international research into cellulosic fibers and composites. *J.Mater. Sci.ence*, 36, 2107–2131

[227] Page D. H., El-Hosseiny F., Winkler K..(1971).Behaviour of single wood fibers under axial tensile strain. *Nature*,229, 252–253

[228] Matos G., Wagner L. (1998) Consumption of materials inthe united states 1900–1995. *Ann.Rev.Ener.Env.*, 23, 107–122

In: Lignin
Editor: Ryan J. Paterson

ISBN 978-1-61122-907-3
© 2012 Nova Science Publishers, Inc.

Chapter 4

OBTAIN LIGNINS FOR SPECIFIC APPLICATIONS

Luis Serrano, Ana Toledano, Araceli García, and Jalel Labidi[*]

Chemical and Environmental Engineering Department,
University of the Basque Country, Plaza Europa, 1, 20018,
Donostia-San Sebastian, Spain

ABSTRACT

The reduction in the emission of gaseous pollutants, the diversification of energy sources, and the dependence on the volatile oil market, all make bio-refinery a good candidate for the production of consumer goods and energy. The valorization of lignin is central to the implementation of the bio-refinery as an alternative to the traditional petrochemical industry. The bio-refinery is the comprehensive use of the total lignocellulosic biomass through the production of high added value products.

Lignin is a polyphenolic polymer with very complex branching structure. Its aromatic structure makes it the only natural polymer with these features. Hence, the huge interest to obtain lignin with homogeneous properties to head production for chemicals, fuels, and/or energy.

There are several treatments applicable for lignocellulosic materials and the fractionation for its major components, cellulose, hemicellulose, and lignin. These treatments are, in general, more focused to achieve good quality solid fraction (cellulose) and pursue the delignification regardless of how it affects lignin structure. However, the different extraction processes strongly affect the structure of the lignin removed, making it difficult to produce homogeneous fractions with industrial interest.

The separation of the liquid fraction after the treatment of the raw material may be a solution to the problem of heterogeneity.

The ultra filtration is an efficient separation process that allows the good separation and purification of lignin to dissolve in the pulping liquor. This method also allows differentiated lignin fractions with diverse physical-chemical characteristics.

Ultra filtration separation method has been compared with another simple method, selective precipitation; in order to establish the effect of the separation process on the properties of lignin fractions produced by both methods. Selective precipitation is a very

---

[*] E-mail: jalel.labidi@ehu.es, tel.:+34-943017178; fax: +34-943017140.

low energy consumption and easy separation method that intends to solve the no-homogeneity of the lignin samples.

Lignin is known to be a natural antioxidant. This capacity is influenced by the properties of the lignin, mainly by the phenol content. As previously mentioned the separation method can reach homogenous fractions and so, enhance their antioxidant capacity. For this purpose, ultra filtrated fractions were studied to know how the final properties of the different fractions affect the antioxidant capacity of the lignin.

This work presents the results of the study and provides a comparison for the physical-chemical characterization of different fractions of lignin obtained by two methods of separation, ultra filtration and selective precipitation. Finally, an antioxidant capacity test was done to confirm the effect of the ultra filtration on the lignin properties.

INTRODUCTION

The use of renewable energy sources (wind power, solar energy, geothermal, hydropower, biomass, etc.) reduces the emissions of greenhouse gases, diversifies energy sources, and reduces our dependence on the volatile and distrustful fossil fuels (including crude oil and gas) markets [1]. Biomass is an important natural reserve for renewable fuels, chemicals, and energy production. The European Union received 66.1% of its renewable energy from biomass, which thus surpassed the total combined contribution from hydropower, wind power, geothermal energy, and solar power. In addition to energy, the production of chemicals from biomass is also essential; indeed, the only renewable source of liquid transportation fuels is currently obtained from biomass. [2].

Biomass is defined as the biodegradable fraction of products, waste and biological waste from agricultural activities (including vegetal and animal substances), forestry and related industries, including the fisheries and aquaculture, as well as the biodegradable fraction of industrial and local wastes [1, 3].

Lignocellulosic biomass comes from agricultural activities (vegetal) and also from forestry. Lignocellulose is the most abundant biomass representing near 70% of the total plant biomass. Classical uses of this raw material are for pulp and paper, building and textile industries that are using only 2% of this type of biomass, leaving a 98% as waste. Important advantages to consider, lignocellulose as feedstocks offers a great abundance, and the fact that there is no concurrency with food industries. There are approximately 200 billion tonnes per year available to be used [4] in the world. However, as a logical consequence, lignocellulosic materials can be interestingly used efficiently by turning them into fuels, chemicals, and energy.

Lignocellulosic materials are mainly composed by hemicelluloses, cellulose and lignin in different proportion essentially depending on the species. There are also other minority components as oils, extracts, mineral compounds.Cellulose is the major component for this type of material. It is made exclusively by β-D-glucose molecules in the form of pyranoses, linked together by β-1, 4-O-glycosidic bonds. Cellulose usually presents a polymerization degree between 8000 and 10,000. It is currently used mostly for the manufacture of paper but it is also used in the production of artificial fibers (cellulose acetate), plastics (cellulose nitrate), explosives (nitrocellulose), thickeners and gelling agents (cellulose ethers such as carboxymethylcellulose, hydroxiethylcellulose and hydroxipropylmethylcellulose)[5]. The main function of the cellulose in the plant cell is as a structural component.

Hemicelluloses (also called polioses) are composed by pentoses (xylose and arabinose) and hexoses (glucose, galactose, manose) linked together forming short chains of low molecular weight and, generally, branching. The hemicelluloses often present degrees of polymerization between 80 and 200 [6] and in the plant cell act as compatibilizer between different components of the cell wall.

Lignin is a phenolic polymer derived from the oxidative combination of three major units derived from the phenylpropanoic acid: syringyl alcohol (S), guaiacyl alcohol (G) and p-coumaryl alcohol (H), which form a randomized structure in a three-dimensional network inside the cell wall [4].

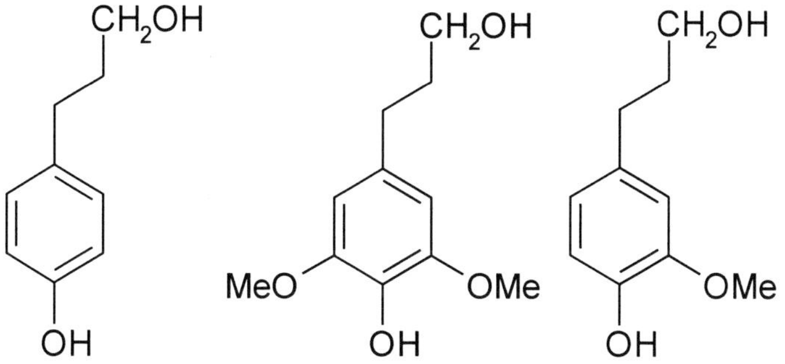

Figure 1. Phenylpropanic units form lignin structure. (H) p-coumaric alcohol (S) syringyl alcohol (G) guaiacyl alcohol.

The components derived from the p-coumaryl, guaiacyl, and syringyl alcohol are linked by different types of carbon-carbon bonds or carbon-oxygen-carbon, as e.g., β-O-4, 5-5, β-5, 4-O-5, β-1, dibenzodioxocine, β –β, of which the β-O-4 linkage is dominant, consisting of more than half of the linkage structures of lignin [2]. The structure of lignin is very complex and changes depending on the species, growing conditions, and employed extraction technique.

The lignin role in the plant cell is like cement between cellulose and hemicelluloses providing rigidity to the cell wall and also as a barrier against biological agents. Nowadays, lignin is burned to produce energy in the pulp and paper industry; however, lignin can be used as a precursor to generate large number of high value added products as chemical compounds [7], polymer formulations [8], dispersant in mixing cement [9]. In addition, the valorization of all components of the lignocellulosic biomass is essential for the bio-refinery to be economically viable.

One of the major drawbacks for lignin to be used in an industrial scale process is the heterogeneity of the isolated samples. This heterogeneity results from the extraction method used causes the random bond linkage breakdown of lignin in the cell wall which is not homogeneously dissolved, causing liquid fraction of different lignin fragments with different molecular weights.

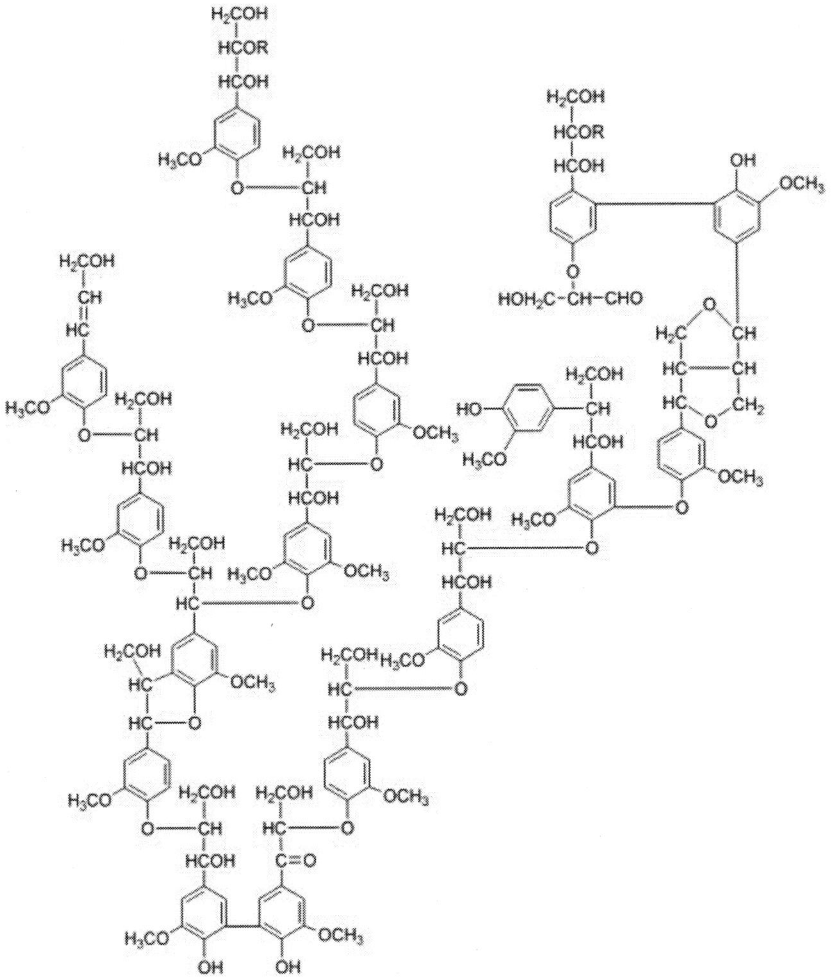

Figure 2. Softwood lignin structure proposed by Adler (1977).

Starting from this point, lignin separation method can play a key role in the properties, structure, and functionalities of obtained lignin, and more importantly, in obtaining marketable products from lignin. Two separation techniques namely – ultra-filtration and selective precipitation have been studied for the production of different lignin fractions.

SEPARATION METHODS

Lignin presents a branched structure that varies depending on the species, extraction method, and age of the plant. High structural diversity, as well as broad weight molecular distributions makes its introduction to industry difficult due to the heterogeneity of lignin fractions. The size of lignin molecules can vary between 1,000 and 100,000 Da within the same sample; this aspect therefore lays the importance of separation and fractionation processes.

There are different methods to separate the lignin dissolved in the liquid fraction that result from the treatment of the raw material, such as extraction with different solvents, selective precipitation, processes involving membrane technology. In this study, two of these techniques were selected and their results from both have been compared.

Membrane separation processes have been studied extensively in recent years since its implementation in the industry is of great interest in different sectors, such as e.g., food, chemical, pharmaceutical industry...

Membrane technology allows four objectives:

- Separation
- Concentration
- Purification
- Fractionation

Using membrane technology for separation processes presents the following advantages:

- Separation can be carried out continuously
- Low energy consumption
- Membrane technology can be easily combined with other separation processes
- Separation can be carried out under mild conditions
- Up-scaling is easy
- Membrane properties are variable and can be adjusted
- No additives are required

The drawbacks must also be taken into account:

- Concentration polarization / membrane fouling
- Low membrane lifetime
- Low selectivity or flux
- Up-scaling factor is more or less linear

The low energy consumption, the no-need to add reagents that can change the structure of the dissolved compounds and provide easy up-scaling, make membrane technology an optimal separation method to be used in a bio-refinery. The disadvantages are important to take into account but can be overcome as confirmed with the existence, nowadays, of production processes that apply membranes technology in various fields [10].

There are different types of processes that involve membrane technology depending on the separation principle. The effectiveness or efficiency of membrane technology based on the separation by molecular size strongly depends on the type of membranes, as in the pore or size of particle that can be retained [12]. Taking these factors into account, four types of processes can be distinguished: microfiltration (MF), ultra filtration (UF), nano filtration (NF), and reversible osmosis (RO). These methods differ in the membrane type and conditions of operation required, and as a result, the application field. In Table 1, process conditions, types of membranes, pore size are shown.

In this work, ultra filtration has been often used since this type of membrane process is suitable for the separation of macromolecules.

Table 1. Common operating conditions and applications of different membrane separation processes [11]

Process	Membrane type and pore size	Membrane material	Driving force (bar)	Applications
MF	Symmetric microporous (0.1 - 10 μm)	Ceramics, Metal oxides (Aluminium, Titanium, Zirconium), Graphite, Polymers (Cellulose nitrate or acetate, PVDF, Polyamides, Polysulfone, PTFE,..)	1 - 5	Sterile filtration, Clarification
UF	Asymmetric microporous, (1 - 10 nm)	Ceramics, Polysulfone, Polypropylene, Nylon 6, PTFE, PVC, Acrylic Copolymer	1 - 10	Separation of macromolecular solutions
NF	Thin-film membranes	Cellulosic Acetate and Aromatic Polyamide	10 - 30	Removal of hardness and desalting
RO	Asymmetric skin-type (0.5 - 1.5 nm)	Polymers, Cellulosic acetate, Aromatic Polyamide	up to 200	Separation of salts and microsolutes from solutions

Ultra filtration membranes are rated on the basis of nominal molecular weight cut-off, but the shape of the molecule to be retained has a major effect on the extent of its retentivity. The membranes discriminate between dissolved macromolecules of different sizes and are usually characterized by their molecular weight cut-off, a loosely defined term generally taken to mean the molecular weight of the globular protein molecule that is 90% rejected by the membrane [13].

Ultra filtration has also been widely studied for the concentration of the resulting black liquor from Kraft pulping and effluent treatment in paper industries processes [14-21]. Jönnson et al. [22] performed a study on the influence of operating conditions in a system of ultra filtration and nano filtration designed to concentrate and purify the lignin from black liquor for the result from the treatment of hardwood species. In general, the main objective for this type of research is the application of UF and/or NF to increase the capacity of the recovery boiler, in order to obtain more energy in the Kraft process. These authors have considered several filtration methods, operating parameters and costs associated with the implementation and application of membrane technology. The high amount of water used in pulp and paper industries and the high cost associated to the recovery process consisting of the contaminated water evaporation of the black liquor before recovering of reagents, make UF/NF systems a promising alternative to the usual methods of black liquor concentration. Schlesinger et al. [23] considered the combination of UF and NF processes for separation of hemicelluloses from the black liquor, concluding that the system suggested, could recover most of the hemicelluloses present in the black liquor. Vegas et al. [24] proved the efficiency of different membranes for the fractionation and purification of xylooligosacharides from the hydrolysis of the rice bran husk.

As mentioned, previous studies been carried out on the efficiency for the use of ultra filtration systems for the concentration of the black liquor, but none of them has focused on the characteristics, properties, and structure of different fractions of the obtained lignin fractions.

Precipitation is the classic separation method of lignin from the liquid fraction after the treatment of the raw material. It is a simple and effective method with very low energy consumption, which is why it has been used traditionally. Precipitation is an easy technique which consists of the simple addition of a mineral acid to reach the desired pH (usually 2 [25, 26]) at which precipitates the lignin present. According to Fengel et al. [27], lignin is highly soluble in alkaline medium but its solubility in liquid fraction decrease dramatically when the pH decreases. Once lignin is precipitated, the slurry is vacuum filtrated. This last step is a bottleneck since the lignin precipitated in the liquor is more or less a viscous gel that makes this stage costly in terms of time of the complete filtration process of liquid fraction.

Mohamad Ibrahim and Chuah [26] isolated alkaline lignin using different mineral acids: sulphuric acid, hydrochloric acid, phosphoric acid, and nitric acid, at different concentrations -20%, 60% (v/v) and concentrated. The objective of this study was to observe the influence of the nature of the mineral acids in the precipitated lignin. They could observe that the percentage in which the soda lignin was recovered is strongly influenced by pulping conditions, the pH value that precipitates the lignin and the nature of the acid employed. However, no significant differences in the structure of the different obtained lignins or in their molecular weight distributions were detected. Taking into account the conclusions reached by Mohamad Ibrahim et al. [26], in the study presented here, concentrated sulphuric acid was used to precipitate the lignin.

Sun and Tomkinson [28] developed a separation process using precipitation for the lignin in two consecutive stages, as a replacement for the traditional precipitation method in one stage. The results showed that separation by precipitation in the two-stage procedure was a fast and adequate method to isolate relatively free of contamination lignin. The obtained fraction in the second stage, after removing in the first stage at pH 7 polysaccharides degradation products, resulted in relatively more free of no-lignin compounds such as ash, salts, extractives, and, as already commented, polysaccharides degradation products, reaching a purity of 98.8% for that lignin fraction. This work attempted to sort out the major disadvantages of the precipitation process, the contamination associated with the dragging of other compounds considered as contaminants in lignin fractions.

Mussato et al. [29] studied different fractions of lignin resulted after the addition of different amounts of sulphuric acid to the black liquor from brewer's spent grain pulping. The study focuses on the determination of the amount of precipitated lignin at each pH, color evaluation of the black liquor after lignin precipitation and determining the concentration of soluble lignin. The main objective was based on the evaluation of the pH influence in the lignin precipitation. Lowering the pH from 12.56 to 2.15, they observed an increase of the precipitated lignin mass, recovering the 81.43% of the lignin that was solubilized in the liquor. It also suggested that each component derived from lignin was affected differently by the pH alteration [29].

ANTIOXIDANT CAPACITY

Due to the high content of diverse functional groups (phenolic and aliphatic hydroxyls, carbonyls, carboxyls ...) and its phenylpropanoic structure, lignin can act as a neutralizer or inhibitor in oxidation processes, stabilizing reactions induced by oxygen radicals and their derived species [30]. However, this antiradical activity depends greatly on lignocellulosic material from which lignin is obtained, the method used for its extraction, and the possible treatments applied during its isolation and purification.

The antioxidant capacity of lignin has been widely studied, since it represents a natural free radical scavenger. Košiková et al. [31] studied the potential medicinal application of the lignin obtained from pulp and paper production (Kraft and prehydrolysis processes), and found that it provided a protective effect on DNA and inhibited mutagenicity process, resulting in a potential antimutagenic and anticarcinogenic agent. Recent works [32, 33] have confirmed that lignin can be used in cosmetics and pharmaceutical preparations without harmful effects to human cells. These studies have also demonstrated that the origin of the lignins strongly influences their antiradical activity.

Many authors [34-38] have analyzed the effect of using different lignin types on the photo-thermal stability of diverse natural or synthetic polymers. The high content and variety of functional groups present in lignin make it a very compatible additive in polymeric materials that allow modifying it in order to obtain different properties of these blends [39, 40]. Modified Kraft lignin using acid treatment proved that the change in the chemical properties of the resulting lignin affected their compatibility with the propylene. An increase in the solubility was observed, but there also was an improvement in their antioxidant capacity (tested by thermogravimetric analysis) due a lower content of hydroxyl groups. A similar study, concerning the use of lignin as stabilizer of natural rubber, was also reported [41]. This studied the possibility of replacing the N-phenyl-N-isopropyl-p-phenylene diamine (IPPD), a commercial antioxidant, with lignin to improve the response of natural rubber against thermo-oxidative aging processes. This study found that lignin gave better results over time than the IPPD, but also managed to slightly improve the mechanical properties of rubber.

Kurosumi et al. [42] affirmed that steam explosion conditions (temperature and time) affected obtaining low molecular weight lignins. This showed higher contents of phenolic groups, and therefore high antioxidant activity since lengthy isolation times can promote lignin repolymerization into insoluble materials. Ultra filtration can allow for the control of molecular weight in lignin fractions and therefore, control the antioxidant capacity.

MATERIALS AND METHODS

Miscanthus Sinensis

Miscanthus sinensis originated in Eastern Asia and since the 1930s in central Europe was used as an ornamental plant [43]. As a perennial crop, it is usually cultivated for 15-20 years. *Miscanthus* needs quite a large water supply (approximately 380 mmha^{-1}) and relatively good soil. During the first 3 years the yields are relatively small and thereafter can range from 10 to 25 t ha^{-1} dry matter (DM) at very good sites up to 30 t DM ha^{-1} [20]. Only the stems are harvested, from November to March, and the foliage remains as much layer on the field. Apart from the utilization of *Miscanthus* as fuel (combined heat and power generation facilities or biogas plants), non-energy use of the fibers is possible. The fibers can be used to replace peat, to produce composites or insulating materials. The latter alternative could enter the market of natural insulations which have established themselves besides conventional products from mineral wool or polystyrene (PS) [44]. *M. sinensis*, elephant grass, belongs to the monocots group. It presents interesting properties such as a C4 metabolism that allows it to photosynthesize and to use water and nutrients with high efficiency, easily adaptable, high resistance at low temperatures, pesticides and diseases and high growing yields. These characteristics make *M. sinensis* suitable for new applications.

The European Community (EC) has shown great interest in *M. sinensis* as a possible new crop to be grown in rural areas which, until now, had been exploited with agriculture food-crops, of which the EC has excess production. That is the reason why the new Common Agricultural Policy imposes a change in the strategy in order to find and exploit species to produce energy and due to specific market purposes. Within these new species, there is a special interest in Miscanthus *sinensis* based on the previously discussed factors: easily adaptable, resistance to temperatures and C4 metabolism allowing it to photosynthesize, and use water and nutrients effectively [45].

The *Miscanthus sinensis* used to develop this work was supplied by Straw Pulp Engineering (SPE) (Zaragoza, España).

Alkaline Treatment

Pre-Treatment Conditions

The preparation of the raw material is of great importance since it allows the liquid to penetrate deeply into the raw material structure, and thus, the delignification process takes place with higher yields.

Raw material is subjected to grounding carried out at SPE with a hammer mill to improve the surface contact with the alkaline solution. After this pretreatment process to the *M. sinensis*, the following distribution of particle size was obtained:

- 0–7mm 36.95%
- 7–18mm 59.68%
- 18–37mm 1.61%
- 37–59mm 1.76%

The raw material was grounded in the laboratory with a Retsch mill using a sieve of 0.4 cm. In taking into account the morphology of the *M. sinensis*, the slices ranged from 1.5 cm up to 0.1 cm. Subsequently, the raw material was sieved to remove particles smaller than 0.1 cm that are difficult to recover.

Alkaline Treatment

Environment protection is one of the main reasons for the pushed development of sulphur-free chemical pulping processes as alternatives to Kraft and sulfite pulping. The various processes that meet these premises are [27]:

- The traditional soda process and its Anthraquinone (AQ) modification
- The two stage soda-oxygen process
- Single-stage oxygen pulping
- The alkaline-AQ-peroxide process

The applied alkaline treatment uses NaOH solution for the delignification of the raw material. It is a classic treatment but unlike the Kraft process, as already mentioned, it does not use sulphur compounds. The advantages for application of the soda treatment are high-speed delignification, which leads to short treatment times, high delignification efficiency and the flexible possibility to be applied to any raw material, woody or non-woody materials [5].

It was found that the soda pulping process is one of the most relevant and used for the delignification of non-woody raw materials [46], such as *M. sinensis*. During soda treatment, lignin is dissolved constituting a rich phenolic compound liquid fraction that represents the process effluent [27]. There are some similarities between soda process and the traditional Kraft process and reactions, such as cleavage of lignin-carbohydrate linkages and depolymerization of the lignin and its recondensation take place. Lignin depolymerization during soda pulping also occurs; principally by the cleavage of α and β aryl ether bonds, first in phenolic units and, in a later phase of delignification, in non-phenolic units. The generation of free phenolic groups in such reactions results in lignin fragments that are soluble in the alkaline environment prevalent during the digestion [46].

The research developed for the study of the selective precipitation as a separation method was done in collaboration with SPE. Soda pulping liquors of *M. sinensis* were provided by SPE. The operation conditions of pilot plant used by SPE are shown in Table 2. The separation of the liquor from solid fraction was carried out by filtration under pressure.

Table 2. Operation conditions used by SPE

Operation Conditions	
NaOH Concentration	7.5 %
Solid: liquid ratio	1 : 10
Temperature	100 °C
Pressure	Atmospheric
Time	3 h

The liquors for the ultra filtration separation were produced by the alkaline treatment of the raw material in a laboratory glass reactor of 20 L. The separation of the liquid fraction from the solid one was carried out by gravity filtration. Table 3 shows the conditions used in the laboratory in the soda treatment of *M. sinensis*.

Table 3. Operation conditions used in the laboratory

Operation Conditions	
NaOH Concentration	7.5 %
Solid: liquid ratio	1 : 18
Temperature	90 °C
Pressure	Atmospheric
Time	1.5 h

The solid:liquid ratio used was greater than the one used in the SPE pilot plant since the objective in the laboratory was to obtain high volume of liquid fraction to apply the ultra filtration.

Ultrafiltration

The liquid fraction obtained after alkaline treatment was introduced in the ultra filtration system to separate lignin fractions depending on the pore size. The ultra filtration module used for this work was provided by IBMEM - Industrial Biotech Membranes (Frankfurt, Germany). Tubular and multichannel ceramic type membranes (active layer of zirconium oxide, titanium oxide support) (Figure 6) were used at different cut-offs. The characteristics of the employed membranes were: hydraulic diameter 2 mm, external diameter of 10 mm, area of 110 cm^2 and a length of 250 mm.

To carry out the ultra filtration process, two stainless steel filter modules were display in series. The liquid fraction was recirculated through the membrane module (Figure 3).

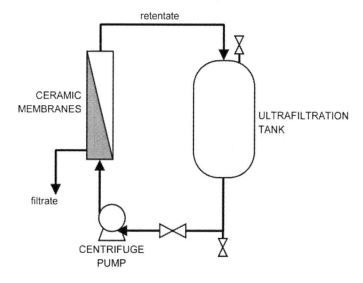

Figure 3. Ultrafiltration system employed.

Lignin dissolved fragments in the liquid fraction with a size lower than the pore size came out of collection constituting the permeate fraction, the remaining liquid flow was recirculated to the tank. The temperature is maintained at 60 °C. Membranes chosen for this study had a pore size of 5, 10, and 15 kDa. The order followed was from the smallest pore size to the biggest allowing the collection of 4 fractions:

- Fraction with molecular size less than 5 kDa (5 kDa fraction)
- Fraction with molecular size between 10 kDa and 5 kDa (10 kDa fraction)
- Fraction with molecular size between 15 kDa and 10 kDa (15 kDa fraction)
- Fraction with molecular size greater than 15 kDa (> 15 kDa fraction)

The resulting permeates were treated with H_2SO_4 (95-98% purity, Panreac) until reaching a pH below 2 to precipitate the present lignin. The obtained mixture was vacuum filtered with a glass fiber filter of 1.2 μm (FVC-110, Albet) washed twice with acidified water (approximately pH 2) with the purpose of dragging possible impurities (hemicelluloses, silicates, rest of reagents, etc.) and vacuum dried at a temperature of 60 °C.

Selective Precipitation

Obtaining different lignin fractions at different pH values was achieved by taking 20 mL of the resulting liquid fraction in *M. sinensis* soda treatment, and adding different amounts of sulphuric acid (98%, Panreac purity) (Table 4). After the addition of the acid, the samples were centrifuged (4000 rpm for 10 minutes) to separate the precipitated lignin in each fraction. Samples were washed with acidified water (approximately pH 2) to remove any possible impurities and/or contamination and then vacuum dried at 50 °C until constant weight.

After removing the precipitated lignin, the supernatant was kept at -27 °C for 20 minutes [29]. After this time, the samples were again centrifuged at 4000 rpm for 10 minutes. The lignin thus precipitated washed twice with acidified water (approximately pH 2) to remove impurities. Finally, the different lignin fractions were dried under vacuum at 50 °C until constant weight.

The total precipitated lignin mass was considered as the sum of the mass recovered after the acid addition and after cooling.

Table 4. Obtained fractions by selective precipitation

Fraction	1	2	3	4	5	6	7	8	9
pH	12.64	11.08	10.41	9.16	6.50	5.40	4.55	2.57	0.72

Lignin Characterization

These lignin fractions obtained apply both separation methods and are physico-chemically characterized.

Fourier Transform Infrared Spectroscopy

The Fourier transform infrared spectroscopy (FT-IR) study was developed in a Nicolet spectrometer of direct transmittance. Samples were analyzed in the region between 4000-400 cm^{-1} with a resolution of 4 cm^{-1} and 20 scans were recorded. Each sample was prepared according to the potassium bromide (KBr) technique, in a proportion of 1:100 (200 mg of KBr, approximately). The mixture was pressed and the resulting pellets were used for the FT-IR test. In order to avoid bands derived by the moisture of the sample, the pellets were dried for 48 hours in an oven at 105° C. Later on, the pellets were cooled in a desiccator until room temperature.

Thermal Behavior

To study the thermal behavior of the samples, two types of tests were developed, differential scanning calorimetry and thermogravimetric analysis.

The differential scanning calorimetry (DSC) analyses were carried out in a thermal analyzer DSC 821 of Mettler Toledo, making dynamic scans of temperature from 30 to 250 °C with a constant heating rate of 10 °C/min. The tests were made under nitrogen atmosphere, and using a sample size of approximately 5 mg to minimize the mass and heat transference differences of the results. A thermogravimetric analysis (TGA) was carried out in a TGA/SDTA RSI analyzer of Mettler Toledo. The samples of ~5 mg were heated of 25 °C up to 800 °C at a rate of 10 °C/min. A constant nitrogen flow was used, which provided an inert atmosphere during the pyrolysis and allowed extracting the gaseous and condensable products that could cause secondary interactions in vapor phase.

Size Exclusion Chromatography

Size exclusion chromatography (SEC) was used to determine lignin number-average (Mn), weight-average (Mw) molecular weight and polydispersity (Mw/Mn) in a Perkin-Elmer instrument equipped with an interface (PE Series 900). Three Waters Styragel columns (HR 1, HR 2, and HR 3) ranging from 100 to 5×105 and a refractive index detector (Series 200) were employed, with a flow rate of 1mL/min. Lignin samples were subjected to acetylating before the analysis in order to enhance their solubility in THF [47]. Calibration was made using polystyrene standards.

Nuclear Magnetic Resonance

NMR spectra were recorded at 25 °C on a Bruker Avance 500 Mz equipped with a z-gradient BBI probe. Typically, 40 mg of sample were dissolved in DMSO-d6. 2D-NMR spectra were recorded with a delay of 5 s. Spectral widths were 4500 Hz for ^1H dimension and 22,000 Hz (HSQC, HSQC-TOCSY) or 30,000 Hz(HMBC) for ^{13}C dimension.

Antioxidant Capacity

Determination of Total Phenol Content

The Folin–Ciocalteu method was used for total phenols content determination [48]. The Folin–Ciocalteu reactive (2.5 mL) was diluted with water (1:10, v/v), and mixed with 2mL of 75 g/L aqueous solution of sodium carbonate. The resultant solution was added to 0.5mL of an aqueous solution of lignin. The mixture was kept for 5 min at 50 °C before measuring the absorbance at 760 nm. The total phenols content was determined from the calibration curve of gallic acid standard solutions (1–20 mg/L) and expressed as mg gallic acid equivalent (GAE)/100 mg of lignin (on dry basis).

Determination of the Antioxidant Capacity

The spectrophotometric method employed is based on the use of the free radical 2,2-diphenyl-1-picrylhydrazyl (DPPH) and was developed by Brand-Williams et al. [49] and modified by Dizhbite et al. [50] for the antiradical activity of lignin. Lignin samples dissolved in dioxane/water (90:10, v/v) were mixed with 3.9 mL of a $6 \cdot 10^{-5}$ mol/L DPPH solution. The absorbance at 518 nm of the mixture was immediately measured using a spectrophotometer Jasco V-630. A reference synthetic antioxidant (Trolox) was tested under the same conditions. The absorbance measurements were made without sample (A_0), and at 60 and 120 min after sample addition (A_{60} and A_{120}). The results were expressed as TAC% (total antioxidant capacity), i.e., as the percentage respect to the reduction in absorbance observed for the DPPH with Trolox just after mixing (t = 0min).

RESULTS AND DISCUSSION

Raw Material Characterization

The results obtained after the characterization of the *M. sinensis* according to the TAPPI standards [51] are shown in Table 5.

Table 5. Characterization of *Miscanthus sinensis* by TAPPI standards

Analysis (%)	Standards	*Miscanthus sinensis*
Moisture	TAPPI T264 om-97	8.13 ± 0.06
Ash	TAPPI T211 om-93	4.06 ± 0.23
Solubility in hot water	TAPPI T207 om-93	12.48 ± 0.21
Solubility in NaOH 1%	TAPPI T212 om-98	36.83 ± 0.47
Extractives	TAPPI T204 om-97	7.72 ± 0.43
Lignin	TAPPI T222 om-98	15.52 ± 1.23
Holocellulose	Mét. Wise et al. [52]	66.59 ± 0.37
α-cellulose	Mét. Rowell [53]	36.06 ± 0.20
Hemicelluloses		30.78 ± 0.16

Table 6 shows a comparison with the results obtained by other authors. As observed, the results vary between authors the reason for such differences are that *M. sinensis* is a biennial plant.

This condition causes greater chance of variations in its structure, depending on the quality of the soil where it was cultivated, growth weather conditions, collection, place and time of storage... All these factors cannot be controlled easily.

Table 6. Characterization of *M. sinensis* comparison with other authors

| | Authors | | | |
	Ye *et al.* [43]	Iglesias *et al.* [45]	Velásquez *et al.* [54]	Obtained results
Holocellulose	72.5 %	65.8 %	-	66.6 %
Cellulose	42.2 %	-	42.6 %	36.1 %
Lignin	19.9 %	23.7 %	19.9 %	15.5 %
Hemicelluloses	-	-	21.1 %	30.8 %
Ash	0.7 %	5.7 %	0.7 %	4.1 %
Extractives	9.1 %	3.2 %	1.6 %	7.7 %
Solubility NaOH 1%	-	44.9 %	3.1 %	36.8 %
Solubility in hot water	3.1 %	9.6 %	-	12.5 %

Liquid Fraction Characterization

After the alkaline treatment of raw material, the liquid fraction was separated from the solid fraction. In this work, there were two different liquid fractions since two different operating conditions were applied to *M. sinensis*. The liquid fraction that subsequently was going to be subjected to ultra filtration as separation method was called "liquid fraction 1." The liquid fraction that subsequently was going to be subjected to selective precipitation was called "liquid fraction 2." As a result of the application for different conditions in the alkaline treatment, liquid fractions have different physico-chemical properties (Table 7).

As it can be observed in the table, the liquid fraction 1 despite having higher content of total dissolved solids, the lignin content is lower than the liquid fraction 2. The reason is that the conditions applied by SPE are optimized.

Inorganic matter content is lower in the fraction liquid 2 than in fraction 1, that in principle, it can be beneficial to have removed lignin with less contamination. The solid:liquid ratio used (1:18) to produce the liquid fraction 1 is greater than the one (1:10) used to generate the liquid fraction 2, that involves higher inorganic content of liquid fraction 1.

Table 7. Characteristics of obtained liquid fractions (the results are presented in mass percentage, w/w)

Liquid fraction	Density (g/mL)	pH	TDS[a] (%)	IM[b] (%)	OM[b] (%)	Lignin (%)
Liquid fraction 1	1.0800	12.68	10.25	7.89	2.36	0.40
Liquid fraction 2	1.0125	12.64	4.17	2.84	1.33	0.99

a: Total dissolved solids.

b: Inorganic and Organic matter referred to TDS content.

Ultra Filtrated Lignin Fractions Characterization

Obtained Permeates

The obtained liquid fractions after applying ultra filtration process to the rough liquid fraction were characterized and the results are shown in Table 8.

Table 8. Characteristics of the different fractions (initial liquor, permeates, and retentate) from the UF process of the alkaline liquor. Results as mass percentage (w/w)

Liquid fraction		Density (g/mL)	pH	TDS[a] (%)	IM[b] (%)	OM[b] (%)	Lignin (%)
Liquid fraction 1		1.0800	12.68	10.25	77.0	23.0	17.0
Permeates	5 kDa	1.0614	12.66	6.71	92.0	8.1	81.5
	10 kDa	1.0705	12.68	8.93	87.5	12.5	34.1
	15 kDa	1.0746	12.68	9.48	82.6	17.4	16.9
Filtrate		1.0830	12.68	10.69	76.5	23.5	20.3

a: Total dissolved solids (dry matter).
b: Inorganic and organic matter, referred to TDS content.
c: Referred to organic matter content.

Physico-chemical properties of the liquid fraction 1 were further influenced by the structures present in more than 15 kDa fraction in comparison to the others. The reason that explains this fact is that many of the components of the raw material present high molecular weight and high steric volume that prevent those components from crossing through the membrane pore.

As expected, the percentage of total dissolved solid decreased as the cut-off was smaller. The use of small pore size membranes means that few dissolved compounds can pass through it, so ultra filtration allows the purification of the compounds in the permeates. Density data agreed with the previous affirmation.

As it was mentioned before, the percentage of inorganic matter was much higher than the organic matter; this fact can be related to the solid/liquid ratio used in the alkaline treatment of the *M. sinensis*.

The percentage of lignin in the obtained fractions, referred to organic matter, decreased as the cut-off was greater. For the obtained permeate using the lowest pore size membrane (5 kDa), most of the organic matter was lignin. In the other obtained permeates (10 kDa and 15 kDa) the percentage of lignin is low being connected to the presence of lignin–carbohydrate complex (LCC), large size complex could only pass through higher cut-offs.

Fourier Transform Infrared Spectroscopy

The FT-IR spectra of the different lignin fractions obtained by ultra filtration are shown in Figure 4. It was found a typical O–H stretching at 3400 cm^{-1} and C–H stretching in methyl and methylene groups at 2930 and 2850 cm^{-1}, respectively. The band at 1700 cm^{-1} can be associated to carbonyl stretching in unconjugated ketone or/and to conjugated carboxylic groups. At 1630 cm-1 there was a band that corresponded to carbonyl stretching in γ-lactone. Finally, the vibrations corresponding to C-H out of plane deformation (trans) and aromatic C–H out of- plane deformation appear at 980cm^{-1} and 835cm^{-1}, respectively.

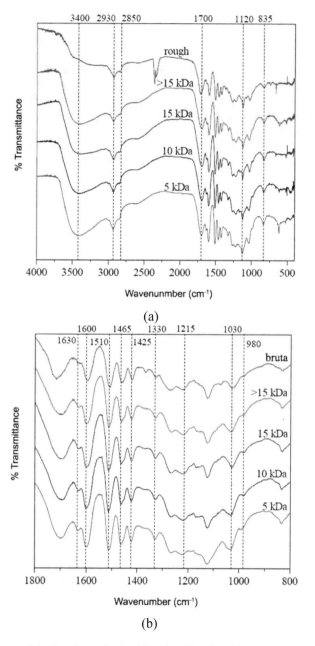

Figure 4. a) FTIR spectra of the fractions obtained by ultrafiltration. b) Magnified region of FTIR spectrums of the fractions obtained by ultrafiltration.

The magnified spectra, as shown in Figure 4b, indicate the vibrations caused by the main subunits in lignin structure for guaiacyl and syringyl type units. The aromatic skeletal vibrations appear at 1600, 1510, and 1425 cm^{-1}. At 1465 cm^{-1} there is a vibration associated to C–H asymmetric deformations. Syringyl ring breathing with C–O stretching occurs at 1330cm^{-1} while guaiacyl ring breathing with C–O stretching appears at 1215cm^{-1}. At 1030cm^{-1} there is a vibration that corresponded to C–H in-plane deformation in guaiacyl or/and C–O deformation in primary alcohol. The band at 1120 cm-1 can be associated to C–H inplane deformation in syringyl.

The bands assignment of the spectra for the different ultra filtrated lignin fractions confirms that the main component is lignin in all cases, according to previous studies [25] where three different types of lignin were analyzed. However, significant structural differences between the different fractions were not observed.

Thermal Behavior

DSC allows to determine the thermal effects associated to physical or/and chemical changes of a substance, when increasing or decreasing the temperature at constant rate. Lignin glass transition temperature (Tg) is difficult to determine with reliability because of the strong electrostatic interactions and because it depends on diverse factors as the extraction process followed [25, 55]. All fractions presented a Tg between 105 and 110° C. These results are in agreement with other authors [25, 47].

The thermogravimetric analysis is widely used to know how the organic polymers decompose. TG curves reveal the weight loss of substances in relation to the temperature of thermal degradation, while the first derivative of that curve (DTG) shows the corresponding rate of weight loss.

Figure 5 shows the results of the thermogravimetric analysis of different lignin ultra filtrated fractions. In all cases there are three areas of weight loss with different slopes that can be distinguished, this involves three components decomposition. First weight loss (DTG$_{max}$= 75 °C, approximately) was due to moisture in the sample. Subsequently weight loss (DTG$_{max}$= 235 °C, approximately) could be ascribed to the degradation of contamination due to polysaccharides (hemicelulosas, mainly) [56].

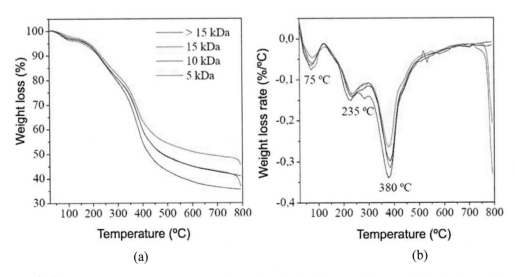

(a)   (b)

Figure 5. a) Thermogravimetric analysis of the samples obtained by ultra filtration. b) Derivative of the TG curves of the ultra filtrated samples.

The 15 kDa fraction presents another weight loss, around 285 °C, which could be associated to heavier hemicelluloses oligomers. Contamination of lignin fractions depends strongly on the applied treatment to the raw material. The greatest loss of mass occurs at DTG_{max} 380 °C (approximately) and was due to the lignin degradation. Actually, the lignin thermal degradation takes place from 185 °C to 500 °C, this fact is because of the complex structure of the lignin with phenolic hydroxyl, carbonyl groups and benzylic hydroxyl, which are connected by straight links [57].

Size Exclusion Chromatography

Gel permeation chromatography was carried out to obtain the molecular weights distribution of the different acetylated lignin fractions obtained by ultra filtration. Number-average (Mn), weight-average (Mw) molecular weight and polydispersity (Mw/Mn) of the different fractions obtained by ultra filtration of the black liquor of alkaline pulping of the *M. sinensis* are shown in Table 9.

Table 9. Results of GPC analysis. Number-average (.Mn), weight-average (.Mw) molecular weight and polydispersity (.Mw/ .Mn) of the different fractions obtained by ultra filtration of the black liquor of alkaline pulping of the *Miscanthus sinensis*

Fraction	$\overline{M}_n$	$\overline{M}_w$	$\overline{M}_w / \overline{M}_n$
Rough	1879	5654	3.01
> 15 KDa	2032	6300	3.10
15 KDa	1891	3544	1.87
10 KDa	946	2022	2.14
5 KDa	940	1806	1.92

Clearly, it can be observed that the objective of the ultra filtration process as a separation method was reached and it was also possible to fractionate the dissolve liginin into fractions with defined molecular weight. The weight average molecular weight decreased as the cut off was smaller. Polydispersity (Mn/Mw) also presented small values. The relatively low polydispersity found indicated the high fraction of low molecular weight (LMW) present in lignin samples. The number of C–C bonds between units is connected to the lignin molecular weight, mainly to the structures involving C5 in the aromatic ring. The most abundant aromatic rings in lignin are guaiacyl-type unit and syringyl-type unit (Figure 6).

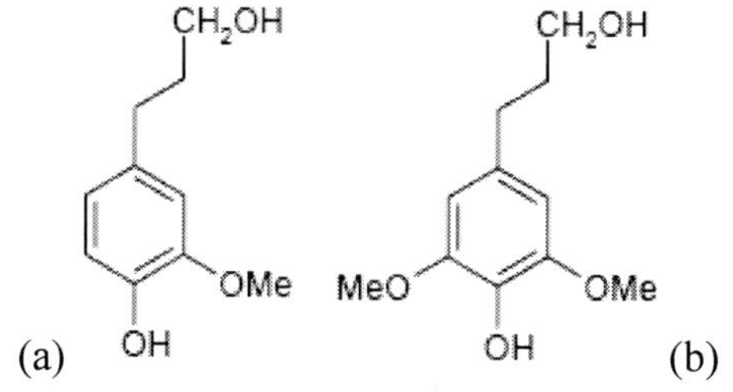

Figure 6. a) Guaiacyl alcohol b) Syringyl alcohol.

Guaiacyl-type units are able to form those kind of bonds, but this is not possible in syringyl-type units, as they have both C3 and C5 positions substituted by methoxy groups. Lignins mostly composed by guaiacyl units are expected to show higher fractions of low molecular weight (LMW) than those presenting high contents of syringyl units that will present high content of high molecular weight fractions (HMW) [58].

According to this explanation, 5 kDa and 15 kDa fractions had high fractions of LMW because of their low polydispersity. It has been reported that this kind of lignins with high fraction of LMW is suitable to be used as an extender or as component of phenol-formaldehyde resins because of their high reactivity, in comparison with lignins with high percentages of high molecular weight molecules [59, 60].

Nuclear Magnetic Resonance

The ^{1}H-NMR experiment was carried out to obtain the chemical structure of the different fractions. In the spectra there are large zones grouping a family of protons, the zone from 6.0 to 7.5 ppm belongs to aromatic protons, from 4.0 to 3.5 ppm belongs to protons next to ether groups. These two areas are very important in the lignin characterization because it is the most influential lignin structures on other properties. Finally, between 2.0 and 0.8 ppm, the signal is due to protons in aliphatic groups such as in the side chains of lignin or in acetyl groups. The main structures are identified in the Table 10.

Obtained spectra shown in Figure 7 illustrate some of the structures present in samples that were identified according to other authors [61-63]. At 4.8 ppm the signal belongs to polysaccharide contamination (LCC) confirming what it was observed in the thermogravimetric analysis. In the spectra of the 5 kDa fraction, it can be pointed that the area from 6.0 to 7.5 ppm is not intense since the cut-off is small and big structures such as aromatic ring substituted with alkyl, hydroxyl, or acetyl chains (syringyl and guaiacyl groups) cannot pass through.

Table 10. Structure correspondence for ^{1}H-NMR

δH (ppm)	Structure
1.90	Protons of aliphatic acetates
2.25	Protons of aromatic acetates
3.40	Moisture in DMSO
3.60	OCH$_3$
4.15	γ(β–β)
6.70	S2,6
7.20	G2,6

Nuclear magnetic resonance in two dimensions allows for researchers to obtain detailed knowledge about the structure of the analyzed compound. In order to find more information about the structure of the lignin fractions obtained by ultra filtration, HSQC (Heteronuclear Single Quantum Coherence) spectra was done for all lignin fraction(Figure 8).The correlated signals of ^{1}H and ^{13}C spectra of side-chains and aromatic structures areas were observed [62, 64]. The main structures are shown in Figure 9.

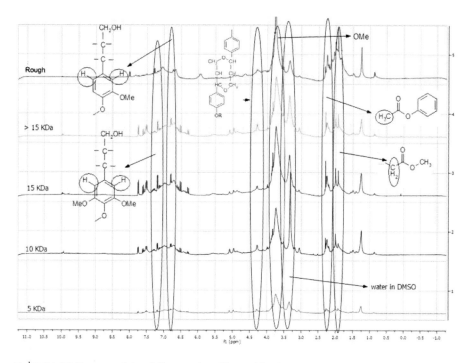

Figure 7. ^{1}H-RMN Spectra of the different ultra filtrated lignin fractions.

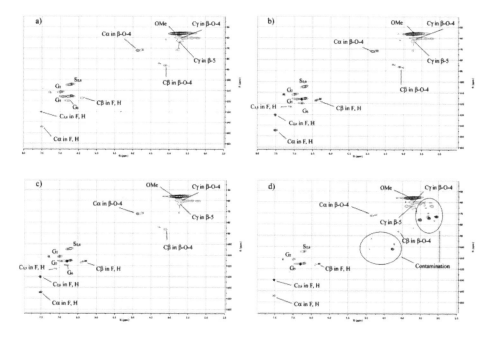

Figure 8. Magnified HSQC spectra of the different ultra filtrated fractions. a) 5 KDa fraction b) 10 KDa fraction c) 15 KDa fraction d) more than 15 KDa fraction.

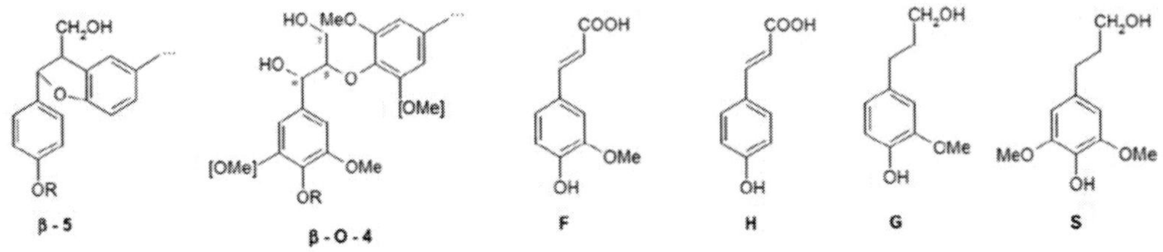

Figure 9. Main structures present in alkali ultra filtrated lignin fractions. β-5 structures; β – O – 4 structures; F: ferulates; H: p-coumaric acid; G: guaiacyl alcohol; S: syringyl alcohol.

Clearly, it can be observed that the lignin obtained from *M. sinensis* by alkali pulping was H:G:S type, with higher content in guaicayl, followed by syringyl content. The proportion of guaiacyl structures in the different lignin fractions increased as the cut-off was greater. This is connected with the polydispersity of the samples if the content type guaiacilo structures increases, the polydispersity also increases due to the possibility that have these structures form carbon-carbon bonds with phenylpropanic units because of having free the C5 position of the aromatic ring, as discussed earlier. This is in agreement with size exclusion chromatography results.

In the side-chain region, different signals could be observed in the ultra filtrated fractions spectra that corresponded to classical lignin substructures such as β-O-4 and β-5. These substructures are connected between them with reactive monolignols, as ferulates and p-coumaric acids. The cross-signals with δ_H/δ_C 7.5/145 ppm; δ_H/δ_C 7.5/130 ppm; δ_H/δ_C 6.35/115ppm revealed the presence of these structures.

The spectrum of the more than 15 kDa fraction suggested that it was a highly contaminated fraction. This spectrum had cross-signals that correspond to polysaccharide contamination (Lignin–carbohydrate complex). Therefore, this finding validated the suggested results of the lignin content and thermogravimetric analysis.

Lignin Fractions Obtained by Selective Precipitation Characterization

Obtained Fractions

Initial pH of the black liquor, produced in the pulping process by the company Straw Pulping Engineering, was 12.64. The progressive lowering of the pH from 12.64 to 0.72, proportioned by the sulphuric acid addition (98%, w/w), produced an increase in the precipitate fraction (Table 11 and Figure 10) and a lightening of black liquor's color (Figure 11).

In Figure 13, two zones of significant precipitation were observed; one in the range of pH 12 - 10 that corresponds to silicates present in the raw material, and another zone in the range of pH 7 - 4 where higher amount of precipitate was obtained. Low percentages obtained in the first fractions at high pH values were as a result of most abundant compounds present in the liquid fraction did not precipitated at those pH values.

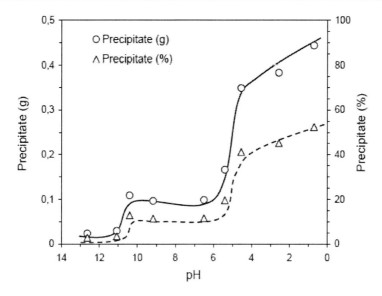

Figure 10. Precipitated mass and precipitated percentage as function of pH.

For the fraction obtained at pH 12.64 (liquid fraction without addition of sulphuric acid), the recovered mass was formed mainly by small suspended solids from the alkaline treatment. From fraction 3 to 5, the majority of the recovered mass corresponds to the presence of silicates [28] linked to organic compounds. In principle, these organic compounds can be low molecular weight compounds and, therefore, more influenced by changes in the medium such as lower pH. Entering more details, the percentage of total dissolved solids increased from 11.62% up to 41.16% when the pH decreases from 6.50 to 4.55. At this pH value, the recovered mass percentage is high and remains around 50% for obtained fractions at lowest pH values.

Table 11. Lignin fractions obtained by selective precipitation

Fraction	pH	Precipitate (g)	% Precipitate
1	12.64	0.0232	2.74
2	11.08	0.0289	3.41
3	10.41	0.1087	12.82
4	9.16	0.0963	11.36
5	6.50	0.0985	11.62
6	5.40	0.1662	19.60
7	4.55	0.3490	41.16
8	2.57	0.3833	45.20
9	0.72	0.4440	52.36

In Figure 11, the color change of the liquid fraction after separation of the precipitated lignin is illustrated. In the original liquid fraction (sample 1), the liquor was dark brown, turning progressively to a pale yellow at the lowest pH value (0.72). Similar liquor colors were obtained by Mussatto *et al.* [29] when the pH of the black liquor of the brewer's spent grain (BSG) was decreased from pH 12.56 to pH 2.15.

Figure 11. Color changes in the treated black liquor from 12.64 (sample 1) to 0.72 (sample 9).

The dark color of the liquid fraction is derived from the chromophoric functional groups, including quinones, carbonyl groups, carboxylic acids, hydroperoxy radicals, phenolic hydroxyl groups, among others, generated during the lignin degradation, which are soluble in alkaline medium [27].

It can also be noted in Figure 11 that in sample 4 the color of the liquid fraction was not altered, being as dark as in the original liquid fraction. Probably, the compounds mentioned here that confer color to the liquor were not removed in this pH range (between 12.64 and 9.16). In sample 5 (pH 6.50), an alteration in the liquor color was observed. The pH decrease from 6.50 to 2.57 (sample 8) also caused a strong influence on the liquor color, which slowly became yellow and almost did not modify with the pH reduction to 0.72 (sample 9). It is evident, thus, that the liquor color was strongly affected by acid pH, suggesting that the chromophore compounds are removed in this pH range, mainly between 9.16 and 2.57.

Fourier Transform Infrared Spectroscopy

The achieved FT-IR spectra of the different fractions obtained at different pH values are shown in Figures 12a and 12b. It also characterized a commercial alkaline lignin (Aldrich) for comparison.

Characteristic vibrations of lignocellulosic materials can be observed, such as the band at 3420 cm^{-1} corresponded to the vibration of the hydroxyl group; stretching at 2925 and 2850 cm^{-1} associated to C-H bond vibrations in methyl and methylene groups; and, the band that appeared at 1715 cm^{-1} is due to the stretching of non-conjugated C=O.

The spectra shown in Figure 12b corresponds with the magnified region of the spectra (1700 - 550 cm^{-1}). Between 1600 and 1515 cm-1, observed bands corresponded to characteristics vibration associated to the aromatic ring present in lignin structure. The band at 1325 cm^{-1} was due to C-O vibration close to syringyl type units while the band to 1265 cm - 1 corresponded to C-O vibration next to guaiacyl type units. At 1130 cm^{-1} the vibration that appeared was because of the in-plane deformation of C-H bond in syringyl. These bands that are associated to substructures present in the lignin structure became most evident in fractions obtained at low pH values, showing the fraction obtained at pH 2.57 the greatest intensity.

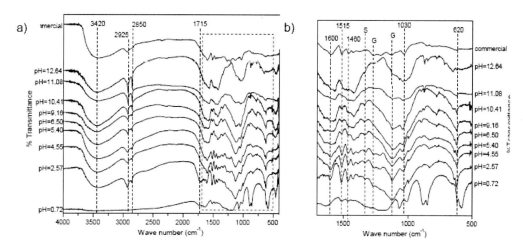

Figure 12. FT-IR spectra for obtained precipitates at several pH values. (a) Wave number from 4000 to 400 cm−1; and, (b) magnification of 1700–550cm−1 region.

The bands that appeared between 1030 and 620 cm^{-1} were attributed to the presence of hemicelluloses and silicates. By the presence of these components, indicate that fractions are contaminated.

Thermal Behavior

The results obtained by the thermogravimetric analysis are shown in Table 12. In Figure 13, details of the obtained curves for some fractions are given.

Table 12. Values of maximal mass loss rate temperature and solid residue for different precipitates in function of pH

pH	9.16	6.50	5.40	4.55	2.57	0.72	Commercial Lignin
TG (°C)	283.5	279.5	291.3	291.3	369.6	369.6	369.6
Solid residue (%)	39.32	32.38	33.52	33.52	35.94	47.12	46.70

Fractions obtained at the pH values of 12.14, 11.92, and 10.46 could not be analyzed in the first two cases due to the limited sample amount of obtained; for the sample fraction at 10.46 was because of the high presence of silicates. Indeed, the low percentage of recovered mass in fractions obtained at higher pH values was expected since there were few components that precipitated at those pH values.

In Figure 13, as in Table 12, it can be observed that fractions obtained at high pH values presented maximum degradation temperatures (sample weight loss in relation to the thermal degradation temperature) between 200 – 300° C. This means that most of the fractions are formed by hemicelluloses in addition to lignin.

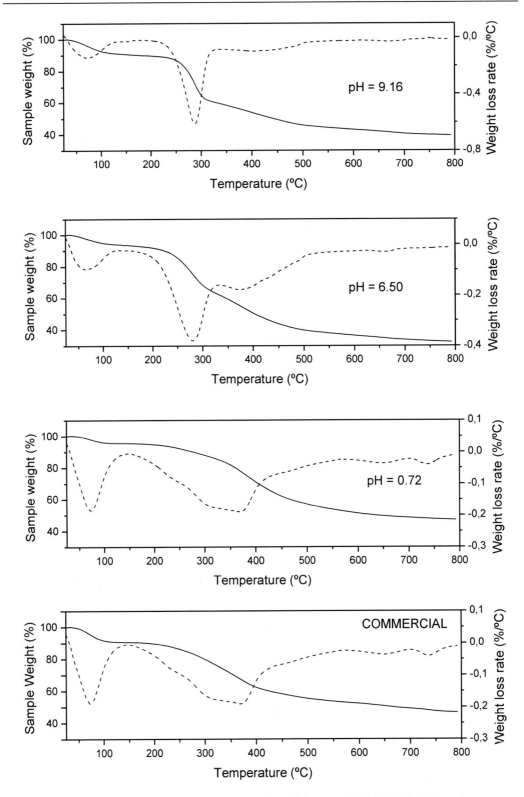

Figure 13. DTG (derivative curves) and TG curves of precipitates at pH 9.16, 6.50, 0.72, and commercial lignin.

Table 13. Values of glass transition temperature (*T*g) of precipitates in function of pH

pH	12.64	11.08	10.61	9.16	6.50	5.40	4.55	2.57	0.72	Com. lignin
Tg (°C)	114.0	84.8	114.0	108.2	109.3	110.5	108.2	105.8	107.0	109.3

This behavior happens until the 8 fraction (pH 2.57), where the maximum speed of degradation was approximated to 400 ° C, the temperature where lignin degradation takes place. Lignin degradation occurs slowly due to the complex structure causing that the derivative of the TG curve displays wide peaks.

Residues, in general, for all fractions are high. It should be noted, that the fraction 9 and 10 (pH 2.57 and 0.72) presented the same maximum degradation temperature than alkaline commercial lignin and fraction obtained at pH 0.72 which also presented similar residue than commercial lignin.

The results obtained after applying differential scanning calorimetry to the samples can be observed in Table 13.

Significant changes were not observed between the different precipitates glass transition temperature (*T*g) values. Strong electrostatic interactions between lignin molecules can convert lignins, in most of cases, into non-melted materials in spite of not presenting very high molecular weights.

For this same reason, the values of glass transition temperature (*T*g) are not frequently imperceptible to determine with any high reliability [65, 66]. At the same time lignins are very influenced by water presence.

In agreement with other published data, lignins presented *T*g between 90 and 180 °C [65, 67]. This *T*g was associated to hydrogen bonds between hydroxyl groups [68] and to the lignin aromatic nature [69].

In the present work it was possible to observe certain similarity between *T*g value of commercial alkaline lignin and lignin obtained at lower value pH.

Size Exclusion Chromatography

Molecular weight of acetylated samples was analyzed through THF-eluted GPC. Table 4 shows the molecular weight-average (Mw), number-average (Mn) and polydispersity (Mw/Mn) of the precipitates and commercial alkaline lignin for comparison.

The fraction obtained at pH 9.16 presented a relative low weight average-weight; this is consistent with the presence of organic compounds of low molecular weight.

In general, all samples presented low polydispersity. As mentioned for the ultra filtrated lignin samples, the low polydispersity indicates greater proportion of fractions LMW which in turn indicates presence of higher amount of syringyl type units unable to establish carbon-carbon bonds in the C5 position of the aromatic ring. As noted in the results obtained by using infrared spectroscopy, all the obtained samples there were guaiacyl and syringyl groups in the structure of the lignin.

However, the data contained in Table 14 suggest greater proportion of type syringilo to guaiacilo-type units.

Table 14. Weight average (Mw), number average (Mn) and polydispersity (Mw/Mn) of acetylated precipitate samples at pH 9.16, 6.50, 0.72 and commercial alkaline lignin

Fraction	$\overline{M}_w$	$\overline{M}_n$	$\overline{M}_w / \overline{M}_n$
Commercial lignin	3135	1886	1.662
pH = 0.72	3501	1908	1.835
pH = 2.57	2432	1311	1.855
pH = 5.40	2120	1142	1.856
pH = 6.50	1990	1430	1.392
pH = 9.16	2160	1550	1.394

Nuclear Magnetic Resonance

The previous analysis concluded all obtained fractions with the exception of fractions obtained at pH 2.57 and 0.72 were very contaminated with hemicelulosas and other components. To enter more details about the chemical structure of the purest samples, proton NMR was carried out testing the obtained lignin precipitates after lowering the pH to 2.57 and 0.72. Commercial alkaline lignin was tested for comparison. The results are shown in Figure 14 and signals assignment in the Table 15.

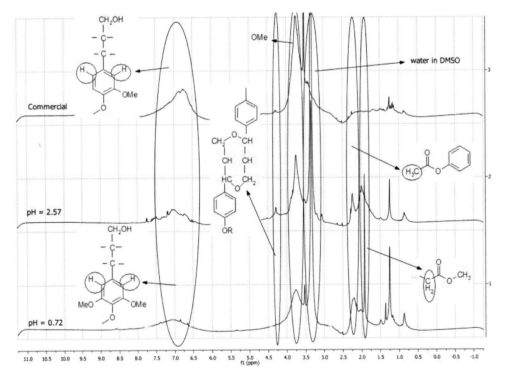

Figure 14. 1H NMR spectra of acetylated precipitate samples at pH 2.57, 0.72 and commercial alkaline lignin.

Table 15. Signal assignment for 1H NMR spectrometry of acetylated precipitate samples at pH 2.57, 0.72 and commercial alkaline lignin

Signal (ppm)	Assignment
8.0-6.0	Aromatic H in G and S units
6.9	G2,6
6.6	S2,6
4.2-3.1	Methoxyl H
2.5-2.2	H in aromatic acetates
2.2-1.9	H in aliphatic acetates
1.5-0.8	Aliphatic protons

^{1}H-NMR spectra revealed the presence of guaiacyl and syringyl groups, as well as acetate and methoxy groups. The corresponding signal to the methoxy group protons is higher in the fraction obtained at pH 2.57 than in fraction at pH 0.72, which might be understood as a greater presence of syringyl units in the first one and therefore a reduction in the average mass molecular weight since these units can not establish carbon-carbon bonds because of having C5 position of the aromatic ring substituted with a methoxy groups. However, nuclear magnetic resonance analysis was carried out qualitatively, so no definitive conclusions can be obtained.

The fraction obtained at pH 2.57 presented structural characteristics similar to commercial lignin as it can be observed in ^{1}H-NMR spectra.

SEPARATION METHODS COMPARISION

Fourier Transform Infrared Spectroscopy

As mentioned in corresponding sections, the infrared spectra of the ultra filtered fractions and the precipitated fractions it can be identified by vibrations associated with functional groups present in lignin structure.

Figure 15 shows the magnified region for the obtained spectra and the different lignin fractions results from the separation application methods of ultra filtered and differential precipitation. The magnified region (1700 - 600 cm^{-1}) is where bands due to characteristic vibrations of lignin substructures linkages can be identified.

In the magnified region (Figure 15b) of the fractions obtained by differential precipitation, bands at 1030–620cm−1 were observed, and attributed to hemicelluloses and silicates contribution. These bands were not as intense as in the magnified region of the fractions obtained by ultra filtration (Figure 15a) which suggest that fractions obtained by acid precipitation were more contaminated than ultra filtered fractions. Suspended lignin drags hemicelluloses and lignin–carbohydrate complex (LCC) when precipitating. This fact was more obvious when using selective precipitation as a fractionation process since membrane technology allows a better purification, where the fraction of more than 15 kDa as most contaminated one.

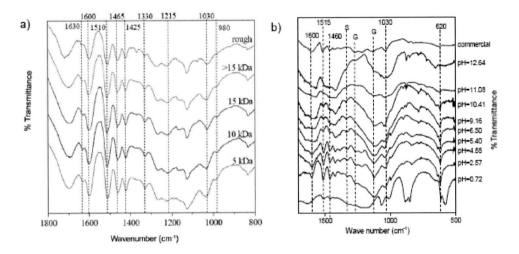

Figure 15. a) Magnified FT-IR spectra of the lignin fractions obtained by ultra filtration. b) Magnified FT-IR spectra of lignin fractions obtained by selective precipitation.

Thermal Behavior

As discussed earlier, the lignin glass transition temperature is difficult to accurately determine because of the strong electrostatic interactions and also due to its dependence on other factors such as the extraction process applied [25, 40, 52]. At the same time lignins are very much influenced by water presence. Fractions obtained by both separation methods presented Tg values between 90 and 110 °C. These results are in agreement with those obtained by other authors [47, 56].

The results of the thermogravimetric analysis applied to the different fractions obtained after subjecting them to the corresponding separation method are presented in Table 16. In all fractions, the weight loss with a DTG_{max} around 70 °C was attributed to the loss of moisture present in the different samples.

It can be observed that in the case of the ultra filtered lignins two different weight loss areas (other than that of the moisture) can be distinguished while there was only a zone of weight loss in fractions obtained at different pH values. This can be explained considering that, just as proved in the physico-chemical characterization of lignin samples obtained by selective precipitation; these fractions are highly contaminated with lignin-carbohydrates complexes (LCC), being so strongly linked that the components of LCC degrade together in a wide temperature range.

In Figure 16, it can be observed that the derivative of the TG curve for all fractions obtained by selective precipitation, presented a very broad peak which corroborates what was explained earlier. In addition, the maximum DTG_{max} for all fractions obtained by selective precipitation, except for fractions obtained at pH 2.57 and 0.72, pH is around 300 °C. These observations suggested the presence of hemicelluloses–lignin complexes in the selective precipitated fractions.

Table 16. Thermogravimetric analysis results of the lignin fractions obtained by both separation methods

Ultrafiltration			Selective precipitation		
Fraction			Fraction		
5 KDa	DTG_{max} (°C)	70.86	4 (pH 9.16)	DTG_{max} (°C)	71.15
		236.94			283.50
		386.33		Residue (%)	39.32
	Residue (%)	41.27	5 (pH 6.50)	DTG_{max} (°C)	66.08
10 KDa	DTG_{max} (°C)	71.5			279.50
		233.14		Residue (%)	32.38
		386.33	6 (pH 5.40)	DTG_{max} (°C)	63.75
	Residue (%)	41.07			291.30
15 KDa	DTG_{max} (°C)	72.00		Residue (%)	33.52
		228.67	7 (pH 4.55)	DTG_{max} (°C)	65.27
		283.5			291.30
		377.5		Residue (%)	33.52
	Residue (%)	35.53	8 (pH 2.57)	DTG_{max} (°C)	73.27
>15 KDa	DTG_{max} (°C)	72.00			369.60
		228.67		Residue (%)	35.94
		377.50	9 (pH 0.72)	DTG_{max} (°C)	70.52
	Residue (%)	48.33			369.60
				Residue (%)	47.12

Decomposition temperature of these products was found to be intermediate between hemicelluloses one (around 230 °C) and lignin (around 380 °C), explaining the low DTG obtained value.

All fractions have a high percentage in the final residue due to lignin aromatic polycondesations.

Size Exclusion Chromatography

Table 17 shows the results after applying size exclusion chromatography to fractions obtained by ultra filtration and selective precipitation.

It can be observed that the separation method based on membrane technology reached not only the lignin separation, but also the fractionation. For the ultra filtration fractionation process, there was a clear trend in the decrease of the weight-average molecular weight as the cut-off used to obtain the fractions was smaller. As a consequence, the polydispersity also decreased with the pore size.

However, fractions obtained by selective precipitation, despite the fact of showing low polydispersity, there was no big differences between fractions in the average weight molecular weight.

It can be concluded that the ultra filtration, as a separation method, produces fractions with more differentiated molecular weights than selective precipitation.

Table 17. SEC results of ultra filtrated lignins and selective precipitated lignins

ULTRA FILTRATION			
Fraction	Mn	Mw	Mw/Mn
Rough lignin	1879	5654	3.01
> 15 KDa	2032	6300	3.10
15 KDa	1891	3544	1.87
10 KDa	946	2022	2.14
5 KDa	940	1806	1.92
SELECTIVE PRECIPITATION			
Fraction	Mn	Mw	Mw/Mn
pH = 0.72	3501	2908	1.20
pH = 2.57	2432	1311	1.86
pH = 5.40	2120	1142	1.62
pH = 6.50	1990	1430	1.40
pH = 9.16	2160	1550	1.41

Nuclear Magnetic Resonance

The obtained spectra from the analysis of different fractions show the presence of lignin substructures protons such as guaiacilo, syringilo, acetates, methoxy groups... Ultra filtration ^{1}HNMR spectra (Figure 16a) showed peaks that did not exist in selective precipitation spectra (Figure 16b) because ultra filtration technology slightly depolymerized lignin. This is due to the alkaline treatment may cause some lignin depolymerization and as the ultra filtation fractions are more purified than fractions obtained by selective precipitation, the resulted compounds can be observed. As previously observed in the other analysis lignin obtained by selective precipitation was more contaminated than the one obtained by using membrane technology. That is the reason why the ^{1}H NMR spectra of both fractionation processes have different shape.

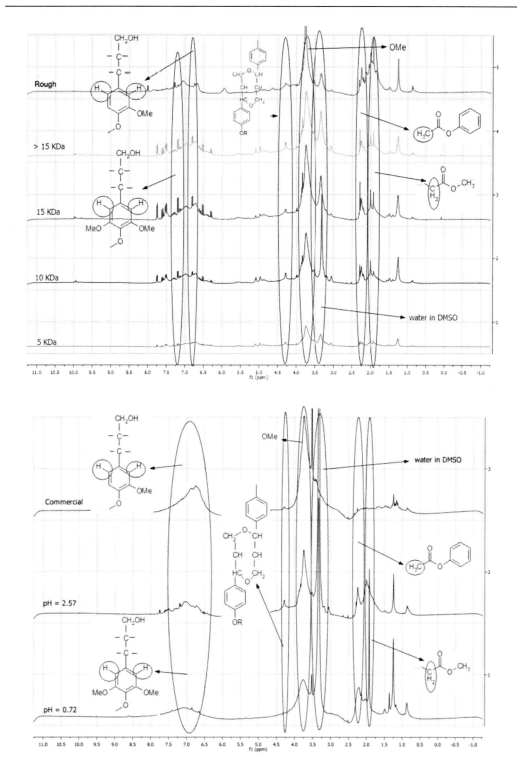

Figure 16. Proton NMR Spectra. a) Lignin fractions obtained by ultra filtration b) Lignin fractions obtained by selective precipitation.

STUDY OF THE ANTIOXIDANT CAPACITY
OF ULTRA FILTRATED LIGNINS

Obtaining of Different Ultra Filtrated Lignins

M. sinensis was treated by two different methods (soda and organosolv treatments) in order to establish the influence of the treatment and the ultra filtration fractionation process on the resulting lignin characteristics.

The alkaline delignification (aqueous solution of 7.5% NaOH in weight) was carried out in a 20 L glass reactor during 90 min at 90 °C, using a solid/liquid ratio of 1:18 (SL lignins). Organosolv treatment consisted in the fractionation of *M. sinensis* with a water/ethanol mixture (40:60, v/v) with a solid/liquid ratio of 1:10, in a 4 L pressurized reactor (180 °C, 90 min) with constant stirring (OL lignins).

Organosolv and soda resulting liquors were ultra filtrated in order to obtain different liquid fractions that contained lignin with specific molar weight distributions. For this purpose, tubular ceramic membranes (provided by IBMEM – Industrial Biotech Membranes, Germany) with cut-offs of 5, 10, and 15 kDa were used. Liquors permeate between 5 and 10 kDa and between 10 and 15 kDa were collected.

Crude liquors and ultra filtration of the liquid fractions were treated to isolate the lignin by precipitation. Alkaline samples were acidified with sulphuric acid until reaching a pH below 2, filtered and finally the precipitate was washed twice with acidified water. To precipitate the lignin from organosolv liquors, two volumes of acidified water were added [70] after that the solid was filtered and washed. In all cases, the isolated lignins were vacuum dried at 40 °C before characterization.

Thermogravimetric Analysis

TG analysis results (Figure 17) showed three steps during samples degradation. The first one corresponded to the moisture removal, and the second one was related by several authors [57] with the degradation of hemicelluloses (between 200 and 300 °C) that can be present in lignin samples. The hemicelluloses degradation peaks were more pronounced for the alkaline lignins which entail high contamination. Lignin degradation, the third step, occurred progressively from 200 up to 450 °C. The maximum weight loss rate depended strongly on the lignin structure (functional groups and linkages), and therefore on the origin and the applied treatments. Minor differences could be observed for the lignins obtained by the diverse used treatments. The residue obtained after thermal degradation was also related with the structural complexity of the lignin molecule and its grade of linkages, i.e., aromatic polycondensations. In this way, high residue percentages imply high thermal stability of the sample [57], i.e., a complex three-dimensional structure. However, high residue values also indicate the presence of inorganic matter. For the analyzed samples it was found that lignin obtained from organosolv treatment showed a residue up to 20–25%, while alkaline samples reached 30–40%, due to the use of sodium hydroxide during raw material delignification.

As mentioned by Dizhbite *et al.* [50], one factor that strongly affected the radical scavenging activity of lignin was its purity and heterogeneity. Thus the presence of non-lignin components as hemicelluloses could diminish the antioxidant capacity of samples, since carbohydrates can generate hydrogen bonding with lignin phenolic groups, interfering in antioxidant properties of lignin. In this way, the alkaline processes [46] produce lignins with higher hemicelluloses contamination than organosolv treatments, because they presented more lignin–carbohydrate linkages. However, for ultra filtrated samples a decrease in the intensity of this peak in both soda and organosolv fractions were observed.

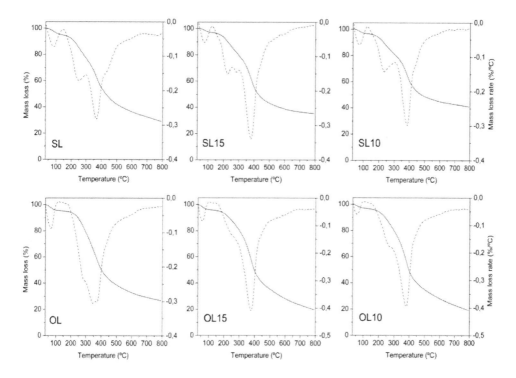

Figure 17. TGA (straight line) and DTG (dotted line) thermograms for the analyzed samples.

Size Exclusion Chromatography

The average number average (Mn), molecular weight (Mw) and polydispersity (Mw/Mn) of studied lignins are shown in Table 18. Significant differences in molecular weight distributions were found for the processes studied for the *M. sinensis* treatment. In this way, the Mw of crude lignins results were directly related to the treatment severity [30] since the soda treatment provided lignin with a higher average molecular weight and wider distribution (5654 g/mol and 3.0, respectively).

By using ultra filtration membranes, lignin fractions with low polydispersity were obtained for both organosolv and soda processes. According to previous studies [25, 70, 71], the molecular weight of lignin samples is related to the degree of condensation of lignin structure, i.e., to the amount of C–C bounds that conforms the aromatic skeleton. In this way, it is expected that lignins with high molecular weight contain more guaiacyl type units.

Table 18. Number-average ($\overline{M}_n$), weight-average ($\overline{M}_w$) molecular weight and polydispersity ($\overline{M}_w / \overline{M}_n$) of the different fractions obtained by ultra filtration

Lignin samples	$\overline{M}_n$	$\overline{M}_w$	$\overline{M}_w / \overline{M}_n$
SL	1885	5654	3.0
SL 10	963	2022	2.1
SL 15	3222	3544	1.1
OL	1147	2180	1.9
OL 10	1346	1750	1.3
OL 15	1357	1900	1.4

Phenolic Content Determination by Folin–Ciocalteu Method

In Table 19, the phenolic group content for the different lignin samples is shown. The result was expressed as equivalent grams of gallic acid per 100 g of analyzed lignin. It can be seen that the highest value of phenolic group corresponded to the lignin obtained by organosolv fractionation process. This fact can be related, as in the molecular weight determination, with the severity of the applied treatment during lignin isolation.

A slightly higher content of phenolic structures was detected for ultra filtrated samples in comparison with the crude ones. This could be explained by the large presence of impurities, dragged during the precipitation process, which can disguise the results of the analysis.

According to the study developed by Pan et al. [72], the use of delignification processes with high proton concentration promoted the loss of terminal groups in the lignin structure, increasing the phenolic hydroxyl group content. This functional group is the one that most affects the antioxidant capacity in lignins [73]. However, Pouteau et al. [40] found that low total hydroxyl content enhanced the antioxidant activity (measured by thermogravimetry) of lignin in polypropylene blends, due to a higher compatibility with the thermoplastic matrix.

Table 19. Phenolic content and antioxidante capacity of the studied lignins

Lignin samples	GAE / 100 g lignin	TAC (%)[a]		
		t = 0 min	t = 60 min	t = 120 min
SL	14.19	12.71	16.07	18.05
SL 10	16.05	14.18	18.64	18.56
SL 15	14.93	26.15	25.22	26.98
OL	17.62	33.38	31.97	43.05
OL 10	18.24	38.79	38.71	49.89
OL 15	19.68	43.93	46.79	54.83

[a] Percentage with respect to the reduction in absorbance observed for the DPPH with Trolox.

Thus, it might be expected that alkaline lignins, which presented the lowest phenolic group content (determined by the Folin–Ciocalteu method) than the other analyzed lignins, could be used as thermal stabilizer additive in polymer blends. The accuracy for this application was corroborated by the great stability observed during thermal behavior analysis of soda lignins, presented higher degradation temperatures.

Antioxidant Capacity of the Isolated Lignins

The results of the antioxidant capacity of the different lignins obtained at different analysis times are shown in Table 19 and Figure 18.

Concerning the isolation process used to obtain the lignins, it was found that the highest antioxidant capacity values obtained corresponded to the OL lignins; followed by SL samples. For the ultra filtrated lignin fractions it could be seen that lignins obtained with a cut-off of 15 kDa reached higher reduction in the DPPH radical absorbance than the 10 kDa samples.

In Figure 19, a dependence of the antiradical activity for the lignin on its molecular weight and polydispersity was found, corresponding to higher TAC values to less polydispersed lignins. Pan et al. [72] reported in their work that low molecular weight resulted from extensive depolymerization of lignin which led to the formation of new aromatic hydroxyl groups, the center to trap radicals.

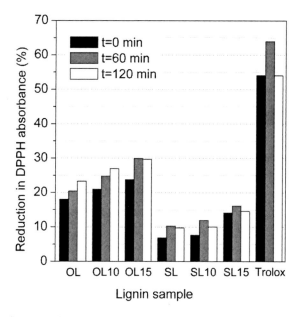

Figure 18. Reduction of DPPH mixture absorbance at 518nm at different reaction times for studied lignins and for the commercial antioxidant Trolox.

Taking into account that in the present work the antioxidant activity of ultra filtrated lignins was higher than the crude lignin, it could be affirmed that membrane technology is a suitable process for the production of high antioxidant capacity lignin fractions, since ultra filtration allowed for obtaining lignins with a narrow molecular weight distribution and lower carbohydrate contamination, as demonstrated by SEC and TGA analysis.

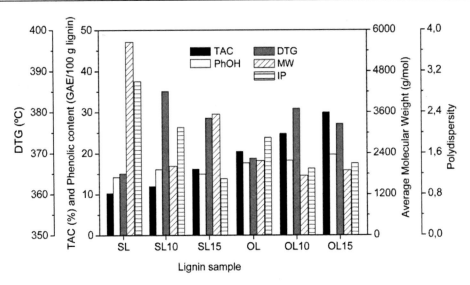

Figure 19. Influence of the diverse parameters studied (phenolic OH content, maximum degradation temperature, average molecular weight and polydispersity) on the antioxidant activity (TAC% at 60 min) of the analyzed lignins.

CONCLUSION

Using ultra filtration as a separation method, besides lignin separation from the other components of the liquid fraction, allows the fractionation into different lignin fractions with defined molecular weight that can be observed in the results by the size exclusion chromatography; 5 kDa and 15 kDa lignin fractions presented greater proportion of LMW.

Different structural analysis showed that the different lignin obtained fractions from *M. sinensis* by alkaline treatment are H:G:S type, with a higher content of G. Structurally, there is no clear differences between fractions since substructures such as β-O-4, β-5, guaiacyl units, syringyl units, monolignols (p-cumaric and ferulic acid) can be observed in all fractions.

The obtained fraction of more than 15 kDa is the most contaminated (mainly with lignin-carbohydrate complex, LCC). This was expected since more than 15 kDa fraction was the remainder liquid when dissolved compounds could not cross through the small pores of the membranes, and remained and precipitated after acid addition. Therefore, it can be concluded that ultra filtration is also a good purification technique for lignin fractions.

Lignin separation and purification was achieved by selective precipitation reaching the purest fractions at low pH values. While decreasing pH from 12.64 to 0.72, the amount of precipitate increased and black liquor color attenuated, giving a maximum value of precipitate at pH 0.72.

The lignin precipitated at pH 2.57 and 0.72 exhibited similar characteristics (FT-IR spectra, TG thermograms, Mw and ^{1}H-RMN) as commercial alkaline lignin. At other pH values obtained, lignin had more impurities mainly due to the presence of hemicelluloses and silicates in the raw material. It is possible to conclude that there are clear differences between the precipitates obtained at the different pH intervals, referring to its composition and, especially, to the content and the properties of obtained lignin.

Comparing both separation methods, it can be said that selective precipitation and ultra filtration were found to be effective techniques to fractionate and to extract lignin from the liquid fraction.

Ultra filtration showed better results as the lignin obtained was less contaminated with hemicelluloses as it was confirmed by the thermal analysis and FTIR spectra. Selecting the right cut-off of the membrane, the weight-average molecular weight could be controlled. Also slightly depolymerization was reached. Smaller lignin fractions (5 and 10 kDa) could be used as adhesives where low molecular weight and purity are important features. These lignin fractions could also be considered to be used as an antioxidant.

Differential precipitation was an easier and simpler technique, less energy consuming than ultra filtration and it does not require maintenance. However, this lignin had lower quality since its weight-average molecular weight cannot be controlled easily and the presence of lignin–carbohydrate complex (LCC) was higher. This type of lignin can be considered for use as dispersant and chelating agent.

Depending on the future use of the lignin, the right technique to obtain the fractions has to be chosen in order to obtain lignin with the appropriate characteristics suitable for the concrete application.

The antioxidant capacity of different ultra filtrated lignin was studied. Between the different lignin samples studied in this work, those obtained by organosolv fractionation process of *M. sinensis* presented the highest antiradical activity (15 kDa > 10 kDa > crude lignin), followed by alkaline samples. Thus, the effect of lignin extraction process and fractionation process –ultra filtration– on its antioxidant and physical properties was confirmed [74]. It has been proved that different extraction processes can affect the antioxidant capacity of the lignin obtained from the same raw material. The pulping process influence several properties of the obtained lignin, mainly its molecular weight distribution, the content on phenolic hydroxyl groups and its purity; all of them are factors that can determine the antioxidant activity of the lignin. The evaluation of different properties has allowed estimating their contribution on the antiradical scavenging of the lignin samples.

In this way, it affirms that, even though phenolic content is the main parameter that establishes the antioxidant capacity of a lignin sample, the molecular weight and the polydispersity are very influential features. The study has revealed that the antioxidant capacity of lignins, obtained by same treatment increased as the polydispersity of the lignin samples decreased. Using ultra filtration as a fractionation and separation method, the molecular weight and the polydispersity of lignin samples can be controlled and as a consequence, the antioxidant capacity of the ultra filtrated lignins.

REFERENCES

[1]   Directiva 2009/28/CE del Parlamento Europeo y del Consejo.

[2]   J. Zakzeski, P.C.A. Brujinincx, A.L. Jongerius, and B.M. Weckhuysen. The catalytic valorization of lignin for the production of renewable Chemicals. DOI: 10.1021/cr900354u.

[3]   S. Octave and D. Thomas (2009). Biorefinery: Towards an industrial metabolism. *Biochimie 91*, 659-664.

[4] A. Garcia, A. Toledano, L. Serrano, I. Egües, M. Gonzalez, F. Marin, and J. Labidi (2009). Characterization of lignins obtained by selective precipitation. *Separation & Purification Technol 68,* 193-198.

[5] M. Gonzalez (2009). Diseño de procesos de bio-refinería. Thesis Doctoral Universidad del País Vasco.

[6] J.M. Fang, R.C. Sun, J. Tomkinson, and P. Fowler (2000). Acetylation of wheat straw hemicellulose B in a new non-aqueous swelling system. *Carbohydrate Polymers 41,* 379-387.

[7] R.A. Northey (1992). Low-cost uses of lignin. *Proceedings of the Materials and Chemicals from Biomass. Symposium Series, Amer Chem Soc 476,* 146-175.

[8] S. Kubo and J.F. Kadla (2004). Poly(ethylene oxide)/organosolv lignin blends: Relationship between thermal properties, chemical structure, and blend behavior. *Macromolecule 37,* 6904-6911.

[9] D. Yang, X. Qiu, M. Zhou, and H. Lou (2007). Properties of sodium lignosulfonate as dispersant of coal water slurry. *Energy Conversion & Mgmt 48,* 2433-2438.

[10] M. Mulder (1996). *Basic Principles of Membrane Technology.* Kluwer Academic Publishers ISBN 0-7923-4247-X (HB), ISBN 0-7923-4248-8 (PB).

[11] *www.ibmem.com.*

[12] Z. Kovacs, M. Discacciati, and W. Samhaber (2009). Modeling of batch and semi-batch membrane filtration processes. *J Membrane Sci 327,* 164-173.

[13] R.W. Baker (2004). *Membrane Technology and Applications.* Wiley ISBN 0-07-135440-9.

[14] I. Tanistra and M. Bodzek (1998). Preparation of high-purity sulphate lignin from spent black liquor using ultra filtration and dia-filtration. *Desalination 115,* 111-120.

[15] A. Keyoumu, R. Sjodahl, G. Henriksson, M. Ek, G. Gellerstedt, and M.E. Lindstrom (2004). Continuous nano- and ultra filtration of Kraft pulping black liquor with ceramic filters. A method for lowering the load on the recovery boiler while generating valuable side-products. *Indust Crops & Products 20,* 143-150.

[16] A. Dafinov, J. Font, and R. Garcia-Valls (2005). Processing of black liquors by UF/NF ceramic membranes. *Desalination 173,* 83-90.

[17] M. Zabkova, E.A. Borges da Silva, and A.E. Rodrigues (2007). Recovery of vanillin from lignin/vanillin mixture by using tubular ceramic ultra filtration membranes. *J Membrane Sci 301,* 221-237.

[18] P.K. Bhattacharjee, R.K. Todi, M. Tiwari, C. Bhattacharjee, S. Bhattacharjee, and S. Datta (2005). Studies on ultra filtration of spent sulphite liquor using various membranes for the recovery of lignosulphonates. *Desalination 174,* 287-297.

[19] S. Bhattacharjee, S. Datta, and C. Bhattacharjee (2006). Performance study during ultra filtration of Kraft black liquor using rotating disk membrane module. *J Cleaner Production 14,* 297-504.

[20] O. Wallberg, A.-S. Jonsson, and R. Wimmerstedt (2003). Ultra filtration of Kraft black liquor with a ceramic membrane. *Desalination 156,* 145-153.

[21] O. Wallberg and A.S. Jonsson (2003). Influence of the membrane cut-off during ultra filtration of Kraft black liquor with ceramic membranes. *Trans. IChemE 81,* 1379-1384.

[22] A.-S. Jönnsson, A.-K. Nordin, and O. Wallberg (2008). Concentration and purification of lignin in hardwood Kraft pulping liquor by ultra filtration and nano filtration. *Chem Engng Res & Design 86* (11), 1271-1280.

[23] R. Schlesinger, G. Gotzinger, H. Sixta, A. Friedl, and M. Harasek (2006). Evaluation of alkali resistant nano filtration membranes for the separation of hemicellulose from concentrated alkaline process liquors. *Desalination 192*, 303-314.

[24] R. Vegas, A. Moure, H. Dominguez, J.C. Parajo, J.R. Alvarez, and S. Luque (2008). Evaluation of ultra- and nano filtration for refining soluble products from rice husk xylan. *Bioresource Technol 99*, 5341-5351.

[25] A. Tejado, C. Peña, J. Labidi, J.M. Echeverria, and I. Mondragon (2007). Physico-chemical characterization of lignins from different sources for use in phenol–formaldehyde resin síntesis. *Bioresource Technol 98*, 1655-1663.

[26] M.N. Mohamad Ibrahim, and S.B. Chuah (2004). Characterization of lignin precipitated from the soda black liquor of oil palm empty fruit bunch fibers by various mineral acids. *ASEAN J Sci & Technol Develop 21* (1), 57-67.

[27] D. Fengel, G. Wegener, Walter de Gruyter (Eds.), (1989). *Wood: Chemistry, Ultrastructure, Reactions*, Berlin and New York.

[28] J. RuncCang Sun, and J. Tomkinson (2001). Fractional separation and physico-chemical analysis of lignins from the black liquor of oil palm trunk fiber pulping. *Separation & Purification Technol 24*, 529-539.

[29] S.I. Mussatto, M. Fernandez, and I.C. Roberto (2007). Lignin recovery from brewer's spent grain black liquor. *Carbohydrate Polymer 70,* 218-223.

[30] R. Randhir, Y-T Lin, and K. Shetty (2004). Stimulation of phenolics, antioxidant and antimicrobial activities in dark germinated mung bean sprouts in response to peptide and phytochemical elicitors. *Process Biochem 39*, 637-646.

[31] B. Košiková, D. Slameňová, M. Mikulášová, E. Horváthová, and J. Lábaj (2002). Reduction of carcinogenesis by bio-based lignin derivatives, *Biomass & Bioenergy 23*, 153-159.

[32] M.P. Vinardell, V. Ugartondo, and M. Mitjans (2008). Potential applications of antioxidant lignins from different sources, *Indust Crops & Products 27*, 220-223.

[33] V. Ugartondo, M. Mitjans, and M.P. Vinardell (2008). Comparative antioxidant and cytotoxic effects of lignins from different sources, *Bioresource Technol 99,* 6683-6687.

[34] C. Pouteau, P. Dole, B. Cathala, L. Averous, and N. Boquillon (2003). Antioxidant properties of lignin in polypropylene, *Polymer Degrad & Stability 81*, 9-18.

[35] M. Canetti, F. Bertini, A. De Chirico, and G. Audisio (2006). Thermal degradation behavior of isotactic polypropylene blended with lignin. *Polymer Degrad & Stability 91*, 494-498.

[36] D.M. Fernándes, A.A. Winkler Hechenleitner, A.E. Job, E. Radovanocic, and E.A. Gómez Pineda (2006). Thermal and photochemical stability of poly(vinyl alcohol)/modified lignin blends. *Polymer Degrad & Stability 91*, 1192-1201.

[37] J.M. Cruz, E. Conde, H. Dominguez, and J.C. Parajó (2007). Thermal stability of antioxidants obtained from wood and industrial wastes. *Food Chem 100*, 1059-1064.

[38] J.S. Fabiyi, A.G. McDonald, M.P. Wolcott, and P.R. Griffiths (2008). Wood plastic composites weathering: Visual appearance and chemical changes. *Polymer Degrad & Stability 93*, 1405-1414.

[39] L. Rosu, C.N. Cascaval, and D. Rosu (2009). Effect of UV radiation on some polymeric networks based on vinyl ester resin and modified lignin. *Polymer Testing 28*, 296-300.

[40] C. Pouteau, B. Cathala, P. Dole, B. Kurek, and B. Monties (2005). Structural modification of Kraft lignin after acid treatment: Characterization of the apolar extracts

and influence on the antioxidant properties in polypropylene. *Indust Crops & Products 21*, 101-108.

[41]   A. Gregorová, B. Košiková, and R. Moravčik (2006). Stabilization effect of lignin in natural rubber. *Polymer Degrad & Stability 91*, 229-233.

[42]   A. Kurosumi, C. Sasaki, K. Kumada, F. Kabayashi, G. Mtui, and Y. Nakamura (2007). Novel extraction method of antioxidant compounds from *Sasa palmata* (Bean) Nakai using steam explosion. *Process Biochem 42*, 1449-1453.

[43]   D. Ye, D. Montané, and X. Farriol (2005). Preparation and characterization of methylcelluloses from *Miscanthus sinensis*. *Carbohydrate Polymer 62*, 258-266.

[44]   L. Serrano, I. Egües, M. González Alriols, R. Llano-Ponte, and J. Labidi (2010). *Miscanthus sinensis* fractionation by different reagents. *Chem Engng J 156*, 49-55.

[45]   G. Iglesias, M. Bao, J. Lamas, and A. Vega.(1996). Soda pulping of *Miscanthus sinensis*. Effects of operational variables on pulp yield and lignin solubilization. *Bioresource Technol 58*, 17-23.

[46]   J. Lora (2008). *Industrial Commercial Lignins: Sources, Properties and Applications. Monomers, Polymers and Composites from Renewable Resources.* Elsevier ISBN 978-0-08-045316-3.

[47]   W.G. Glasser, V. Dave, C.E. Frazier (1993). Molecular weight distribution of semi-commercial lignin derivatives, *J Wood Chem & Technol 13* (4), (1993) 545–559.

[48]   G. Vazquez, E. Fontenla, J. Santos, M.S. Freire, J. Gonzalez-Alvarez, and G. Antorrena (2008). Antioxidant activity and phenolic content of chestnut (*Castanea sativa*) shell and eucalyptus (*Eucalyptus globulus*) bark extracts. *Indust Crops & Products 28*, 279-285.

[49]   W. Brand-Williams, M.E. Cuvelier, and C. Reset (1995). Use of a free radical method to evaluate antioxidant activity. *Lebensm Wiss Technol 28*, 25-30.

[50]   T. Dizhbite, G. Telysheva, V. Jurkjane, and U. Viesturs (2004). Characterization of the radical scavenging activity of lignins—natural antioxidants. *Bioresource Technol 95*, 309-317.

[51]   Technical Association of the Pulp and Paper Industry, TAPPI. TAPPI Standard Test Methods. *www.tappi.org.*

[52]   L.E. Wise, M. Murphy, and A.A. D'Adieco (1946). Chlorite holocellulose, its fractionation and bearing on summative wood analysis and on studies on the hemicelluloses. *Paper Trade J 122*, (2), 35-43.

[53]   R. Rowell (1983). The chemistry of solid wood. Based on short course and symposium sponsored by the *Division of Cellulose, Paper and Textile Chemistry at the 185th meeting of the American Chemical Society*, Seattle, Washington, pp. 70-72.

[54]   J.A. Velásquez, F. Ferrando, X. Farriol, and J. Salvadó (2003). Binderless fiberboard from steam exploded *Miscanthus sinensis*. *Wood Sci & Technol 37*, 269-278.

[55]   J.H. Lora and W.G. Glasser (2002). Recent industrial applications of lignin -a sustainable alternative to nonrenewable materials. *J Polymers & Envir 10*, 39-48.

[56]   A. Tejado (2007). Modificación de matrices novolacadas por uso de compuestos naturales: caracterización físico-química de ligninas y estudio del curado y de las propiedades de resinas lignofenólicas. Thesis Doctoral, Universidad del País Vasco.

[57]   J.C. Dominguez, M. Oliet, M.V. Alonso, M.A. Gilarranz, and F. Rodriguez (2008). Thermal stability and pyrolysis kinetics of organosolv lignins obtained from *Eucalyptus globulus*. *Indust Crops & Products 27,* 150-156.

[58] E. Sjostrom (1981). Wood polysaccharides.In: Wood *Chemistry, Fundamentals and Applications*, Academic Press, New York pp. 51-67.

[59] N.E. El Mansouri and J. Salvado (2006). Structural characterization of technical lignins for the production of adhesives: Application to lignosulfonate, Kraft, sodaanthraquinone, organosolv and ethanol process lignins. *Indust Crops & Products 24*, 8-16.

[60] A.R. Gonçalves and P. Benar (2001). Hydroxymethylation and oxidation of organosolv lignins and utilization of the products. *Bioresource Technol 79*, 103-111.

[61] N. Vivas, M.F. Nonier, I. Pianet, N. Vivas de Gaulejac, and E. Fouquet (2006). Structure of extracted lignins from oak heartwood (*Quercus petraea Liebl., Q. robur L.*). *Comptes Rendus Chimie 9*, 1221-1233.

[62] J.C. del Río, J. Rencoret, G. Marques, A. Gutierrez, D. Ibarra, J.I. Santos, J. Jimenez-Barbero, L. Zhang, and A.T. Martinez (2008). Highly acylated (acetylated and/or p-coumaroylated) native lignins from diverse herbaceous plants. *J Agricult & Food Chem 56* (20), 9525-9534.

[63] F. Xu, J.-X. Sun, R.C. Sun, P. Fowler, and M.S. Baird (2006). Comparative study of organosolv lignins from wheat straw. *Indust Crops & Products 23*, 180-193.

[64] A.T. Martinez, J. Rencoret, G. Marques, A. Gutierrez, D. Ibarra, J. Jimenez-Barbero, and J.C. del Rio (2008). Monolignol acylation and lignin structure in some nonwoody plants: A 2D NMR study. *Phytochemistry 69*, 2831-2843.

[65] W.G. Glasser and R.K. Jain (1993). Lignin derivatives I. Alkanoates. *Holzforschung 47*, 225-233.

[66] A. Abacherli, R.J.A. Gosselink, E. de Jong, S. Baumberger, B. Hortling, C. Bonini, M. D'iAuria, F. Zimbardi, D. Barisano, J.C. Duarte, G. Sena-Martins, B. Ribeiro, E. Koukios, D. Koullas, E. Avgerinos, C. Vasile, G. Cazacu, R. Mathey, D. Ghidoni, G. Gellerstedt, J. Li, P. Quintus-Leino, S. Piepponen, A. Laine, P. Koskinen, J. Gravitis, J. Suren, and M. Fasching (2005). Intermediary status of the round robins in the Eurolignin network, *Proceedings of the ILI 7th Forum*, Barcelona, pp. 119-124.

[67] D. Feldman, D. Banu, J. Campanelli, and H. Zhu (2001). Blends of vinylic copolymer with plasticized lignin: Thermal and mechanical properties. *J Appl Polymer Sci 81*, 861-874.

[68] H. Yoshida, R. Mörck, K.P. Kringstad, and H. Hatakeyama (1987). Kraft lignin in polyurethanes I. Mechanical properties of polyurethanes from a Kraft lignin-polyether triol-polymeric MDI system. *J Appl Polymer Sci 34*, 1187-1198.

[69] T. Hatakeyama, K. Nakamura, and H. Hatakeyama (1978). Differential thermal analysis of styrene derivatives related to lignin. *Polymer 19*, 593-594.

[70] M. González, A. Tejado, M. Blanco, I. Mondragon, and J. Labidi (2009). Agricultural palm oil tree residues as raw material for cellulose, lignin, and hemicelluloses production by ethylene glycol pulping process. *Chem Engng J 148*, 106-114.

[71] A. García, A. Toledano, L. Serrano, I. Egües, M. González, F. Marin, and J. Labidi (2009). Characterization of lignins obtained by selective precipitation. *Separation & Purification Technol 68*, 193-198.

[72] X. Pan, J.F. Kadla, K. Ethara, N. Gilkes, and J.N. Saddler. Organosolv ethanol lignin from hybrid poplar as a radical scavenger: Relationship between lignin structure, extraction conditions, and antioxidant activity. *J Agricult & Food Chem 54*, 5806-5813.

[73] B. Verma, P. Huclb, and R.N. Chibbar (2009). Phenolic acid composition and antioxidant capacity of acid and alkali hydrolyzed wheat bran fractions. *Food Chem 116*, 947-954.

[74] H. Nadji, P.N. Diouf, A. Benaboura, Y. Bedard, B. Riedl, and T. Stevanovic (2009). Comparative study of lignins isolated from Alfa grass (*Stipa tenacissima L.*). *Bioresource Technol 100*, 3585-3592.

In: Lignin
Editor: Ryan J. Paterson
ISBN 978-1-61122-907-3
© 2012 Nova Science Publishers, Inc.

Chapter 5

PRODUCTION OF BIO-PHENOLS AND BIO-BASED PHENOLIC RESINS FROM LIGNIN AND LIGNOCELLULOSIC BIOMASS

Chunbao (Charles) Xu,[1,4,] Shuna Cheng,[2] Zhongshun Yuan,[1] Mathew Leitch,[2] and Mark Anderson[3]*

[1]Department of Chemical Engineering, Lakehead University,
Thunder Bay, ON, Canada
[2]Faculty of Natural Resources Management, Lakehead University,
Thunder Bay, ON, Canada
[3]Research and Technology , Arclin, Mississauga,
ON, Canada [4]Department of Chemical and Biochemical Engineering, University of
Western Ontario, London, ON, Canada

ABSTRACT

Lignin is the second most abundant polymer in nature and a major by-product of the paper industry. Crude lignin is also generated as a waste stream in the organosolv delignification process and the steam explosion process for cellulosic ethanol production. There is an increasing interest in utilizing lignin as a potential raw material for the chemical industry in the past two decades, due to the its immense quantities produced annually and the depletion of fossil fuels as well as the environmental concerns associated with the use of fossil resources. Lignin is an amorphous macromolecule of three phenyl-propanols i.e., p-hydroxyl-phenyl propanol, guaiacyl-propanol and syringyl-propanol. These phenyl-propanols are linked mainly by condensed linkages (e.g., 5-5 and β-1 linkages), and more dominantly by ether linkages (e.g., β-O-4 and α-O-4) between the three main lignin building blocks. This macromolecule can decompose/degrade into oligomeric and monomeric phenolic compounds via thermochemical technologies (such as pyrolysis, hydrothermal liquefaction, and hydrolysis) and some biological processes. Lignin is thus attractive to many industries such as the manufacture of phenol-formaldehyde (PF) resins and adhesives, as the bio-phenols derived from lignin could

---

* Corresponding author: Fax: 1-807-343-8928; E-mail: cxu@lakeheadu.ca.

potentially substitute for petroleum-based phenol in the resin synthesis. This chapter provides an overview of chemistry of PF resin synthesis, liquefaction of lignocellulosic materials and production of bio-oil-based PF resins, and the recent advances in lignin extraction, modification and de-polymerization.

1. INTRODUCTION

Lignin is the second most abundant naturally synthesized polymer after cellulose, present in forest/agricultural biomass. Agricultural biomass is typically composed of 40 wt.% cellulose, 20-25 wt.% hemicellulose and 10-20 wt.% lignin, and forestry biomass contains 40-45 wt.% cellulose, 15-35 wt.% hemicellulose and 20-35 wt.% (40-50 wt.% for barks) lignin (S'anchez, 2009; Research Note FPL-091, 1971). Lignin is produced in large quantities (estimated at 50 million tons per year) as a by-product (e.g., the alkali or kraft lignin, and lignosulfonates) from the chemical pulping processes in the pulp/paper industry. Crude lignin is also generated as a waste stream in the organosolv deligninfication process and the steam explosion process for cellulosic ethanol production. Lignin has traditionally been viewed as a waste material or used predominantly as a solid fuel at a value of ~$200/t for heat generation in the recovery boilers in pulp/paper mills (Stewart, 2008). Roughly only 1-2% of lignin is isolated from pulping liquors and used for specialty products (Lora and Glasser, 2002). However, 60-70% of North American kraft mills are recovery boiler limited, so precipitating and utilizing lignin for the production of value-added bio-products can enhance the mill capacity and the mill economy. In North America alone, the potential of precipitated lignin was estimated at 1.5 Mt/year (Schmidt and Laberge, 2008).

Lignin is a natural polymer of three phenyl-propanols, i.e., p-coumaryl alcohol (p-hydroxyl-phenyl propanol), coniferyl-alcohol (guaiacyl-propanol) and sinapyl-alcohol (syringyl-propanol) (Tejado et al, 2007). Lignin is an amorphous three-dimensional phenyl-propanol polymer as illustrated in Figure 1 (Desch, and Dinwoodie, 1996; Tsoumis, 1991). As displayed in the Figure, the phenyl-propanols are linked mainly by two types of linkages between the three main lignin building blocks: condensed linkages (e.g., 5-5 and β-1 linkages), and more dominantly ether linkages (e.g., α-O-4 and β-O-4) (Chakar and Ragauskas, 2004). Under certain conditions (such as hydrolysis, pyrolysis, and liquefaction), lignin can degrade or depolymerise into oligomers and even the respective mono-phenolic monomers.

In today's society, petroleum prices are high, petroleum resources are depleted coupled together with an increased demand for petroleum from the developing economies and the increasing environmental concerns over the fossil-based resources. There is an intensified interest in researching and developing alternative renewable (non-petroleum) resources for energy and chemical production. Lignin can be a renewable source for a variety of valuable chemicals/products currently produced from petroleum, in particular phenolic products (resins, adhesives, polyesters, polyurethanes, carbon fibres, plastics, surfactants) (Sudan, 2003; Angles et al., 2003). Replacing petroleum-based phenol totally or partially with lignin or solid lignocellulosic materials for the manufacture of phenol-formaldehyde (PF) resins has attracted significant interests due to the economic and environmental benefits (Alma and Basturk, 2006).

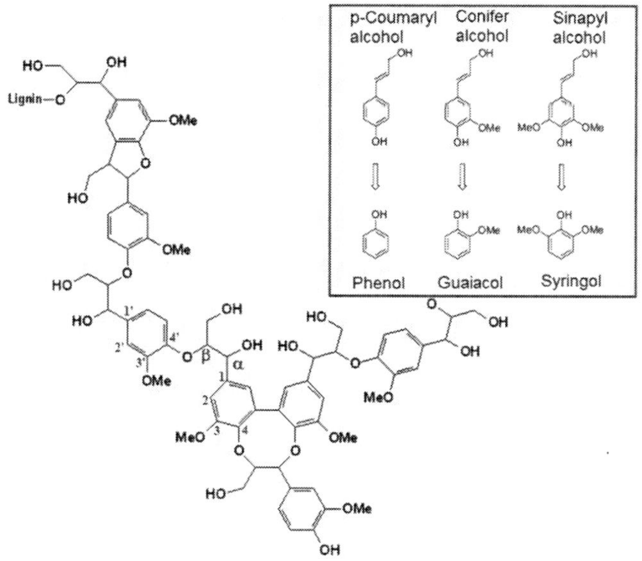

Figure 1. Chemical structure of a piece of lignin molecule and three monomers of lignin.

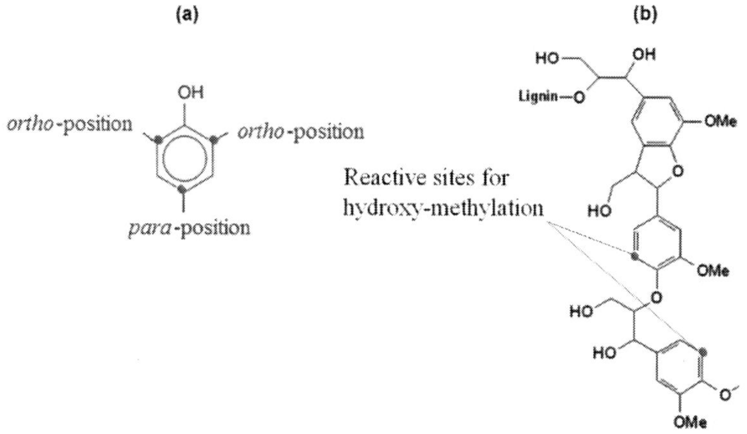

Figure 2. The reactive sites of phenol (a) and lignin fragments (b) for the PF resin synthesis reactions.

Phenol formaldehyde (PF) resins are most widely used as wood adhesives in the manufacturing of engineered wood products such as plywood, laminated veneer lumber, particle boards, oriented strand board (OSB) and fiberboards. Phenol formaldehyde (PF) resins are also used in different industrial products such as insulation, coated abrasives, paper saturation, and floral foam. The PF resin manufacture is a $ 2.3 billion industry in North America. Phenol is primarily produced from petroleum-derived benzene through the cumene process, which makes it the most costly feedstock for the production of PF resins. The production of phenol-formaldehyde resins consumes approximately 35-40% of the phenol production in the USA. Due to the fluctuating price of petroleum-based phenol (currently at about $700-1,500 per metric ton), the cost of PF resins varies from US$ 1,500 to US$ 2,000 per metric ton. The production of phenolic compounds from less expensive and renewable sources such as biomass is therefore of great interest.

Lignin has been successfully used to directly replace petroleum-based phenol in phenolic resin synthesis, but the replacement ratio is generally less than 30-50% due to the lower reactivity of lignin compared with pure phenol (Cetin and Ozmen, 2002(a); Effendi et al, 2008; Wang et al., 2009 (a)). Direct use of lignin as a substitute for phenol in PF resins is however limited because lignin has a lower reactivity due to its fewer reactive sites and the steric hindrance effects resulting from its complex chemical structure (Alonso et al, 2004; Çetin and Özmen, [2002](a)), as illustrated in Figure 2. Therefore, although being a valuable, inexpensive and abundant renewable source for phenol, lignin has found limited practical applications so far.

In the past 10 to 15 years, extensive research efforts were made in utilizing lignin for the synthesis of PF resins as wood adhesives (Wang et al., 2009 (a); Vázquez et al., 1997), and coating/molding materials (Park et al., 2008; Tejado et al., 2007; Perez et al., 2009). Lignin-modified resins, with up to 35% phenol replacement by lignin, have been widely used in the USA to bond fibreboard and plywood. Research work continues to find different routes to produce adhesive resins with 50% or a greater replacement of phenol with biomass derived phenolic compounds (Peng and Reidl, 1994). Among different routes investigated, the following two proved to be very effective:

1) Degrade the macromolecular lignin to smaller molecules of oligomers or monomers by liquefaction and de-polymerization (Thring, 1994);
2) Introduce reactive sites into lignin molecules by such process as phenolation (Matsushita and Yasuda, 2003).

It was found that β-ether and α-ether bonds in the β-O-4 and α-O-4 linkages in lignin are fairly easy to cleave, while the 5-5 (biphenyl)-type and aromatic rings structures were more stable (Tsujino et al., 2003; Minami et al., 2003). Several chemical degradation methods including oxidative (with the presence of oxidants of O_2 or H_2O_2), reductive (with the presence of reducing agents such as H_2 and/or formic acid), acidic (Matsushita et al., 2008[)] and basic hydrolytic process have been studied to break down lignin polymers to oligomers or monomers, as will be overviewed in details in the following sections of this chapter. Irrespective of what de-polymerization process was employed, although the molecular weight and steric hindrance were reduced significantly, the degraded lignin products were mainly phenol derivatives whose ortho and para positions are occupied by alkyl substitute groups such as methoxy groups. This greatly reduced the chance for the phenol derivatives reacting with formaldehyde in the synthesis of PF resins. Accordingly, modification of the degraded lignin products is still necessary to incorporate reactive sites into the phenolic molecules for the synthesis of bio-based PF resins with a high substitution ratio for phenol. Matsushita and Yasuda (2003) employed phenolation by introducing phenol to acid lignin to improve its reactivity at 60°C for 6 h for the purpose of synthesis of ion exchange resin. Okuda et al. (2004) and Wang et al. (2009b) employed direct de-polymerization/liquefaction of lignin/biomass in a water–phenol mixture at elevated temperatures (300-400°C), in which phenol acted as a capping agent to react with the degraded lignin intermediates to prevent re-polymerization. In a recent research by the authors' group, a basic catalyst (such as NaOH) was used to hydrolytically degrade alkaline lignin of a very high molecular weight in a hot-compressed water-ethanol medium at a much lower temperature (220-300°C) with phenol as the capping agent (Yuan et al., 2010). The degradation products, oligomers of significantly lower molecular weights with more reactive functionality, could be a promising phenolic feedstock for the synthesis of bio-based PF resins with a high substitution ratio of phenol. More recently, the authors' group de-polymerized an alkali lignin (M_n ~ 10,000) in a hot compressed alcohol solution at approx. 300 °C, producing bio-phenols with M_n <500 at a yield of >90 wt.%. These results will be discussed in more details in the later part of this chapter.

2. CHEMISTRY OF PHENOL-FORMALDEHYDE RESIN SYNTHESIS

Phenol-formaldehyde (PF) resin was the first synthetic polymer commercialized (Pizzi and Walton, 1992). It has become one of the most widely utilized synthetic polymers since Baekeland developed a commercial manufacturing process in 1907 (Wendler and Frazier, 1995). The global production and consumption of PF resins in 2009 were approximately 3.0 Mt. It is expected an average growth of 3.9% per year from 2009 to 2014, and 2.9% per year from 2014 to 2019 (SRI consulting, 2010). The global PF resin market value is about $4.5-6 billion per year. The PF resin manufacture is an important industry valued approx. $10 billion in the world, and $ 2.3 billion in North America.

By varying the catalyst type and the formaldehyde (F) and phenol (P) molar ratio, two classes of PF resin can be synthesized: resoles (resols) and novolaks (novolacs). Resols are synthesized under basic conditions with excess formaldehyde (i.e. F/P>1); novolacs are synthesized under acidic conditions with excess phenol (i.e. F/P<1) (Pizzi and Walton, 1992). Resols and novolacs are inherently different: resols are heat curable (thermosetting) while novolacs (thermoplastic) require addition of a crosslinking agent such as hexamethylenetetramine (HMTA) to cure. For most novolacs, this additional step results in slower cure rates and lower crosslinking than resols (Wendler and Frazier, 1996 (a)).

PF resins were first introduced as binders for particleboard and plywood in the mid 1930's (Wendler and Frazier, 1996 (b)); they have since become one of the most important thermosetting adhesives in the wood composites industry, especially for exterior applications. In 1998, PF resins comprised approximately 32% of the total 1.78 million metric tons of resin solids consumed in the North American wood products industry (Ni and Frazier, 1998). Almost all PF resins currently used in wood bonding applications are resols. PF resols are desirable for exterior applications due to their rigidity, weather (moisture) resistance, chemical resistance and dimensional stability (Wendler and Frazier, 1996 (b)). PF resols, in either a liquid or a spray-dried form, are currently used mainly as binders for the manufacture of plywood and oriented strand board (OSB) (Wendler and Frazier, 1996 (a)). Compared to polymeric diphenylmethane diisocyanate (PMDI), another binder currently used in OSB manufacturing in North America, PF resols have the advantage of low cost, good thermal stability and reasonably fast cure.

2.1. Chemistry of PF Resol/Novolac Resin Synthesis

PF resols are poly-condensation products of phenol and formaldehyde in an alkaline aqueous medium with excess formaldehyde (F/P >1.0). Formaldehyde is often used in the form of an aqueous solution (formalin) during commercial production of PF resols. PF resols used as wood binders are typically synthesized at 80-90°C with a formaldehyde/phenol (F/P) ratio of 1.6 to 2.5, with a final resin solids content of 40-60% and a final pH value of 9-12 (Bao et al., 1999; Rosthauser et al., 1997). The most commonly used catalyst in the manufacture of commercial resols is sodium hydroxide (NaOH). Besides its catalytic effect, sodium hydroxide also improves the solubility of PF resols in aqueous solution, which allows resols to be synthesized with a high degree of advancement for fast curing, while maintaining good processability (less viscous). PF resol synthesis is a step growth polymerization process comprising two steps: addition and condensation.

Addition Reactions

In an aqueous solution, formaldehyde is present in its hydrated form, i.e., methylene glycol. In an alkaline medium, phenols are deprotonated to form phenoxide ions as shown in Figure 3a. The electron rich ortho and para positions in phenoxide ions are susceptible to electrophilic aromatic substitution (often referred to as addition). It was found that the para position is more reactive toward electrophilic aromatic substitution than is the ortho position ascribing to the less severe stereo hindrance at the para position (Lin-Gibson and Riffle, 2003).

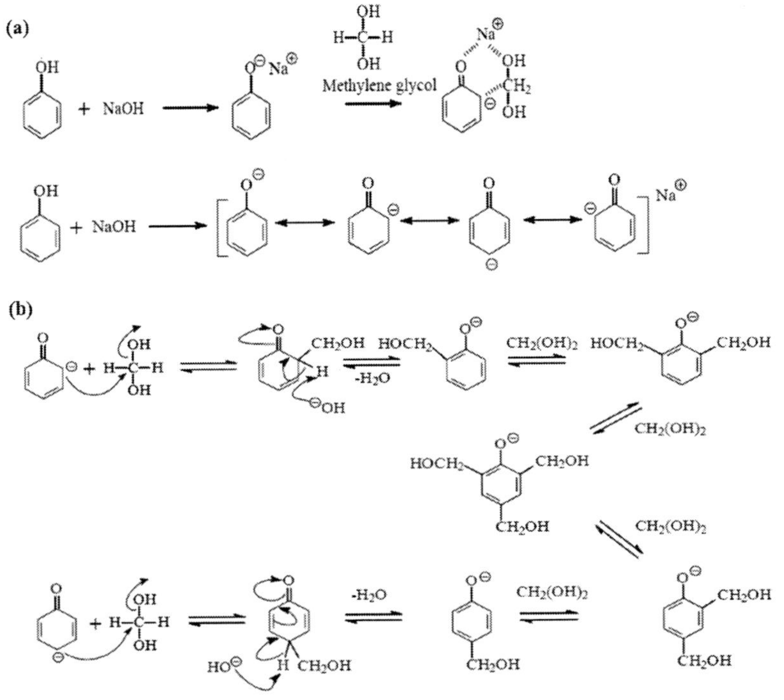

Figure 3. Addition reactions for the PF resol resin synthesis: formation of mono-, di- and tri-substituted HMPs (Lin-Gibson and Riffle, 2003).

Also, the hydroxymethylation at the ortho positions was found to further increase the reactivity of the para positions (Zhou and Frazier, 2001). This was explained by the formation of intra-molecular hydrogen bonds that lead to a higher negative charge on the para position. When sodium hydroxide is used as the catalyst, a chelate structure may form, which favors ortho-substitution (Haider et al., 2000). It was also found that the mono-substituted hydroxymethylated phenols (HMPs) react with methylene glycol to form di- and tri-substituted HMPs before the slower condensation reaction could occur (Lin-Gibson and Riffle, 2003). The formation of mono-, di- and tri- substituted HMPs is illustrated in Figure 3b.

Condensation Reactions

The step growth polymerization of HMPs is a water-producing condensation. However, as will be shown, this condensation occurs through electrophilic aromatic substitution. It is generally agreed that HMPs condense through quinone methide (QM) intermediates (Papa and Critchfield, 1979; Zhuang and Steiner, 1993; Haider et al., 2000). The formation of a quinone methide from HMP is shown in Figure 4a. Quinone methides are strong electrophiles and will readily substitute onto other electron rich phenoxides to form methylene bridges (Figure 4b). The remaining hydroxymethyl groups will be linked via methylene bridges upon curing (Papa and Critchfield, 1979). In contrast to novolaks, dimethylene ether-type of bridges are scarce in resols, and may only be present as short-lived intermediates that will convert to the more stable methylene linkages upon cure via heating (Figure 4c) (Zhuang and Steiner, 1993; Kim and Watt , 1996).

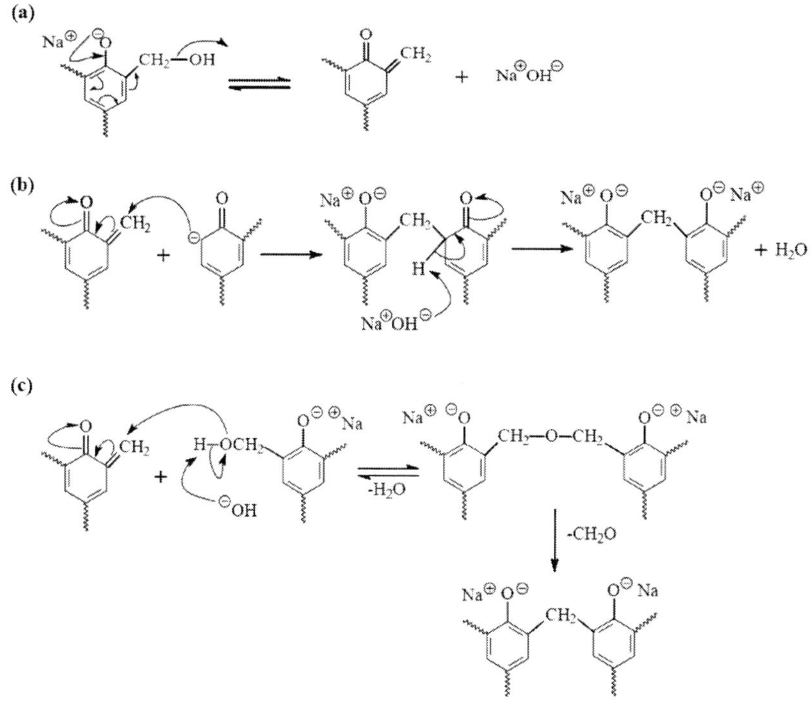

Figure 4. The step growth polymerization of HMPs via water-producing condensation (Haider et al, 2000).

Similarly to the above mechanism of the PF resol resin synthesis, novalac PF resin synthesis is also via electrophilic aromatic substitution, as illustrated in Figure 5.

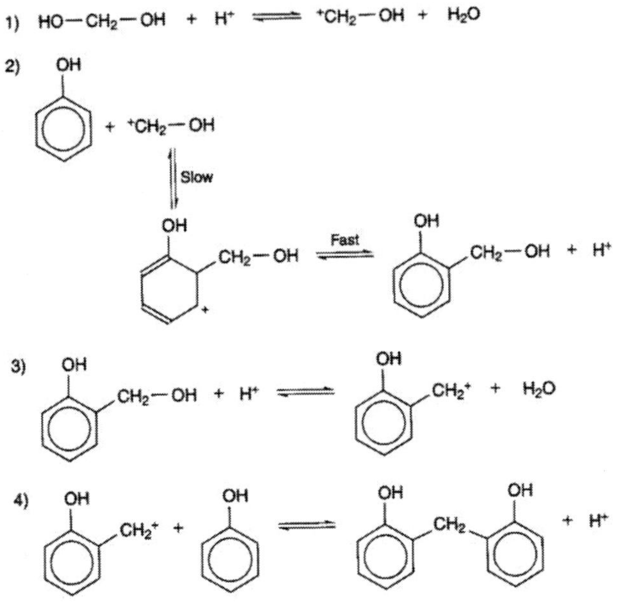

Figure 5. Mechanism of novalac PF resin synthesis (Lin-Gibson and Riffle, 2003).

2.2. PF Resol Resin Cure Chemistry

In the wood-based composites industry, PF resols are usually cured to the insoluble and infusible resitol stage (the intermediate stage of cured phenolic resins) by applying heat and pressure. The resol cure process is essentially a continuation of the electrophilic aromatic substitution through quinone methide intermediates as it occurs in the condensation stage (Zhuang and Steiner, 1993; Schaefer et al, 1977). Dimethylene ether- bridges were found to be negligible in the cured PF resols (Zhuang and Steiner, 1993; Larsson et al, 1997; McBrierty and Packer, 1993). However, besides methylene bridges, additional cross-linking mechanisms have been proposed for the cured resols. These additional crosslinking reactions were hypothesized to directly involve phenolic hydroxyls (phenoxides) and methylene bridges (McBrierty and Packer, 1993; McBrierty and Douglass, 1981). Evidence suggests that the ether structures form between phenolic hydroxyls (phenoxides) and hydroxymethyl groups (Figure 10a), the crosslinks generated between methylene bridges by free formaldehyde (Figure 10b), and the crosslinks formed from the condensation between methylene carbons and hydroxymethyl groups (Figure 10c) are all of importance in cured resols (Schaefer et al., 1977).

In summary, the chemistry of resol synthesis and cure is very complex, and a full mechanistic understanding is still lacking. However, the wood adhesives industry has been successful for tailoring PF resols to make their properties suitable for various wood bonding applications. Presently, the cure rate of resols can be dramatically increased to compete with other fast-curing adhesives such as PMDI. Some improvements developed include modifying resin cook procedures and adding cure accelerators.

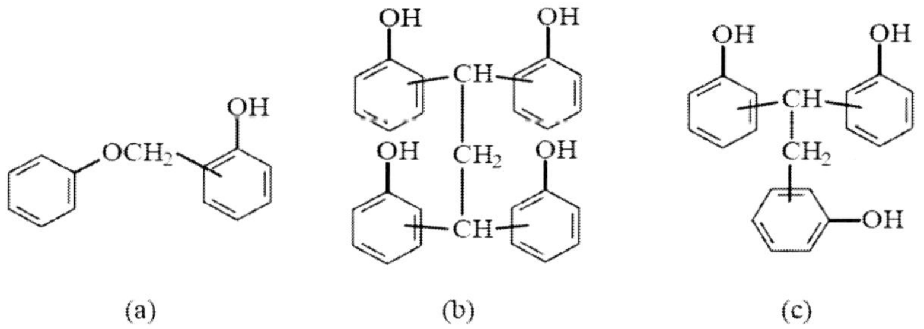

Figure 6. Cross linked structures in cured resols besides methylene bridges (Schaefer et al., 1977).

3. Liquefaction of Lignocellulosic Materials and Production of Bio-Oil-based PF Resins

3.1. Pyrolysis Technologies and Applications of Pyrolysis Oils in PF Resins

In order to overcome the lower reactivity of lignin and to further increase the ratio of lignin/biomass substitution, an effective and promising approach may be liquefaction of biomass/lignin through thermo-chemical processes such as pyrolysis. Depending on the operating conditions, pyrolysis processes could be classified into two major types: fast pyrolysis and vacuum pyrolysis. Fast pyrolysis is so far the only industrially realized thermo-chemical process for biomass liquefaction. A fast pyrolysis process typically operates at > 500°C in an inert atmosphere under a very high heating rate with a short vapor residence time of less than 2 seconds, producing liquid products (pyrolysis oil or bio-oil) at a yield as high as 70-80% (Bridgwater and Peacocke, 2000; Bridgwater, 2004). Compared to fast pyrolysis, vacuum pyrolysis, operating under reduced pressure with a much longer vapor residence time, produces a much lower liquid yield (Effendi et al., 2008). Pyrolysis processes cleave the primary and secondary bonds in both lignin and high molecular weight polysaccharides (cellulose and hemicellulose) into lower molecular weight compounds, e.g. guaiacyl compounds (4-hydroxyl-3-methoxy-phenylpropanones) and syringyl compounds (4-hydroxy-3, 5-dimethoxyphenylpropanones).

Pyrolysis oil is a complex mixture of water (15-30 wt.%) and organics of a high oxygen-content (45-50 wt.%). This oxygen content distributes in more than 300 identified compounds which consist of hydroxyaldehydes, hydroxyketones, sugars, carboxylic acids and phenolics (Bridgwater, 2004). The majority of the phenolics are present as oligomers containing varying numbers of acidic, phenolic and carboxylic acid hydroxyl groups as well as aldehyde, alcohol and ether functions. Typical chemical compositions of pyrolysis oils obtained from vacuum pyrolysis of softwood bark residues are shown in Table 1 (Chan et al., 2002). Pyrolysis oils are unstable upon storage and require further upgrading, which will result in a high cost. They have mainly been used as an energy source by direct combustion. The liquid product from pyrolysis of lignocellulosic materials, i.e., bio-oil or pyrolysis oil, can be separated into two fractions, water-soluble fraction and water-insoluble fraction, based on the differences in water solubility and density by decanting or solvent extraction: The water-insoluble fraction that usually constituies 25-30 wt.% of the whole bio-oil is often called pyrolytic lignin because it is composed of much oligomeric fragments derived from thermal degradation of the native lignin (Radlein et al., 1987; Meier and Scholze, 1997). Although without being commercialized so far, the use of pyrolytic lignin as a source of phenol in PF resins seems to be promising. Applications of bio-oil in the production of bio-based phenolic resins are briefly overviewed as follows.

Table 1. Typical chemical compositions of pyrolysis oils obtained from vacuum pyrolysis of softwood bark residues (Chan et al, 2002)

Family of compounds	Concentration (wt.%, d.b.)
Hydrocarbons	3
Sugars	9
Low-molecular weight-acids	1.5
High-molecular weight-acids	10
Alcohols	2.5
Esters and ketones	4
Phenols	10
Steroids and triterpenoids	4
Lignin/tannin-based compounds	48
Labile compounds	8
Total	100

The U.S. National Renewable Energy Laboratory (NREL) has been actively exploiting phenolic compounds from sawdust and bark through a controlled pyrolysis technology. Research at the NREL demonstrated that it is possible to produce phenolic resins from wood for about half the cost of those produced from fossil fuels (NREL 2003). Chum and co-workers (Chum et al., 1993; Chum et al., 1989; Chum and Kreibich, 1992) synthesized resins of both novolac and resole types using fast pyrolysis bio-oils derived from softwood, hardwood and bark residue to partially substitute phenol. The reactive phenolic components and neutral fractions were fractionated from fast pyrolysis oil for resin synthesis. The bio-based resins were evaluated and tested for wood panel production. The reactive fraction separated from pyrolysis oil consists of mainly phenolic compounds and aldehyde groups. Accordingly, the reactive fraction was not only to replace phenol but also some of the formaldehyde used during the resin synthesis process. Himmelblau from Biocarbon Corporation patented a technology of producing PF resins using fast pyrolysis bio-oils from mixed hardwood (maple, birch, beech) after removing of non-reactive fraction from the oils (Himmelblau, 1991; Himmelblau and Grozdits, 1999). The whole phyrolysis oil from wood and bark was also investigated as feedstock for phenolic resins (Nakos et al., 2001). Although the wood adhesive properties of the bio-based resins were normally inferior to the commercial resins with 100% petroleum-based phenol, higher phenol substitution was considered to likely be possible with a fraction enriched in reactive phenolics. Ensyn Group Inc. developed patented methods of preparing a natural resin from wood, bark, forestry residues and wood industry residues using a patented rapid pyrolysis process (Giroux et al, 2001, 2003), as described below. To produce the natural resin, pyrolyis oil may partly replace phenol, or both phenol and formaldehyde. The properties of wood adhesives using synthesized PF resin comprising up to 60 wt.% the natural resin precursor were demonstrated to be comparable to those of a commercial control.

A vacuum pyrolysis oil was used to replace 25-35 wt.% of phenol in PF resin formulations (Chan et al. 2002). The resin was tested as adhesive for OSB. The mechanical properties such as modulus of elasticity (MOE), modulus of rupture (MOR) and internal bond (IB) strength, were investigated using 3-layer boards (11.1mm thick) bonded with the bio-based phenol formaldehyde resin, and it was found that the bio-based phenolic resin (25-35wt.% phenol replacement) exceeded the minimum requirements set by CSA 0437 Series 93, both in dry and wet (2-hr. boil) tests. Higher concentrations of the bio-based resins (50 wt.% phenol substitutions) in the formulations lowered the performance of the bio-based phenolic resin in terms of cure kinetics and the extent of condensation (Amen-Chen et al, 2002 (a)). The curing behavior of bio-based resins could be improved by addition of a small amount of polypropylene carbonate (at 0.5-1.5 wt.%). The addition of polypropylene carbonate however did not have an apparent effect on the mechanical properties of OSB (Amen-Chen et al, 2002 (b)). Mourant et al. (2009) reported a vacuum pyrolysis process, performed at 450°C and at a total pressure of 20 kPa, to produce vacuum pyrolysis bio-oil from a bark mixture (balsam fir-70% and white spruce-30%), and the obtained vacuum pyrolysis bio-oil was employed as a phenol replacement in PF resins at a high ratio up to 50-85 wt.%.

3.2. Hydrothermal/Solvolytic Liquefaction and Application of Bio-Oils in PF Resins

Direct liquefaction of lignocellulosic materials in a suitable solvent at relatively lower temperatures (<400°C), is known as solvolytic liquefaction or hydrothermal liquefaction (if water is used as the reaction medium). The hydrothermal/solvolytic liquefaction processes are considered to be more advantageous than the pyrolysis processes with respect to the energy efficiency and the quality of the liquid products. As such, the hydrothermal/solvolytic liquefaction technologies for direct liquefaction of lignocellulosic materials have attracted an increasing interest for the production of bio-phenol precursors for the synthesis of bio-based phenolic resins and heavy oil (bio-crude) for the bio-fuel production. In terms of the type of solvent and operating conditions employed in the liquefaction processes, solvolytic liquefaction processes may be grouped into two categories: (1) solvolysis under low temperatures and atmospheric pressure, and (2) solvolysis in a hot-compressed and sub-/supercritical fluid.

3.2.1. Solvolysis under Low Temperatures and Atmospheric Pressure

Lignocellulosic materials can be efficiently liquefied at a low temperature and atmospheric pressure in the presence of phenol with a proper catalyst such as an acid (hydrochloric acid, sulfuric acid, phosphoric acid, oxalic acid) (Alma et al., 1995; Lin et al., 2000; Lin et al., 1994; Lin et al., 1995; Alma and Basturk, 2006; Alma et al., 1998; Alma and Kelley, 2000), an alkaline solution or a medal salt (NaOH, Cu(OH)$_2$, NaHCO$_3$, NH$_4$Cl, AlCl$_3$, CuSO$_4$, FeSO$_4$, NaH$_2$PO$_4$) (Maldas and Shiraishi, 1997(a), Maldas et al, 1997(b)). The resulting phenolated products were widely used as feedstocks for the synthesis of phenolic resins (Lin et al, 1995; Alma and Basturk, 2006; Alma and Kelley, 2000; Maldas and Shiraishi, 1997(a)). Strong acids, such as concentrated sulfuric acid has been used widely as an effective catalyst in the phenolysis liquefaction of lignocellulosic materials, and the resulting liquefied products could be used as matrix resins with satisfactory mechanical properties (Hassan and Mun, 2002; Russell and Riemath, 1985). However, disadvantages of the processes using strong acids include partial carbonization of biomass during the liquefaction process and corrosion of the equipment. Moreover, the use of phenol as the liquefaction solvent could cause some problems, such as a high cost of the solvent, and the difficulty in recycling of phenol from the liquefied products as well as some environmental concerns.

Other solvents instead of phenol, such as alcohols (Shiraishi et al, 1992; Hassan and Shukry, 2008) and cyclic carbonates (Yamada and Ono, 1999) and co-solvent (Pasquini et al, 2005) (e.g. alcohols and water) were tested as solvents for low-temperature liquefaction of biomass. These low boiling-point solvents could be recycled easily for reuse by distillation/evaporation after liquefaction and are much cheaper than phenol. Hassan and Shukry (2008) investigated the liquefaction of bagasse and cotton stalks by using polyhydric alcohols (polyethylene glycol and glycerin) in the presence of sulfuric acid as a catalyst at temperatures ranging from 140 to 170°C under atmospheric pressure. The residue for both bagasse and cotton stalks were less than 10 wt.%. In addition, ionic liquids (ILs) - liquids at or below 100°C that are comprised entirely of cations and anions, have been developed over the past decades as green solvents with tunable solvation properties (Earle and Seddon, 2000). Recently, ionic liquids were employed as solvents to liquefy natural polymeric materials of cellulose and starch (Swatloski et al., 2002), lignin (Jiang et al., 2007) and wood (Xie and Shi, 2006). Xie and Shi (2006) reported that *Metasequoia glyptostroboides* (Dawn Redwood) could be effectively liquefied at 120°C for 25 min with imidazole-based ionic liquid as the liquefaction solvent, being more efficient than the conventional phenol/H_2SO_4 system. However, in a more recent study by Jiang et al. (2007), only up to 20% of lignin could be dissolved in the 1-butyl-3-methylimidazolium-based ILs. Another major challenge for the application of ILs for biomass conversion may be the difficulty in recovering of the expensive ILs from the reaction products.

3.2.2. Solvolysis in Hot-Compressed and Sub-/Supercritical Fluids

The liquid bio–oil products from the above mentioned low-temperature solvolytic processes using phenol or alcohols are however much less reactive than petroleum-based phenol in the synthesis of phenolic resins due to their high molecular weights and shortage of mono-phenolic compounds with unoccupied ortho and para positions (Effendi et al, 2008). The liquid products from the low-temperature solvolytic processes are rich in large molecules of oligomers derived from lignin and cellulosic components. In this regard, liquefaction of lignocellulosic materials at a high temperature is preferable as the oligomers derived from lignin and cellulosic components could be degraded further into products of smaller molecular weights at a higher temperature with the aids of suitable solvents and catalysts. Numerous studies have been reported on the liquefaction of biomass at a relatively higher temperature and under high pressure. For instance, agricultural residues, peat moss, sawdust and wood chips were liquefied at 290-350°C and 10-20 MPa with water as the solvent and sodium carbonate or calcium carbonate as the catalyst, and the extracted reactive phenolic content was used for preparing phenolic resin (Russell and Riemath, 1985). The bond strength of birch plywood using their bio-based phenol formaldehyde resins was found to be superior to that of a commercial control. Wood wastes (Monarch birch, *Betula maxim owiczina Regel*) were liquefied with phenol in the presence of NaOH at 250°C in a pressure-proof tube placed in an oil bath for 1h (Alma et al, 1997; Alma et al, 2001). The liquefied phenolic oil product was resinified with formaldehyde, and the resulting resol-type resins were applied as adhesives to the production of plywood. The bonding results indicated that almost all of these resin adhesives could meet the Japanese Industrial Standard as far as dry shear adhesive strengths of plywood were concerned.

Generally, the properties of liquefied wood are dependent on the liquefaction conditions such as reaction temperature, time, and catalyst used (Alma et al, 1995; Lin et al, 1994; Alma et al, 1996). More recently, there has been a huge surge of interest in using hot-compressed and sub-supercritical fluids for biomass liquefaction. Supercritical fluids have found applications for the chemical conversion of lignocellulosic materials due to their unique properties, e.g., they possess unique transport properties (gas-like diffusivity and liquid-like density) and supercritical fluids have the ability to dissolve materials not normally soluble in either liquid or gaseous phase of the solvent, and hence to promote the gasification/liquefaction reactions. Hot-compressed or sub-critical water was used by many researchers for biomass liquefaction (Matsumura et al, 1999; Karagoz et al, 2004; Karagoz et al, 2005; Xu and Lad, 2008). Karagoz et al. (2005) studied the oil compositions produced from pine sawdust, rice husk, lignin and cellulose which were liquefied at 280°C for 15 min in water. The main compounds of oils from pine sawdust and rice dusk were phenolic compounds, while rice husk derived oil consisted of more benzenediols than sawdust derived oil. Tymchyshyn and Xu (2010) investigated hydrothermal liquefaction of lignocellulosic wastes (sawdust and cornstalks) and two model biomass compounds (pure organosolv lignin and pure cellulose as references) in hot compressed water at temperatures from 250°C to 350°C in the presence of 2 MPa H_2, for the production of phenolic compounds that may be suitable for the production of green phenol-formaldehyde resins. The liquefaction operations at 250°C for 60 min produced the desirable product of phenolic/neutral oil at a yield of about 53 wt.%, 32 wt.%, 32 wt.% and 17 wt.% for lignin, sawdust, cornstalk and cellulose, respectively, as displayed in Figure 7. Significant quantities of phenolic compounds such as 2-methoxy-phenol, 4-ethyl-2-methoxy-phenol, and 2, 6-dimethoxy-phenol, were present in the resulting phenolic/neutral oils from the two lignocellulosic wastes and the pure lignin.

Sub-/supercritical phenol (Lee and Ohkita, 2003) was used for liquefaction of wood meal (Birch, *Betula maximomawiczii Regel*) in a pressure-proof autoclave. A very high yield (up to about 95%) of liquid products was achieved within 2 min at 421°C with phenol/wood weight ratio of 3, where combined phenol content was about 75%. Sub-/super-critical alcohols have been tested for liquefaction of lignocellulosic materials (Xu and Etcheverry, 2008; Miller et al, 1999). An advantage of using alcohols as the solvent for biomass liquefaction is that these alcohols themselves are renewable as they can be produced from biomass. In addition, alcohols have a higher solubility for the liquid products derived from cellulose, hemicelluloses, and lignin due to their lower dielectric constants when compared with that of water (Yamazaki et al, 2006). Xu and Etcheverry (2008) reported that a high yield of bio-crude (rich in phenolic compounds) up to 60 wt.% was obtained by hydro-liquefaction of Jack pine sawdust in sub-/super-critical ethanol at 200-350°C under H_2 of cold pressure of 2.0-10.0 MPa without and with iron-based catalysts (FeS or $FeSO_4$). Co-solvent of alcohol and water was found to be more effective in liquefaction of lignocellosic materials at temperatures about 300°C, as reported by Minami and Saka (2005) and Cheng et al. (2010).

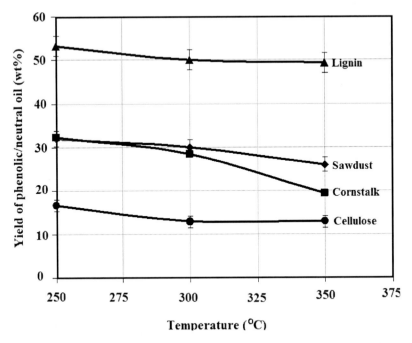

Figure 7. Phenolic/neutral oil yield from various lignocellulosic materials during hydrothermal liquefaction in hot-compressed water for 60 min at various temperatures (modified from Tymchyshyn and Xu (2010)).

More than 95 wt.% of Japanese beech (*Fagus crenata* Blume) wood was liquefied in water-methanol (10%/90% v/v) at 350°C for 5 min. Guaiacyl and syringyl lignin were decomposed to coniferial alcohol-γ-methyl ethers and sinapyl alcohol-γ-methyl ethers and further converted to isoeugenol and 2,6-dimethoxy-4-(1-propenyl) phenol after a long treatment, respectively. More recently, the authors' group demonstrated that the use hot-compressed co-solvent of water-ethanol (50%/50%, w/w) achieved a very high biomass conversion (up to 95 wt.%) and a high yield of bio-oil (66 wt.%) by liquefaction of Eastern white pine (*Pinus strobus L.*) sawdust at 300°C for 15 min (Cheng et al., 2010). The obtained phenolic bio-oil produced was further used to substitute for phenol in the synthesis of bio-oil phenol formaldehyde (BPF) resol resins as adhesives for 3-ply yellow birch plywood. The addition of bio-oil to substitute for phenol at a high level (up to 75 wt.%) could still produce bio-based resol resins useful as plywood adhesives, ascribing to the low molecular weights of the obtained phenolic bio-oil (with an M_w of 1072 g/mol and an M_n of 342 g/mol). All the obtained BPF resins possess broader molecular weight distributions, but similar chemical/thermal properties compared with a conventional phenol formaldehyde (PF) resol resin (or 0 wt.% BPF resin) as the reference. As displayed in Figure 8, the dry tensile strengths of the plywood samples bonded with the BPF resins at up to 75 wt.% phenol substitution exceeded or were comparable to that of the conventional pure PF resin adhesive. All the BPF resin-bonded plywood samples showed comparable wet tensile strengths to that with the conventional PF adhesive too.

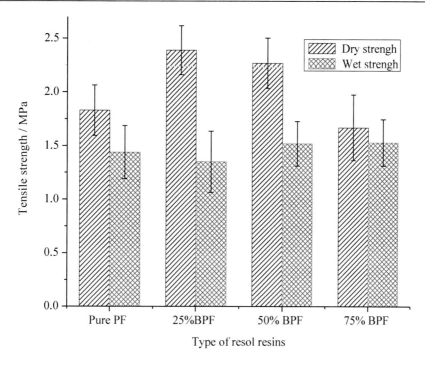

Figure 8. Tensile strength of plywood using BPF and pure PF resol resins as adhesives.

4. LIGNIN EXTRACTION, MODIFICATION AND DEPOLYMERIZATION

4.1. Lignin Extraction

Two types of lignins have been produced from lignocellulosic materials: sulfur-containing lignins and sulfur-free lignins. The sulfur-containing lignins include Kraft lignin and lignosulfonates which are by-products present in the black liquor from the pulping processes. The sulfur-free lignins, including soda, organosolv, and steam-explosion, oxygen delignification and hydrolysis lignins, are mainly from the bio-fuel production processes. 80% of chemical pulping lignins come from the widely used Kraft process (El Mansouri and Salvado, 2006; Kirk-Othmer, 1999-2008). Lignin can be separated from the black liquor through the acid precipitation approaches using H_2SO_4, CO_2 or by some electrochemical methods. For example, the following experimental procedure for lignin extraction from black liquor was proposed by Olivares et al. (1988). The black liquor normally with pH >13 was first acidified using H_2SO_4 to pH 8-9.5 at temperature 75-85°C and filtrated while hot. The filtrate was further acidified to pH 2, followed by final filtration and washing and drying to give highly pure lignin. This procedure is advantageous since it allows recycling of a solution rich in Na^+ and SO_4^{2-} to the kraft process. However, the Kraft pulping process has long-standing problems with the air/water pollution and the odour issues (Wegener, 1992).

A better method to extract lignin from lignocellulosic materials should have following characteristics: less environmental pollution, low energy consumption, easy operation, easy solvent recovery and ease of product separation. To this end, many organic solvents have been tested to fractionate wood into its components: lignin and carbohydrates. These include alcohols (methanol, ethanol, propanol, etc.), alcohol-water, organic acid (formic acid and acetic acid, and other organic acid and salts of organic acid like), phenol and cresols, ethyl acetate, amines and amine oxides (ethylenediamine, hexamethylene diamine, methylamine, ethanolamine, amine oxides), ketones and diaxane, etc. (Muurinene, 2000). Aqueous ethanol (or ethanol-water mixture) could penetrate readily into the structure of wood resulting in uniform delignification (Kleinert, 1975). As reported by Li et al. (1988), red spruce was effectively de-lignified in ethanol-water mixture at 190°C and 29 MPa. Pasquini et al. (2005) also described an effective organosolv delignification process at around 190°C using ehanol-water mixture and CO_2 for *Pinus taeda* wood chips. In addition to alcohol-water mixture, formic acid was found to be a potential good solvent for separating lignin and extractives from wood, and it could also result in hydrolytic breakdown of lignin polymers into smaller and more soluble molecules (Sundquist, 1996).

4.2. Lignin Modification for the Production of PF Resins

Crude lignin has a much lower reactivity than pure phenol because it has less reactive sites for reacting with formaldehyde (Alonso et al, 2004; Clarke and Dolenko, 1978; Lee et al, 2000). Many researchers have attempted to modify lignin in order to improve its reactivity through various approaches such as methylolation (Clarke and Dolenko, 1978; Lawson and Kleln, 1985), phenolation (Cetin and Ozmen, 2002(a); 2002(b)), and demethylation (Olivares et al, 1988; Narayamusti and George, 1954; Hayashi et al, 1967; Shashi et al, 1982). Clarke and Dolenko (1978) reported a highly cross linked methylolated Kraft lignin was obtained by stirring a mixture of Kraft lignin and formaldehyde with pH value 12-12.5 at room temperature for a period of at least three days. Then the methylolated Kraft lignin and phenol formaldehyde (with the addition amount of at least 5 wt.% of the total resin solids) were combined to produce the lignin-phenol-formaldehyde (LPF) at a pH of 6-7 for a sufficient time. The LPF resin was applied as adhesive to 3-ply poplar plywood. The plywood samples bonded with the methylolated LPF resins had dry bond strength of 2.1 MPa (wood failure 40%) and wet strength of 1.7 MPa (wood failure 60%), compared with a dry strength of 1.6 MPa (0% wood failure) and no wet strength for those with the LPF resin using unmethylolated Kraft lignin. Cetin and Ozmen (2002 a and b) described the production of LPF resin using phenolated organosolv lignin, and the properties of particle board using this LPF resin. In the first step, phenolated lignin was prepared by adding lignin into a mixture of phenol and industrial methylolated spirit (IMS) at desired weight ratios at 70°C under stirring for a sufficient time. The IMS was removed by evaporation under vacuum at 40-50°C. The remaining product, i.e., the phenolation lignin, was subsequently mixed with formaldehyde (35%) solution to synthesize LPF resole resin at pH 10.2 at 50°C for 1h. Then the temperature was increased up to 80°C and kept at the temperature until the desired viscosity was achieved. The final viscosity (200-250 cP) of LPF resin was obtained by removing excess water under vacuum at 40-50°C. The LPF resin was used for the production of particleboards whose physical and mechanical properties were tested, such as internal bond strength (IB), modules of rupture (MOR), modulus of elasticity (MOE)). The results showed that the mechanical properties of phenolated-lignin-formaldehyde resins with 30% phenol replacement were comparable to those of a commercial PF resin. Lignin demethylation can be an effective measure to increase lignin reactivity, and the demethylation of lignin could be achieved by reaction with potassium dichromate ($K_2Cr_2O_7$) (Narayamusti and George, 1954; Hayashi et al, 1967; Shashi et al., 1982). Using $K_2Cr_2O_7$ in the presence of acetic acid, the concentration of methoxyl groups in a lignin decreased from 11.9 to 2.8%, accompanied by an increase in the phenolic hydroxyl groups from 2.4 to 11.2% (Shashi et al., 1982). As per Olivares et al. (1988), for demethylation of 100 g of Kraft lignin or alkali lignin, the lignin were dispersed in 2000 ml water in a reactor at 40°C with moderate agitation; 750 ml of 0.033M $K_2Cr_2O_7$ together with 30 ml of acetic acid was added, in order to maintain a pH of 4. After 2 h, lignin was separated by precipitation at pH 2. As such, the lignin demethylation process using $K_2Cr_2O_7$ has a big challenge in wastewater disposal and the high acid consumption in the process.

4.3. Lignin Depolymerization for the Production of PF Resins

Crude lignin has a much lower reactivity not only because it has less reactive sites compared with pure phenol for reacting with formaldehyde as discussed before (Alonso et al., 2004; Clarke and Dolenko, 1978; Lee et al., 2000), but also because the reactive positions of the macromolecules of lignin or its fragments have less accessibility to formaldehyde due to the steric hindrance.

It is thus an effective way to improve lignin reactivity for the synthesis of bio-based phenolic resins through degradation/de-polymerization of lignin into monomeric bio-phenols or oligomers of reduced molecular weights (Alma et al., 2001; Lee and Ohkita, 2003). Some model lignin compounds, such as, guaiacol, guaiacol-β-guaiacol, veratrote (or diphenyl ether), 2,6-dimethoxyphenol, and 1,2,3-trimethoxybenzen, biphenyl and 5-5, β-1, β-O-4, and α-O-4 types of dimeric lignin fragments, were used as a probe into the depolymerization mechanism of lignin in supercritical fluids of solvents such as water and methanol (Lawson and Kleln, 1985; Tsujino et al., 2003; Minami et al., 2003). The influence of supercritical water on guaiacol pyrolysis was examined at 383°C. Neat guaiacol pyrolysis produced catechol and char, and a small amount of phenol and o-cresol. However, guaiacol hydrolysis led to catechol and methanol, and the selectivity to which was found to be an increasing function of water density (Lawson and Kleln, 1985). Reaction behavior of lignin in supercritical methanol (250-270 °C, 24-27 MPa) was investigated by using lignin model compounds (Tsujino et al., 2003; Minami et al., 2003). It was found that β-ether linkage in β-O-4 model compounds and α-ether linkage in α-O-4 model compounds could be easily cleaved, but 5-5 (biphenyl)-type and aromatic rings structures were much stable as expected. Recombination of obtained highly reactive radicals to form solid residues would be the major challenge for this approach.

Several chemical degradation processes have been developed to break down lignin polymers to oligomers or monomers. These include oxidative processes (with the presence of oxidants such as O_2 or H_2O_2), reductive processes (with the presence of reducing agents such as H_2 and/or formic acid), acidic (Matsushita et al., 2008[)] and basic hydrolytic processes.

4.3.1. Oxidative/Reductive Depolymerization

Oxidative degradation of lignin in the presence of a catalyst such as a noble metal (Crestini et al, 2006) or peroxidise (Hofrichter, 2002) and an oxidant (H_2O_2) usually produce benzaldehydes, ketones, and benzoic acids which are not suitable for the synthesis of PF resins, but they can be used for perfumes and pharmaceutical intermediates.

In order to prevent the recombination, hydrogenation using hydrogen as an reductive agent could be a solution based on Kleinert's work (Kleinert and Barth, 2008 (a)), suggesting that hydrogenation led to a higher yield of monomeric phenols and less char formation. The reactive hydrogen could be obtained from solvents medium or gaseous hydrogen in combination with a suitable catalyst. Attempts have been made to hydro-treat lignin over alumina-supported NiMo and CoMo catalysts (Piskorz et al., 1989; Oasmaa et al, 1993; Thring et al, 2000). It was recently reported by Yan et al. (2008) that wood lignin (using pine birch sawdust as the lignin source) could be effectively degraded into monomers of guaiacylpropane, guaiacylpropanol, syringylpropane and syringylpropanol in hot-compressed water at 200°C for 4 h under 4MPa (cold pressure) H_2 with carbon-supported Pt or Ru catalysts, where the biomass-to-solvent ratio was fixed at about 1:15 (wt/wt). The yield of total mono-phenols was as high as 45% of the total amount of C_9 units in the lignin when co-solvent of dioxane/H_2O (1:1 wt/wt) with 1 H_3PO_4 (1 wt.% of the solvent) was used combined with Pt/C catalyst (5wt.% of the sawdust). The application of hydrogen-donating solvents such as Tetralin, 9,10-dihydroanthracene (AnH$_2$) and its derivatives and 1,4,5,8,9,10-hexahydroanthracene, have been widely used as effective hydrogen donors for the liquefaction of coal (Shen and Iino, 1994; Li et al, 2003), but they are yet to be tested for lignin de-polymerization. Recently, the authors' group performed research on depolymerization of lignin using hot-compressed solvents of ethanol, water and ethanol-water mixtures. In the study, an alkali lignin was depolymerized with ethanol-water co-solvents of varying ethanol contents from 0 to 100% at 300 °C for 4hr in the presence of H_2. The yields of degraded lignin (DL), solid residue (SR), gas products, and the GPC results of DLs were listed in Table 2. As shown in Table 2, the mono-solvent of either pure ethanol or pure water was less effective for lignin conversion and DL yield, compared with the ethanol-water co-solvents. 50% ethanol solution was found to be the most effective solvent for lignin degradation in terms of the DL yield and lignin conversion (referring to the SR yield). The lower activity of the 100% ethanol for lignin depolymerization than that of 100% water may be predictable due to the poor solubility of the alkali lignin in pure ethanol and probably the limited hydrolysis reactions in the system with pure ethanol. The addition of water as a co-solvent into ethanol would promote lignin de-polymerization by increasing the lignin solubility in the reaction medium and hydrolysis of the lignin by water. As seen from Table 2 the alkali lignin was greatly depolymerized from M_w 60,000 and M_n 10,000, to a degraded lignin product with M_w 1010 and M_n 415 in the 50% aqueous ethanol solution, and to a DL product with M_w 631 and M_n 260 in 100% ethanol.

Table 2. Effect of composition of ethanol-water co-solvents on Alkali lignin de-polymerization at 300°C in the presence of H$_2$

Degradation conditions	Solvent	DL[1] / %	SR[2] / %	GP[3] / %	GPC results[4]		
					M$_n$	M$_w$	M$_w$/M$_n$
Raw lignin	–	–	–	–	~10,000	~60,000	~6.0
300 °C, 240min, 5 MPa H$_2$, 1g lignin, 10 ml solvents	100% water	48.6 (±4.4)	41.7 (±2.6)	1.6 (±0.3)	NA[5]	NA	NA
	25% water-75% ethanol	48.6 (±1.7)	35.7 (±1.8)	8.0 (±0.1)	NA	NA	NA
	50% water-50% ethanol	87.8 (±2.7)	5.7 (±3.3)	6.4 (±0.7)	415	1010	1.95
	75% water-25% ethanol	52 (±3.8)	34.3 (±1.5)	6.5 (±2.6)	NA	NA	NA
	100% ethanol	15.3 (±1.6)	64.1 (±0.8)	3.6 (±0.7)	260	631	2.43

[1]DL– Degraded lignin products;
[2]SR – Solid residue;
[3]GP – Gaseous products;
[4]The GPC results were from the THF-soluble portion of the DLs;
[5]NA – not analyzed.

There is another family of solvent such as formic acid and 2-propanol, which are less thermally stable and will decompose to give hydrogen upon being heated at elevated temperatures. For example, formic acid decomposes completely into hydrogen, and carbon dioxide, and 2-propanol can be decomposed into hydrogen and acetone by heating. Recently, this type of hydrogen-donating solvent has found special applications in hydro-conversion of biomass and lignin. As reported by Kleinert et al. (Kleinert and Barth, 2008 (a), 2008 (b); Kleinert et al, 2009; Kleinert and Barth, 2008 (c); Kleinert and Barth, 2007 (a); 2007 (b)), formic acid and 2-propanol were used as hydrogen donators for de-polymerize and hydrogenate lignin. Kleinert and co-workers obtained a high yield (25-35 wt.%) of phenolic fraction, exclusively composed of monoaromatic phenols with alkylation ranging from C$_1$–C$_7$ in the side chain(s), by one-step conversion of lignin to oxygen-depleted bio-fuels and phenols using co-solvent of formic acid and ethanol at about 400°C. The yield was 2 or 3 times that of an earlier work by Dorrestijn et al. (1999) using AnH$_2$ for de-polymerization of wood lignin at 352°C. A recent preliminary research by the authors' research group demonstrated that at >350°C hot-pressed ethanol-formic acid solvent with a Pt or Ni catalyst could effectively de-polymerize lignin into low molecular phenols (Mn: 143, Mw: 247 determined by GPC analysis). The research work in this regard continues in the authors' research lab.

4.3.2. Hydrolytic Depolymerization

As a common process of biomass acid hydrolysis, hydrolytic degradation of lignin can be achieved using 72% sulfuric acid at 60-80°C (Matsushita and Yasuda, 2005) to produce sulfuric lignin. This lignin can then be used to synthesis ionic exchange resins. The main drawback of acidic hydrolysis includes unavoidable repolymerization between phenol reactive sites and the α-position of phenol propanol and the waste disposal problem for sulfuric acid. To address the above problems of the acid hydrolysis, aqueous alkaline de-polymerization of lignin was carried out in a 5% NaOH solution at a temperature of 180°C for 6 hours with anthraquinone as a co-catalyst (0.5 wt %) (Nenkova et al., 2008). A variety of monophenol and phenyl carbonyl compounds in the liquid products were identified by GC-MS. The de-polymerization of Kraft- and organosolv-derived lignins by KOH in supercritical methanol or ethanol at 290°C was studied in rapidly heated batch micro-reactors (Miller et al, 1999). High conversions were realized and the conversion was fast, reaching the maximum value within 10–15 min. The dominant de-polymerization route is the solvolysis of the ether linkages. A drawback of this method was its high yield (up to 7%) of solid residues due to re-polymerization and condensation of the reaction intermediates and the liquid products. The Kraft process is also an alkaline de-polymerization process with re-polymerization occurring (Chakar and Ragauskas, 2004). Usually a capping agent such as sulfur or sulfite is used to react with degraded intermediates.

In a recent research by the authors' group (Yuan et al., 2010), alkaline lignin of a very high molecular weight was successfully degraded into oligomers in a hot-compressed water-ethanol medium with NaOH as the catalyst and phenol as the capping agent at 220-300°C. Under the optimal reaction conditions, i.e., 260°C, 1 hour, with the lignin/phenol ratio of 1:1 (w/w), almost complete degradation was achieved, producing <1% solid residue and negligible gas products. The obtained degraded lignin had a number-average molecular weight M_n and weight-average molecular weight M_w of 450 and 1,000 g/mol respectively, significantly lower than the M_n and M_w of 10,000 and 60,000 g/mol of the original lignin. A higher temperature and a longer reaction time favoured phenol combination, but increased the formation of solid residue due to the condensation reactions of the degradation intermediate/products.

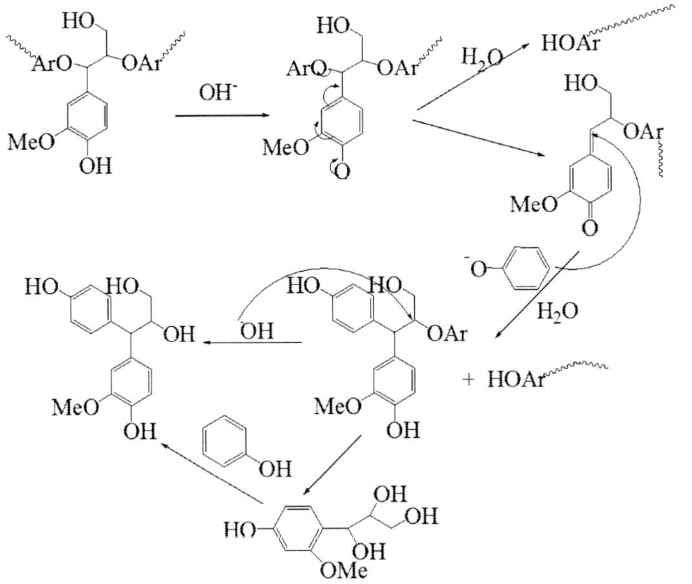

Figure 9. Possible hydrolytic degradation mechanisms of alkaline lignin catalyzed by NaOH (Yuan et al., 2010).

The degraded lignin products were soluble in organic solvents (such as THF), and were characterized by HPLC/GPC, IR and NMR. A possible mechanism for lignin hydrolytic degradation was also proposed in this study.

SUMMARY

Lignin, an important component of lignocellulosic biomass, can be a potential feedstock for the production of aromatic/phenolic chemicals, and resins/adhesives. Conversion of lignin and lignocellulosic biomass into valuable chemicals such as bio-phenols for the production of bio-based phenolic resins have attracted increasing interest, motivated by the abundant resources of natural lignocellulosic biomass and lignin and the relatively high price of petroleum-based phenol.

Direct use of lignin to substitute for phenol in PF resins is however limited because lignin has a much lower reactivity than phenol due to its fewer reactive sites and the steric hindrance effects resulting from its complex chemical structure. Degradation of the macromolecular lignin to smaller molecules of oligomers or monomers by liquefaction and de-polymerization proved to be an effective approach to improve the reactivity of the lignin for the production of bio-based phenolic resins. Biomass and lignin can be effectively liquefied by two thermochemcial processes: pyrolysis process and hydrothermal/sovolytic liquefaction process.

To date, fast pyrolysis is the only industrially realized process for biomass liquefaction. However, pyrolysis oil is highly complex, unstable, and it requires further upgrading. The hydrothermal/solvolytic liquefaction processes, using solvents of water, alcohols, ionic liquids and phenol, etc., are considered to be more advantageous than the pyrolysis processes with respect to the energy efficiency and the quality of the liquid products.

As such, the hydrothermal/solvolytic liquefaction technologies have attracted an increasing interest for the production of bio-phenol precursors for the synthesis of bio-based phenolic resins and heavy oil (bio-crude) for the bio-fuel production. It has been demonstrated that the use hot-compressed co-solvent of water-ethanol (50%/50%, w/w) achieved a very high biomass conversion (up to 95 wt.%) and a high yield of bio-oil (66 wt.%) by liquefaction of pine sawdust at 300°C for 15 min.

The obtained phenolic bio-oil produced was successfully used to substitute for phenol in the synthesis of bio-oil phenol formaldehyde resol resins as adhesives for 3-ply yellow birch plywood. The addition of bio-oil to substitute for phenol at a high level (up to 75 wt.%) could still produce bio-based resol resins useful as plywood adhesives, ascribing to the low molecular weights of the obtained phenolic bio-oil.

As another effective way to improve lignin reactivity for the synthesis of bio-based phenolic resins, degradation/de-polymerization of lignin into monomeric bio-phenols or oligomers could be achieved chemically via the oxidative processes (with the presence of oxidants such as O_2 or H_2O_2), reductive processes (with the presence of reducing agents such as H_2 and/or formic acid), and acidic/basic hydrolytic processes.

ACKNOWLEDGMENTS

The authors are grateful for the financial support from the Ontario Ministry of Agriculture, Food and Rural Affairs (OMAFRA) through the New Directions Research Program Grant, and from the industry partners of Arclin Limited and GreenField Ethanol. The authors would also like to acknowledge the financial support from FedNor, NOHFC and NSERC.

REFERENCES

Alma, M.H., Basturk, M.A. (2006). Liquefaction of grapevine cane (Vitis vinisera L.) waste and its application to phenol–formaldehyde type adhesive. *Ind. Crops Prod.* 24, 171–176.

Alma, M.H., Yoshioka, M., Yao, Y., Shiraishi, N. (1995). Characterization of the phenolated wood using hydrochloric acid as a Catalyst. *Wood Sci. Technol.* 30, 39–47.

Alma, M.H, Yoshioka, M., Yao, Y., Shiraishi, N. (1998). Preparation of sulfuric acid-catalyzed phenolated wood resin. *Wood Sci. Technol.* 32: 297–308.

Alma, M.H., Kelley, S.S. (2000). Thermal stability of novolak-type thermosettings made by the condensation of bark and phenol. *Plymer Degradation and Stability.* 68: 413–418.

Alma, M.H, Maldas, D., Shiraishi, N. (1997). Liquefaction of several biomass wastes into phenol in the presence of various alkalies and metallic salts as cataysts. *J. Polym. Eng.* 18, 163

Alma, M.H., Basturk, M.A., Shiraishi, N. (2001). Cocondensation of NaOH-catalyzed liquefied wood wastes, phenol, and formaldehyde for the production of resol-type adhesives. *Ind. Eng. Chem. Res.* 40: 5036–5039.

Alma, M.H., Yoshioka, M., Yao, Y., Shiraishi, N. (1995). Some characterizations of hydrochloric acid catalyzed phenolated wood -based materials. *Mokuzai Gakkaishi.* 41,741–748.

Alma, M.H., Yoshioka, M., Yao, Y., Shiraishi, N. (1996). The preparation and flow properties of HCL catalyzed phenolated wood and its blend with commercial Novolak resin. *Holzforschung.* 50: 85–90.

Alonso, M.V., Oliet, M., Perez, J.M., Rodriguez, F., Echeverria, J. (2004). Determination of curing kinetic parameters of lignin-phenol-formaldehyde resole resins by several dynamic differential scanning calorimetry methods. *Thermochim. Acta.* 419, 161–167.

Amen-Chen, C., Riedl, B., Roy, C. (2002 (a)). Softwood bark pyrolysis oil–PF resols–Part2. Thermal analysis by DSC and TG. *Holzforschung.* 56, 273–280.

Amen-Chen, C., Riedl, B., Wang, X.M., Roy, C. (2002 (b)). Softwood bark vacuum pyrolysis oil–PF resols for bonding OSB Panels. Part III. Use of propylene carbonate as resin cure accelerator. *Holzforschung.* 56: 281–288.

Alonso, M.V, Oliet, M., Perez, J.M, Rodriguez, F., Echeverria, J. (2004). Determination of curing kinetic parameters of lignin-phenol-formaldehyde resol resins by several dynamic differential scanning calorimetry methods. *Thermochimica Acta.* 419: 161–167.

Angles, M.N., Reguant, J., Garcia-Valls, R., Salvado, J. (2003). Characteristics of lignin obtained from steam exploded softwood with soda/anthraquinone pulping. *Wood Sci. Technol.* 37, 309–320.

Bao, S., Daunch, W.A., Sun, Y., Rinaldi, P.L., Marcinko, J.J., Phanopoulos, C. (1999). Solid state NMR studies of methylene diisocyanate (PMDI) derived species in wood. *J. Adhesion*. 71:377–394.

Bridgwater, A.V., Peacocke, G.V.C. (2000). Fast pyrolysis processes for biomass. *Sustainable Renew Energy Rev*. 4(1):1–73.

Bridgwater, A.V. (2004). Biomass fast pyrolysis. *Thermal Science*. 8(2):21–49.

Cetin, N. S., Ozmen, N. (2002 (a)). Use of organosolv lignin in phenol-formaldehyde resins for particleboard production: I. Organosolv lignin modified resins *Int. J. of Adhesion and Adhesives*. 22: 477–480.

Cetin, N.S, Ozmen, N. (2002 (b)). Use of organosolv lignin in phenol–formaldehyde resins for particleboard production−II. Particleboard production and properties. *Int. J. of Adhesion and Adhesive*. 22: 481–486.

Chan, F., Riedl, B., Wang, X.M., Lu, X., Amen-Chen, C., Roy, C. (2002). Performance of pyrolysis oil-based wood adhesives in OSB. *Forest products Journal*. 56: 273–280.

Chakar, F.S., Ragauskas, A.J. (2004). Review of current and future softwood kraft lignin process chemistry. *Ind. Crops Prod*. 20, 131–141.

Cheng, S., Dcruz, I., Wang, M., Leitch, M., Xu, C. (2010). Highly efficient liquefaction of woody biomass in hot-compressed alcohol-water co-solvents. *Energy Fuels* (In press).

Chum, H.L., Black, S.K., Diebold, J.P., Kreibich, R.E. (1993). Resole resin products derived from fractionated organic and aqueous condensates made by fast-pyrolysis of biomass materials. *U.S. Patent* 5 235 021,.

Chum, H., Diebold, J., Scahill, J., Johnson, J., Black, S.D., Schroeder, H., Kreibich, R.E. (1989). Biomass pyrolysis oil feedstocks for phenolic adhesives. In Adhesive from Renewable Resources; Hemingway E W, Connor A J, Branham S J, Eds.; ACS Sympsium Series 385; American Chemical Society: Washington D C. 135.

Chum, H.L., Kreibich, R.E. (1992). Process for preparing phenolic formaldehyde resole resin products derived from fractionated fast-pyrolysis oils. *US Patent* 5 091 499, Midwest Research Institute.

Clarke, M.R., Dolenko, J.D. (1978). Methylolated kraft lignin polymer resin. *U.S. Patent* 4 113 675

Crestini, C., Caponi, M.C., Argyropoulos, D.S., Saladino, R. (2006). Immobilized methyltrioxo rhenium (MTO)/H_2O_2 systems for the oxidation of lignin and lignin model compounds, *Bioorg. Med. Chem*. **14**, 5292–5302.

Desch, H.E., Dinwoodie, J.M. (1996). Timber: structure, properties, conversion and use. 7th edition. 40.

Dorrestijn, E., Kranenburg, M., Poinsot, M., Mulder, P. (1999). Lignin depolymerization in hydrogen-donor solvents. *Holzforschung*. 53 (6): 611–616.

Effendi, A., Gerhauser, H., Bridgwater, A.V. (2008). Production of renewable phenolic resins by thermochemical conversion of biomass: A review. *Renewable Sustainable Energy Rev*. 12: 2092–2116.

Earle, M.J., Seddon, K.R. (2000). Ionic liquids. Green solvents for the future, *Pure Appl.Chem*, 72, 1391–1398.

EI Mansouri N-E, Salvado, J. (2006) Structural characterization of technical lignins for production of adhesives: Application to lignosulfonate, kraft, soda-anthraquinone, organosolv and ethanol process lignins. *Industrial Crops and Products*. 24: 8–16.

Giroux, R., Freel, B., Graham, R. (2001). Natural resin formulations. *US Patent* 6 326 461, Ensyn Group Inc.

Giroux, R., Freel, B., Graham, R. (2003). Natural resin formulations. *US Patent* 6 555 649, Ensyn Group Inc.

Haider, K.W., Wittig, M.A., Dettore, J.A., Dodge, J.A., Rosthauser, J.W. (2000). On the trail to isocyanate/phenolic hybrid wood binders: model compound studies. Wood Adhesives 2000, Extended Abstracts. *Forest Products Society*, Tahoe, 85–86.

Hassan, M.E., Mun, S.H. (2002). Liquefaction of pine bark using phenol and lower alcohols with methanesulfonic acid catalyst. *J. Ind. Chem.* 8:359–364.

Hassan, E.M., Shukry, N. (2008). Polyhydric alcohol liquefaction of some lignocellulosic agricultural residues. *Industrial crops and products*. 27: 33–38.

Hayashi, A., Namura, Y., Urkita, T. (1967). Demethylation of lignosulphonate during the gelling reaction with dichromate. *Mokozai Gakkaishi*. 13: 194–197.

Himmelblau, D.A. (1991). Method and apparatus for producing water-soluble resin and resin product made by that method. US Patent 5 034 498 Bicarbons Co.

Himmelblau, D.A., Grozdits, G.A. (1999). Production and performance of wood composite adhesives with air-blown, fluidized-bed pyrolysisoil. In: Overend R P, Chornet E, editors. Proceedings of the 4th biomass conference of the Americas, Oxford, UK: Elsevier Science. 1:541–547.

Hofrichter, M. (2002). Review: lignin conversion by manganese peroxidase (MnP). *Enzyme Microb. Technol.* 30, 454–466.

Karagoz, S., Bhaskar, T., Muto, A., Sakata, Y. (2004). Effect of Rb and Cs carbonates for production of phenols from liquefaction of wood biomass. *Fuel.* 83, 2293–2299.

Karagoz, S., Bhaskar, T., Muto, A., Sakata, Y. (2005). Comparative studies of oil compositions produced from sawdust, rice husk, lignin and cellulose by hydrothermal treatment. *Fuel.* 84, 875–884.

Kim, M.G., Watt, C. (1996). Effects of urea addition to phenol-formaldehyde resin binders for oriented strandboard. *Journal of Wood Chemistry and Technology*. 16(1): 21–34.

Kirk-Othmer Encyclopedia of Chemical Technology [electronic resource], Web ed., John Wiley and Sons, (1999-2008), New York, Vol.21.

Kleinert, T.N. (1975). Ethanol-water delignification of sizable pieces of wood, disintegration into stringlike fiber bundles. *Holzforschung*. 29(3): 107–109.

Kleinert, M., Barth, T. (2008 (a)). Phenols from lignin. *Chem. Eng. Technol.* 31(5): 736–745.

Kleinert, M., Barth, T. **(2008 (b)).** One-step conversion of lignin to oxygen-depleted bio-fuels and phenols, In Proceedings of the 16th European Biomass Conference (Eds: Schmid J., Grimm H-P, Helm P, Grassi A), Valencia, 1993-1997. ISBN 978-88-89407.

Kleinert, M., Gasson, J.R., Barth, T. (2009). Optimizing solvolysis conditions for integrated depolymerisation and hydrodeoxygenation of lignin to produce liquid biofuel. *J. Anal. Appl. Pyrolysis,* 85:108–117.

Kleinert, M., Barth, T. (2008 (c)). Towards a lignocellulosic biorefinery: Direct one-step conversion of lignin to hydrogen-enriched bio-fuel. *Energy Fuels.* 22, 1371–1379.

Kleinert, M., Barth, T. (2007 (a)). One-step conversion of solid lignin to liquid products. *European Patent application*.

Kleinert, M., Barth, T. **(2007 (b)).** Production of biofuel and phenols from lignin by hydrous pyrolysis. In Proceedings of the 15th European Biomass Conference (Eds: Maniatis K, Grimm H -P, Helm P, Grassi A), Berlin. 1371–1379. ISBN 978-88-89407-59-X.

Larsson, P.T., Wickholm, K., Iversen, T. (1997). A CP/MAS[13] C NMR investigation of molecular ordering in celluloses. *Carbohydr. Res.* 302: 19–25.

Lawson, J R, Klein, M.T. (1985). Influence of water on guaiacol pyrolysis. *Ind. Eng. Chem. Fundam.* 24, 203–208.

Lee, S.H, Ohkita, T. (2003). Rapid wood liquefaction by supercritical phenol. *Wood Sci. Technol.* 37: 29–38.

Lee, S.H., Yoshioka, M., Shiraishi, N. (2000). Preparation and properties of phenolated corn bran (CB)/phenol/formaldehyde cocondensed resin. *Journal of Applied Polymer Science.* 77(13), 2901–2907.

Li, L., Kiran, E. (1988). Interaction of supercritical fluids with lignocellulosic materials. *Ind. Eng. Chem. Res.* 27, 1301.

Li, C.Q., Takanohashi, T., Saito, I., Iino, M. (2003). Role of N-methyl-2-pyrrolidinone in hydrogen donation from 9,10-dihydroanthracene to coal at 300 °C. *Energy Fuels.* 17 (5):1399–1400.

Lin-Gibson, S., Riffle, J.R. (2003). Chemistry and properties of phenolic resins and network. In synthetic methods in step-growth polymers. Edited by Martin E. Rogers and Timothy E. Long. John Wiley and Sons Inc. Hoboken, New Jersey: 375–430.

Lin, L., Yao, Y., Shiraishi, N. (2000). Liquefaction mechanism of β-O-4 lignin model compound in the presence of phenol under acid catalysts—Part 2. Reaction behavior and pathways. *Holzforschung.* 55: 625–630.

Lin, L.Z., Yoshioka, M., Yao, Y.G., Shiraishi, N. (1994). Liquefaction of wood in the presence of phenol using phosphoric acid as a catalyst and the flow properties of the liquefied wood. *J. Appl. Polym. Sci.* 52: 1629–1636.

Lin, L.Z., Yoshioka, M., Yao, Y.G., Shiraishi, N. (1995). Preparation and properties of phenolated wood/phenol/ formaldehyde cocondensed resin. *J. Appl. Polym. Sci.* 58:1297–1304.

Lora, J.H., Glasser, W.G. (2002). Recent industrial applications of lignins: A sustainable alternative to non-renewable materials, *J. Polym. Environ.* **10**, 39–48.

Maldas, D., Shiraishi, N., Harada, Y. (1997(a)). Phenolic resol resin adhesives prepared from alkali catalyzed liquefied phenolated wood and used to bond hardwood. *J. Adhesion Scitechnol.* 11:305–316.

Maldas, D., Shiraishi, N. (1997(b)). Liquefaction of Biomass in the presence phenol and water using alkalies and salts as the cataysts. *J. Biomass and Bioenergy.* 12(4): 273–279.

Matsushita, Y., Imai, M., Iwatsuki, A., Fukushima, K. (2008). The relationship between surface tension and the industrial performance of water-soluble polymers prepared from acid hydrolysis lignin, a saccharification by-product from woody materials. *Bioresour. Technol.* 99, 3024–3028.

Matsushita, Y., Yasuda, S. (2003). Preparation of anion-exchange resins from pine sulfuric acid lignin, one of the acid hydrolysis lignins. *J. Wood Sci.* 49, 423–429.

Matsumura, Y., Nonaka, H., Yokura, H., Tsutsumi, A., Yoshida, K. (1999). Co-liquefaction of coal and cellulose in supercritical water. *Fuel.* 78, 1049–1056.

Matsushita, Y., Yasuda, S. (2005). Preparation and evaluation of lignosulfonates as a dispersant for gypsum paste from acid hydrolysis lignin. *Bioresour. Technol.* 96, 465–470.

McBrierty, V.J., Douglass, D.C. (1981). Recent advances in the NMR of solid polymers. Journal of Polymer Science. *Macromolecular Reviews.* 16: 295–366.

McBrierty, V.J., Packer, K.J. (1993). Nuclear magnetic resonance in solid polymer, Chapter 6, Cambridge, UK; Cambridge University Press.

Meier, D., Scholze, B. (1997). Fast pyrolysis liquid characteristics. In: Biomass Gasification and Pyrolysis - State of the Art and Future Prospects. (eds. Kaltschmitt M and Bridgwater AV), CPL Press Newbury, U.K. 431–441.

Miller, J.E., Evans, L., Littlewolf, A., Trudell, D.E. (1999). Batch microreactor studies of lignin and ligninmodel compound depolymerization by bases in alcohol solvents. *Fuel.* 78: 1363–1366.

Minami, E., Saka, S. (2005). Decomposition behavior of woody biomass in water-added supercritical methanol. *J. Wood Sci.* 51: 395–400.

Minami, E., Kawamoto, H., Shiro, S. (2003). Reaction behaviour of lignin in supercritical methanol as studied with lignin model compounds. *J. Wood Sci.* 49, 158–165.

Mourant, D., Yang, D.Q., Lu, X., Riedl, B., Roy, C. (2009). Copper and boron fixation in wood by pyrolytic resins. *Bioresource Technology.* 100:1442–1449.

Muurinene, E. (2000). Organosolv pulping, a review and distillation study related to peroxyacid pulping. Oulu University Library. (URL:http://herkules.oulu.fi/issn 03553213).

Nakos, P., Tsiantzi, S., Athanassiadou, E. (2001). Wood adhesives made with pyrolysis oils. In: Proceedings of the 3rd European wood-based panel symposium. November 12–14. Hanover, Germany: European Panel Federation and Wilhelm Klauditz Institute.

Narayamusti, D., George, J. (1954). *Composite Wood.* 1: 51–55.

Nenkova, S., Vasileva, T., Stanulov, K. (2008). Production of phenol compounds by alkaline treatment of technical hydrolysis lignin and wood biomass. *Chem. Nat. Compd.* 44.

Ni, J.W., Frazier, C.E. (1998). 15N CP/MAS NMR study of the isocyanate/wood adhesive bondline. effects of structural isomerism. *J. Adhesion.* 66:89–116.

Olivares, M., Guzman, J.A., Natho, A., Saavedra, A. (1988). Kraft lignin utilization in adhesives. *Wood Sci. Technol.* 22: 157–165.

Okuda, K., Umetsu, M., Takami, S., Adschiri, T. (2004). Disassembly of lignin and chemical recovery-rapid depolymerization of lignin without char formation in water-phenol mixtures. *Fuel Process. Technol.* 85, 803–813.

Oasmaa, A., Alen, R., Meier, D. (1993). Catalytic hydro-treatment of some technical lignins. *Bioresour Technol.* 45(3): 189–194.

Papa, A.J., Critchfield, F.E. (1979). Hybrid Phenolic/Urethane Foams. *Journal of Cellular Plastics.* 15: 258–266.

Park, Y., Doherty, W.O.S., Halley, P.J. (2008). Developing lignin-based resin coatings and composites. *Ind. Crops Prod.* 27, 163–167.

Pasquini, D., Pimenta, M.T.B., Ferreira, L.H, Curvelo, A.A.S. (2005). Extraction of lignin from sugar cane bagasse and Pinus taeda wood chips using ethanol–water mixtures and carbon dioxide at high pressures. *J. of Supercrit. Fluids.* 36:31–39.

Peng, W., Reidl, B. (1994). The chemorheology of phenol-formaldehyde thermoset resin and mixtures of resin fillers. *Polymer* 35, 1280–1286.

Perez, J.M., Oliet, M., Alonso, M.V., Rodriguez., F. (2009). Cure kinetics of lignin–novolac resins studied by isoconversional methods. *Thermochim. Acta.* 487, 39–42.

Piskorz, J., Majerski, P., Radlein, D., Scott, D.S. (1989). Conversion of lignins to hydrocarbon fuels. *Energy Fuels.* 3(6), 723–729.

Pizzi, A., Walton, T. (1992). Non-emulsifiable, water-based diisocyanate adhesives for exterior plywood, Part I, *Holzforschung*, 46(6): 541–547.

Pu, Y., Jiang, N., Ragauskas, A.J. (2007). Ionic liquid as a green solvent for lignin, *Journal of Wood Chemistry and Technology*, 27: 23–33.

Radlein, D., Piskorz, J., Scott, D. (1987). Lignin derived oils from the fast pyrolysis of polar wood. *J. Anal. Appl. Pyrol.* 12, 51.

Rasthauser, J.W., Haider, K.W., Hunt, R.N., Gustavich, W.S. (1997). Chemistry of PMDE wood binders: Model studies. In: M. Wolcott (ed.) Proceedings of the 31st International Particleboard/Composite Materials Symposium. Pullman,WA: Washinton State University Press. 161–176.

Russell, J.A., Riemath, W.F. (1985). Method for making adhesive from biomass. US Patent 4 508 886, USA as represented by the United States Department of Energy,

Schaefer, J., Stejskal, E.O., Buchdahl, R. (1977). Magic-angle 13C NMR analysis of motion in solid glassy polymers. *Macromolecules*. 10, 384–405.

S'anchez, C. (2009). Lignocellulosic residues: Biodegradation and bioconversion by fungi. *Biotechnology Advances*. 27(2): 185-194.

Shen, J.L., Iino, M. (1994). Heat treatment of coals in hydrogen-donating solvents at temperatures as low as 175-300 °C. *Energy Fuels*. 8 (4): 978–983.

Schmidt, J., Laberge, S. (2008). Presentation of "Chemicals and Fuels from Lignocellulosic Materials", 09/09/2008.

Shashi, Jolli S.P., Singh, S.V., Gupta, R.C. (1982). Kinetics and mechanisms of lignin formaldehyde resinification reaction. *Cellul. Chem. Technol.* 16: 511–522.

Shiraishi , N., Shirakawa, K., Kurimoto, Y. (1992). Japanese Pat. Appl. 106128.

SRI consulting (2010), World Petrochemical (WP) report on PF Resins. http://www.sriconsulting.com/WP/Public/Reports/pf_resins (retrieved Jan 14, 2011)

Stewart, D. (2008). Lignin as a base material for materials applications: Chemistry, application and economics. *Ind. Crops Prod.* 27, 202–207.

Sudan, V. (2003). Process for preparing a black liquor-phenol formaldehyde thermoset resin. *US Patent 6632912B2*, Oct 14.

Sundquist, J. (1996). Chemical pulping based on formic acid, summary of Milox research. *Paperi ja Puu*. 78(3): 92–95.

Swatloski, R.P., Spear, S.K., Holbrey, J.D., Rogers, R.D. (2002). Dissolution of cellulose with ionic liquids. *J. Am. Chem. Soc.* 124, 4974–4975.

Tejado, A., Pena, C., Labidi, J., Echeverria, J.M., Mondragon, I. (2007). Physico-chemical characterization of lignins from different sources for use in phenol–formaldehyde resin synthesis. *Bioresour. Technol.* 98, 1655–1663.

Thring, R.W. (1994). Alkaline degradation of ALCELL® lignin. *Biomass Bioenergy*. 7, 125–130.

Thring, R.W., Katikaneni, S.P.R., Bakhshi, N.N. (2000). Production of gasoline range hydrocarbons from alcell lignin using HZSM-5 Catalyst. *Fuel Processing Technology*. 65, 17–30.

Tsoumis, G. (1991). Science and technology of wood: structure, properties, utilization. 36.

Tsujino, J., Kawamoto, H., Saka, S. (2003). Reactivity of lignin in supercritical methanol studied with various lignin model compounds. *Wood Sci. Technol.* 37, 299–307.

Tymchyshyn, M., Xu, C. (2010). Liquefaction of biomass in hot-compressed water for the production of phenolic compounds. *Bioresource Technology*. 101, 2483–2490.

US National Renewable Energy Laboratory (NREL). (2003). http://www.nrel.gov.

USDA Forest Service Research Note FPL-091 Bark and its possible uses. Revised 1971. Available on http://www.fpl.fs.fed.us/documnts/fplrn/fplrn091.pdf.

Vázquez, G., González, J., Freire, S., Antorrena, G. (1997). Effect of chemical modification of lignin on the gluebond performance of lignin-phenolic resins. *Bioresour. Technol.* 60, 191–198.

Wang, M., Leitch, M., Xu, C. (2009 (a)). Synthesis of phenol-formaldehyde resol resins using organosolv pine lignins. *European Polymer Journal.* 45, 3380–3388.

Wang, M., Xu, C., Leitch, M. (2009 (b)). Liquefaction of corn stalk for the production of phenol-formaldehyde resole resin. *Bioresource Technology*, 100, 2305-2307.

Wegener, G. (1992). Pulping innovations in German. *Ind. Crops Prod.* 1, 113–117.

Wendler, S.L., Frazier, C.E. (1995). The ^{15}N CP/MAS NMR characterization of the isocyanate adhesive bondline for cellulosic substrates. *Journal of adhesion.* 50: 135–153.

Wendler, S.L., Frazier, C.E. (1996 (a)). Effect of moisture content on the isocyanate/wood adhesive bondline by 15N CP/MAS NMR. *Journal of Applied Polymer Science.* 61:775–782.

Wendler, S.L., Frazier, C.E. (1996 (b)). The effects of cure temperature and time on the isocyanate-wood adhesive bondline by 15N CP/MAS NMR. *Int. Journal of Adhesion and Adhesives.* 16:179–186.

Xie, H. L., Shi, T. J. (2006). Wood liquefaction by ionic liquids. *Holzforschung.* 60: 509–512.

Xu, C., Lad, N. (2008). Production of heavy oils with high caloric values by direct-liquefaction of woody biomass in sub-/near-critical water. *Energy Fuels.* 22, 635–642.

Xu, C., Etcheverry, T. (2008). Hydro-liquefaction of woody biomass in sub- and super-critical ethanol with iron-based catalysts. *Fuel.* 87 (3): 335–345.

Yamada, T., Ono, H. (1999). Rapid liquefaction of lignocellulosic waste by using ethylene carbonate. *Bioresour. Technol.* 70: 61–67.

Yamazaki, J., Minami, E., Saka, S. (2006). Liquefaction of beech wood in various supercritical alcohols. *J. Wood Sci.* 52, 527–532.

Yan, N., Zhao, C., Dyson, P.J., Wang, C., Liu, L., Kou, Y. (2008). Selective degradation of wood lignin over noble-metal catalysts in a two-step process. *ChemSusChem.* 1, 626–629.

Yuan, Z., Cheng, S., Leitch, M., Xu, C. (2010). Hydrolytic degradation of alkaline lignin in hot-compressed water and ethanol. *Bioresource Technology* (In press).

Zhou, X., Frazier, C.E. (2001). "Double labeled isocyanate resins for the solid-state NMR detection of urethane linkages to wood" *international Journal of Adhesion and Adhesives.* 21(3): 259–264.

Zhuang, M., Steiner, P.R. (1993). Thermal reactions of diisocyanate (MDI) with phenols and benzylalcohols: DSC study and synthesis of mdi adducts. *Holzforschung.* 47: 425–434.

In: Lignin
Editor: Ryan J. Paterson

ISBN 978-1-61122-907-3
© 2012 Nova Science Publishers, Inc.

Chapter 6

REACTIVITY AND REACTION MECHANISM OF CELLULOSE, LIGNIN AND BIOMASS IN STEAM GASIFICATION AT LOW TEMPERATURES

Chihiro Fushimi and Atsushi Tsutsumi*

Collaborative Research Center for Energy Engineering,
Institute of Industrial Science, The University of Tokyo, Tokyo, Japan

1. INTRODUCTION

Biomass has been important resource because of its renewable and carbon neutral characteristics. Steam gasification is a very promising technology for energy conversion and hydrogen production with high efficiency. In steam gasification of biomass, initially pyrolysis takes place, producing volatiles (tar and gases) and solid residue (char). Then, steam reacts with the residual char and tar, producing gases such as H_2, CO and CO_2. Moreover, a large amount of tar is evolved in biomass gasification. Tar causes blockages and corrosion of pipes, and also reduces the overall thermal efficiency of the process [Bridgewater, 1995; Devi et al., 2003]. It is, therefore, imperative to rapidly convert the char and tar into gases to achieve high overall efficiency of steam gasification of biomass.

In conventional gasification processes, heat required for gasification reaction is supplied by partial oxidation of biomass/coal. This reduces the energy efficiency of biomass/coal conversion. In fact, even the integrated coal gasification combined cycle (IGCC) power generation remains around 80% of cold gas efficiency (ratio of the heating value of product gases to that of original fuel).

Thus, we have proposed exergy recuperation concept [Kuchonthara et al., 2003a,b, 2005, 2006; Tsutsumi, 2004; Hayashi et al., 2006], which utilizes the exhaust heat of the gas turbine or solid fuel cell instead of partial oxidation of biomass/coal as a heat source for endothermic gasification reactions, so as to increase cold gas efficiency. The steam gasification with exergy recuperation requires that the reactions be carried out at low temperatures, which are

---

* Corresponding author: fushimi@iis.u-tokyo.ac.jp, 4-6-1 Komaba, Meguro-ku, Tokyo, 153-8505, Japan, Tel: +81-3-5452-6293 Fax: +81-3-5452-6728.

not generally preferred for rapid and complete conversion. Thus, it is necessary to investigate reaction rate at low temperatures.

2. EFFECT OF HEATING RATE ON THE REACTIVITY AND REACTION MECHANISM OF BIOMASS MEASURED BY RAPID-HEATING THERMOGRAVIMETRIC-MASS SPECTROMETRIC (TG-MS) ANALYSIS

So far, many kinetic studies have been conducted mainly with thermobalance reactors at various heating rates to study pyrolytic reactivity of cellulose [Antal et al., 1980, 1995, 1998; Várhegyi et al., 1994; Conesa et al., 1995; Milosavljevic et al., 1995, 1996; Grφnli, 1999; Völker and Reickmann, 2002], lignin [Caballero et al., 1996; Ferdous et al., 2002], and biomass [Cooley and Antal, 1988; Antal et al., 1990, 1997; Font et al., 1991; Narayan and Antal, 1996; Raveendran et al. 1996; Várhegyi et al.1997, 2002; Lanzetta and Blash, 1998; Reynolds et al., 1997; Caballero et al., 1997; Teng and Wei, 1998; Rao and Sharma, 1998; Reina et al., 1998; Órfão et al., 1999; Burnham and Braun, 1999; Conesa et al., 2001; Garcìa-Pèrez et al., 2001; Grφnli et al., 2002; Manyà et al., 2003]. Several researchers have investigated steam gasification kinetics of biomass [Antal et al., 1980; Encinar et al., 2001,2002; Branca et al., 2003; Müller et al, 2003]. However, in most of the studies, the heating rates were at most 2-3 K s^{-1}. Chen et al. (1997) reported that rapid-heating pyrolysis of the birch wood char possessed higher reactivity in reactions than char formed with a slow heating rate. Marcilla et al. (2000a) and Iniesta et al. (2001) also reported that the larger reactivity with carbon dioxide observed in the samples obtained with a flash treatment, as compared to a low heating rate treatment. Thus, it is inferred that rapid heating is effective for enhancement of steam gasification of biomass char at low temperatures by pyrolysis. However, few studies have been conducted to investigate reactivity of steam gasification of biomass char in situ formed up to high heating rates in a thermobalance reactor.

In addition, evolution profiles of gaseous products during steam gasification have to be investigated to explore the reaction mechanism for steam gasification of biomass. At present, several studies have reported biomass pyrolysis [Antal, 1983; Zaror et al., 1985; Piskorz et al., 1986; Cooley and Antal, 1988; Alves and Figueiredo, 1989; Pouwels et al., 1989; Boroson et al., 1989; Banyasz et al., 2001a,b; Li et al., 2001; Brown et al., 2001; Völker and Riekmann, 2002; Ferdous et al., 2001,2002]. In the case of lignin, Py-GC/MS has been generally used to measure the evolution profiles of volatiles [Evans et al., 1986; Faix et al., 1987; Genuit et al., 1987; Pouwels and Boon, 1990; Jakab et al., 1995; Camarero et al., 1999; Río et al., 2001; Greenwood et al., 2002]. Serio et al. (1994) analyzed gaseous products in pyrolysis of lignin using Thermogravimetric, Fourier Transform Infrared (TG-FTIR) spectroscopy. However, few studies have investigated gas evolution profiles in the presence of steam, especially in the case of rapid heating.

Hence, we have developed a novel rapid-heating thermobalance reactor, which can heat biomass sample at the heating rate up to 100 K s^{-1} [Fushimi et al., 2003a,b]. By using this reactor, the reactivity of biomass in steam gasification at a low temperature (973 K) is examined at the heating rates of 1, 10 and 100 K s^{-1}. In addition, 1) the effect of heating rates on production of low-molecular-weight gases such as H_2, CH_4, CO, and CO_2 and 2) the reaction mechanisms in steam gasification of biomass were examined by coupling the quadropole-mass spectrometer with the rapid-heating themobalance reactor (=thermogravimetric-mass spectrometric (TG-MS) analysis).

2.1. Experimental

2.1.1. Apparatus

A schematic diagram of the experimental apparatus and sample basket are shown in Figure 1 [Fushimi et al., 2003a,b]. The system mainly consists of a quartz thermobalance reactor of 25 mm in inner diameter, an infrared gold image furnace and a balance sensor (HP-TG-9000; ULVAC-RIKO, Inc.). Length of the furnace's isothermal zone is approximately 210 mm. A ceramic basket of 8 mm in diameter and 10 mm in length is suspended in the thermobalance. Temperature is measured by an R-type thermocouple placed near the sample. Water is fed by a chemical pump (NP-KX-100; Nihon Seimitsu Kagaku, Inc.) and heated by a steam generator. The steam is fed through a coil above the sample basket. The lower part of the quartz reactor is cooled by a water-cooling jacket to prevent secondary gas-phase reaction.

2.1.2. Procedure

A 10-20 mg sample was placed into the ceramic basket. Ar gas of 2.72 $Ncm^3 s^{-1}$, which is 0.55 cm s^{-1} at the standard state, was fed into the thermobalance reactor. Subsequently, temperature was increased and was kept at 473 K to prevent steam from condensing in the reactor. Then, steam was introduced into the reactor with carrier gas Ar (50:50 vol%).Then, the reactor was heated up to a desired temperature. The heating rate was variable up to 100 K s^{-1}. Temperature and weight loss of the sample during reaction were recorded on a personal computer at time intervals of 0.2 or 0.5 s. Pyrolysis was also carried out without introducing steam for comparison with steam gasification.

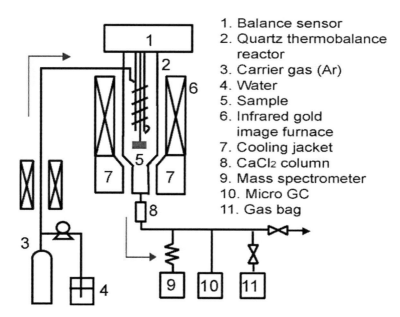

1. Balance sensor
2. Quartz thermobalance reactor
3. Carrier gas (Ar)
4. Water
5. Sample
6. Infrared gold image furnace
7. Cooling jacket
8. $CaCl_2$ column
9. Mass spectrometer
10. Micro GC
11. Gas bag

Figure 1. Schematic diagram of thermobalance reactor. [Fushimi et al., 2003a].

The produced tar and water were eliminated in a $CaCl_2$ column, then, the gaseous products (H_2, CH_4, CO and CO_2) were sampled and analyzed with a quadrupole mass spectrometer (Standum; Ulvac, Inc.). In conjunction with mass spectrometry, the gaseous products were also analyzed by a TCD-micro gas chromatograph (model M-200H; Hewlett Packard, Co.) to verify accuracy of the mass spectrometer data. H_2, O_2, N_2, CH_4, CO and CO_2 were measured at time intervals of approximately 90 s.

After steam gasification or pyrolysis was completed, char was burned by introducing oxygen to calculate the conversion of sample to volatile matter. All of the experiments were conducted at atmospheric pressure.

2.1.3. Sample

Cellulose (Merck Co. Ltd.) and kraft lignin made of softwood with sulfonic acid (Kanto Chemical Co. Ltd.) were purchased and used. Bagasse was also used as sample. Bagasse was ground with a pestle and a mortar for 1 h prior to an experiment. Table 1 lists their elemental compositions and ash content.

Table 1. Elemental compositions and ash content of biomass samples

	C	H	N	S	O	
	Elemental compositions [wt%, d.a.f. basis]					Ash (d.b.)
cellulose	44.44	6.17	0	0	49.39	0
lignin	64.47	5.60	0.15	2.83	26.95	17.55
bagasse	46.20	5.74	0.15	0	47.91	2.38

2.2. Results and Discussion

2.2.1. Time Profiles of Mass Change in Pyrolysis and Steam Gasification

Time profiles of temperature and relative mass change of cellulose in pyrolysis and steam gasification at the heating rates of 1, 10, and 100 K s^{-1} are shown in Figures 2-4. The solid lines and dotted lines represent the results in pyrolysis and steam gasification, respectively. The final temperature was 973 K. The initial time (0 s) was defined as the start of heating from 473 K. Relative mass was recorded on a dry ash-free basis (d.a.f.). In the case of 1 K s^{-1} (Figure 2(a)), relative mass was observed to decrease rapidly above 600 K (127 s). Approximately 85wt% of cellulose was converted into volatiles in pyrolysis up to 700 K (227 s). Above 700 K, the relative mass continued to decrease gradually. Pyrolysis was almost completed at 973 K (500 s), reaching conversion of 93wt%. This trend is agreement with the previous reports [Antal et al., 1980, 1998; Antal and Várhegyi, 1994, 1995, 1997; Grønli et al., 1999; Shafizadeh, 1982; Piskorz et al., 1989; Milosavljevic et al., 1995, 1996; Raveendran et al., 1996; Reynolds and Burnham, 1997; Völker and Rieckmann, 2002]. However, the relative mass of char was slightly decreased at 973 K due to condensation. By adding steam, further reduction of relative mass was observed above 700 K. This indicates that steam gasification occurs above 700 K for char produced in pyrolysis. Steam gasification was completed in 1000 s after the temperature reached 973 K, reaching final conversion of 98wt%.

In the case of 10 K s^{-1} (Figure 2(b)), relative mass was observed to decrease above 650 K and finish at 973 K in pyrolysis. Steam gasification of char took place above 880 K, reaching final conversion of 98wt%.

When the heating rate was 100 K s^{-1} (Figure 2(c)), rapid reduction of relative mass was observed above 700 K. With rapid heating, most of cellulose was converted into volatiles during heating irrespective of steam. The relative mass of cellulose gradually decreased due to steam gasification at 973 K and was completed in 1000 s.

Figure 3 shows time profiles of temperature and relative mass change of lignin at the heating rates of 1, 10, and 100 K s^{-1}. When the heating rate was 1 K s^{-1} (Figure 3(a)), a rapid drop in the relative mass of lignin was observed above 550 K (77 s). The rate of relative mass reduction was observed to be slow at 773-923 K (300-450 s), reaching approximately 40wt% of conversion to volatiles. Above 923 K, the relative mass of lignin char gradually decreased. Relative mass reduction continued after the temperature reached 973 K, indicating that pyrolysis continued. Finally, 44wt% of lignin remained as char in pyrolysis. In the case of 10, and 100 K s^{-1} (Figures 3(b), (c)), pyrolysis was not completed during heating and continued at 973 K, implying the occurrence of condensation and carbonization of char.

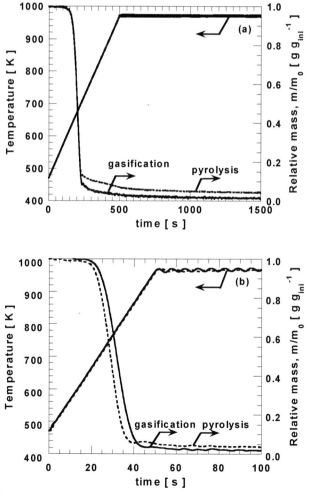

Figure 3. (Continued).

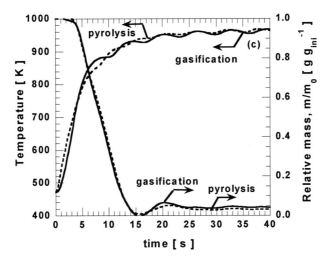

Figure 2. Profiles of temperature and relative mass of cellulose: (a) heating rate = 1 K s^{-1} (b) heating rate = 10 K s^{-1} (c) heating rate = 100 K s^{-1} [Fushimi et al., 2003a].

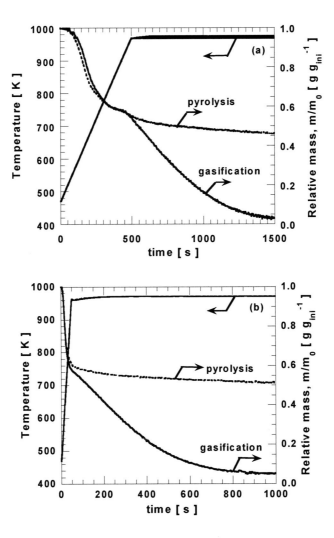

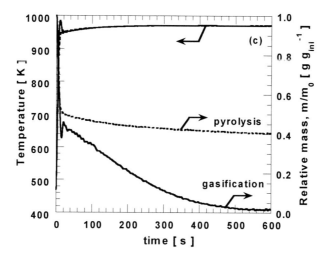

Figure 3. Profiles of temperature and relative mass of lignin: (a) heating rate = 1 K s^{-1} (b) heating rate = 10 K s^{-1} (c) heating rate = 100 K s^{-1} [Fushimi et al., 2003a].

By adding steam, the steam gasification reaction of lignin char produced in pyrolysis occurred above 923 K (450 s) and finished approximately 900 s after the temperature reached 973 K in the case of 1 K s^{-1} (Figure 3(a)).

In the case of 10 K s^{-1} (Figure 3(b)), steam gasification of nascent char took place above 703 K (23 s). Steam gasification of char finished approximately 800 s after the temperature reached 973 K.

With rapid heating, steam gasification started after the temperature reached 973 K and was completed in approximately 600 s. It was found that rapid heating substantially shortened the time to complete steam gasification of lignin char and that some amount of unreacted char remained after steam gasification in the case of lignin.

Figure 4 shows the time profile of relative mass change of bagasse in pyrolysis and steam gasification at the heating rate of 1, 10 and 100 K s^{-1}. The profiles of bagasse were observed to be similar to those of cellulose because the bagasse contains 49wt% of cellulose. In the case of 1 K s^{-1} (Figure 4(a)), rapid decrease of relative mass due to devolatilization was observed above 550 K (77 s).

Approximately 75wt% of bagasse was converted into volatiles in pyrolysis up to 700 K (227 s) and pyrolysis was almost completed at 973 K. Steam reacted with nascent char of bagasse above 700 K. With increasing heating rate, the rapid reduction of the relative mass in pyrolysis was observed at higher temperature (Figures 4(b), (c)). The relative mass slightly decreased at 973 K.

This implies that lignin component in bagasse still repolymerized and carbonized. Steam was observed to react with nascent char at 770 K in the case of 10 K s^{-1} and 870 K in the case of 100 K s^{-1}. We found that most of bagasse is converted into volatiles during heating in pyrolysis and that steam reacts with nascent char of biomass after the devolatilization of volatile.

Table 2 lists the final conversion of biomass in pyrolysis and steam gasification. It is shown that the higher heating rates increase final conversion of biomass in pyrolysis and steam gasification.

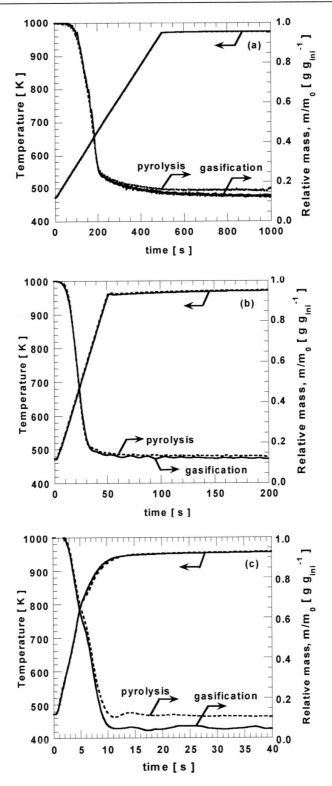

Figure 4. Profiles of temperature and relative mass of bagasse: (a) heating rate = 1 K s^{-1} (b) heating rate = 10 K s^{-1} (c) heating rate = 100 K s^{-1}.

Several researchers reported that rapid heating of brown coal to high temperatures enhances the probability of simultaneous bond-breaking, leading to a release of volatiles within a very short period.

It is inferred that higher heating rates reduce the chance of repolymerization of volatiles inside the char in pyrolysis and steam gasification. This results in the increase of final conversion to volatiles.

Ferdous *et al.* (2002) also reported that a higher heating rates gives higher conversion and higher synthesis gas production for both Alcell and kraft lignin. The present results agree with these reports.

Table 2. Final conversions of biomass in pyrolysis and steam gasification (at 973 K) [Fushimi et al., 2003a]

Sample	Reaction	Heating rate [K s^{-1}]	Conversion [wt%]
cellulose	Pyrolysis	100	99
		10	96
		1	96
	Gasification	100	100
		10	99
		1	98
bagasse	Pyrolysis	100	91
		10	87
		1	84
	Gasification	100	96
		10	90
		1	88
lignin	Pyrolysis	100	60
		10	52
		1	55
	Gasification	100	98
		10	95
		1	98

2.2.2. Rate Constants of Char Gasification

Assuming the first-order reaction, the Arrhenius parameter is obtained by:

$$-\frac{d(M - M_f)}{dt} = k(M - M_f) \qquad (1)$$

$$k = k_0 \exp\left(\frac{-E_a}{RT}\right) \qquad (2)$$

where M [g g$_{ini}^{-1}$] is the relative mass at each time, M_f [g g$_{ini}^{-1}$] is the relative mass remaining after steam gasification, t [s] represents time, k [s^{-1}] is the first-order rate constant of steam gasification, k_0 [s^{-1}] is the frequency factor, E_a [J mol^{-1}] is the activation energy of the reaction, R [J mol^{-1} K^{-1}] is the gas constant, and T [K] is the temperature. The initial reaction rate constants of steam gasification of char at 973 K (after heating) were also obtained by solving eq (1)

$$\ln(M - M_f) = -kt + C \qquad (3)$$

where C is a constant. Thus, plotting $ln(M-M_f)$ against time gives a straight line with a slope of $(-k)$. Figure 5 shows rate constants of steam gasification of char at heating rates of 1, 10, and 100 K s^{-1}. This result confirms that rapid heating substantially accelerates the initial steam gasification rate of biomass char.

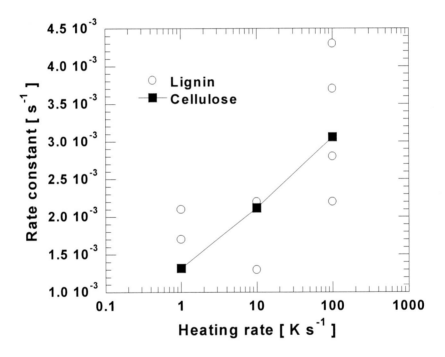

Figure 5. Rate constants of steam gasification of cellulose and lignin [Fushimi et al., 2003a].

By SEM photographs of original lignin and char heated at the heating rates of 1 and 100 K s^{-1}, we observed that formation of needle-like fragments on the surface of the char and the agglomeration of char in the case of slow heating. These needle-like fragments were considered to be produced by condensation or cross-linking of fragments of the side chain. On the other hand, in the case of rapid heating, formation of large pores of several micrometers' diameter of was observed on the surface of lignin char. These large pores are considered to be active sites which increases the reactivity of biomass in steam gasification. Agglomeration of char and needle-like fragments were not observed up to 973 K [Fushimi et al., 2003a].

2.2.3. Gas Evolution Profiles in Pyrolysis and Steam Gasification with Slow Heating

Figure 6 shows time profiles of temperature, relative mass of char, gases and tar and gas evolution rates at the heating rate of 1 K s^{-1} in pyrolysis and steam gasification of cellulose. Since the water and tar were eliminated in the $CaCl_2$ column before the MS and GC measurement, the uncollected volatiles including pyrolytic water will be referred as tar. Thus, tar yield was calculated by subtracting the relative mass of char and produced gases from that of original cellulose. The dotted line of relative mass of tar in Figure 6(b) represents the difference of relative mass of char and gas after steam gasification of char started. In the initial stage of cellulose pyrolysis, depolymerization of cellulose began above 523 K (50 s) to produce CO_2 and depolymerizing cellulose [Banyasz et al., 2001a,b]. Evolution of CO and H_2 proceeded. Above 600 K (127 s), evolution rates of CO_2, CO and H_2 increased rapidly with decrease of relative mass of char. Evolution of these gases peaked at 673 K (200 s) and then rapidly decreased up to 700 K (227 s), in correspondence with relative mass change. As a result, 81wt% of cellulose was converted to tar (including pyrolytic water) at 600-700 K by decomposition of depolymerizing cellulose [Piskorz et al., 1986; Cooley and Antal, 1988, Alves and Figueiredo, 1989; Pouwels et al., 1989; Antal and Várhegyi, 1997; Várhegyi et al., 1998, 2002; Banyasz et al., 2001a,b; Li et al., 2001]. Above 700 K, second peaks of evolution of CO_2, CO and H_2 were coincident with completion of pyrolysis around 973 K. Second peaks of gas evolution are attributable to further decomposition and condensation of nascent char to yield CO_2, CO, H_2, and char.

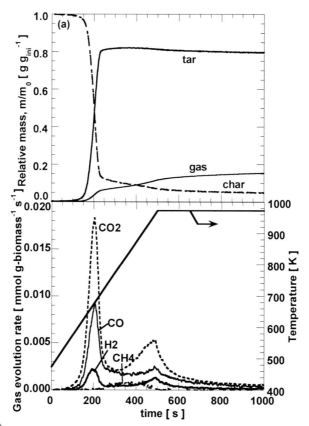

Figure 6. (Continued).

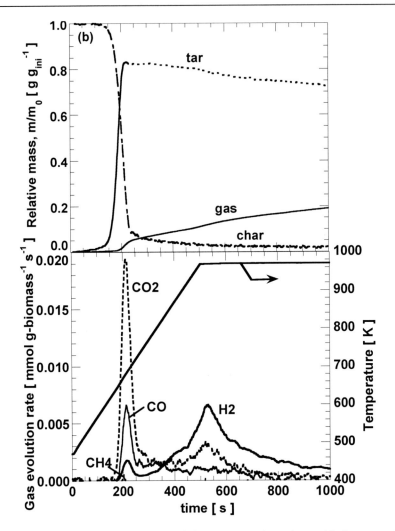

Figure 6. Profiles of temperature, relative mass of char, gases, and tar along with the gas evolution rate of cellulose with the heating rate of 1 K s^{-1}: (a) pyrolysis (b) steam gasification [Fushimi et al, 2003b].

As Figure 6(b) shows, the presence of steam did not have significant effect on gas evolution profiles and reduction in relative mass of char below 700 K (227 s). Subsequently, an increase in H$_2$ evolution was observed above 700 K. However, no significant increases of evolution of CO, CO$_2$ and CH$_4$ were observed with addition of steam. Gas evolution peaked at 973 K; then, it decreased. These results indicate that cellulose is decomposed mainly into tar at 600-700 K; subsequently, steam gasification of nascent char produced by decomposition occurs above 700 K to emit H$_2$.

Figures 7 and 8 show time profiles of temperature, relative mass and gas evolution rate at the heating rate of 1 K s^{-1} in pyrolysis and steam gasification of lignin and bagasse, respectively.

In the case of lignin pyrolysis (Figure 7(a)), CO$_2$ evolution was observed above 500 K (27 s), followed by evolution of CO and CH$_4$. Above 550 K (77 s) depolymerization and decomposition of lignin occurred, yielding CO$_2$, CO, CH$_4$, tar and 60wt% of nascent char. Evolution of CO$_2$ peaked at 673 K (200 s). A significant increase in H$_2$ evolution was observed above 773 K (300 s) and peaked at 873 K (400 s). However, no pronounced evolution of CO and CO_2 was observed in this temperature range. These results imply that aromatization and carbonization of the lignin-nascent char proceed to yield char, evolving excess hydrogen. Evolution of CO_2 and CO exhibited a weak peak at 973 K in accordance with cellulose pyrolysis.

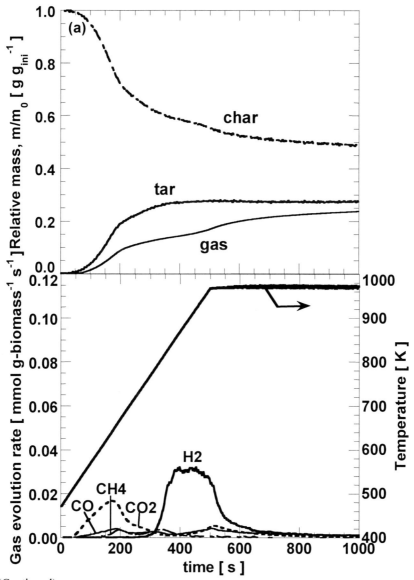

Figure 7. (Continued).

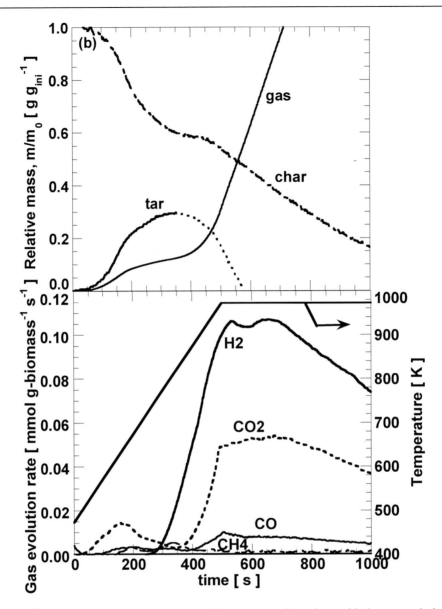

Figure 7. Profiles of temperature, relative mass of char, gases, and tar along with the gas evolution rate of lignin with the heating rate of 1 K s^{-1}: (a) pyrolysis (b) steam gasification [Fushimi et al., 2003b].

No significant difference between pyrolysis and steam gasification of lignin was observed up to 823 K (350 s). Above 823 K (350 s), H_2 and CO_2 evolution increased drastically. Then, a steep rise in CO evolution was observed. As mentioned, relative mass change of lignin char in steam gasification exhibited a similar curve to that of char in pyrolysis up to 923 K. These suggest that water-gas-shift reaction takes place above 823 K and that steam gasification of nascent char takes place above 923 K to form H_2, CO_2 and CO in parallel with repolymerization and carbonization. The evolution of H_2, CO, and CO_2 was observed until steam gasification of char was completed [Fushimi et al., 2003b].

In the case of bagasse pyrolysis (Figure 8(a)), CO_2 evolved above 500 K (27 s); it was followed by evolution of CO and H_2. The shoulders of CO_2, CO, and H_2 evolution were observed at 623 K (150 s). These may be due to the depolymerization and devolatilization of hemicellulose that decomposed at lower temperature than cellulose [Raveendran *et al*, 1996; Teng and Wei, 1998; Marcilla *et al*., 2000]. The peaks of the three gases evolution were observed at 673 K (200 s). This is attributable to cellulose decomposition along with tar evolution. Above 673 K, similar evolution profiles of CO_2, CO, H_2 and CH_4 to those of cellulose were observed. The second peaks of CO_2, CO and H_2 evolution were observed at 973 K. By adding steam, insignificant difference in gas evolution was observed until 873 K (400 s). After that, steam gasification of char takes place, increasing evolution of H_2.

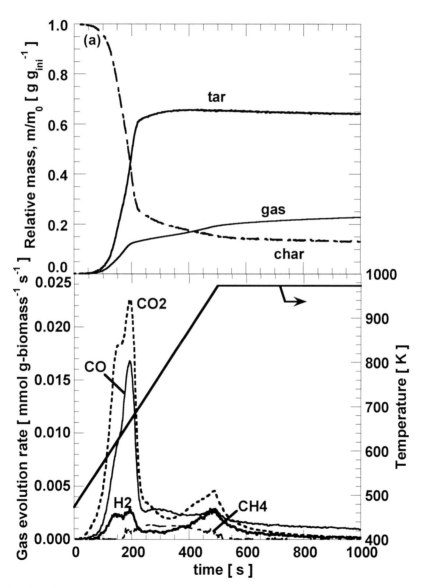

Figure 8. (Continued).

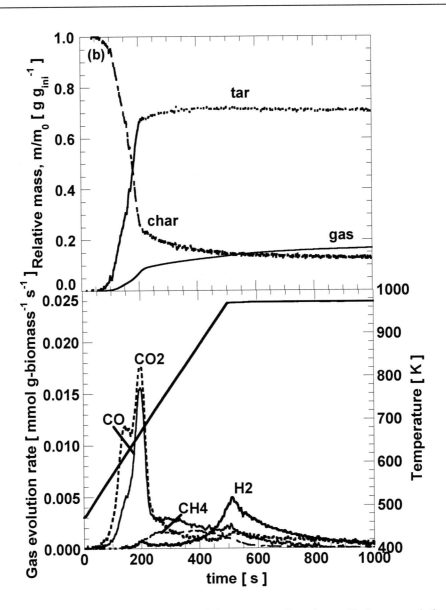

Figure 8. Profiles of temperature, relative mass of char, gases, and tar along with the gas evolution rate of bagasse with the heating rate of 1 K s^{-1}: (a) pyrolysis (b) steam gasification.

2.2.4. Gas Evolution Profiles in Pyrolysis and Steam Gasification at Higher Heating Rates

Figures 9-11 show time profiles of temperature, relative mass and gas evolution rate in pyrolysis and steam gasification of cellulose, lignin, and bagasse at the heating rate of 100 K s^{-1}, respectively.

In all cases, all gases were evolved almost simultaneously in pyrolysis and exhibited single peaks. In the case of cellulose (Figure 9(a)), CO was the main product. By adding steam (Figure 9(b)), the evolution of H_2 and CO_2 increased and that of CO decreased, indicating steam reforming and water-gas-shift reaction take place during rapid heating.

The evolution of H_2 due to steam gasification of char was observed after 40 s. In steam gasification of lignin (Figure 10(b)), evolution of the four gases significantly increased just after temperature reached 973 K with rapid heating. Evolution of H_2 and CO_2 due to steam gasification of char was observed after 15 s [Fushimi et al, 2003b].

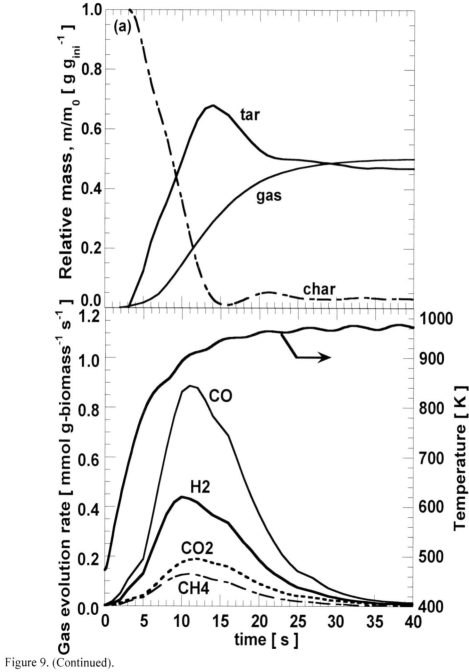

Figure 9. (Continued).

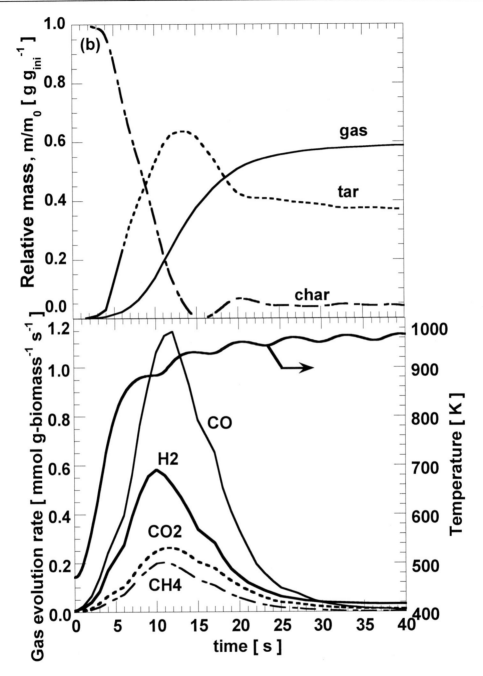

Figure 9. Profiles of temperature, relative mass of char, gases, and tar along with the gas evolution rate of cellulose with the heating rate of 100 K s^{-1}: (a) pyrolysis (b) steam gasification [Fushimi et al., 2003b].

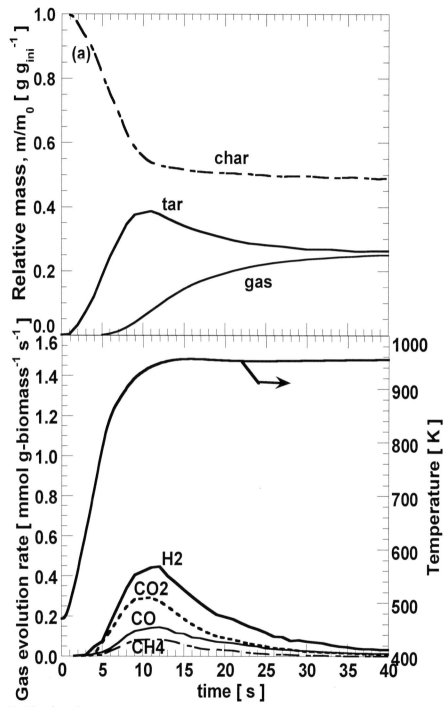

Figure 10. (Continued).

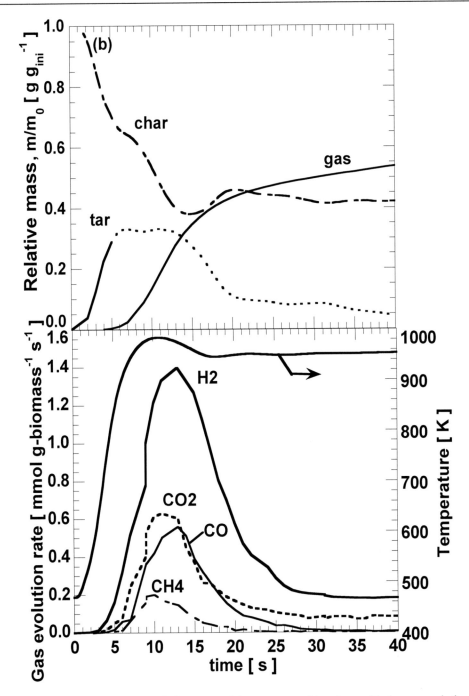

Figure 10. Profiles of temperature, relative mass of char, gases, and tar along with the gas evolution rate of lignin with the heating rate of 100 K s^{-1}: (a) pyrolysis (b) steam gasification [Fushimi et al., 2003b].

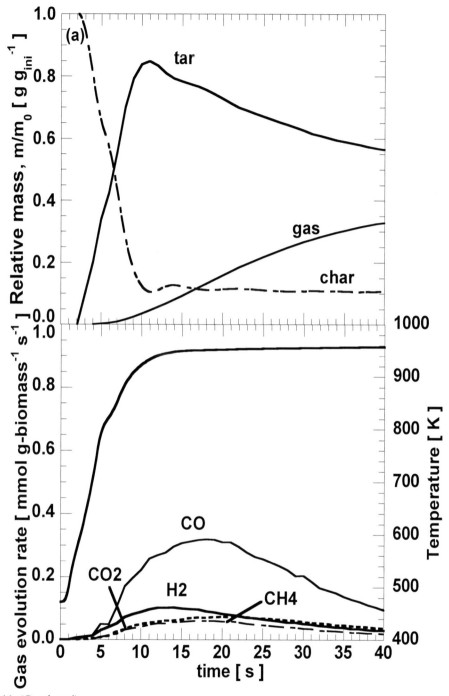

Figure 11. (Continued).

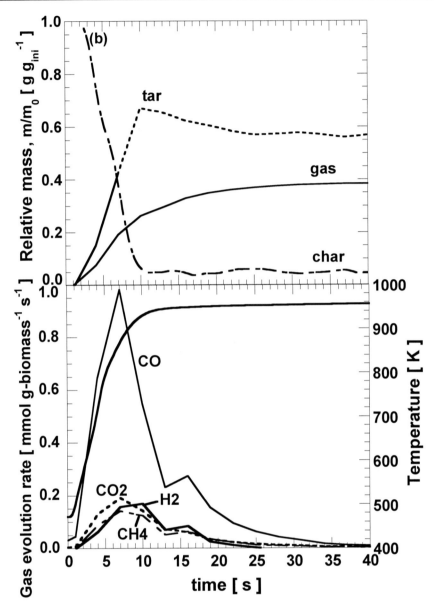

Figure 11. Profiles of temperature, relative mass of char, gases, and tar along with the gas evolution rate of bagasse with the heating rate of 100 K s^{-1}: (a) pyrolysis (b) steam gasification.

2.2.5. Summary of Reaction Mechanism of Pyrolysis and Steam Gasification

Figures 12 and 13 present summaries of reaction mechanisms of pyrolysis/steam gasification of cellulose and lignin, respectively. When cellulose is heated, depolymerization takes place. Devolatilization of volatiles follows depolymerization, leading to evolution of CO_2, H_2, and CO. 79wt% of cellulose is converted into tar (including pyrolytic water). The remaining nascent char condenses and is converted into char, evolving small amount of CO_2, CO and H_2. Devolatilization occurs at higher temperature in the case of rapid heating. In this temperature range, secondary tar cracking takes place, leading to significant increase of CO, H_2, and CH_4 evolution.

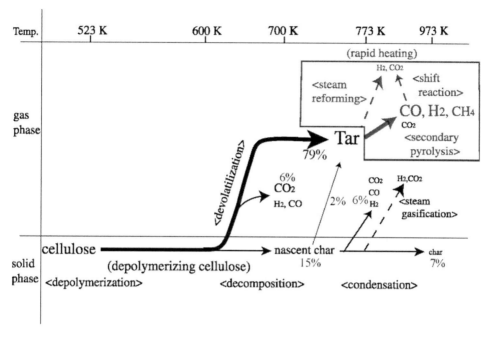

Figure 12. Reaction mechanism of pyrolysis and steam gasification of cellulose (dotted line represents the reaction with steam) [Fushimi et al., 2003b].

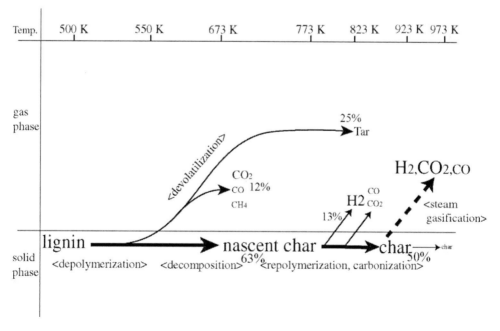

Figure 13. Reaction mechanism of pyrolysis and steam gasification of lignin (dotted line represents the reaction with steam) [Fushimi et al., 2003b].

By adding steam, steam reforming of volatiles and water-gas-shift reaction take place, decreasing CO evolution and producing H_2 and CO_2. On the other hand, when lignin is heated, depolymerization and decomposition occur, evolving tar along with CO_2, H_2 and CO. Since lignin has large amount of aromatic carbons that has low reactivity, the main product in pyrolysis is nascent char.

The repolymerization and carbonization of the nascent char follows depolymerization, evolving large amount of excess hydrogen. The carbonization of lignin is not completed at 973 K.

Rapid heating does not enhance devolatilization. This may be due to the weak bond energy of the peripheral groups of lignin, leading to relatively rapid bond cleavage (depolymerization) during heating. Steam gasification of lignin char takes place, producing large amount of H_2 and CO_2.

2.3. Conclusion

Time histories of weight change and gas evolution rate of biomass samples (cellulose, lignin, and bagasse) in steam gasification and pyrolysis were studied with a rapid-heating thermobalance reactor at heating rates 1, 10 and 100 K s^{-1}. Effect of heating rate on the conversion to volatiles and reactivity of char were investigated. The following conclusions are drawn;

1) Rapid heating substantially increase the reactivity of lignin char in steam gasification as a result of the development of many macropores, which are considered to have active sites, on the char surface by rapid evolution of volatiles. Rapid heating increased reactivity of cellulose and bagasse char in steam gasification.

2) The higher heating rate increases final conversion of the biomass to volatiles as a consequence of the reduced chance of repolymerization of volatiles to the char.

3) Steam reacted with nascent char of bagasse after devolatilization above 700 K. Steam gasification of bagasse char took place above 873 K.

4) In pyrolysis of cellulose, rapid heating significantly increased evolution of H_2, CO and CH_4 through enhancement of secondary pyrolysis (tar cracking). Addition of steam decreased CO and increased H_2 and CO_2 evolution, suggesting that steam reforming and water-shift reactions take place.

5) Formation of nascent char at 500-773 K was predominant with evolution of CO_2, CO, and CH_4 in the case of lignin pyrolysis. The nascent char is converted into char by repolymerization and carbonization, significantly emitting excess hydrogen. Steam gasification of char takes place above 823 K, increasing in H_2 and CO_2 production significantly.

6) Time profiles of gas production of bagasse are similar to those of cellulose because bagasse contains about 50% of cellulose. Rapid heating remarkably increased gas production in steam gasification of bagasse, indicating that steam gasification of biomass can proceed sufficiently by rapid heating even at low temperatures.

3. INTERACTION OF CELLULOSE, LIGNIN AND HEMICELLULOSE DURING GAS AND TAR EVOLUTION IN THE PYROLYSIS/GASIFICATION MEASURED WITH A CONTINUOUS CROSS-FLOW MOVING BED TYPE DIFFERENTIAL REACTOR (CCDR)

In the previous section, we studied the reaction rate of char and evolution rates of volatiles in pyrolysis and steam gasification of biomass (cellulose, lignin and bagasse) using thermogravimetric-mass spectrometric (TG-MS) analysis with slow and rapid heating. The effect of heating rate and steam on time profiles for weight change of samples during pyrolysis and steam gasification was investigated in detail. The reaction mechanism of pyrolysis and steam gasification from the view points of gas evolution and char reactivity in semi batch operation was explained. However, it is very difficult to investigate the time profile of tar and gas evolution during biomass gasification with a differential method of kinetic analysis in a continuous feeding condition with a drop tube reactor, a fixed bed reactor, a fluidized bed reactor and a thermobalance reactor. Thus, we have newly developed a continuous cross-flow moving bed type differential reactor (CCDR) [Yamaguchi et al, 2006, Fushimi et al., 2009a,b], in which the biomass sample is continuously fed and the products (tar, gas, and char) can be fractionated from each compartment according to the reaction time. The fractionated volatile matter and char were separated immediately and quenched. Under this experimental condition, secondary reaction between volatile matter and char can be minimized because of the low temperature and short residence time. Therefore, the reaction mechanism of biomass, especially tar evolution, in the initial stage of gasification can be investigated by CCDR.

So far, many researchers have examined product distribution and kinetics of real biomass and proposed reaction models [Evans and Milne, 1987; Di Blasi and Lanzetta, 1997; Várhegyi et al., 1997; Di Blasi, 1998; Miller and Bellan, 1997a,b; Morf et al.,2002; Manyà et al., 2003; Müller-Hagedorn et al., 2003; Gómez et al, 2004; Svenson and Pettersson, 2004; Kersten et al., 2005; Wang et al., 2005; Yang et al., 2006]. In many studies, it was reported that pyrolysis of the three major components of real biomass (cellulose, hemicellulose, and lignin) takes place independently without interaction.

On the other hand, some researchers reported the product yield and kinetics of biomass in pyrolysis cannot be explained by the superposition of the values of the three components because of the interaction among the three components during tar production and evolution [Sagehashi et al.,2006; Hosoya et al., 2007a,b,c; Worasuwannarak et al., 2007]. However, so far the works on the interaction among cellulose, lignin and xylan on the gas and tar evolution during pyrolysis and gasification are not sufficient.

In this section, the time profile of tar and gas evolution was investigated in steam gasification of real biomass (pulverized eucalyptus) and its major components (cellulose, xylan as hemicellulose, and lignin) by using CCDR and the interaction between cellulose, xylan and lignin in the initial stage of pyrolysis and steam gasification is explained.

3.1. Experimental

3.1.1. Apparatus.

Figure 14 shows the schematic diagram of the CCDR. The reactor consists of a quartz-glass half-tube covered with a quartz-glass plate and a belt-conveyor system. The reactor is divided into six compartments (W 90 mm × D 80 mm × H 40 mm), where gas flows are independent. The reactor is heated using an infrared gold image furnace (Ulvac Riko, Co. Ltd.).

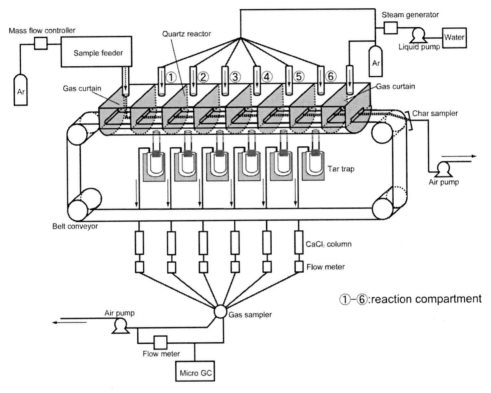

Figure 14. Schematic image of continuous cross-flow moving bed type differential reactor [Yamaguchi et al, 2006; Fushimi et al., 2009a].

The temperatures of each zone are measured by K-type thermocouples and are controlled to be constant. Reaction temperature was set to 673 K. The biomass sample is fed out of a feeder onto the conveyor belt, which carries the sample across the six compartments. The initial time is defined as the time when the sample is fed into the preheating zone. Tar and gases produced in each compartment are sampled with a carrier gas and fractionated according to the reaction time. At the end of the belt conveyor, char is collected using the char-sampling system. The residence time of each compartment can be varied by changing belt speed. Steam was fed into the reactor with an Ar carrier gas. Tar evolved in each compartment went through heated sampling lines and then was collected separately in six cold tar traps. Water was eliminated in $CaCl_2$ columns. The flow rate of the effluent gas was measured with a mass flow-meter. We analyzed H_2, O_2, N_2, CH_4, CO, CO_2, C_2H_4, and C_2H_6 using a micro-gas chromatograph (micro-GC, M-200H; Hewlett-Packard Co.).

3.1.2. Biomass Samples

Pulverized wood biomass (Chilean eucalyptus, <500 μm), Avicel microcrystalline cellulose (Merck Co. Ltd., < 160 μm, Avicel), xylan from birch wood (Sigma-Aldrich Co.), kraft lignin (Kanto Chemical Co. Ltd.) were used. The elemental compositions and ash content of these samples were shown Table 3. In addition, two kinds of model biomass samples were used. Sample A was a mixture of cellulose (65wt%) and lignin (35wt%). Sample B was a mixture of cellulose (50wt%), xylan (23wt%) and lignin (27wt%). Note that the ratio was based on the composition of the eucalyptus wood and the ratio of cellulose to lignin in the sample A is the same as that in sample B. The samples A and B were mixed using a ball mill for 5 h prior to an experiment.

Table 3. Elemental compositions and ash content of samples [wt%]

| | Elemental compositions (dry, ash-free) | | | | | Ash | | | |
	C	H	O (diff.)	N	S	(dry basis)	cellulose	lignin	xylan
Cellulose	44.44	6.17	49.39	0	0	0	100	0	0
Lignin	64.47	5.57	26.90	0.15	2.91	17.48	0	100	0
Xylan	40.16	6.12	52.96	0.14	0.62	0.28	0	0	100
Eucalyptus	48.87	6.05	44.17	0.19	0.72	0.43	50	27	23
Sample A	51.45	5.96	41.52	0.05	1.02	6.12	65	35	0
Sample B	48.86	6.00	44.14	0.07	0.93	4.78	50	27	23

3.1.3. Procedure

The biomass sample was placed in the feeder, and the feeder box and the reactor were purged with Ar. The steam generator, steam feeding lines, and sampling lines were preheated to avoid the condensation of steam and tar. After the reactor was heated up to 673 K, steam and biomass were fed. The belt speed was 750 mm min^{-1}; the corresponding residence time of biomass sample in each compartment was 6.4 s.

In each compartment of CCDR the gas evolution rate was defined as follows:

$$gas\ evolution\ rate\ (weight\ basis) = \frac{M_G}{F \cdot \tau}$$

where M_G, F and τ represent molar amount of gases in product [mol], the amount of fed sample [g], and residence time [s], respectively.

3.1.4. Analysis of Tar

After the experiment, the thimble filters were removed from the tar traps and placed in glass jars. At first, distilled water was added in the glass jars and the jars were heated to extract water-soluble tar. Next, acetone was added in the glass jars and heated to extract water-insoluble tar. Then, part of the extracted water-soluble and water-insoluble tar was placed in a dry evaporation dish and held in an oven at 383 K. The amount of each tar was calculated from the difference of the weight of the evaporation dish.

238

Chihiro Fushimi and Atsushi Tsutsumi

Then, distilled water was introduced in the reactor and sampling lines. The weight of water-soluble and water-insoluble tar was measured with the same method as mentioned above.

3.2. Results and Discussion

3.2.1. Trend of Gas and Tar Evolution in Gasification of Each Sample

In the case of cellulose, it can be seen from Figure 15 the tar evolution rapidly increased with reaction time and peaked at 22 s. The evolution rate of CO and CO_2 increased slightly with an increase in reaction time to 35 s. These results are attributed to the existence of intermediates [Yamaguchi et al.; 2006]. The evolution rates of H_2, CH_4, C_2H_4 and C_2H_6 had similar trends to those of CO and CO_2 although their values were much smaller. In the case of lignin, it can be seen from Figure 16 that the evolution rates of CO and CO_2 were large at first and decreased with reaction time. This implies CO and CO_2 are released from lignin and/or nascent char. The evolution rate of CH_4 was much larger than that of cellulose (cf. Figure 15). This is probably due to methoxyl groups in lignin. For xylan, as shown in Figure 17, tar evolution decreased monotonically. On the other hand, the time profile of gas evolution had two peaks, initially and at 29 s. This result agrees with the reports that stated hemicellulose has two decomposition steps [Müller-Hagedorn et al., 2003].

In the case of real biomass (Figure 18), a similar trend for tar and gas evolution with xylan gasification was observed, indicating a larger amount of gaseous products is derived from xylan in real biomass.

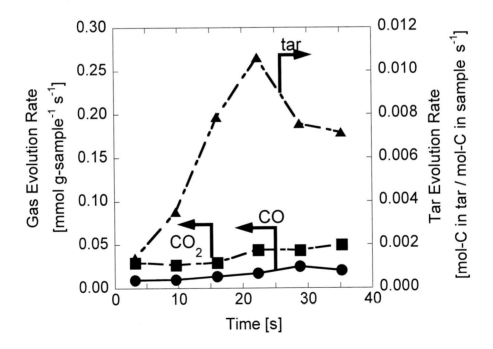

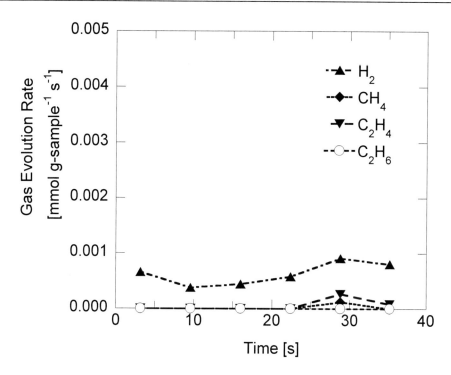

Figure 15. Time profile of gas and tar evolution (cellulose) ; (a) CO, CO_2 and tar (b) H_2, CH_4, C_2H_4 and C_2H_6. [Fushimi et al., 2009a]

Figure 19 shows the evolution rate predicted by a superposition of evolution rates for cellulose, lignin, and xylan in the ratio of 50:27:23 by weight.

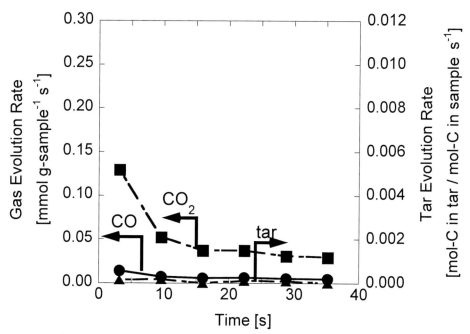

Figure 16. (Continued).

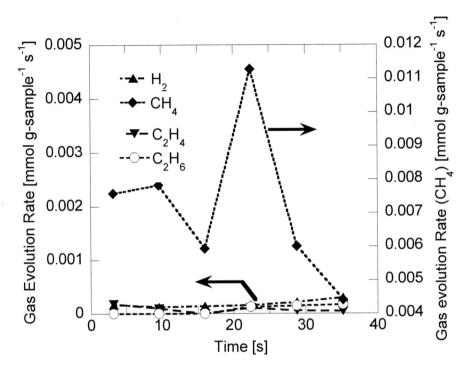

Figure 16. Time profile of gas and tar evolution (lignin) ; (a) CO, CO_2 and tar (b) H_2, CH_4, C_2H_4 and C_2H_6. [Fushimi et al., 2009a]

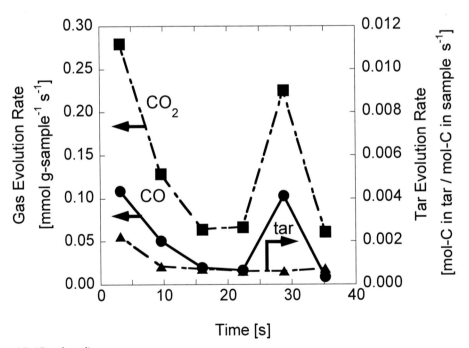

Figure 17. (Continued).

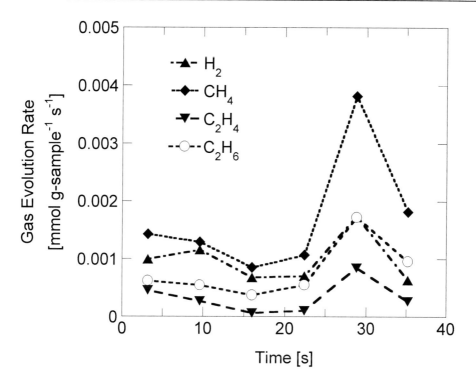

Figure 17. Time profile of gas and tar evolution (xylan) ; (a) CO, CO$_2$ and tar (b) H$_2$, CH$_4$, C$_2$H$_4$ and C$_2$H$_6$. [Fushimi et al., 2009a]

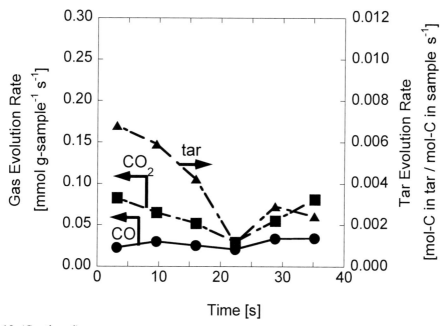

Figure 18. (Continued).

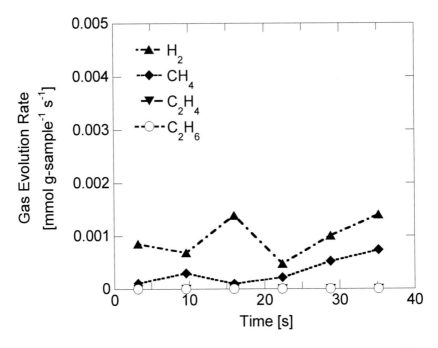

Figure 18. Time profile of gas and tar evolution (biomass : experimental result) ; (a) CO, CO_2 and tar (b) H_2, CH_4, C_2H_4 and C_2H_6. [Fushimi et al., 2009a]

The predicted time profile of the gas evolution is similar to the time profile of experimental data. However, the tar evolution profile was found to differ for experimental and estimated values. This result indicates an interaction among the tar components from cellulose, lignin and xylan.

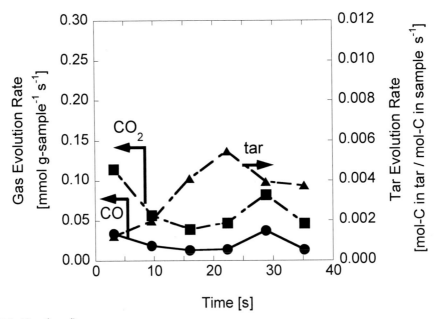

Figure 19. (Continued).

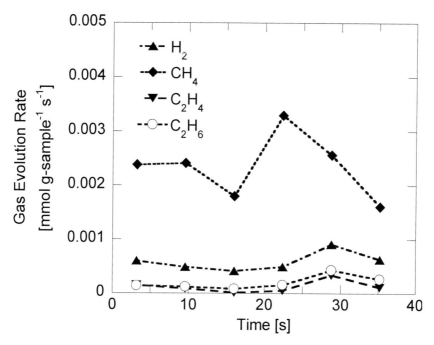

Figure 19. Time profile of gas and tar evolution (biomass : calculated by superposition of cellulose (50 %), lignin (27 %) , and xylan (23 %)); (a) CO, CO_2 and tar (b) H_2, CH_4, C_2H_4 and C_2H_6 [Fushimi et al., 2009a].

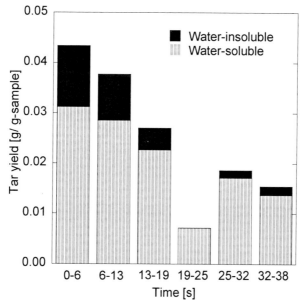

Figure 20. (Continued).

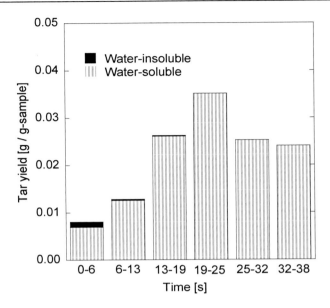

Figure 20. Amount of tar in biomass gasification (a) experiment using real biomass (b) calculated by the superposition of cellulose, lignin, and xylan. [Fushimi et al., 2009a].

Figures 20 (a) and (b) show the amounts of water-soluble and water-insoluble tar in the gasification of real biomass and the predicted amounts calculated by the superposition of the amounts for the three components, respectively. In the case of real biomass (Figure 20 (a)), both water-soluble and water-insoluble tar evolved initially and the ratio of water-insoluble tar decreased with reaction time. On the other hand, insignificant water-insoluble tar production was predicted from the superposition of the three components (Figure 20 (b)). We surmised that tar was derived mainly from cellulose because little evolved from lignin and xylan.

These results suggest that in gasification of real biomass, the amount of water-insoluble tar that mainly evolves from lignin increases and that the evolution of water-soluble tar from cellulose occurs earlier than in the gasification of pure lignin or cellulose.[Fushimi et al., 2009a]

3.2.2. Gas and Tar Evolution from Mixture of Cellulose and Lignin (Sample A)

Figure 21 shows the gas evolution rate from the sample A. The predicted values obtained from the superposition of the results for each component by assuming that the each component reacts independently in gasification, are also shown. Thus, the difference between experimental values (closed symbols and solid lines) and predicted values (open symbols and dotted lines) indicate the interaction between the two components during gas evolution. The CO_2 evolution rates in the experiment were smaller than the prediction at 3.2 and 9.6 s, indicating the suppression of initial evolution of CO_2 by the interaction between cellulose and lignin. Then, the evolution of CO_2 became larger than the prediction. The evolution rates of CO in experiment were smaller than the predicted values.

These imply the mechanism of CO_2 and CO evolution is different in the mixture of cellulose and lignin. The evolution rates of H_2, CH_4 and C_2H_4 in the experiments were smaller than the predicted values, implying that the interaction between cellulose and lignin suppresses the evolution of these gases (Figure 21b).

Figure 22 shows the amount of produced water-soluble and water-insoluble tars from the sample A. In the experiment, the evolution of water-soluble tar monotonically increased with the rise in time. The trend is similar to the gas evolutions (cf. Figure 21).

The evolutions of water-soluble tar evolution were much smaller and peaked later than the prediction. In contrast, the evolution rates of water-insoluble tar in the experiment were larger and peaked earlier than the estimated values.

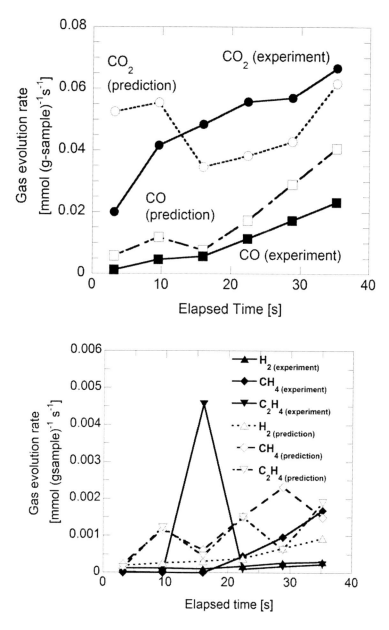

Figure 21. Time profiles of gas evolution rate from mixture of the two components (cellulose 65%, lignin 35%) and predicted values obtained from the superposition of the results for each component by assuming there is no interaction (a) CO and CO_2 (b) H_2, CH_4 and C_2H_4 [Fushimi et al., 2009b].

These suggest that the interaction between cellulose and lignin enhances evolution of water-insoluble tar while substantially suppressing the evolution of gases (CO, H_2, CH_4 and C_2H_4) and water-soluble tar derived mainly from cellulose. In addition, it was found that the interaction between cellulose and lignin delays the evolution peaks of water-soluble tar and accelerates those of water-insoluble tar.[Fushimi et al., 2009b]

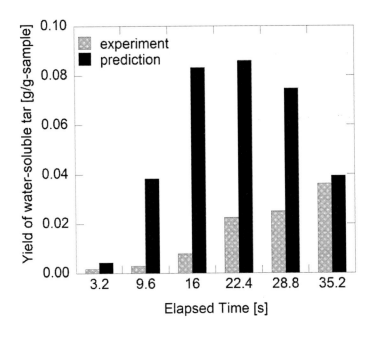

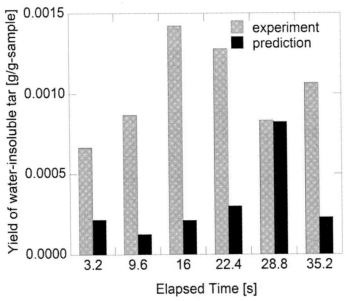

Figure 22. Time profiles of yields of produced tar evolved from mixture of the two components (cellulose 65%, lignin 35%) and predicted values obtained from the superposition of the results for each component by assuming there is no interaction (a) water-soluble tar (b) water-insoluble tar [Fushimi et al., 2009b].

3.2.4. Gas and Tar Evolution from a Mixture of the Three Components (Sample B)

Figures 23 shows the gas evolution rate and tar yield from the mixture of the three biomass components (sample B) in steam gasification. The predicted amounts of gas and tar evolution were calculated by a superposition of those for cellulose (50wt%), xylan (23%) and lignin (27wt%). Figure 23 shows the yields of water-soluble and water-insoluble tars from the mixture of the sample B in steam gasification. Compared with the predicted values, the evolution rate of CO_2 from the sample B was smaller at 3.2 s. However, the CO evolution rate and production of water-soluble and water-insoluble tars from the sample B was almost the same at 3.2 s. This implies that the primary decomposition of lignin is hindered by the interactions with pyrolysates of cellulose and xylan and that the CO evolution from xylan component in the sample B is not affected by other components. These results agree with the report [Hosoya et al., 2007c].

The evolution rates of CO and CO_2 from the sample B show much larger peaks after 3.2 s than the predicted values. However, the evolution rates of water-soluble tar were much smaller than the predicted values. The evolution of water-soluble tar peaked earlier. In addition, slight increase in production of water-insoluble tar was observed compared with the predicted values. These may imply that after 3.2 s cellulose-derived tar decomposes into low-molecular weight gases, especially CO_2 and promotes the evolution of water-insoluble tar from xylan by the interaction with other pyrolysates.

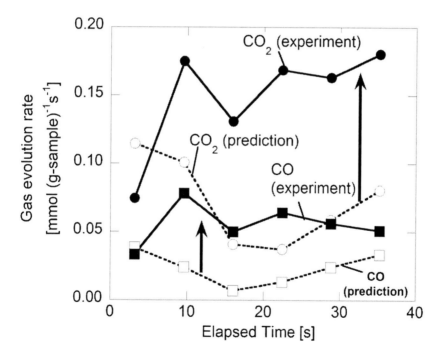

Figure 23. (Continued).

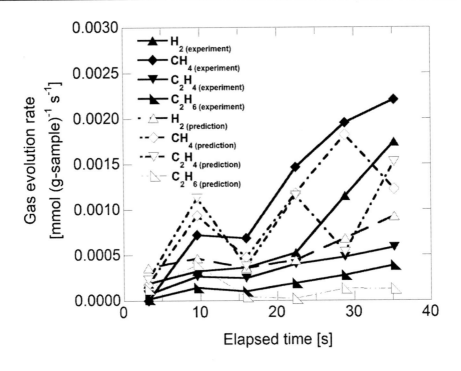

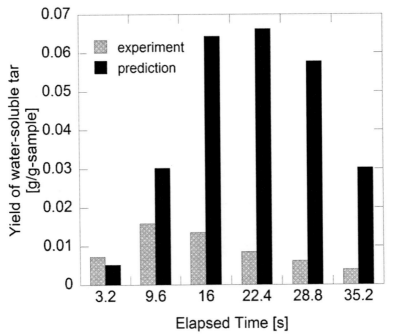

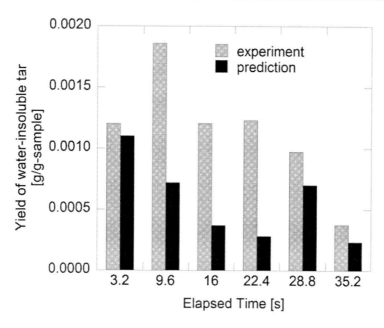

Figure 23. Time profiles of tar yield from mixture of the three components (cellulose 50%, xylan 23%, lignin 27%) and predicted values obtained from the superposition of the results for each component by assuming there is no interaction (a) water-soluble tar (b) water-insoluble tar [Fushimi et al., 2009b].

3.3. Summary of Interaction

Figure 24 summarizes the interaction of cellulose-lignin and cellulose-lignin-xylan from the viewpoint of evolutions of gas, water-soluble tar and water-insoluble tar. The x-axis means the reaction time and the y-axis means the enhancement (positive value) or suppression (negative value) of gas and tar evolution by the interaction. In the case of cellulose and lignin (Figure 24a), the evolutions of water-soluble tar and gas (CO, H_2, CH_4 and C_2H_4) are significantly suppressed and those of water-insoluble tar is enhanced by the interaction.

This would suggest the cellulose pyrolysate adsorb on lignin and char very rapidly and that deoxygenating reactions are accelerated. The evolution of CO_2 is initially (< 9.6 s) suppressed, indicating the primary decomposition of lignin is hindered by interaction with cellulose. After 9.6 s, the CO_2 evolution is substantially enhanced. The evolution of water-soluble tar is delayed by the interaction. These may imply that the polymerization of water-soluble tar derived from cellulose is enhanced by lignin while suppressing the volatilization and then the decomposition of char derived from lignin and polysaccharide takes place. This suggests that the tar evolution can be greatly reduced by adding biomass char which is mainly derived from lignin.

In the case of mixture of cellulose, xylan and lignin (Figure 24b), the evolutions of CO_2, CO, H_2 and CH_4 are enhanced and those of water-soluble tar and levoglucosan are suppressed after 3 s. It was found that the addition of xylan greatly enhances the gas evolution such as CO_2, CO, CH_4 and H_2 and accelerates evolution of water-soluble tar and CO_2. These results imply that xylan enhances evolution and decomposition of water-soluble tar into gases and that xylan decomposes into gases without significant interaction with cellulose or lignin. [Fushimi et al., 2009b]

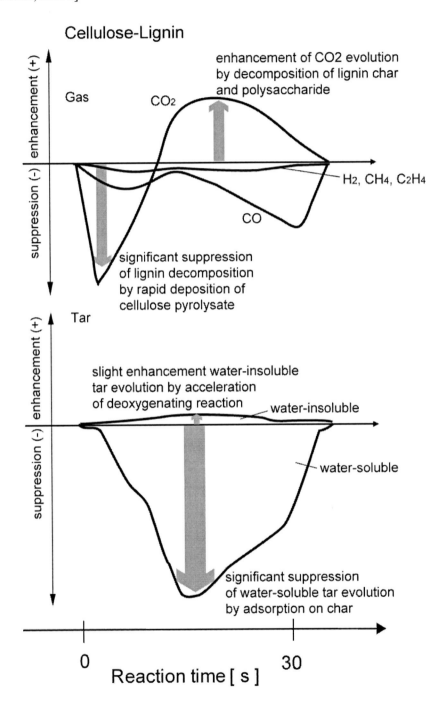

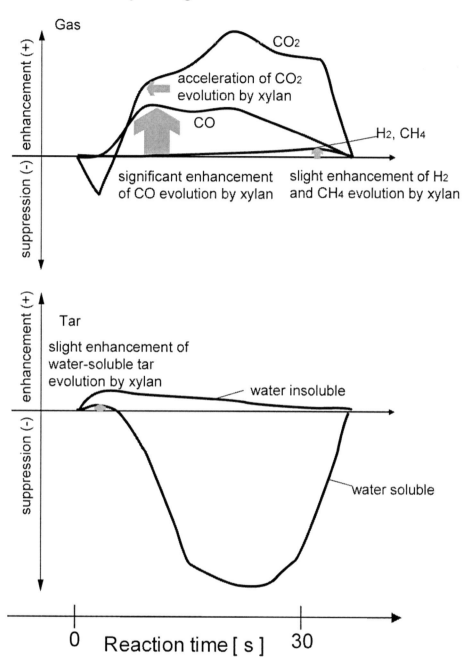

Figure 24. Transitions of the non-dimensional gas and tar yields; (a) shows the effect of the interactions between cellulose and xylan, and (b) shows the effect of the interactions among cellulose, xylan and lignin. [Fushimi et al., 2009b].

CONCLUSIONS

Using the CCDR, steam gasification of cellulose, lignin, xylan and real biomass (pulverized eucalyptus) was investigated at 673 K. In order to clarify the interaction, the steam gasification of the mixtures of cellulose and lignin (sample A) and the mixures of cellulose, lignin and xylan (sample B) was also conducted. From the experimental results, following conclusions are drawn:

1) In the gasification of real biomass, the evolution of water-soluble tar evolved from cellulose occurs earlier than in the gasification of pure cellulose and that the evolution of water-insoluble tar is enhanced.

2) In steam gasification of the mixture of cellulose and lignin, the evolutions of water-soluble tar and gas are significantly suppressed by the interaction between cellulose and lignin. In contrast, the interaction enhances the evolution of water-insoluble tar. This suggests the cellulose pyrolysate adsorb on lignin and char very rapidly and that deoxygenating reaction is accelerated.

3) The evolution of CO_2 is initially suppressed, indicating the primary (initial) decomposition of lignin is hindered by interaction with pyrolysate of cellulose.

4) The CO_2 evolution is then substantially enhanced by the interaction between cellulose and lignin. The interaction delays the evolution of water-soluble tar. These results may imply that volatilization of water-soluble tar derived from cellulose is suppressed by lignin and then the decomposition of char occurs.

5) The addition of xylan greatly enhances the gas evolution such as CO_2, CO, CH_4 and H_2 and accelerates evolution of water-soluble tar and CO_2, implying that the enhancement of decomposition of water-soluble tar into gases and/or xylan decomposes into gases without significant interaction with cellulose or lignin.

ACKNOWLEDGMENTS

The financial supports provided by the 1) "Core Research for Evolutional Science and Technology" grant from the Japan Science and Technology Agency (JST), 2) the New Energy and Industrial Technology Development Organization (NEDO) and 3) the Grant-in-Aid for Young Scientists (B) (number 20760520) from the Japanese Society for Promotion of Science (JSPS) are acknowledged. The authors also thank Mitsubishi Paper Mills Co. Ltd. for providing biomass samples.

REFERENCES

Alves, S. S.; Figueiredo, J. L. *J. Anal. Appl. Pyrolysis*. 1989, 17, 37-46.
Antal, M. J.; Friedman, H. L.; Rogers, F. E. *Combust. Sci. Technol*. 1980, 21, 141-152.
Antal, M. J. *Ind. Eng. Chem. Prod. Res. Dev.,* 1983, 22, 366-375.
Antal, M. J.; Mok, W. S. L.; Várhegyi, G.; Szekely, T. *Energy Fuels*. 1990, 4, 221-225.
Antal, M. J.; Várhegyi, G. *Energy Fuels*. 1994, 8, 1345-1352.
Antal, M. J.; Várhegyi, G. *Ind. Eng. Chem. Res*. 1995, 34, 703-717.

Antal, M. J.; Várhegyi, G. *Energy Fuels*. 1997, 11, 1309-1310.

Antal, M. J.; Várhegyi, G.; Jakab, E. *Ind. Eng. Chem. Res.* 1998, 37, 1267-1275.

Banyasz, J. L.; Li, S.; Lyons-Hart, J.; Shafer, K. H. *J. Anal. Appl. Pyrolysis*. 2001a, 57, 223-248.

Banyasz, J. L.; Li, S.; Lyons-Hart, J.; Shafer, K. H. *Fuel*, 2001b, 80, 1757-1763.

Boroson, M. L.; Howard, J. B.; Longwell, J. P.; Peters, W. A. *AIChE J.,* 1989, 35, 120-128.

Branca, C.; Blasi, C. D. *J. Anal. Appl. Pyrolysis*.2003, 67, 207-219.

Bridgwater. A. V. Fuel 1995, 74, 631–653.

Brown, A. L.; Dayton, D. C.; Daily, J. W. *Energy Fuels*. 2001, 15, 1286-1294.

Burnham, A. K.; Braun, R. L. *Energy Fuels*. 1999, 13, 1-22.

Caballero, J. A.; Font, R.; Marcilla, A. *J. Anal. Appl. Pyrolysis*. 1996, 36,159-178.

Caballero, J. A.; Conesa, J. A.; Font, R.; Marcilla, A. *J. Anal. Appl. Pyrolysis*. 1997, 42, 159-178.

Camarero, S.; Bocchini, P.; Galletti, G. C.; Martínez, A. T. *Rapid Commun. Mass Spectrom.* 1999, 13, 630-636.

Chen, G.; Yu, W; Sjöström, K. *J. Anal. Appl. Pyrolysis*. 1997, 40-41, 491-499.

Conesa, J. A.; Caballero, J. A.; Marcilla, A. Font, R *Thermochimica Acta*, 1995, 254, 175-192.

Conesa, J. A.; Marcilla,A.; Caballero, J. A.; Font, R. *J. Anal. Appl. Pyrolysis*. 2001, 58-59, 617-633.

Cooley, S.; Antal, M. J. *J. Anal. Appl. Pyrolysis*. 1988, 14, 149-161.

Devi, L.; Ptasinski, K. J.; Janssen, F. J. J. G. *Biomass and Bioenergy* 2003, 24, 125–140.

Di Blasi, C.; Lanzetta, M.; *J. Anal. Appl. Pyrolysis*. 1997, 40-41, 287-303.

Di Blasi, C. J. *Anal. Appl. Pyrolysis*. 1998, 47, 43-64.

Encinar, J. M.; González, J. F.; Rodríguez, J. J.; Ramiro, M. J. *Fuel*. 2001, 80, 2025-2036.

Encinar, J. M.; González, J. F.; González, J. *Fuel Process. Technol.* 2002, 75, 27-43.

Evans, R. J.; Milne, T. A.; Soltys, M. N. *J. Anal. Appl. Pyrolysis*, 1986, 9, 207-236.

Evans, R. J.; Milne, T. A. *Energy Fuels*. 1987, 1, 123-137.

Faix, O.; Meier, D. Grobe, I. *J. Anal. Appl. Pyrolysis*. 1987, 11, 403-416.

Ferdous, D.; Dalai, A. K.; Bej, S. K.; Thring, R. W.; Bakhshi, N. N. *Fuel Process. Technol.* 2001 ,70, 9-26.

Ferdous, D.; Dalai, A. K.; Bej, S. K.; Thring, R. W. *Energy Fuels*. 2002, 16, 1405-1412.

Fushimi, C.; Araki, K.; Yamaguchi, Y.; Tsutsumi, A. *Ind. Eng. Chem. Res.* 2003a, 17, 3922-3928.

Fushimi, C.; Araki, K.; Yamaguchi, Y.; Tsutsumi, A. *Ind. Eng. Chem. Res.* 2003b, 17, 3929-3936.

Fushimi, C.; Katayama, S.; Tasaka, K.; Suzuki, M.; Tsutsumi, A. *AIChE J.* 2009a, 55, 529-537.

Fushimi, C.; Katayama, S.; Tsutsumi, A. *J. Anal. Appl. Pyrolysis*, 2009b, 86, 82-89.

Font, R.; Macilla, A.; Verdú, E.; Devesa, J. *J. Anal. Appl. Pyrolysis*. 1991, 21, 249-264.

Garcìa-Pèrez, M.; Chaala, A.; Yang, J.; Roy, C. *Fuel*. 2001, 80, 1245-1258.

Genuit, W.; Boon, J. J.; Faix, O. *Anal. Chem.*, 1987, 59, 508-513.

Gómez, C. J.; Manyà, J. J.; Velo, E.; Puigjanaer, L. *Ind. Eng. Chem. Res.* 2004, 43, 901-906.

Grønli, M.; Antal, M. J.; Várhegyi, G. *Ind. Eng. Chem. Res.* 1999, 38, 2238-2244.

Grønli, M.; Várhegyi, G.; Blasi, C. D. *Ind. Eng. Chem. Res.* 2002, 41, 4201-4208.

Greenwood, P. F.; van Heemst, J. D. H.; Guthrie, E. A.; Hatcher, P. G. *J. Anal., Appl., Pyrolysis.* 2002, 62, 365-373.

Hayashi, J.-i.; Hosokai, S. Sonoyama, N. *Trans. IChem E: Part B.* 2006, 84(B6), 409-419.

Hosoya, T.; Kawamoto, H.; Saka, S. *J. Anal. Appl. Pyrolysis.* 2007a, 78, 328-336.

Hosoya, T.; Kawamoto, H.; Saka, S. *J. Wood. Sci.* 2007b, 53, 351-357.

Hosoya, T.; Kawamoto, H.; Saka, S. *J. Anal. Appl. Pyrolysis.* 2007c, 80, 118-125.

Iniesta, E.; Sánchez, G.; García, A. N.; Marcilla, A *J. Anal. Appl. Pyrolysis.* 2001, 58-59, 967-981.

Jacab, E.; Faix, O.; Till, F.; Székely, T. *J. Anal. Appl. Pyorlysis.* 1995, 35, 167-179.

Kersten, S. R. A.; Wang, X.; Prins, W.; van Swaaij, W. P. M. *Ind. Eng. Chem. Res.* 2005, 44, 8773 -8785.

Kuchonthara, P.; Tsutsumi, A.; Bhattacharya, S. *J. Power Sources.* 2003a, 117, 7-13.

Kuchonthara, P.; Bhattacharya, S.; Tsutsumi, A. *J. Power Sources.* 2003b, 124, 65-75.

Kuchonthara, P.; Bhattacharya, S.; Tsutsumi, A. *Fuel,* 2005, 84, 1019-1021.

Kuchonthara, P.; Tsutsumi, A. *J. Chem. Eng. Jpn.* 2006, 39, 545-552.

Lanzetta, M.; Blasi, C. D. *J. Anal. Appl. Pyrolysis.* 1998, 44, 181-192.

Li, S.; Lyons-Hart, J.; Banyasz, J.; Shafer, K. H. *Fuel,* 2001, 80, 1809-1817.

Mae, K. *J. Jpn. Inst. Energy* (in Japanese), 1996, 75, 167-177.

Manyà, J. J.; Velo, E.; Puigjaner, L. *Ind. Eng. Chem. Res.* 2003, 42, 434-441.

Marcilla, A.; García-García, A.; Asensio, M.; Conesa, J. A. *Carbon,* 2000, 38, 429-440.

Miller, R. S.; Bellan, *J. Combust. Sci. and Tech.* 1997, 126, 97-137.

Miller, R. S.; Bellan, *J. Combust. Sci. and Tech.* 1997, 127: 97-118.

Milosavljevic, I.; Suuberg, E. M. *Ind. Eng. Chem. Res.* 1995, 34, 1081-1091.

Milosavljevic, I.; Oja, V.; Suuberg, E. M. *Ind. Eng. Chem. Res.* 1996, 35, 653-662.

Morf, P.; Hasler, P.; Nussbaumer, T. *Fuel.* 2002, 81, 843-853.

Müller-Hagedorn, M.; Bockhorn, H.; Krebs, L.; Müller, U. J. *Anal. Appl. Pyrolysis.* 2003, 68-69, 231-249.

Müller, R.; Zedtwitz, P.; Wokaun, A.; Steinfeld, A. *Chem. Eng. Sci.* 2003, 58, 5111-5119.

Narayan, R.; Antal, M. J. *Ind. Eng. Chem. Res.* 1996, 35, 1171-1721.

Órfão, J. J. M.; Antunes, F. J. A.; Figueiredo, J. L. *Fuel.* 1999, 78, 349-358.

Piskorz, J.; Radlein, D.; Scott, D. S. *J. Anal. Appl. Pyrolysis.* 1986, 9, 121-137.

Piskorz, J.; Radlein, D.; Scott, D. S.; Czernik, S. *J. Anal. Appl. Pyrolysis.* 1989, 16, 127-142.

Pouwels, A. D.; Eijkel, G. B.; Boon, J. J. *J. Anal. Appl. Pyrolysis.* 1989, 14, 237-280.

Pouwels, A. D.; Boon, J. J. *J. Anal. Appl. Pyrolysis,* 1990, 17, 97-126.

Rao, T. R.; Sharma, A. *Energy.* 1998, 23, 973-987.

Raveendran, K.; Ganesh, A.; Khilar, K. C. *Fuel,* 1996, 75, 987-998.

Reina, J.; Velo, E.; Puigjaner, L. *Ind. Eng. Chem. Res.* 1998, 37, 4290-4295.

Reynolds, J.; Burnham, A. K. *Energy Fuels.* 1997, 11, 88-97.

Río, J. C.; Gutiérrez, A.; Romero, J.; Martínez, M. J.; Martínez, A. T. *J. Anal. Appl. Pyrolysis.* 2001, 58-59, 425-439.

Sagehashi, M.; Miyasaka, N.; Shishido, H.; Sakoda, A. *Bioresource Technology.* 2006, 97, 1272-1283.

Serio, M. A.; Charpenay, S.; Bassilakis, R.; Solomon, P. R. *Biomass Bioenergy.* 1994, 7, 107-124.

Shafizadeh, F. *J. Anal. Appl. Pyrolysis.* 1982, 3, 283-305.

Svenson, J.; Pettersson, J. B. C. *Combust. Sci. and Tech.* 2004, 176, 977-990.

Teng, H.; Wei, Y.-C. *Ind. Eng. Chem. Res.* 1998, 37, 3806-3811.

Tsutsumi, A. *Clean Coal Technology J.* (in Japanese). 2004, 11, 17-22.

Várhegyi, G.; Jakab, E.; Antal, M. *J. Energy Fuels.* 1994, 8, 1345-1352.

Várhegyi, G.; Antal, M. J.; Jakab, E.; Szabo, P. *J. Anal. Appl. Pyrolysis*, 1997, 42, 73-87.

Várhegyi, G.; Szabó, P.; Till, F.; Zelei, B.; Antal, M. J.; Dai, X. *Energy Fuels.* 1998, 12, 969-974.

Várhegyi, G.; Szabó, P.; Antal, M. *J. Energy Fuels.* 2002, 16, 724-731.

Völker, S.; Rieckmann, T. J. *Anal. Appl. Pyrolysis.* 2002, 62, 165-177.

Wang, X.; Kersten, S. R. A.; Prins, W.; van Swaaij, W. P. M. *Ind. Eng. Chem. Res.* 2005, 44, 8786 -8795.

Worasuwannarak, N.; Sonobe, T.; Tanthapanichakoon, W. J. *Anal. Appl. Pyrolysis*, 2007, 78, 265-271.

Yamaguchi, Y.; Fushimi, C.; Tasaka, K.; Furusawa, T., Tsutsumi, A. *Energy Fuels.* 2006, 20, 2681-2685.

Yang, H.; Yan, R.; Chen, H.; Zheng, C.; Lee, D. H.; Liang, D. T. *Energy Fuels.* 2006, 20, 388-393.

Zaror, C. A.; Hutchings, I. S.; Pyle, D. L.; Stiles, H. N.; Kandiyoti, R. *Fuel*, 1985, 64.

In: Lignin
Editor: Ryan J. Paterson

ISBN 978-1-61122-907-3
© 2012 Nova Science Publishers, Inc.

Chapter 7

ISOLATION OF LIGNIN FROM ALKALINE PULPING LIQUORS

Himadri Roy Ghatak
Chemical Technology Department, S.L.I.E.T.,
Longwal, Punjab, India

ABSTRACT

Alkaline pulping is the most widely used pulping method for the pulping of lignocelluloses. The spent liquor generated, called the black liquor, contains almost all the lignin present in the original raw material. Most of this lignin finds its way, as a fuel, to a power boiler for captive power generation. Methods, however, exist for the isolation of lignin from the black liquor. Commercial isolation of lignin from spent pulping liquors amounts to about 1 million tonnes annually, which is less than 2% of the total technical potential. A huge opportunity, thus, exists for tapping this potential to put lignin to better uses and niche applications. Precipitation of lignin by acidification is the current most practiced separation method. Another method is by membrane separation in its different forms. Recently, some electrochemical methods have been used for the isolation of lignin from black liquor. This chapter provides an insight into the different methods for the separation of lignin from alkaline pulping liquors.

1. INTRODUCTION

Lignin is an amorphous, complex polyphenolic plant constituent present in the cell wall of plants [1-2]. Its main function is to cement the cellulose fibres in the plant. It is one of the major polymers occurring in the plant kingdom; the second most abundant terrestrial polymer after cellulose [2-5]. Traditionally, lignosulfonates (lignins resulting from the acidic sulfite pulping process) have been the only type of lignin extensively used in the industry. On the other hand kraft lignin, which accounts nearly 85% of total lignin production in the world, has usually been seen as waste product [3]. The limited use of lignin compared to cellulose is attributable to its molecular structure, i.e., the lignin molecule lacks stereo-regularity and the repeating units of the polymer chain are heterogeneous [6-7]. As a result it is mainly used as fuel to recover part of the energy of the pulping process [2-3]. In a recent review, Suhas et al., have estimated this amount to be 50 million tonnes/year, which is a substantial quantity [5]. By another estimate, more than 26 million tonnes of Kraft lignin are generated as byproducts of pulping operations every year in the United States only [8]. This makes it a potentially abundant and relatively low cost material. In recent times, however, there has been a paradigm shift vis a vis the utility of lignin mainly because of three reasons.

First, as a highly abundant, renewable raw material that is currently underutilized, lignin has attracted increasing interest [7]. Over the last 50 years, advances in petroleum processing and industrial organic chemistry have enabled petroleum to become a primary source of energy and the major feedstock for industrial organic chemicals, accounting for over 90% (by tonnage) of all organic chemicals produced [9]. This, however, is an unsustainable practice owing to the non-renewable finite nature of these resources. Moreover, the pressing need to arrest catastrophic environmental degradation also nedds to be addressed. Consumption of large amounts of limited fossil resources such as petroleum and coal pollute the global environment. There is a growing interest in reducing dependence on petroleum and returning to the use of renewable resources to meet at least some of the demand for fuel and organic chemicals. For a sustainable society in future, reproducible natural materials must be utilized as resources for the chemical industry instead of fossil sources [6, 10]. The most obvious reason to reduce dependence on petroleum feedstocks is that petroleum is a finite resource. Although projections vary, most estimates indicate that over the next 10–40 years petroleum will become considerably less abundant and more expensive than it is today [11]. A conservative estimate puts the availability of fossil fuels at 50000 EJ for coal, 12000 EJ for oil, and 10000 EJ for natural gas [12]. It has been reported that the reserve to production ratio of fossil fuels for North America, Europe and Eurasia, and Asia Pacific is 10, 57, and 40 years, respectively [13]. In addition to being finite, petroleum resources are also isolated to certain geographic regions of the world. Many countries have biomass resources that could be converted into renewable feedstocks for fuels, chemicals, and materials. In fact, in some cases these 'resources' are nuisance wastes that pose a disposal problem. Synthetic chemical products are also undergoing scrutiny for a variety of reasons [9]. Safety, health and environmental concerns are mandating change in the chemical process industry. The ever increasing uses of synthetic materials (polymers and plastics) that are non-biodegradable pose a serious risk to the environment. Although lignin is one of the most durable biopolymers, it is perfectly biodegradable in nature in the presence of some enzymes [14-16] unlike synthetic polymers. With right process chemistries, lignin could be a good renewable raw material for oxygenated aromatic compounds [9]. Due to its phenolic nature, many chemical modifications have been studied [17-18].

Second, straw and other agricultural residues are slowly regaining their importance as candidate raw materials for the pulp and paper industry in Europe and USA [19]. Non-wood fiber pulp mills are normally small in capacity, since the feedstock is bulky, difficult to store, and normally is not available all year around. Technology developed for much larger wood pulping operations may not be technically or economically feasible when pulping non-woods. This is particularly true concerning the processing of the black liquor to recover energy and pulping chemicals [20-21]. In the case of wood chemical recovery, it is normally performed with a highly engineered and very capital-intensive process. To be economically sustainable, such a process requires a scale of operation well beyond the average non-wood pulp mill [22].

Disposal of black liquor poses environmental problems which have been correlated with mutagenic and carcinogenic activity [23-24], and constitutes a threat for human health [25]. Black liquor contributes only 10–15% of the total wastewater, but accounts for nearly 80% of color, 30% of the biochemical oxygen demand (BOD) and 60% of the chemical oxygen demand (COD) of the total pollution load of pulp and paper mill effluent [26]. Discharge of coloured effluents from the pulp and paper mills is not only a serious aesthetic problem, but also interferes with algal and aquatic plant productivity by limiting light transmittance [27]. The primary contributors to the color and toxicity of these effuents are high-molecular-weight lignin and its derivatives [28]. Rather complex chemical structure of lignin resists traditional biological treatment processes due to their non-biodegradable nature [29]. Lignin recovery from black liquor can significantly reduce the amount of environmental pollution. Furthermore, the remaining effluent (after lignin removal) can be more easily degraded by biological treatment [30]. Moreover, many large kraft pulp mills desiring incremental capacity expansion often find the recovery boiler as the major bottleneck. One approach is to retrofit the recovery boiler to handle the increased load [31], but this approach is often expensive and involves many engineering complexities. A simpler approach would be to keep the load of the recovery boiler constant by extracting lignin in proportion to the desired production increase [32-34].

Third, biomass is attractive for energy purposes in at least three respects: contribution to sustainable development, reduction of fossil fuel related anthropogenic greenhouse gas (GHG) emissions, and security of supply [35]. In this context, integrated pulp and paper industries are increasingly being seen as biorefineries. The focus is gradually but surely shifting from the earlier notion of paper as the product from such industries. The modern view looks at these industries as producers of power and chemicals besides paper [36-39]. Several studies of more energy efficient pulp mills have been performed. In the spirit of the biorefinery concept, an increasing interest in further utilization of technical lignins outside the pulp mills has arisen [40]. In the scenario of future pulp mills, lignin and hemicelluloses will replace fossil fuels in other furnaces and will be converted into speciality chemicals [41]. The Swedish 'EcoCyclic Pulp Mill Project' which was a Swedish national research and development programme carried out by STFI, has defined the energy system for a reference mill incorporating best available technology. According to this project, future pulp mills should not only be energy self sufficient using internal biofuel resources, but should also be able to export excess biofuels and lignin. Identifying potential energy efficiency measures for retrofitting of a pulp mill requires the use of appropriate process integration tools, including both traditional pinch technology tools, e.g. [42] and advanced tools specially suited to retrofit situations, e.g. [43-44]. With energy savings measures a part of the lignin present in the black liquor can be extracted and sold as biofuel or chemical feedstock. Loutfi et al. presented a case study for the removal of 57 t/day of product lignin in a mill producing 770 ADt pulp/day [32]. This amounted to 9.5% lignin originally present in the black liquor. They estimated the payback period to be 4 years. Axelsson and co-workers, through computer simulations, have examined the possibility of a 25% capacity expansion in a model 1000 t/day capacity mill through energy saving measures and lignin separation from black liquor [34]. With this approach, they predicted the availability of 0.15 tonnes of lignin for every tonne of pulp produced [34, 45]. Simulations by Wising et al. showed that 37 – 50% lignin removal from black liquor should be possible without jeopardizing the recovery boiler operation [46]. Several other studies have also illustrated the possibility of substantial energy savings in existing pulp mills [47-51].

2. LIGNIN ISOLATION

Isolation of lignin from lignocelluloses follows two general pathways. The first category is based on the hydrolysis and/or solubilisation of carbohydrates (Cellulose and hemicelluloses) leaving behind lignin as an insoluble material. The second, and the commercially important one known as pulping in industrial parlance, uses chemical agents to dissolve the lignin to separate the cellulose fibres. The spent liquor, known as the black liquor, contains the lignin. The overall annual world production of black liquor is approximately 500 million tons [52]. Alkaline pulping is currently the most widely used method having two variations, kraft process and the soda process. In the first, a combination of sodium hydroxide and sodium sulphide is used for lignin dissolution, whereas the second one uses sodium hydroxide alone. In recent years, some mills have also used anthraquinone as a pulping catalyst. The dark brown colour of the black liquor results from the lignin degradation products formed during pulping [53-54]. While the ether bonds of lignin cleave efficiently under alkaline pulping conditions, the formation of condensed structures decrease its reactivity and diminish the solubility [55]. Among the reactions that are known to interfere with the process of alkaline delignification are those involving the formation of carbon–carbon bonds [55]. From conservative estimates, every tonne of pulp produced by alkaline pulping results in 350 – 500 kgs of lignin in black liquor.

When discussing lignins in general and alkali lignins in black liquor specifically, one has to remember that they are polydisperse in many senses [56]. The kraft lignin has components with a wide range of molar masses. Reported values of the number average molar mass for lignin are about 1000–2000 g/mole, while the corresponding weight average molar mass is 2–4 times greater [57]. The apparent pK_a value of kraft lignin has been shown to increase with the molecular weight due to an increased electrostatic counterion attraction, including hydrogen ions, as the size of the fragments increase [58]. The latter might also influence the overall solubility and the colloidal stability of aqueous alkaline lignin solutions. Rudatin et al. declared that the presence of large lignin macromolecules facilitates self-association of lower molecular weight lignin [59]. Like Lindstrom [60], they found that factors influencing the solution conditions, such as the pH value and ionic strength, were important to consider in controlling the stability of kraft lignins. One other important parameter is the influence of temperature on the kraft lignin solution behaviour. Dissociation, solubility, and conformation behaviour of polymers in solution change with change in temperature.

2.1. Acid Precipitation

Lignin can be isolated from black liquor through acidification. Lignin present in the black liquor precipitates out when the pH of the black liquor is brought down. This is the oldest and by far the most widely used method for the isolation of lignin from black liquor and forms the basis of several commercial technologies. Carbon dioxide and sulphuric acid are two acidifying agents commonly used [61-62]. Lignin Precipitation System (LPS) patented and commercialized by Granit SA, of Switzerland, is specifically developed for non-wood soda black liquors. The high purity, lack of sulfur, and other characteristics of the soda non-wood lignins offer new opportunities of speciality applications. In this process the black liquor is treated to precipitate lignin using mineral acid or carbon dioxide, the choice depending on the liquor characteristics. The resulting slurry is conditioned in a maturation step, and is then dewatered and washed in a filter. The lignin cake obtained is dried to generate a high purity lignin powder at about 5 % moisture. By recovering lignin, about 50% of the Chemical Oxygen Demand (COD) in the liquor is removed [63-64]. LignoBoost process was developed by STFI-Packforsk in collaboration with Chalmers University of Technology, Sweden, and sold to Metso which reports a maximum of 70% lignin precipitation with carbon dioxide [65]. In this process, partially evaporated black liquor at about 30% concentration is acidified to a pH of 10.

Initially, the use of sulphuric acid as the acidifying agent for lignin precipitation was thought to be encouraging as part of the acid requirement could be met with the spent acid stream from old chlorine dioxide generators. However, with the advent of methanol based chlorine dioxide generation this option does not look attractive anymore. Fresh sulphuric acid could be used, but the quantity of sulphuric acid required would upset the liquor cycle chemical balance with excess sulphur [32]. Carbon dioxide, therefore, is the preferred acidifying agent. It is also reported to yield an easier to filter precipitate [66]. The possibility of obtaining the carbon dioxide from the boiler flue gases could further improve the economics [32, 34]. Alen et al. found that lignin recovery from pine kraft black liquor at a carbonation pressure of 1500 kPa was maximized at a solid content of approximately 27% [66], whereas for birch liquor, optimal solids concentration was between 30% and 35% [67]. With dilute liquors the carbon dioxide consumption increases and the lignin yield is somewhat decreased while concentrated liquors posed handling problems due to high viscosity [32]. In many laboratory studies, hydrochloric acid has been used as the acidifying agent [2, 68] with the resulting lignin reported to be more reactive [2]. Use of hydrochloric acid, however, would not be industrially viable as it would introduce unacceptable amounts of chloride into the system. Ibrahim et al. used phosphoric and nitric acid besides sulphuric and hydrochloric acid for the precipitation of lignin from soda black liquor of oil palm empty fruit bunch [69] by acidifying to pH 2. No yield values were, however, reported. In mill applications, however, use of nitric acid is likely to cause high NOx emission. In a variation of laboratory procedure, Zhang et al. separated lignin with 80% yield by bringing the black liquor pH down to 2 – 3 by absorption of sulphur dioxide [70]. In an interesting Chinese study lignin was precipitated from black liquor by acidification with biologically formed acids [71]. Villar et al. have proposed a calcium salt in aqueous methanol as the precipitating agent giving 90% lignin recovery [72]. In a recent study, Kim et al. have reported 93% lignin recovery from softwood kraft black liquor by acidifying in the presence of magnetite followed by application of a magnetic field [73].

Precipitation of lignin from black liquor is basically coagulation of a colloid. The initial nucleation phenomenon involves self association of lignin fragments in black liquor resulting in entities of colloidal dimensions. Gradually the number of such colloidal particles increases. These colloidal particles then grow in size due to adhesive interaction finally leading to precipitation. Kinetics of lignin precipitation is believed to follow two well defined regimes. The diffusion limited cluster – cluster aggregation is generally rapid whereas the reaction limited cluster – cluster aggregation is slow [74]. The latter also produces relatively denser and compact aggregates [40]. The precipitation process is strongly influenced by the solution condition. Early studies established the end pH of acidification as one important parameter. At pH 2 – 3, obtained using mineral acids, lignin yield approaching 90% could be possible [61, 66]. With carbon dioxide as the acidifying agent, the optimum end pH would be around 8.5 providing 65 – 75% yield with easy filterability [66, 75]. In recent studies, Ohman found that the end pH and temperature were the two most critical parameters during acid precipitation which influenced the yield as well as the filtration characteristics of the separated lignin [76]. In their experiments with softwood black liquor 70% lignin could be precipitated at pH 8.5 – 9. However, their yield calculations were based on UV absorbance rather than direct gravimetric measures. In a very early study with eucalyptus kraft black liquor, Merewether found the optimum carbonation temperature to be 80 – 85 ^{0}C for best filterability [75]. Wallmo and co-workers investigated the kinetics of lignin precipitation by carbon dioxide acidification and found the rate of decrease of pH to be governed by the CO_2 partial pressure in the gas stream [77]. These investigators have further reported the process to be controlled principally by the reactions and chemical composition of the black liquor rather than the mass transport rate of CO_2. Self association of lignin fragments in solution is reportedly affected by ions of the so-called Hofmeister series [78]. At slightly alkaline pH lignin starts to aggregate into large clusters in the presence of high concentrations of monovalent metal ion salts [79]. Norgren and Mackin studied the use of additives to improve lignin precipitation during acidification. The study revealed that additions of sodium sulfate increased the yield of precipitation at elevated temperatures. Similarly, cationic surfactants gave rise to fast aggregation and relatively high yields mainly due to attractive electrostatic interactions between the cationic surfactant headgroup and the oppositely charged groups on the kraft lignin macromolecules [40]. For non ionic surfactants the effect was not so generic. Gilarranz et al. were able to recover lignin from wheat straw soda black liquor with 80% yield, and 99.5% post washing purity, by acidifying with sulphuric acid. They found the optimum end pH to be 3.5 [80]. Ibrahim et al. extracted lignin from soda black liquor of oil palm empty fruit bunch by acidifying with sulphuric acid and found that decreasing the end pH from 3 to 1 had little impact [81]. From statistical analysis they concluded that extraction temperature had significant effect on yield. Sun et al., in their study on black liquor from oil palm empty fruit bunch, fractionated lignin by precipitating at progressively lower pH values of 4.8, 4.0, 3.0, 2.0, and 1.5. Their results showed that the lignin yield and purity increased with decreasing pH if the carbohydrate degradation products are first removed from the black liquor [82]. Garcia et al. also reported higher precipitate amount when the black liquor is acidified to a lower pH [83].

Low end pH, though yielding more lignin, produced a gelatinous mass which was difficult to filtrate and wash [61]. With kraft black liquors, a very low end pH resulted in emission of hydrogen sulphide in large amounts [32]. Average specific filtration resistance of the precipitated lignin increased with the increase in mixing speed during carbon dioxide precipitation [77]. For hardwood black liquor, the study also illustrated that the presence of hemicelluloses also affected the filtration rate of the precipitated mass with better filtration rates obtained when the hemicelluloses were separated through ultrafiltration and nanofiltration prior to precipitation. Allen et al. found the heating of the suspension necessary to avoid fine dispersion [66]. In later studies Wienhaus et al., also found the carbonation temperature to be an important parameter for the ease of filtration of the precipitated slurry

[84]. However, precipitation temperature in excess of 85 ^{0}C was found to produce a soft tacky lignin [61]. Recently, Ohman et al., conducted elaborate studies on the filtration properties of softwood lignin suspensions obtained through acidification of black liquor [76]. Their quantitative measurements have established the specific filtration resistance to depend on the pH of precipitation and the filtering temperature. In the pH range of 8.5 – 10.5 studied by them, the specific filtration resistance steadily decreased with the decrease in pH. They further suggested a narrow temperature window of 80 – 85 ^{0}C for the optimum filtration results and also related the filterability characteristics of lignin suspensions to the different particle size distributions of the precipitate at different pH. Washing the precipitated lignin post filtration is equally important as some of the sodium present in the black liquor binds to the precipitated lignin. Sodium is bound to the carboxylic and phenolic groups in lignin [32]. High sodium content in the final lignin product is not acceptable in many applications. Washing with dilute sulphuric acid solution is more effective in sodium removal than simple water wash [61]. Hassan proposed treatment of black liquor with selected polymeric flocculants, together with surfactant treatment [85-86]. This resulted in immediate formation of dense biomass aggregates upon acidification to pH 3-4, which could be easily collected. In the commercial LignoBoost process [65], instead of washing the lignin directly after filtration, the filter cake is re-dispersed once again. The new slurry is filtered and, finally, washed using displacement washing. Arguably, when the filter cake is re-dispersed in liquor where the pH and temperature are controlled to approximately that of the final wash liquor, the gradients in pH and ionic strength during the washing stage will be small. The pH change and most of the change in ionic strength, and thus the resulting change in lignin solubility, will then take place in the re-suspension stage instead of in the filter cake during washing. This results in significantly improved wash liquid flow during washing.

The filtrate obtained after lignin precipitation and filtration deserves equal attention as it often requires additional treatment. Bulk of the inorganic chemicals in the black liquor is retained in the filtrate. More than 90% of the sodium in the original black liquor samples can be retained in the filtrate when lignin is precipitated with sulphuric acid at final pH 2.0 [87]. Its viscosity is reported to be significantly lesser than the original black liquor. Moosavifar found the viscosity of the filtrate to be between 2 to 100 times lower than the original samples of softwood and hardwood black liquors after separation of 70% lignin through acid precipitation [88]. At the same time the change in boiling point elevation was not found to be appreciable. Ohman et al. reported progressive reduction in the viscosity of the filtrate with the decrease in precipitation pH and increase in filtration temperature [76]. The return of the filtrates from lignin washing to the liquor recovery cycle would disrupt the chemical balance of the liquor cycle at some mills if sulphuric acid was the acidifying agent [32].

Co-precipitation of silica is one of the technical problems associated with acid precipitation of lignin from black liquor; especially from non-wood black liquors. Selective precipitation of silica has been investigated by several researchers [89-91]. Myreen and co-workers reported that precipitation of silica starts at pH 10.6 and at pH 10.0 almost all silica gets precipitated [21]. However, in the same pH interval lignin also starts to precipitate. For wheat straw soda black liquor, they observed that 90% silica and 35% lignin is separated at pH 9.0 when acidified with carbon dioxide.

2.2. Membrane Based Techniques

As a unit operation, membrane separation has the necessary prerequisites for separating lignin from black liquors, in which separation is accomplished without addition of chemicals. Schematically, in membrane separation, a solution containing macromolecular solute is allowed to pass through a semi-permeable membrane. The membrane allows the solvent and some of the solutes to pass while blocking the rest; the driving force being the pressure difference across the membrane. The process is further classified into microfiltration, ultrafiltration, and nanofiltration based on the cutoff size of the macromolecule to be separated. A diafiltration step may sometimes follow to further purify the retentate and reduce its inorganic content. Another variation of the membrane separation process is reverse osmosis. The two main advantages of using ultrafiltration for lignin recovery are that no adjustment of the pH or temperature is needed, and that the concentration of the liquor to be treated is not crucial [92]. This fact is important when extracting lignin from black liquor taken directly from the digester to improve delignification during pulping. This gives advantages in terms of pulp quality and bleachability. However, Woerner and McCarthy [93] had previously suggested that to produce purified high molecular weight lignin, UF should be operated at low pressure and under high alkalinity. One key aspect of black liquor ultrafiltration is that the lignin gets fractionated in the process; high molecular weight fraction retained while low molecular weight ones permeating through the membrane [94]. Fractionated high molecular weight lignins are more valuable as chemical feedstock for specialty chemicals. Membranes used are either polymeric or ceramic in nature. Polymeric membranes used for this purpose include, cellulose acetate [95], cellulose triacetate [96], polyacrylonitrile, polyaryletherketone, and polyethersulphone [97]. Among the ceramic membranes studied for black liquor are composites of α-alumina coated with an inner skin layer of either TiO_2 or ZrO_2 [52, 92, 97].

The frictional pressure drop is the pressure difference between the inlet and outlet of the module.

$$\Delta p_f = \Delta p_{in} \; \square \; \Delta p_{out} \tag{1}$$

The driving force for the membrane separation, the transmembrane pressure is the pressure difference across the membrane. For cross flow membrane operation, it can be computed as

$$\Delta p = (\Delta p_{in} + \Delta p_{out})/2 - p_{perm} \tag{2}$$

where, p_{perm} is the pressure at the permeate side. The rate of separation is measured in terms of the permeation flux which is defined as

$$J = (1/A) \, (dV/dt) \tag{3}$$

where, A is the area of the membrane and V is the permeate volume. As expected for a pressure driven separation process, the rate of separation, measured as permeation flux, is strongly dependent on the transmembrane pressure. Since a mechanical device, such as a pump, would be needed to maintain this pressure, it governs the energy consumption in the process. Bhattacharjee et al., for cellulose acetate membranes, found the flux to linearly increase with the increase in transmembrane pressure up to a certain level beyond which the process became mass transfer controlled [96]. This pressure independent limiting flux phenomenon was explained in terms of formation of gel/polarized layer of solute to form on the membrane surface. Transmembrane pressure also affects the extent of lignin recovery with higher pressures giving higher yields [96]. Tanistra and Bodzek found that the lignin purity also increased with the increase in transmembrane pressure [94]. Also, in their work, the permeate flux did not level off with increasing pressure from 0.5 MPa to 1.0 MPa. For ceramic membranes, Dafinov et al., could not get a definite relationship between permeate flux and transmembrane pressure [52]. While membranes with 1000 Da and 5000 Da cut-offs had the expected flux increase with pressure, the one with 15000 Da cut-off had permeate flux independent of pressure. This behaviour was explained by pore blocking due to specific adsorption of lignin entities inside the pore. These researchers could not observe any limiting flux phenomenon in the studied transmembrane pressure range of 3 – 7 bar. Wallberg and Jonsson found linear dependence of flux on tranmembrane pressure for ceramic membrane, as shown in figure below [98]. In addition, Volume reduction (VR), and volume reduction factor (VRF), are two important parameters. VR is defined as the ratio between the permeate volume withdrawn and the initial feed volume, and VRF as the ratio between the initial feed volume and the retentate volume.

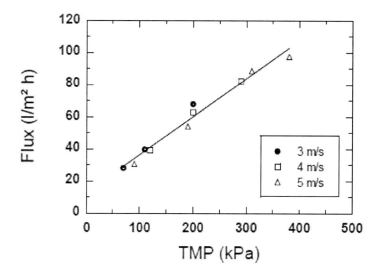

Figure 1. Linear flux dependence on transmembrane pressure for ultrafiltration of softwood kraft black liquor through ceramic membranes at different cross flow velocities [98].

Early attempts were made to separate water and inorganic substances from black liquor through reverse osmosis or a combination of reverse osmosis and ultrafiltration, leaving behind an organic rich retentate [99-100]. Wallberg et al. investigated the retention of lignin and the flux for three polymeric membranes with cut-offs of 4000, 8000 and 20000 Da, with the retention of lignin found to be 80%, 67% and 45%, respectively [101]. Liu et al. could separate more than 75% organics larger than 60,000 Da from black liquor obtained from a pulp mill in China using ultrafiltration with a 60,000 Da cut-off membrane at pH 10 and 11 [97]. These investigators studied the performance of several polymeric ultrafiltration membranes and ceramic microfiltration membranes for treating wheat straw kraft black liquor. They reported very high lignin retention of nearly 80% in 30 minutes microfiltration; with the membrane pore size between 0.2 and 0.8 μm at 200 kPa transmembrane pressure [97, 102]. With polyethersulphone ultrafiltration membranes of 10000 and 3000 Da cut-off 100% lignin retention could be achieved in same time duration. Significantly, more than 80% of the silica present in the black liquor was reportedly separated out with the lignin. Also, in the microfiltration experiments, the sodium concentration on the two sides of the membrane remained same thus providing a lignin with high ash content [102]. Tanistra and Bodzek used polyacrylonitrile membranes of 20000 Da cut-off for black liquor ultrafiltration. They achieved maximum lignin retention coefficient of 80% with 83% lignin purity with permeate flux of 0.45 m^3/m^2d at 1.0 MPa transmembrane pressure [94]. The lignin purity improved to 88% when the retentate from the ultrafiltration was subjected to five stages of diafiltration. Similarly, Wallberg et al. improved the lignin purity, obtained in ultrafiltration retentate, from 36% to 78% through semi-continuous diafiltration using a 8000 Da cut-off polysulfone membrane at a volume reduction of 1.09 [101]. It is apparent from all studies that molecular weight cut-off plays an important role in lignin retention and lignin purity, which is often a trade off. Higher cut-off membranes give low lignin yield but the product is of higher purity. Overall, ultrafiltration, rather than acid precipitation, should be the preferred route if lignin fractionation and high purity is the objective. Toledano et al. found that ultrafiltration gives lignin that is less contaminated with lignin carbohydrate complexes, and also provides for controlling the molecular weight of the lignin fractions [103].

The chemical and temperature resistance of the ceramic membranes compared to polymeric membranes makes them particularly attractive for the treatment of black liquor. They also have longer operating lives than polymeric membranes and can be used at higher pressures. Using ceramic membranes, lignin and hemicelluloses can even be isolated from kraft cooking liquor taken directly from the digester without cooling or adjustment of the pH [98]. Polymeric membranes are generally not suited for such applications as these membranes have limited temperature and pH resistance. Another important aspect of ceramic membranes is the possibility of regeneration of the fouled membrane through high temperature heat treatment [97]. Jonsson and Wallberg performed pilot scale ultrafiltration studies on hardwood cooking liquors from the digester, as well as partially concentrated black liquor from evaporators, using Al_2O_3-TiO_2 ceramic membrane with a cut-off of 15000 Da [92]. At 90 ^{0}C, they obtained a final volume reduction of 0.90 during ultrafiltration of cooking liquor and 0.66 during ultrafiltration of black liquor equivalent to a volume reduction factor of 10 for the cooking liquor and 3 for the black liquor. This was achieved at transmembrane pressures of 400 kPa and 100 kPa, respectively, for the two liquors. The cross flow velocities were 3.7 m/s and 5.0 m/s. Lignin separation was reported to be 30–50% for the black liquor directly taken from the digester and 15–25% for the semi-concentrated liquor with near zero retention of inorganics. Almost the entire amount of hemicelluloses was also retained. The same investigators reported 70% lignin separation with polymeric membranes at 60 ^{0}C and 800 kPa transmembrane pressure [104]. In subsequent studies these investigators devised a pressurized ceramic ultrafiltration membrane module to treat digester liquor at 140 – 160 ^{0}C [98]. Lignin retention of 10 – 30%, with not so impressive purity, was achieved when operated in a feed and bleed mode. But most significantly, the pilot facility could be run continuously for eight months without any serious membrane fouling, when cleaned once in a week. Dafinov et al., in continuous operation could separate about 50% organic matter from soda antharaquinone black liquor with a 1000 Da cut-off ceramic membrane, with little variation with varying transmembrane pressure [52]. They further reported about 13% inorganic matter, mainly attributed to the association of the ions with the lignin macromolecules retained by the membrane. Retention was 25 – 30% when using membranes of 5000 and 15000 Da cut-off, with about 10% inorganic retention. In discontinuous experiments, with 5000 Da cut-off membrane, an almost steady state flux of 35 l/m^{2}h was obtained up to a volume reduction factor of 1.6. Interestingly, with membrane of 15000 Da cut-off the permeate flux steadily declined to a final value of 17 l/m^{2}h, at a volume reduction factor of 1.9, mainly due to progressive pore blockage. Liu et al., tested the long term suitability of a 0.2 μm α-alumina ceramic microfiltration membrane. At 32 ^{0}C the average permeate flux was 200 l/m^{2}h for 374 hour of operation. It increased to 400 l/m^{2}h at 63 ^{0}C for 999 hours of operation [97].

One of the major drawbacks in the use of UF for treating Kraft black liquor is the decline of flux with time, and is one of the reasons for the lower commercial acceptability of the process.

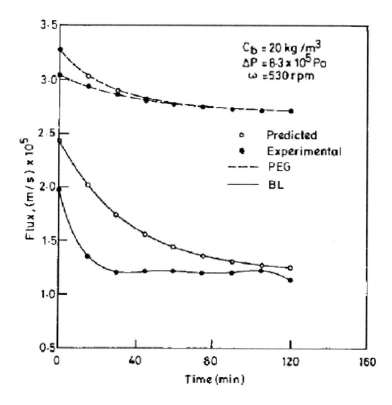

Figure 2. Decline of flux with time in ultrafiltration of hardwood kraft black liquor through cellulose acetate membrane of 5000 Da cut-off [105].

The flux of pure solvent through the membrane increases linearly with the transmembrane pressure according to the well known Darcy's equation. For solutions containing macromolecular solute such as black liquor the permeate flux follows the darcy's law only up to a certain transmembrane pressure. Thereafter, it reaches a limiting value giving rise to a phenomenon called concentration polarization. Under this condition the solute concentration at the membrane surface considerably exceeds the bulk concentration, giving rise to an osmotic pressure working against the externally applied pressure. Flux decline may also be attributed to gel formation or fouling of the membrane resulting from reversible or irreversible pore-plugging [96]. Such complexities do not allow a rigorous quantitative estimation of the flux and its variation from theoretical considerations, though several models have been proposed by different investigators. For an extended knowledge of such mathematical deliberations the reader is referred to the related references given at the end of the chapter [95- 96, 105-108].

The concentration profile in the film layer [105] is fully developed and reached steady state within a few seconds only. Thereafter, the flux decline occurs only due to the formation of a secondary gel-type layer, the thickness of which continuously grows with time, thus giving extra resistance to permeate flow [105-106]. Sridhar, for asymmetric cellulose acetate complex membrane with 5000-molecular weight cut-off, concluded the phenomenon of polarized layer deposition as the key cause that reduces flux of black liquor through the membrane [95]. Jonsson and Wallberg [92] reported steady decline in the ultrafiltration flux through ceramic membrane from 120 l/m^2h to 10 l/m^2h while concentrating to a volume reduction of 0.9 for digester cooking liquor at 20% concentration. Similarly, for 30% concentration black liquor they observed a steady decline in the permeation flux from 50 l/m^2h to about 12 l/m^2h for a volume reduction of 0.66. The calculated average flux was 72 l/m^2h and 33 l/m^2h, respectively. The energy required for pumping was 6.5 and 30.5 kWh/m^3 permeate, respectively, for the two liquors. Increasing the turbulence to improve permeate flux through the membrane is one of the alternatives. The simplest approach would be to let the feed solution to flow parallel to the membrane surface, known as cross flow ultrafiltration. This causes reduction in growth of the boundary layer near the surface and increase in back diffusion from the surface to the bulk of the solution [108]. Satyanarayan et al. studied black liquor ultrafiltration with different flow patterns and concluded that the operating Reynolds number is the primary factor governing performance [109]. The highest flux was in stirred cells for any transmembrane pressure. Flux was lower in rectangular flow cells and the least in radial flow cells. Their calculated values of Reynolds number for the three cells were 5, 925, and 16666, respectively. The highest lignin recovery was also obtained for the stirred cell. Bhattacharjee et al. attempted a stirred rotating disc ultrafiltration module to minimize the flux decline phenomena by minimizing the effect of concentration polarization [96]. They contended that in a stirred rotating disk membrane module, it is possible to create very high shear in the vicinity of the membrane resulting in reduced boundary layer resistance. Further they pre-treated the black liquor through centrifugation and microfiltration through polyether sulphone membrane of 0.45 μm pore size. 23% increase in initial flux was reported compared to fixed membrane at 7 kg/cm^2 transmembrane pressure by rotating the membrane at 450 rpm. This improved to 31% initial flux increase at 600 rpm membrane rotation. The benefits of membrane rotation were even more pronounced after extended period of operation. Similar benefits, synergistic with benefits from membrane rotation, were also obtained by stirring the black liquor. In such arrangements the transmembrane pressure was relatively uniform and they could yield high retentate concentration in single pass. Similar results were also obtained with alkaline sulphite liquor [110]. Drawbacks of such systems, however, are their complexity and high cost of manufacture and limited membrane area, which seldom exceeds 2 or $3m^2$. The effect of membrane molecular weight cut-off on the permeate flux is not very straightforward. While Satyanarayan et al. found the flux to increase with the increase in molecular weight cut-off [109], Liu et al. in their experiments with different polymeric ultrafiltration membranes, and ceramic microfiltration membranes, could not find any relationship between the limiting flux and the molecular weight cut-off or the pore size of membrane [97, 102]. However, their study revealed that the mechanism for flux decline depended, not only on the molecular weight cut-off, but also on membrane material.

A comparison between precipitation and ultrafiltration by Uloth [111] showed acid precipitation to be cost effective. However, the removal of high molecular-weight lignin by ultrafiltration was found to be feasible and economically attractive in simulation studies by Kirkman [112]. In most cases, the capital cost is the dominating cost for an ultrafiltration plant, and the annuity is thus crucial for the economy of an ultrafiltration plant. Membranes also require periodic replacement. Polymeric membranes have a lifetime of the order of about 18 months. In comparison ceramic membranes may last upto 6 years. The operating costs arise from electricity, the replacement of membranes, cleaning, maintenance and labour costs. Electricity is needed mostly for pumping. Jonsson and Wallberg in their cost analysis study reported unit capital cost of € $3300/m^2$ membrane area for ceramic membrane and lignin production cost of € 60/tonne from digester cooking liquor and € 33/tonne for semi-concentrated black liquor [92].

2.3. Electrochemical Methods

Black liquor is an aqueous electrolyte, consisting of Na^+ ions as the principal cation, with negligible amounts of other cations like K^+, Ca^{++} etc. also present. Anions present in the black liquor are mainly lignin degradation products resulting from the reaction of native lignin with alkali during alkaline pulping. Other anions are mainly OH^- ions that come from the alkali left unreacted during pulping. It can, therefore, be a potential candidate for electrochemical separation techniques.

Electrodialysis is a membrane based separation method where the preferential movement of some ions across the membrane is driven by an applied electric potential. Ion exchange membranes are used. In early studies, Radhamohan and Basu [113] attempted the regeneration of sodium hydroxide and lignin from rice straw black liquor in a two stage process. The first stage involved electrodialysis followed by a second stage of electrogravitational separation. In the process developed by Azarniouch and Prahacs [114] the black liquor is on the anodic side of a cation permselective membrane. Upon application of electric potential, the sodium ions present in the black liquor cross the membrane to the cathodic side to be neutralized and converted to the corresponding base sodium hydroxide. Hydronium ion is produced at the anode to maintain the electrical neutrality of the anolyte. This decreases the black liquor pH and causes the lignin to precipitate. About 75% lignin and 80% sodium could be recovered. Hydrogen is produced as a byproduct in the cathode chamber. Oxygen is produced at the anode. Beaudry and Caro have reported the results of pilot plant studies [115] with the optimum conditions being; temperatures 60 °C, current density 1200 A/m^2, final pH of black liquor 8.8, and 10% catholyte caustic strength. Electrical energy consumption was 4.5 kWh/kg NaOH and the membrane used was Nafion 450 of Du Pont make. Cloutier et al. studied the effect of process parameters using two different membranes, Nafion 324 from Du Pont and R 4010 from RAI Corporation, and four different anode materials, graphite, nickel, platinum, and iridium oxide coated titanium [116-117].

Among the two membranes, Nafion 324 provided better energy efficiency due to its better ion selectivity which restricted back diffusion of hydroxyl ions. Among the anode materials iridium oxide coated titanium performed best for energy efficient operation. Increasing current density lowered the energy efficiency but produced higher sodium flux through the membrane. One important finding was that addition of sodium sulphate increased the energy efficiency of the process.

As with ultrafiltration, membrane and electrode fouling is a major problem with electroldialysis. As the membrane resistance increases due to fouling, more and more electrical potential is required to maintain the separation rate. This leads to increased energy consumption and added operating cost. Fouling can also reduce the ion selectivity of the membrane. The problem is further compounded if the fouling is irreversible. This would require periodic replacement of the membrane. Mishra and Bhattacharya [118] studied electrodialysis of black liquor using different membranes and noticed excessive lignin deposition on the membrane in the case of anion exchange membranes. This made the process very diffficult. One way to avoid fouling on the anode is to maintain necessary turbulence in the electrocell [119]. This does not allow the lignin to precipitate on the anode.

Electrolysis can separate lignin from black liquor through electrodeposition at anode [120-121]. In laboratory experiments, partial lignin recovery of about 10-15% could be achieved by electrolysis of black liquor requiring 1.8-2.0 kWh electricity per kg of lignin removed [122]. The process looks attractive due to co-production of hydrogen at the cathode. However, the process was slow because the electrolysis had to be carried out at low interelectrode potential to avoid oxygen evolution [123]. Also, the lignin separated through electrolysis showed spectroscopic and thermochemical characteristics distinct from that obtained by other isolation methods [68, 124].

REFERENCES

[1]   Canetti, M; Bertini, F. Supermolecular structure and thermal properties of poly(ethylene terephthalate)/lignin composites. *Composites Sc. Tech.* 2007, 67(15-16), 3151-3157.

[2]   Tejado, A.; Pena, C.; Labidi, J.; Echeverria, J.M.; Mondragon, I. Physico-chemical characterization of lignins from different sources for use in phenol–formaldehyde resin synthesis. *Bioresource Tech.* 2007, 98(8), 1655-1663.

[3]   Pouteau, C.; Dole, P.; Cathala, B.; Averous, L.; Boquillon, N. Antioxidant properties of lignin in polypropylene. *Poly. Deg. Stab.* 2003, 81(1), 9-18.

[4]   Gosselink, R.J.A.; deJong, E.; Guran, B.; Abacherli, A. Coordination network for lignin—standardisation, production and applications adapted to market requirements (EUROLIGNIN). *Ind. Crops Prod.* 2004, 20(2), 121–129.

[5]   Suhas; Carrot, P.J.M.; Ribeiro Carrott, M.M.L. Lignin – from natural adsorbent to activated carbon: A review. *Bioresource Tech.* 2007, 98(12), 2301-2312.

[6]   Parajuli, D.; Inoue, K.; Ohto, K.; Oshima, T.; Murota, A.; Funaoka, M.; Makino, K. Adsorption of heavy metals on crosslinked lignocatechol: a modified lignin gel. *React. Func. Poly.* 2005, 62(2), 129-139.

[7]   Lu, F; Ralph, J. Derivatization followed by reductive cleavage (DFRC method), a new method for lignin analysis: Protocol for analysis of DFRC monomers. *J. Agric. Food Chem.* 1997, 45(7), 2590-2592.

[8]   Suparno, O.; Covington, A.D.; Philips, P.S.; Evans, C.S. An innovative new application for waste phenolic compounds: Use of Kraft lignin and naphthols in leather tanning. *Resources Conserv. Recyc.* 2005, 45(2), 114-127.

[9]   Embree, H.D.; Chen, T.; Payne, G.F. Oxygenated aromatic compounds from renewable resources: motivation, opportunities, and adsorptive separations. *Chem. Eng. J.* 2001, 84(2), 133-147.

[10] Kubo, S.; Uraki, Y.; Sano, Y. Preparation of carbon fibers from softwood lignin by atmospheric acetic acid pulping. *Carbon.* 1998, 36(7-8), 1119-1124.

[11] Kerr, R.A. The next oil crisis looms large — and perhaps close. *Science.* 1998, 281, 1128–1131.

[12] Salgado, J.R.; Martinez, A.E. Roadmap towards a sustainable hydrogen economy in Mexico. *J. Power Sources.* 2004, 129(2), 255–263.

[13] BP Statistical Review of World Energy. *British Petroleum,* 2003.

[14] Xia, Z.; Yoshida, T.; Funaoka, M. Enzymatic degradation of highly phenolic lignin-based polymers (lignophenols). *European Poly. J.* 2003, 39(5), 909-914.

[15] Alexy, P.; Kosikova, B.; Podstranka, G. The effect of blending lignin with polyethylene and polypropylene on physical properties. *Polymer.* 2000, 41(13), 4901–4908.

[16] Guerra, A.; Mendonca, R.; Ferraz, A.; Lu, F.; Ralph, J. Structural characterization of lignin during Pinus Taeda wood treatment with Ceriporiopsis Subvermispora. *App. Env. Microbiol.* 2004, 70(7), 4073-4078.

[17] Pucciariello, R.; Villani, V.; Bonini, C.; D'Auria, M.; Vetere, T. Physical properties of straw lignin-based polymer blends. *Polymer.* 2004, 45(12), 4159-4169.

[18] Thring, R.W.; Breau, J. Hydrocracking of solvolysis lignin in a batch reactor. *Fuel.* 1996, 75(7), 795-800.

[19] Chaudhary, P.B. A concept of a viable alkali recovery system for a non-wood pulp mill. Tappi 1993 Pulping Conference Proceedings; Tappi Press: Atlanta, 1993; pp 273-279.

[20] Xiao, C.; Bolton, R.; Pan, W.L. Lignin from rice straw Kraft pulping: Effects on soil aggregation and chemical properties. *Bioresource Technol.* 2007, 98(7), 1482-1488.

[21] Myreen, B. A novel recovery process for small-scale non-wood pulp mills. Tappi 1998 International Chemical Recovery Conference Proceedings; Tappi Press: Atlanta, 1998; pp 823-829.

[22] Lora, J.H.; Caro, R.F.; Cloutier, J.N. Treatment of nonwood black liquors by electrolysis and lignin precipitation. Tappi 2005 Engineering, Pulping and Environmental Conference Proceedings; Tappi Press: Atlanta, 2005; pp 267-271.

[23] Helmy, S.M.; El Rafie, S.; Ghaly, M.Y. Bioremediation post-photo-oxidation and coagulation for black liquor effluent treatment. *Desalination.* 2003, 158(1-3), 331-339.

[24] Yang, R.; Pickard, J; Omatani, K. Assessment of industrial effluent toxicity using flow through fish egg/alleviants/fry (EAF) toxicity test. *Bull. Environm. Contam.* 1999, 62(4), 440-447.

[25] Lara, M.A.; Rodriguez-Malaver, A.J.; Rojas, O.J.; Holmquist, O.; Gonzalez, A.M.; Bullon, J.; Penaloza, N.; Araujo, E. Black liquor lignin biodegradation by *Trametes elegans. Int. Biodeterior. Biodegr.* 2003, 52, 167–173.

[26]   Ma, F.; Xiong, Z.; Zheng, Y.; Yu, X.; Zhang, X. Repeated batch process for biological treatment of black liquor using brown-rot basidiomycete Fomitopsis sp. IMER2. *World J. Microbiol. Biotechnol.* 2008, 24(11), 2627–2632.

[27]   Panchapakesan, B. Process modifications, end-of-pipe technologies reduce effluent color. *Pulp Paper* 1991, 65(8), 82–84.

[28]   Grover, R.; Marwaha, S.S.; Kennedy, J.F. Studies on the use of an anaerobic baffled bioreactor for the continuous anaerobic digestion of pulp and paper mill black liquor. *Process Biochem.* 1999, 34(6-7), 653–657.

[29]   Zaied, M.; Bellakhal, N. Electrocoagulation treatment of black liquor from paper industry. *J. Haz. Mat.* 2008, 163(2-3), 995-1000.

[30]   Lora, J.H.; Glasser, W.G. Recent industrial applications of lignin: A sustainable alternative to nonrenewable materials. *J. Pol. Env.* 2002, 10(1), 39-48.

[31]   Williamson, D.; Santyr, G.M. Capacity upgrade of an existing chemical recovery boiler. Tappi Engineering Conference Proceeding; Tappi Press: Atlanta, 1988; pp 381-388.

[32]   Loutfi, H.; Blackwell, B.; Uloth, V. Lignin recovery from kraft black liquor: preliminary process design. *Tappi J.* 1991, 74(1), 203-210.

[33]   Davy, M.F.; Uloth, V.C.; Cloutier, J.N. Economic evaluation of black liquor treatment processes for incremental kraft pulp production. *Pulp Paper Can.* 1998, 99(2), 35-39.

[34]   Axelsson, E.; Olsson, M.R.; Berntsson, T. Increased capacity in kraft pulp mills: Lignin separation and reduced steam demand compared with recovery boiler upgrade. *Nord. Pulp Pap. Res. J.* 2006, 21(4), 485-492.

[35]   Adahl, A.; Harvey, S.; Berntsson, T. Assessing the value of pulp mill biomass savings in a climate change conscious economy. *Energy Policy.* 2006, 34(15), 2330-2343.

[36]   Ragauskas, A.J.; Nagy, M.; Kim, D.H.; Eckert, C.A.; Hallett, J.P.; Liotta., C.L. From wood to fuels: Integrating biofuels and pulp production. *Ind. Biotech.* 2006, 2(1), 55-65.

[37]   Thorp, B., Biorefinery offer industry leaders business model for major change. *Pulp and Paper.* 2005, 79(11), 35-39.

[38]   Thorp, B.; Raymond, D. Forest biorefinery could open door to bright future for PandP industry. *Paper Age.* 2004, 120(7), 16-18.

[39]   Consonni, S.; Katofsky, R.E.; Larson, E.D. A gasification-based biorefinery for the pulp and paper industry. *Chem. Eng. Res. Des.* 2009, 87(9), 1293–1317.

[40]   Norgren, M.; Mackin, S. Sulfate and surfactants as boosters of kraft lignin precipitation. *Ind. Eng. Chem. Res.* 2009, 48(10), 5098-5104.

[41]   Axegard, P; Tomani, P; Backlund, B. The pulp mill based biorefinery. Proceedings of PulPaper 2007; Helsinki, 2007, pp 19–26.

[42]   Linnhoff, B.; Mason, D.R.; Wardle, I. Understanding heat exchanger networks. *Computers Chem. Eng.* 1979, 3(1-4), 295-302.

[43]   Carlsson, A.; Franck, P.A.; Berntsson, T. Design better heat exchanger network retrofits. *Chem. Eng. Prog.* 1993, 89, 87–96.

[44]   Nordman, R.; Berntsson, T. New Pinch Technology based HEN analysis methodologies for cost-effective retrofitting. *Canadian J. Chem. Eng.* 2002, 79(4), 655-662.

[45]   Olsson, M.R.; Axelsson, E.; Berntsson, T. Exporting lignin or power from heat-integrated kraft pulp mills: A techno-economic comparison using model mills. Nord. *Pulp Pap. Res. J.* 2006, 21(4), 476-484.

[46]   Wising, U.; Algehed, J.; Berntsson, T.; Delin, L. Consequences of lignin precipitation in the pulp and paper industry. *Tappi J.* 2006, 5(1), 3-8.

[47] Towers, M. Energy reduction at a kraft mill: Examining the effects of process integration, benchmarking, and water reduction. *Tappi J.* 2005, 4(3), 15-21.

[48] Fouche, E.; Banerjee, S. Improving energy efficiency in the forest products industry. *Tappi J.* 2004, 3(11), 24-26.

[49] Thollander, P.; Ottosson, M. An energy efficient Swedish pulp and paper industry – exploring barriers to and driving forces for cost-effective energy efficiency investments. *Energy Efficiency.* 2008, 1(1), 21-34.

[50] Sarimveis, H.K.; Angelou, A.S.; Retsina, T.R.; Rutherford, S.R.; Bafas, G.V. Optimal energy management in pulp and paper mills. *Energy Conv. Manag.* 2003, 44(10), 1707-1718.

[51] Laaksometsa, C.; Axelsson, E; Berntsson, T.; undstrom, A. Energy savings combined with lignin extraction for production increase: case study at a eucalyptus mill in Portugal. *Clean Tech. Environ. Policy.* 2009, 11(1), 77-82.

[52] Dafinov, A.; Font, J.; Garcia-Valls, R. Processing of black liquors by UF/NF ceramic membranes. *Desalination.* 2005, 173(1), 83-90.

[53] Rodriguez-Mirasol, J.; Cordero, T.; Rodriguez, J.J. High temperature carbons from kraft lignin. *Carbon.* 1996, 34(1), 43-52.

[54] Helmy, S.M.; El Rafie, S.; Ghaly, M.Y. Bioremediation post-photo-oxidation and coagulation for black liquor effluent treatment. *Desalination.* 2003, 158(1-3), 331-339.

[55] Argyropoulos, D.S.; Sun, Y.; Palus, E. Isolation of residual kraft lignin in high yield and purity. *J. Pulp Paper Sc.* 2002, 28(2), 50-54.

[56] Norgren, M.; Edlund, H.; Nilvebrant, N.O. Physiochemical differences between dissolved and precipitated kraft lignin fragments as determined by PFG NMR, CZE and quantitative UV spectrophotometry. *J. Pulp Pap. Sc.* 2001, 27(11), 359-363.

[57] Jacobs, A.; Dahlman, O. Absolute molar mass of lignins by size exclusion chromatography and MALDI-TOF mass spectroscopy. *Nord. Pulp Pap. Res. J.* 2000, 15(2), 120–127.

[58] Norgren, M.; Lindstrom, B. Dissociation of phenolic groups in kraft lignin at elevated temperatures. *Holzforschung* 2000, 54(5), 519–527.

[59] Rudatin, S.; Sen,Y.L.; Woerner, D.L. Association of kraft lignin in aqueous solution. *ACS Symp. Ser.*, 1989, 397, Ch. 11, 144–154.

[60] Lindstrom, T. The colloidal behaviour of kraft lignin. Part II: Coagulation of kraft lignin sols in the presence of simple and complex metal ions. *Colloid Poly. Sc.* 1980, 8(2), 168–173.

[61] Uloth, V.C.; Wearing, J.T. Kraft lignin recovery: acid precipitation versus ultrafiltration. *Pulp Pap. Can.* 1989, 90(9), T310.

[62] Forss, K.G.; Fuhrmann, A. Finnish plywood, particleboard and fiberboard made with a lignin base adhesive. *Paperi Puu.* 1976, (58)11, 817.

[63] Abacherli, A.; Doppenberg, F.; Lora, J.H. Chemical recovery from non-wood fiber black liquors by lignin precipitation and wet oxidation. Tappi 1999 Pulping Conference Proceedings, Volume 1; Tappi Press: Atlanta, 1999; pp 5-11.

[64] Abacherli, A.; Doppenberg, F. Method for preparing alkaline solutions containing aromatic polymers. *US Patent* 6239198 B1, 2001.

[65] Ohman, F.; Wallmo, H.; Theliander, H. Precipitation and filtration of lignin from black liquor of different origin. *Nord. Pulp Pap. Res. J.* 2007, 22(2), 188-193.

[66] Alen, R.; Patja, P.; Sjostrom, E. Carbon dioxide precipitation of lignin from pine kraft black liquor. *Tappi J.* 1979, 62(11), 108-110.

[67] Alen, R.; Sjostrom, E.; Vaskikari, P. Carbon dioxide precipitation of lignin from alkaline pulping liquors. *Cellulose Chem. Tech.* 1985, 19(5), 537-541.

[68] Ghatak, H.R. Spectroscopic comparison of lignin separated by electrolysis and acid precipitation of wheat straw soda black liquor. *Ind. Crops Prod.* 2008, 28(2), 206-212.

[69] Ibrahim, M.N.M.; Chuah, S.B. Characterization of lignin precipitated from the soda black liquor of oil palm empty fruit bunch fibres by various mineral acids. *ASEAN J. Sc. Tech. Dev.* 2004, 21(1), 57-67.

[70] Zhang, S.; Wang, S.; Shan, X.; Mu, H. Influences of lignin from paper mill sludge on soil properties and metal accumulation in wheat. *Biol. Fertil. Soils.* 2004, 40(4), 237-242.

[71] Zhang, X.Y.; Zhang, J.N.; Han, R.L.; Li, Z.H. Method of biological acid precipitation of lignin from alkaline pulp black liquor. *Chin. J. Appl. Environ.* Biol. 1999, 5(6), 618-622.

[72] Villar, J.C.; Caperos, A.; Garcia-Ochoa F. Oxidation of hardwood kraft lignin to phenolic derivatives with oxygen as oxidant. *Wood Sci. Tech.* 2001, 35(3), 245-255.

[73] Kim, D.H.; Pu, Y.; Chandra, R.P.; Dyer, T.J.; Ragauskas, A.J.; Singh, P.M. A novel method for enhanced recovery of lignin from aqueous process streams. *J. Wood Chem. Tech.* 2007, 27(3-4), 219-224.

[74] Norgren, M.; Edlund, H.; Wagberg, L. Aggregation of lignin derivatives under alkaline conditions. Kinetics and aggregate structure. *Langmuir.* 2002, 18(7), 2859-2865.

[75] Merewether, J.W.T. The precipitation of lignin from eucalyptus kraft black liquors. *Tappi J.* 1962, 45(2), 159-163.

[76] Ohman, F.; Theliander, H. Filtration properties of lignin precipitated from black liquor. *Tappi J.* 2007, 6(7), 3-9.

[77] Wallmo, H.; Theliander, H.; Richards, T. Lignin precipitation from kraft black liquors: kinetics and carbon dioxide absorption. *Paperi Puu.* 2007, 89(7-8), 436-442.

[78] Norgren, M.; Edlund, H. Ion specific differences in salt induced precipitation of kraft lignin. *Nord. Pulp Pap. Res. J.* 2003, 18(4), 400-403.

[79] Norgren, M.; Edlund, H.; Wagberg, L.; Lindstrom, B.; Annergren, G. Aggregation of kraft lignin derivatives under conditions relevant to the process. *Part I. Phase behaviour. Colloids Surf., A* 2001, 194(1-3), 85-96.

[80] Gilarranz, M.A.; Rodriguez, F.; Oliet, M.; Revenga, J.A. Acid precipitation and purification of wheat straw lignin. *Sep. Sci. Technol.* 1998, 33(9), 1359-1377.

[81] Ibrahim, M.N.M.; Azian, H. Extracting soda lignin from the black liquor of oil palm empty fruit bunch. *J. Teknol.* 2005, 42(C), 11-20.

[82] Sun, R.C.; Tomkinson, J.; Jones, G.L. Fractional characterization of ash-AQ lignin by successive extraction with organic solvents from oil palm EFB fiber. *Pol. Deg. Stab.* 2000, 68(1), 111-119.

[83] Garcia, A.; Toledano, A.; Serrano, L; Egues, I.; Gonzalez, M.; Marin, F.; Labidi, J. Characterization of lignins obtained by selective precipitation. *Sep. Purif. Tech.* 2009, 68(2), 193-198.

[84] Wienhaus, O.; Bernaczyk, Z.; Pecina, H. Research into the precipitation of lignin from sulphate black liquors, particulary by carbon dioxide. *Papier* 1990, 44(11), 563-571.

[85] Hassan, E. Process for separating lignins and dissolved organic compounds from kraft spent liquor. *US Patent* 5635024, 1997.

[86] Hassan, E. Treating spent, waste, alkaline digestion liquor from paper pulping operations and product. *US Patent* 6632327, 2003.

[87] Bruley, A.J.; Cook, W.G.; Ross, R.A. The pyrolysis of extracted solids from oxidized kraft black liquor after lignin precipitation. *Can. J. Chem. Eng.* 1973, 51(6), 746-750.

[88] Moosavifar, A.; Sedin, P.; Theliander, H. Viscosity and boiling point elevation of black liquor: consequences when lignin is extracted from the black liquor. *Nord. Pulp Pap. Res. J.* 2006, 21(2), 180-187.

[89] Misra, D.K. Selective removal of silica from alkaline spent liquors. Tappi 1982 Pulping Conference Proceedings; Tappi Press: Atlanta, 1982; pp 225-229.

[90] Kopfmann, K.; Hudeczek, W. Desilication of black liquors: *Pilot plant tests.* 1988, 71(10), 139-147.

[91] Mandavgane, S.A.; Paradkar, G.D.; Subramanian, D. Desilication of agro based black liquor using bubble column reactor. *J. Sc. Ind. Res.* 2006, 65(7), 603-607.

[92] Jonsson, A.; Wallberg, O. Cost estimates of kraft lignin recovery by ultrafiltration. *Desalination.* 2009, 237(1-3), 254-267.

[93] Woerner, D.L.; McCarthy, J.L. Ultrafiltration of kraft black liquor. *AIChE Sym. Ser.* 1984, 80: 25-34.

[94] Tanistra, I.; Bodzek, M. Preparation of high-purity sulphate lignin from spent black liquor using ultrafiltration and diafiltration processes. *Desalination.* 1998, 115(2), 111-120.

[95] Sridhar, S.; Bhattacharya, P.K. Limiting flux phenomena in ultrafiltration of Kraft black liquor. *J. Membrane Sci.* 1991, 57(2-3), 187-206.

[96] Bhattacharjee, S.; Datta, S.; Bhattacharjee, C. Performance study during ultrafiltration of kraft black liquor using rotating disk membrane module. *J. Cleaner Prod.* 2006, 14(5), 497-504.

[97] Liu, G.; Liu, Y.; Ni, J.; Shi, H.; Qian, Y. Treatability of Kraft spent liquor by microfiltration and ultrafiltration. *Desalination* 2004, 160(2), 131-141.

[98] Wallberg, O.; Jonsson, A. Separation of lignin in kraft cooking liquor from a continuous digester by ultrafiltration at temperatures above 100°C. *Desalination.* 2006, 195(1-3), 187–200.

[99] Jonsson, A.; Wimmerstedt, R. The application of membrane technology in the pulp and paper industry. *Desalination,* 1985, 53(1-3), 181-196.

[100] Basu, S.; Sapkal, V.S. Membrane technique in simplification of soda recovery process in pulp and paper industry. *Desalination.* 1987, 67, 371-379.

[101] Wallberg, O.; Jonsson A.; Wimmerstedt, R. Fractionation and concentration of kraft black liquor lignin with ultrafiltration. *Desalination.* 2003, 154(2), 187-199.

[102] Liu, G.; Liu, Y.; Shi, H.; Qian, Y. Application of inorganic membranes in the alkali recovery process. *Desalination.* 2004, 169(2), 193-205.

[103] Toledano, A.; Serrano, L.; Garcia, A.; Mondragon, I.; Labidi, J. Comparative study of lignin fractionation by ultrafiltration and selective precipitation. *Chem. Eng. J.* 2010, 157(1), 93-99.

[104] Wallberg, O.; Linde, M.; Jonsson, A. Extraction of lignin and hemicelluloses from kraft black liquor. *Desalination.* 2006, 199(1-3), 413-414.

[105] Bhattacharjee, C.; Bhattacharya, P.K. Prediction of limiting flux in ultrafiltration of kraft black liquor. *J. Memb. Sc.* 1992, 72(2), 137-147.

[106] Bhattacharjee, C.; Bhattacharya, P.K. Flux decline analysis in ultrafiltration of kraft black liquor. *J. Memb. Sc.* 1993, 82(1-2), 1-14.

[107] Bhattacharjee, C.; Sarkar, P.; Datta, S.; Gupta, B.B.; Bhattacharya, P.K. Parameter estimation and performance study during ultrafiltration of kraft black liquor. *Sep. Purif. Tech.* 2006, 51(3), 247-257.

[108] De, S.; Bhattacharya, P.K. Flux prediction of black liquor in cross flow ultrafiltration using low and high rejecting membranes. *J. Memb. Sc.* 1996, 109(1), 109-123.

[109] Satyanarayana, S.V.; Bhattacharya, P.K.; De, S. Flux decline during ultrafiltration of kraft black liquor using different flow modules: a comparative study. *Sep. Purif. Tech.* 2000, 20(2-3), 155-167.

[110] Bhattacharjee, C.; Bhattacharya, P.K. Ultrafiltration of black liquor using rotating disk membrane module. *Sep. Purif. Tech.* 2006, 49(3), 281-290.

[111] Uloth, V.C.; Wearing, J.T. Kraft lignin recovery: acid precipitation versus ultrafiltration. Part II: Technology and economics. *Pulp Pap.* Canada. 1989, 90(10), 34–37.

[112] Kirkman, A.G.; Gratzl, J.S.; Edwards, L.L. Kraft lignin recovery by ultrafiltration: economic feasibility and impact on the kraft recovery system. *Tappi J.* 1986, 69(5), 110–114.

[113] Radhamohan, K.; Basu, S. Electrodialysis in the regeneration of paper mill spent liquor. *Desalination.* 1980, 33(2), 185-200.

[114] Azarniouch, M.K.; Prahacs, S. Recovery of NaOH and Other Values from Spent Liquors and Bleach Plant Effluents. *US Patent* 5061343, 1991.

[115] Beaudry, E.G.; Caro, R.F. Electrolysis of weak black liquor and bleach effluent. Tappi 1996 Engineering Conference Proceedings; Tappi Press: Atlanta, 1996; pp 263-267.

[116] Cloutier, J.N.; Azarniouch, M.K.; Callender, D. Electrolysis of Weak Black Liquor. Part I: Laboratory Study. *J. Pulp Pap. Sc.* 1993, 19(6), J244.

[117] Cloutier, J.N.; Azarniouch, M.K.; callender, D. Electrolysis of weak black liquor part II: Effect of process parameters on the energy efficiency of the electrolytic cell. *J. Appl. Electrochem.* 1995, 25(5), 472-478.

[118] Mishra, A.K.; Bhattacharya, P.K. Alkaline black liquor treatment by continuous electrodialysis. *J. Memb. Sc.* 1987, 33(1), 83-95.

[119] Herron, J.R.; Beaudry, E.G.; Jochums, C.E.; Medina, L.E. Turbulent flow electrodialysis cell. *US Patent* 5334300, 1994.

[120] Kennedy, A.M.; Jernigan, J.M. Process and apparatus for electrolytically treating black liquor. *US Patent* 2905604, 1959.

[121] Ghatak, H.R. Electrolysis of black liquor for hydrogen production: some initial Findings. *Int. J. Hydrogen Energy.* 2006, 31(7), 934-938.

[122] Ghatak, H.R. Reduction of organic pollutants with recovery of value added products from soda black liquor of agricultural residues by electrolysis. *Tappi J.* 2009, 8(7), 4-10.

[123] Ghatak, H.R.; Kumar, S.; Kundu, P.P. Electrode processes in black liquor electrolysis and their significance for hydrogen production. *Int. J. Hydrogen Energy.* 2008, 33(12), 2904-2911.

[124] Ghatak, H.R.; Kundu, P.P.; Kumar, S. Thermochemical comparison of lignin separated by electrolysis and acid precipitation from soda black liquor of agricultural residues. *Thermochim. Acta.* 2010, 502(1-2), 85-89.

In: Lignin
Editor: Ryan J. Paterson

ISBN 978-1-61122-907-3
© 2012 Nova Science Publishers, Inc.

Chapter 8

BIOSORBENTS BASED ON LIGNIN USED IN BIOSORPTION PROCESSES FROM WASTEWATER TREATMENT: A REVIEW

Daniela Suteu,[1]* *Carmen Zaharia*[2] *and Teodor Malutan*[3]

[1]'Gheorghe Asachi' Technical University of Iasi,
Faculty of Chemical Engineering and Environment Protection,
Department of Organic and Biochemical Engineering,
71A Prof.dr.docent D. Mangeron Blvd, 700050, Iasi, Romania
[2]'Gheorghe Asachi' Technical University of Iasi,
Faculty of Chemical Engineering and Environment Protection,
Department of Environmental Engineering and Management,
71A Prof. dr.docent D. Mangeron Blvd, 700050, Iasi, Romania
[3]'Gheorghe Asachi' Technical University of Iasi,
Faculty of Chemical Engineering and Environment Protection,
Department of Natural and Synthetic Polymers,
71A Prof. dr.docent D. Mangeron Blvd, 700050, Iasi, Romania

ABSTRACT

Taking into consideration the advantages of using natural materials as biosorbents in the cleaning and/or clearing wastewater technologies, the specialists' attention has been turned to the development of some new treatment methods, as a part of the environmental biotechnologies. These include, among others, biosorption. A special attention is focused on lignocellulosic materials, due to their fundamental characteristics, such as accessibility and low prices, mechanical resistance, high porosity and specific surface area, hydrophilic character that ensures a rapid sorption rate, tolerance to biological adsorbed solid layers, easy functionalization, the possibility of be used in different forms (particles of different dimensions, fibbers, filters, textile materials for cloths) and regimes (discontinuous and continuous processes).

* Email corresponding author: danasuteu67@yahoo.com.

The paper is a review about our researches and comparisons with literature regarding different types of industrial wasted lignocellulose and lignin materials with sorptive properties that were used into biosorption processes of wastewater treatment.

Keywords: lignocellulosic materials, biosorption, dyes, batch experiments, modeling and optimization sorption process.

INTRODUCTION

All industries discharge potential contaminants into public sewerage systems or different emissaries and have the obligation to report to the local water authority each exceeding of the maximum admissible concentration or hourly load. To avoid the pollution of water resources it is necessary the individual or combined industrial wastewater treatment with municipal wastewater.

The viable procedures of the *'production water' management* into each company take into consideration the following possible options: (1) minimization of water use; (2) the development of new processes for different industrial effluents or final wastewater treatment, and the optimization of existing or new ones with the aim of saving water, materials and energy, and (3) the reuse and recycling of industrial wastewater [1-5].

An industrial effluent treatment can be applied into decentralized treatment systems with or without reuse or central wastewater treatment. For a high wastewater flow rate and a high COD concentration, an estimation of the treatment costs lets us know whether it is better to treat it in an end-of-pipe system or in a decentralized effluent treatment system.

The interest towards an entire production process respecting the requirements of environment protection must direct all efforts to develop new production processes with reduced consumption levels of water and raw materials (i.e. dry processes or other electrochemical, physical ones), with reduced production of wastewater and to treat efficiently the individual or final effluents [3].

By large volume and composition of wastewaters from dyeing processes, the textile industry has a great polluting impact. Within this industry, the industrial processing of textile materials is a complex process, and the chemical textile finishing involves a high water consumption (ca 100 m^3/tone of product) used mainly as transportation media of dyes, auxiliaries and thermal energy. Because of the effluents' volume but also of the fact that ca 90% from the consumption water is used for pretreatment operations, dyeing, printing and finishing, the produced wastewaters of textile industry have a high pollution level so that this industry represents one of the most pollutant economic sectors [6-8]. During the dyeing and finishing operations, 10-15% of the dye is lost in the wastewaters [8, 9].

Concerning the elimination of pollutants from wastewaters produced into textile industry, because of their presence in mixture and extremely various quantities into the polluted environment, it is very difficult to establish or design some general treatment technologies, being proposed technological scheme dependent of the pollutants composition and structure, and also of the origin media. The characteristics of wastewaters produced from the processes of chemical textile finishing depend on both type of the production process (i.e. glueing together, whitening, dyeing, hot alkaline treatment, etc.) and also the nature of textile material for chemical finishing (cotton, proteic fibers, polyamides fibers, etc.). Even in the same textile enterprise can appear important modifications of wastewater composition, determined by the changes interfering in the production structure or restructuring of the technological process [2]. The textile effluents resulted in tinctorial processes contain many types of dangerous chemical organic and inorganic compounds that influence the quality of wastewaters and need to be treated before their disposal in municipal pipe network or other receiving basins [10].

An important category of organic compounds into textile effluents is represented by the *textile dyes* - synthetic compounds with aromatic molecular structures, resistant to light, heat and oxidizing agents and other non-biodegradable materials. It has been suggested that the presence of coloured compounds in aqueous environments can reduce light penetration, thus affecting the photosynthesis process of aquatic plants. In addition to their visual effect (aesthetic impact on receiving waters), many synthetic dyes are toxic, mutagenic and carcinogenic. Due to their aromatic structures these compounds are more stable, no biodegradable and difficult to remove from the industrial effluents before their discharges into the urban wastewater sewage system or different emissaries [15]. One of the most important group of dyes are represented by the *reactive dyes* as industrial compounds for dyeing the cellulose fibres, and can be characterized by low absorbability on a wide range of adsorbent, and limited biodegradability in an aerobic environment [11]. However, many reactive dyes contain in their molecules some azo compounds that transform these dyes into important organic pollutants for aquatic ecosystems because of their potential to form dangerous aromatic amines, other carcinogenic and mutagenic compounds [10].

The methods used for colour removal from industrial effluents (decolourization) are based on [3, 6, 7, 12, 13]:

- Biological processes – that achieve dyes decolourization and decomposition/mineralization in aerobic or anaerobic conditions, under the influence of some bacteria (pseudomonas) or fungi.
- Physical processes – that suppose the dyes elimination by filtration (inverse osmosis, micro-filtration, ultra-filtration, nano-filtration), separation through membrane, flotation and adsorption.
- Chemical processes – that suppose the dyes elimination and mineralization/decomposition by chemical precipitation and complexation, coagulation-flocculation, electro-coagulation, chemical reduction, chemical oxidation, and homogenous or heterogeneous photochemical oxidation. The advanced oxidation processes (AOPs) are based on the action of some oxidants as ozone (O_3), H_2O_2, UV, O_3/UV, H_2O_2/UV, O_3/UV/ H_2O_2, Fe^{2+}/ H_2O_2 (Fenton), UV/H_2O_2/Fe^{2+} (foto-Fenton) or photocatalytic oxidation – UV/TiO_2, UV/ZnO, UV/Ti-MCM-41, UV/Ti-Sn-MCM-41.
- Mixed processes: physical-chemical, biochemical, electrochemical processes.

As a function of their specificity, the non-biological processes present both advantages (i.e. do not produce residual sludge in high quantities, perform decolourization both of soluble and insoluble dyes, a part of secondary products are not dangerous) but also disadvantages (i.e. some methods produce concentrated sludge or secondary products that must be eliminated, applicability only for aminic aromatic compounds, some of them involve high costs or are not efficient for all types of dyes) [2, 3, 7, 12]. Many of these processes are often complicated and time consuming, generate sludge or other toxic wastes and may be ineffective or expensive for diluted dye solutions [15].

Sorption is one of the methods very useful in the treatment of coloured wastewaters because of their design simplicity, inexpensiveness, specific physic-chemical interactions between dyes and the solid sorbent, and low matrix effects [6, 12, 16, 17]. Dye removal by sorption in batch conditions is a relatively simple method which can be carried out without sophisticated equipments. Decolourisation by adsorption is mainly determined in equal extent on the dyes's properties and structure and on the surface chemistry of the sorbent. Interaction of the functional group of the dyes (one or more groups of -OH, -COOH, -SO$_3$H, -N=N-, etc.) with the sorbent surface could be result anywhere from covalent, coulombic, hydrogen bonding or weak van der Waals forces [18]. Several research studies indicate that adsorbents which contain high concentrations of cellulose irreversibly adsorb basic dyes through coulombic attraction and ion exchange processes. Acid dyes are mostly adsorbed in a reversible process involving physical adsorption (combination of van der Waals attraction, hydrogen bonding and coulombic attraction that generates a negative surface charge of the adsorbent in contact with water) [4, 5].

One of the main advantages of sorption process is the possibility to use as sorbents many classes of materials: synthetic to natural low-cost materials (natural as well as wasted materials from different industries and agriculture) as suitable sorbents for decolourization of industrial effluents [6].

Selection of new sorbents is determined by the high efficiency of the sorption process (high affinity and dye binding capacity, sorption kinetics, regeneration properties and cost), cost effectiveness, availability and adsorptive properties. Most sorbents discussed into the scientific literature are synthetic and engineered materials, such as synthetic resins, ion exchange celluloses, chemically modified fibers, activated charcoal and ashes.

The high costs of sorbent preparation, the necessity to avoid some disadvantages of conventional sorbents based on synthetic polymers (high prices, difficulties in obtaining, pollution produced during their synthesis), and the actual tendency of replacing chemically synthesized compounds with others of natural origin [10] reoriented the researches to the testing of low cost materials and easily obtainable, included into the category of 'non-conventional' or 'low cost', having a synthetic (such industrial wastes) and/or natural (such cellulosic and/or lignocellulosic agricultural wastes) structure, such as:

     i.   industrial/ agricultural/ domestic wastes or industrial/agricultural by-products (ash, sludge, sawdust, textile fibbers, mud, bark, straw, etc.)
     ii.   natural materials (peat, seashell, algae, lignite, wood, etc.).

that assure large specific surface areas, high sorption rates, the possibility to be functionalized with organic ligands for increasing their sorption capacity [16].

Recently, one of the most important subjects is to investigate new materials with sorptive potential and to implicate them into wastewater treatment process. Sorption onto activated charcoal is a very useful technique, but the high cost implied to obtain this sorbent stimulated the search of cheaper alternatives. In this context, the overall attention was moved to the non-conventional and low-cost materials which include some agriculture and industrial by-products and wastes, industrial wasted biomass and natural materials to retain and accumulate the textile dyes [3, 16, 19].

The sorption capacities of dyes onto non-biological wastes or by-products as activated carbon, coal ash, fly ash, physicochemical modified coal ashes depend on the surface charge of the adsorbent on contact with water. For activated carbon, the surface charge will be neutral, the physical adsorption will predominate and a high adsorption capacity for both acid and basic dyes will be achieved [20, 21].

A special attention was focused on *lignocellulosic materials*, due to their fundamental characteristics, such as accessibility and low prices, mechanical resistance, high porosity and specific surface area, hydrophilic character that ensures a rapid sorption kinetic, tolerance to biological structures, easy functionalization, the possibility to be used in different forms (particles of different dimensions, fibbers, filters, cloths) and in continuous or discontinuous processes [22]. The removal of some pollutant species by sorbents based on lignin, modified lignin and/or lignocellulosic materials has been intensively studied (Table 1) [23]; the results suggested that sorption on lignin is 'a progression towards a perspective method'.

Table 1. Applicability of lignocellulosic sorbents in the removal of pollutants from aqueous systems

Sorbent	Retained pollutants	Adsorption capacity	Ref.
Modified lignin	Pb (II) Cd (II)	q= (8.2-9.0) mg Pb/ g q= (6.7 – 7.5) mg Cd/g	24
Modified corncob: -with nitric acid - with citric acid	Cd(II) Cd(II)	q= 19.3 mg Cd (II)/ g q= 55.2 mg Cd (II)/g	25
Pine fruit	Cu (II) Fe (II)	q= 4.8 mg Fe(II)/ g q= 14.1 mg Cu (II)/g	26
Wood apple carbon	Methylene Blue	q = 36.9 mg/g	27
Kraft lignin	Cu (II) Cd (II)	q = 87.05 mg Cu (II) /g q = 137.14 mg Cd (II) /g	28
Sawdust	Brilliant Red HE-3B Methylene Blue Rhodamine B Crystal Violet	q = 11.61 mg/g q = 7.215 mg/g q = 7.309 mg/g q= 12.594 mg/g	29
Peat	Brilliant Red HE-3B Methylene Blue Rhodamine B	q = 16.286 mg/g q = 38.314 mg/g q = 16.722 mg/g	30
Rice bran Bagase fly ash of sugarcane *Moninga oleifera* pods Rice husk	Methyl parathion pesticide	q = 0.9 mmol/g q = 1 mmol/g q = 1 mmol/g q = 0.9 mmol/g	31
Corncobs	Cu (II)	q = 26 mg/g	32
Pre-boiled sunflower stem	Cr (VI)	q = 5.34 mg/g	33
Formaldehyde-treated sunflower stem	Cr (VI)	q = 4.81 mg/g	
Black tea waste	Cu (II) Pb(II)	q = 48 mg/g q = 65 mg/g	34

This paper reviews our researches in comparison with scientific literature regarding different types of industrial wasted lignocellulosic and lignin materials with sorptive properties which were used into biosorption processes of wastewater treatment. In the same time, we want to present an overview about effects of various parameters (factors) that influence the biosorption of textile dyes onto industrial wasted lignocellulosic and lignin materials, and also the equilibrium, kinetic and thermodynamic models studied in order to understand the biosorption mechanism.

2. GENERALITIES

2.1. Working Methodology

Biosorption is a process of chemical species sorption on biological or natural materials as sorbents [35]. This process is characterized mainly by: low cost, elimination of ecological risk of the toxic compounds recovery, the possibility to treat large volumes of effluents with low concentrations of pollutants [36]. Biosorption is a process achieved through extracellular and intracellular bonding, interactions that depend of the nature of pollutants, of the structure of biosorptive materials into the microbian metabolism and into the transport process. Similar as the biodegradation of organic compounds, biosorption implies the breaking and creating new chemical bonds that can change the molecular structure of the pollutant.

The biosorption experiments were usually performed using 'batch technique' - a simple and ease to operate methodology (easily control and adjustment of the operating parameters) in the case of small and medium size processes from laboratory practice in order to obtain the preliminary information about parameters or variables (factors) that affect the biosorption process, and that are necessary for the modelling of process and for future simulation of real case study [38].

The main factors influencing the sorption equilibrium can be classified in two groups:

1) process variables such as pH, temperature, sorbent dose, initial dye concentration, sorption time, and
2) variables depending on sorbent and sorbet such as structure of dyes and sorbent, and the particle size of sorbent [1, 38-43].

It was worked with suspending samples of sorbent into volumes of aqueous solution containing knowing dye concentrations placed into 100 mL Erlenmeyer flasks, under an intermittent agitation, at different temperatures (2° - 45°C). The pH of aqueous solutions was adjusted to a desired value by addition of HCl 1M solution and measuring the pH value with a HACH One Laboratory pH meter. After an established sorption time, the concentration of the dyes in supernatant was determined by absorbance measurement with a VIS Spectrophotometer, model SP 830 Plus, Metertech Inc. Version 1.06 (i.e. a digital microprocessor controlled instrument providing photometric absorbance, transmittance and concentration measurements with a wavelength range of studied dyes).

For the evaluation of the potential sorptive capacity of different materials the laboratory experiments, in static conditions, applied for aqueous systems or synthetic wastewaters that contain textile dyes were studied the influence of some operational parameters (temperature, pH, quantity of sorbent, time of equilibrium and initial dye concentration) against the dyes removal.

The biosorption capacity of the sorbents was expressed by:

- the amount of dye sorbed: $q = (C_0 - C) \cdot V \cdot 10^{-3} / G$ (mg of dye/g of sorbent),
- percentage of dye removal (R), $R = [(C_0-C)/C_0] \cdot 100$ (%)

where: C_0 and C are the initial and equilibrium concentration of dye in solution (mg/L), G is the amount of sorbent (g) and V is the volume of solution (mL).

The relationship between the amount of sorbate at constant temperature and its concentration in the solution equilibrium is named *sorption isotherm*. For the interpretation of sorption equilibrium and calculation of the characteristic constants were used the usual models of isotherms (liniar, Freundlich, Langmuir, Dubinin-Radushkevich, Temkin) [37].

The calculation of thermodynamic parameters (ΔG, ΔH and ΔS) [37] and the study of dye sorption kinetic (based on the main models from literature: pseudo-first order (Lagergreen model), pseudo-second order (Ho model) and intraparticle diffusion (Webber and Morris) models [37]) were contributed to the elucidation of sorption mechanism.

2.2. Selected Textile Dyes

The textile dyes represent a category of organic compounds, generally considered as pollutants, presented into wastewaters resulting mainly from processes of chemical textile finishing.

These are aromatic compounds produced by chemical synthesis, and having into their structure aromatic rings that contain electrons delocated and also different functional groups. Their color is due to the chromogene-chromophore structure (acceptor of electrons) and the dyeing capacity is due to auxochrome groups (donor of electrons). The chromogene is constituted from an aromatic structure normally based on rings of benzene, naphthaline or antracene, from which are binding chromofores that contain double conjugated links with delocated electrons. The chromofore configurations are represented by the azo group (-N=N-), etylene group (=C=C=), methin group (-CH=), carbonylic group (=C=O), carbon-nitrogen (=C=NH; -CH=N-), carbon-sulphur (=C=S; ≡C-S-S-C≡), nitro (-NO₂; -NO-OH), nitrozo (-N=O; =N-OH) or chinoide groups. The auxochrome groups are ionizable groups, that confer to the dyes the binging capacity onto the dyeing textile material. The usual auxochrome groups are: -NH₂ (amino), -COOH (carboxyl), -HSO₃ (sulphonate) and -OH (hydroxil).

The reactive dyes, which represent the largest class of dyes used in textile processing industries, are almost azo compounds, i.e. molecules with one or several azo (-N=N-) bridges linking substituted aromatic structures. These dyes are designed to be chemically and photolytically stable; they exhibit a high resistance to microbial degradation and are highly persistent in natural environment.

The *selected dyes* by authors that are used as commercial salts are characterized in Table 2.

Table 2. Structure and characteristics of the studied dyes

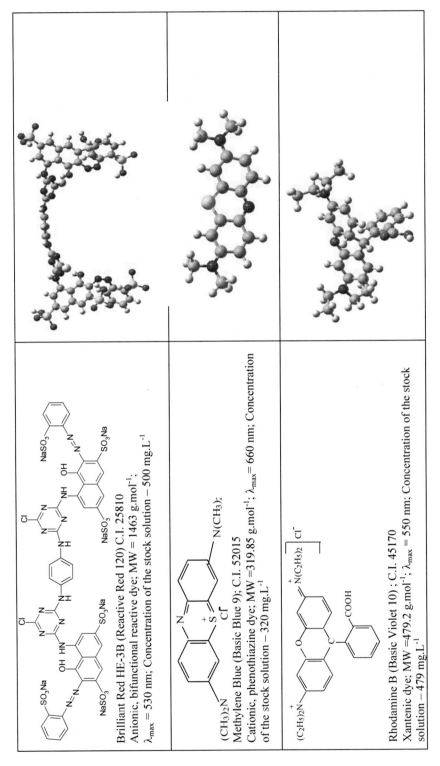

Brilliant Red HE-3B (Reactive Red 120) C.I. 25810
Anionic, bifunctional reactive dye; MW = 1463 g.mol⁻¹;
λ_{max} = 530 nm; Concentration of the stock solution – 500 mg.L⁻¹

Methylene Blue (Basic Blue 9); C.I. 52015
Cationic, phenothiazine dye; MW =319.85 g.mol⁻¹; λ_{max} = 660 nm; Concentration
of the stock solution – 320 mg.L⁻¹

Rhodamine B (Basic Violet 10) ; C.I. 45170
Xantenic dye; MW =479.2 g.mol⁻¹; λ_{max} = 550 nm; Concentration of the stock
solution – 479 mg.L⁻¹

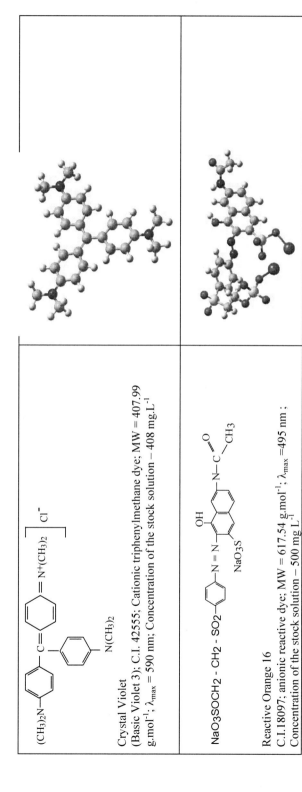

(CH₃)₂N

C=

N⁺(CH₃)₂   Cl⁻

N(CH₃)₂

Crystal Violet
(Basic Violet 3); C.I. 42555; Cationic triphenylmethane dye; MW = 407.99
g.mol⁻¹; λ_{max} = 590 nm; Concentration of the stock solution – 408 mg.L⁻¹

NaO₃SOCH₂ - CH₂ - SO₂

N = N

OH

N – C

O

CH₃

NaO₃S

Reactive Orange 16
C.I.18097; anionic reactive dye; MW = 617.54 g.mol⁻¹; λ_{max} =495 nm ;
Concentration of the stock solution – 500 mg L⁻¹

Abbreviations: BRed - Brilliant Red HE-3B; MB – Methylene Blue; RB – Rhodamine B; CV - Crystal Violet; RO – Reactive Orange.

2.3. Biosorbents Based on Lignin, Modified Lignin and/or Lignocellulose

The following materials can be used for their properties as biosorbents:

- biomass (natural support, generated by plants, mushrooms, seaweed, microorganisms from soil or water is biodegradable and can be integrated in carbon cycles being also compatible with self control environmental processes) which includes: phytomass agriculturally produced such as lignin – cellulose biomass of wooden plants [22,44] ; the aquatic biomass [45]; biomass obtained as agriculture waste; domestic biomass used in technologies for biogas production; microbial biomass from pharmaceutical and food industry [46-48] ; microorganisms easily available in large amounts in nature or especially cultivated or propagated for these process.
- abundant natural materials as peat, charcoal, moss, sawdust, hemp fibers [35, 49-51];
- biosorbents – resulting from the biomass transforming in a form much easier to use that finally leads to high efficiency of the separation process of biomass containing metals from solution. Also, they are effective and sustainable in repeated long-term applications. Different types of support can be used for retaining biomass with various forms and porosity. Biosorbents can be used as filling material for chromatographic columns used for purifying wastewaters. They are efficient in metal-ions and organic compounds removing from aqueous medium and have the sorption feature superior in comparison with conventional materials from final treatment of effluents.

Table 3. Intrinsic properties of lignocellulosic components of
wasted sorptive material [53]

Component	Properties
Cellulose	- linear homo polysaccharide with repeating β-D-glucopyranose units - the degree of polymerization is 10000-15000 - high crystallinity and hydrophilicity - capacity to form intra- and intermolecular hydrogen bonds - insoluble in water and common organic solvents - presents three hydroxyls with differ reactivity in the elementary structure - good chelating properties - the presence of interfibrillar interstices, voids and capillaries - negative charge on the macromolecular chains - non toxic - potential enzymatic degradation
Lignin	- cross-linked aromatic polymer (phenyl propane units) - hydrophobic and amorphous component of biomass - various reactive groups for chemical activation - insoluble in water (except lignosulphonates) - potential degradation made by micro-organisms like fungi and bacteria
Hemicelluloses	- hetero polysaccharide with various sugar units - the degree of polymerization is low (100-200) - amorphous matrix component of vegetal materials - partly or even totally water soluble; soluble in diluted alkali; - low chemical and thermal stability - chelating properties - non toxic - potential enzymatic degradation

A new wasted material, characterized by low cost and huge availability is that with lignocellulosic composition (table 3).

An important constituent of *lignocellulosic materials* is *lignin* (Figure 1) and represent 16% - 33% from composition of plant biomass after celluloses and hemicelluloses. Is an inexpensive natural material, available as a by-product from the paper and pulp industry (50 millions tones /years). Lignin is an aromatic, three-dimensional polymer, containing many oxygen functionalities (phenolic and alcoholic hydroxyl, carbonyl, carboxyl, methoxyl, ether) which can be responsible of the sorptive capacity [53, 54].

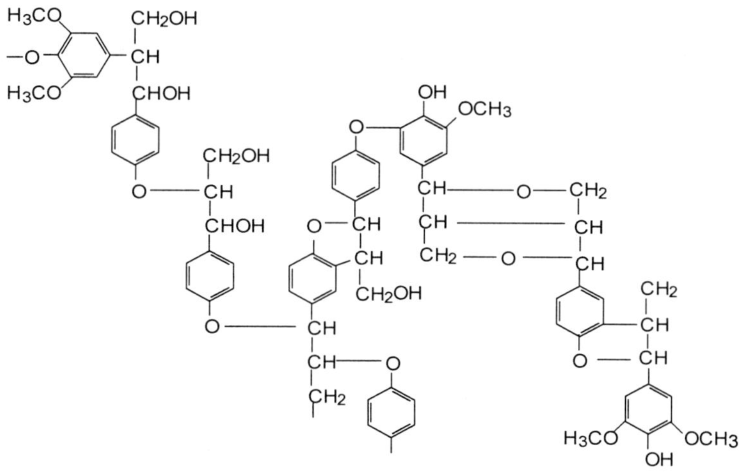

Figure 1. Chemical structure of a portion of a lignin macromolecule.

The wasted lignin is in a huge amount, but less than 10% of them were utilized and this fact is imposing as a serious environmental problem [53, 54].

As a function of the composition of residual vegetal material (table 4) these materials can be valorised by hydrolytic and/or fermentative procedures for obtaining of: etheric oils, furfural, proteic mass, bioethanol, different composts, adhesives, antimicrobian agents, antioxidants, sorptive material for wastewater treatment, alimentary additives, etc.

Table 4. Chemical composition of some vegetal lignocellulosic wastes

Type of agricultural waste	Cellulose	Hexosane	Pentosane	Polyuronic acids	Furfurol	SR obtained from total hydrolysis	Lignin
Creeping stalks of grape vine	29	38	17	9	13	66	35
Corncobs	32	38	33	7.4	25	79	15
Sunflower seed shells	25	28	17	11	14	56	28
Rice straws	28	32	16	4	12	52	19
Hemp	36	-	25	8	15	61	19
Reed	37	39	20	5	15	67	21

The selected lignocellulosic materials for our works (Table 5) are characterized by high capacity for pollutants removal, employing and disposing, and also low cost, accessibility, high inner and outer surface, macro and micro-porous structure, rapid kinetics of the sorption process, the possibility to use some materials in different shapes (particles of different dimensions, fibres, filters, textures) [16].

Table 5. Characteristics of the employed sorbents

Lignocellulosic sorbent	Characteristics
Industrial lignin	It is the main by-product of the pulp industry and also represents a product obtained from renewable resources The characteristics had been: acid insoluble lignin, 90%; acid soluble lignin, 1 %; COOH, 3.8 mmole/g; aromatic OH, 1.7-1.8 mmole/g; OH/C9 groups chemical method = 1.02; Ash, 2.5 %; pH (10% dispersion) = 2.7; MW = 3510; T softening, 170 ^{0}C; Solubility in furfural alcohol, 88.5 %; Solubility in aqueous alkali, pH 12, 98.5%
Sphagnum moss peat	Collected from Poiana Stampei (Romania), sphagnum moss peat is a fibrous natural material. Having as main components: cellulose, lignin, humic and fulvic acids, the peat moss contain different polar functional groups such as carboxyl, carbonyl, phenolic and enolic hydroxide. The employed material has the following physical and chemical characteristics: colour - brown; organic carbon content - 49 w%; total proteins - 7.8 w%; ash 3.44 w%; pH - 3.5; specific surface area - 192 m^2/g.
Sawdust	It is a fibrous waste material produced during the mechanical manufacture of the conifer wood. Sawdust contains various organic compounds (hemicelluloses (10-16%), cellulose (48-57), lignin (27-33%) with polyphenolic groups and, also, ionisable carboxylic groups (uronic acids) which are the most important sorption sites. The humidity of tested material was of 4%.
Cellolignin	Represent a residual product obtained after treating the wood with dilute mineral acid at a temperature of 150-160^{0}C, with following composition: (45-48) % cellulose, (32-35) % lignin, (4-8) % pentosane and (1-1.5) % ash.
Sun flower seed shells	These materials were obtained from local oil industry and used after air drying at room temperature for two days. The seed shells were grounded and sieved to obtain a particle size range of 0.8 mm and stored in plastic bottle for further use. No other chemical or physical treatments were performed. The major constituents of sunflower seed shells are cellulose, lignin and pentose.
Corncob	The material represents an important by-product from local agro-industrial activities and can be included in the lignocellulosic groups of the sorptive materials. The crude material was dried at room temperature, granulated and sieved to obtain different fractions. We used the fractions with size < 800 μm.

3. PERFORMANCE EVALUATION OF LIGNOCELLULOSIC SORBENT

3.1. Influence of Lignocellulosic Sorbent Nature

The sorption capacity of lignocellulosic materials is closely dependent of their composition (especially the lignin percentage into the polymeric structure, that pretends the highest capacity of dye binding) as well of material graining (Table 4). It can be considered that the highest quantity of lignin can be fund into the grape wine stalks close to sunflower seed shells and reed. It must be mentioned the fact that the lignin is higher than 19% in almost all wasted sorptive materials, fact that give to the tested sorbents good sorptive capacity that is clearly presented into the following table (Table 7).

3.2. Influence of Operating Parameters

Biosorption of different dyes on lignocellulosic materials (natural, such as sphagnum moss peat, and by-products, such as sawdust, cellolignin, lignin, sunflower seed shells, and corncob) is strongly dependent on the solution pH. These sorbents are characterized by the pH of zero charge (pH_{pzc}), as neutral pH beyond which the material surface is either positively or negatively charged. The dyes sorption is dependent on both the ionic charge of the dyes and the ionic character of the sorbent. At pH lower that pH_{pzc} the sorbent has affinity for anionic dyes, but at pH higher than pH_{pzc} the sorbent is available to interaction with cationic dyes (Figure 2a, 2b) [43].

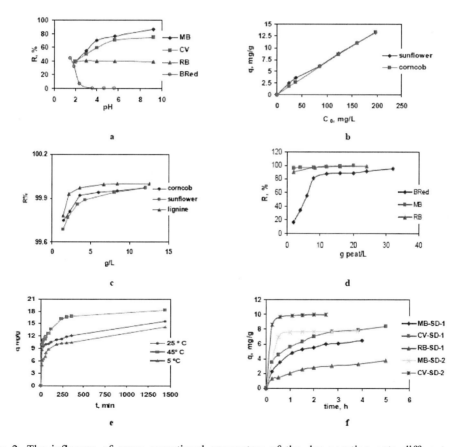

Figure 2. The influence of some operational parameters of the dye sorption onto different wasted materials at room temperature: a) influence of solution pH on the dye sorption on sawdust; b) influence of initial RO dye concentration on sorption onto sunflower shells and corncob; c) influence of dose of corncob, lignin and sunflower seed shell onto sorption the MB dye; d) effect of sphagnum peat dose on sorption the dyes; e) influence of contact time and temperature on the MB dye sorption onto cellolignin; f) influence of sawdust particles size and contact time on sorption of dyes (SD-1 - particle size 1-2 mm, SD-2 – powdered with particle size < 0.1 mm) [10, 43, 55, 56].

Studies about the influence of sorbent dose (Figure 2c) on the decolourization of aqueous dye solutions showed that the sorption of dyes increases with the amount of sorbent. This fact can be attributed to an increase of sorbent surface area availability for more sorption sites than a result of an increase of sorbent amount. In the same time, the amount of sorbed dye per mass unit of sorbent decreased with the increase of sorbent dose.

The results show that the amount of dyes retained increases with increases in initial dye concentration (Figure 2b). Also, the obtained results indicate that the amount of dyes sorbed onto employed sorbents increases with an increasing in temperature, suggesting an endothermic process and, also, the fact that the high temperatures favour the dye molecule diffusion in the internal porous structure of sorbent [10, 21, 43, 55-57]. The effect is more important at higher concentrations.

The influence of sorption time on the decolourization of aqueous solutions containing different textile dyes is different, depending on the sorbent structure (Figures 2e, 2f) [10]. For example, in the case of cellolignin (Figure 2e) the experimental data showed that the removal of dye is faster into initial stages of contact period and then, the amounts of sorbed dye slowly increases near the equilibrium: after 2 hours were removed more than 75 % from the total amount of MB dye corresponding to the dye amount removed after 24 h [58].

3.3. Modelling of the Biosorption Process

3.3.1. Equilibrium Isotherm Models

The equilibrium sorption data were adequately analyzed by the Freundlich, Langmuir, Dubinin-Radushkevich and Temkin sorption isotherm models, in order to calculate the values of sorption constants (Table 6). The best fit equilibrium model was established based on the linear regression correlation coefficients [37]. Figure 3 presents sorption isotherms for dyes removal onto different studied lignocellulosic materials.

**Table 6. Mathematical equation forms, constants and parameters
of the tested isotherm models [56]**

Sorption isotherm model	Linear form of equation	Isotherms parameters, significance	Ref.
Freundlich	$\lg q = \lg K_F + 1/n \lg C$	K_F – adsorption capacity n – adsorption intensity	21
Langmuir	$\dfrac{1}{q} = \dfrac{1}{q_0 \cdot K_L \cdot C} + \dfrac{1}{q_0}$	q_0 - saturation capacity K_L - binding (sorption) energy	21
Temkin	$q = \dfrac{RT}{b_T} \ln K_T + \dfrac{RT}{b_T} \ln C$	b_T – heat of sorption K_T – intensity of sorption (maximum binding energy)	59
Dubinin-Radushkevich	$\ln q = \ln q_D - \beta_D \varepsilon^2$ $\varepsilon = RT \ln(1 + \dfrac{1}{C})$ $E = \dfrac{1}{\sqrt{-2\beta_D}}$	q_D - the maximum amount sorbed β_D – sorption energy ε - Polanyi potential E – mean free energy of sorption	60

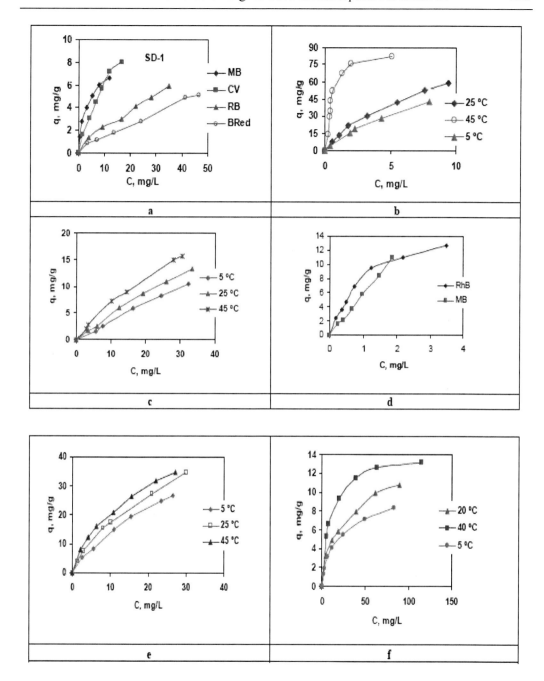

Figure 3. The isotherms of the sorption of some studied dyes onto: a) sawdust; b) MB dye sorption onto cellolignin; c) Orange 16 dye sorption onto corncob; d) peat; e) MB dye sorption onto sunflower seed shell; f) BRed dye sorption onto lignin.

Some of the most important parameters of sorption processes were presented in Table 7, together with the main experimental conditions.

Table 7. The isotherm constants for the biosorption of the studied textile dyes onto lignocellulosic materials

Sorbent	Experimental conditions and sorption constants determined at room temperature (K_L - Langmuir constant; q_0 - the maximum value of sorption capacity; K_F and n – Freundlich constants; C_0 – concentration of dye working solution; E (kJ/mol) - mean energy of sorption	Ref
Sawdust	BRed: (pH = 2; sorbent dose –20 g/L; C_0= (20-150) mg/L) q_0= 11.61 mg/g; K_L= 0.0146 L/mg; K_F= 0.293; n = 1.36;	
	MB:(pH = 5.7; sorbent dose – 4 g/L; C_0= (6.4-38.4) mg/L) q_0= 7.215 mg/g; K_L= 0.479 L/mg; K_F= 2.236; n = 1.99	[41]
	CV: (pH= 5.7; sorbent dose – 4 g/L; C_0= (8.16-48.96) mg/L) q_0= 12.594 mg/g; K_L= 0.0899 L/mg; K_F= 1.151; n = 1.391	
	RB: (pH = 5.7; sorbent dose – 4 g/L; C_0= (9.58-57.5) mg/L) q_0= 7.309 mg/g; K_L= 0.056 L/mg; K_F= 1.95; n = 1.482	
	RO: (pH= 1;sorbent dose – 8 g/L; C_0= (24.7- 159.25) mg/L) q_0= 8.554 mg/g; K_L= 0.0175 L/mg; K_F= 0.307; n = 1.562	[10]
Lignin	BRed: (pH = 1.5; sorbent dose – 14 g/L; C_0= (20-300) mg/L) q_0= 10.173 mg/g; K_L= 10.19 L/mg; K_F= 1.9602; n = 2.621; E = 13.13 kj/mol	[23, 42]
	MB: (pH= 6; sorbent dose – 4 g/L; C_0 = (25.6 - 281.6) mg/L) q_0 =28.01 mg/g; K_L= 0.309 L/mg; K_F= 6.179; n = 1.9; E = 11.47 kj/mol	[61]
	RO: (pH = 1.5; sorbent dose – 12 g/L; C_0= (30-200) mg/L) q = 6.45-90.46 mg/g; K_L=21.38 L/mg; K_F=3.304; n = 2.964; E =16.18 kj/mol	this paper
Sunflower seed shell	RO: (pH= 1; sorbent dose – 8 g/L; C_0= (24.7- 159.25) mg/L) q_0= 8.554 mg/g;K_L= 0.0175 L/mg; K_F= 0.307; n = 1.562; E = 8.276 kj/mol	[10]
	MB: (pH= 6; sorbent dose – 4g/L; C_0 = (25.6 - 281.6) mg/L) q_0= 43.29 mg/g; K_L= 0.0748 L/mg; K_F= 3.49; n= 1.457; E = 9.806 kj/mol	[55]
Corncob	RO: (pH= 1; sorbent dose – 8 g/L; C_0= (24.7-159.25) mg/L) q_0= 8.554 mg/g; K_L= 0.0175 L/mg; K_F= 0.307; n= 1.562; E = 8.28 kj/mol	[56, 65]
	MB: (pH= 6; sorbent dose – 4 g/L; C_0 = (25.6 - 281.6) mg/L) q_0= 28.11 mg/g; K_L= 0.4002 L/mg; K_F= 6.725; n= 1.632; E = 10.66 kj/mol	
Cellolignin	MB: (pH= 6; sorbent dose – 4 g/L; C_0 = (25.6 - 281.6) mg/L) q_0= 92.59 mg/g; K_L= 0.164 L/mg; K_F= 12.963; n= 1.428; E = 10 kj/mol	[58]

Abbreviations: BRed - Brilliant Red HE-3B; MB – Methylene Blue; RB – Rhodamine B; CV - Crystal Violet; RO – Reactive Orange.

3.3.2. Kinetic Studies

Effect of contact time on the biosorption of dyes onto lignocellulosic materials was determined by the 'limited bath' technique, when sample of biosorbent was added under stirring to volume of solution of dye with necessary pH and knowing initial concentration. The temperature of solutions was maintained constant, in general at room value. After different contact times, volumes of supernatant were taken for spectrophotometric measurements of dye content. The extent of sorption was expressed by the fractional attainment of equilibrium: $F = \frac{q_t}{q}$, where q_t and q (mg/g) are the amounts of dye sorbed at time t and at equilibrium (24 h), respectively.

In order to analyse the sorption of dyes onto the lignocellulosic sorbents and also to understand the dynamic of the sorption process in terms of the order of the rate:

- pseudo-first order model (Lagergreen 1898), applied when the sorption is preceded by diffusion through a boundary,
- pseudo-second order kinetic model (Ho model 1999) which assumes that the sorption follows a second order mechanism and the rate limiting step may be chemical sorption involving valence forces or covalent forces between sorbent and sorbate,
- intraparticle diffusion kinetic model (Webber and Morris 1963) which assumes that adsorption is a multi-step process involving transport of sorbate from aqueous solution to the sorption sites of the sorbent and the diffusion into pores is the slow rate determining process.

were applied to the data (Table 8) [23].

Table 8. Mathematical equation forms and constants of the tested kinetic models

Kinetic model	Equation	Kinetic parameters	Ref.
Pseudo-first order (Lagergreen model)	$lg(q_0 - q) = -\dfrac{k_1}{2.303}t + lg\,q_0$	k_1 – rate constant of pseudo-first order model sorption	[24]
Pseudo-second order (Ho model)	$\dfrac{t}{q} = \dfrac{1}{k_2 q_0^2} + \dfrac{1}{q_0} \cdot t$	k_2 – rate constant of pseudo-second order model sorption $k_2 q_0^2 = h$ – initial sorption rate	[25]
Intraparticle diffusion (Webber and Morris model)	$q = k_p t^{0.5} + c$	k_p is the rate constant for intra-particle diffusion $(mg/g.min^{0.5})$ C - constant	[37]

Some of the most important results of kinetic studies about biosorption processes of studied dyes onto selected lignocellulosic sorbents were presented in Table 9, together with the main experimental conditions.

3.3.3. Thermodynamic Studies

The evolution of biosorption vs. temperature modification, together with the calculation of thermodynamic parameters has an important role to determine spontaneity and heat change for the biosorption process.

In order to evaluate the effect of temperature on the dyes biosorption onto lignocellulose sorbents, the apparent thermodynamic parameters (free energy change - ΔG, enthalpy change - ΔH and entropy change - ΔS) were determined using the values of binding Langmuir constant, K_L (L/mol) at temperatures desired, previously determined, and also by the following two equations [37]:

Gibbs equation: $\Delta G = -RT \ln K_L$ (4)

van't Hoff equation: $\ln K_L = -\dfrac{\Delta H}{R \cdot T} + \dfrac{\Delta S}{R}$ (5)

where R is the universal gas constant and T is the absolute temperature.

Table 9. The kinetic evaluation and thermodynamic nature of the biosorption of the studied textile dyes onto lignocellulosic materials

Sorbent	Experimental conditions, kinetic evaluation and thermodynamic nature of biosorption process	Ref
Sawdust	RO: (pH= 1; sorbent dose – 8 g/L; C_0= (24.7- 159.25) mg/L) sorption process followed the pseudo-second order kinetic model (k_2= 0.431×10^{-4} g/ mg·min); ΔG = (-23.352 ÷ - 20.725 kj/mol) - thermodynamic spontaneous sorption process; ΔH = -4.8115 kj/mol shows the exothermic nature of the sorption process; ΔS <0	[15]
Lignin	BRed: (pH = 1.5; sorbent dose – 14 g/L; C_0= (20-300) mg/L) sorption process followed the pseudo-second order kinetic model (k_2= 0.00547 g/ mg·min); ΔG = (-31.516 ÷ - 27.468 kj/mol) - thermodynamic spontaneous sorption process; ΔH = 4.589 kj/mol shows the endothermic nature of the sorption process; ΔS >0	[56]
	MB: (pH= 6; sorbent dose – 4 g/L; C_0 = (25.6 - 281.6) mg/L) ΔG = (-31.390 ÷ - 26.998 kj/mol) - thermodynamic spontaneous sorption process; ΔH = 3.7587 kj/mol shows the endothermic nature of the sorption process; ΔS >0	[61]
Sunflower seed shell	RO: (pH= 1; sorbent dose – 8 g/L; C_0= (24.7- 159.25) mg/L) sorption process followed the pseudo-second order kinetic model (k_2= 0.0105 g/ mg·min); ΔG =(-27.707 ÷ -21.564 kj/mol) - thermodynamic spontaneous sorption process; ΔH = 20.952 kj/mol shows the endothermic nature of the sorption process; ΔS >0	[55]
	MB: (pH= 6; sorbent dose – 4 g/L; C_0 = (25.6 - 281.6) mg/L) sorption process followed the pseudo-second order kinetic model (k_2= 0.00485 g/ mg·min); ΔG < 0 - thermodynamic spontaneous sorption process	this paper
Corncob	RO: (pH= 1; sorbent dose – 8 g/L; C_0= (24.7-159.25) mg/L) sorption process followed the pseudo-second order kinetic model (k_2= 0.045 g/ mg·min); ΔG = (-23.835 ÷ -22.343 kj/mol) - thermodynamic spontaneous sorption process; ΔH = -14.324 kj/mol shows the exothermic nature of the sorption process; ΔS >0	[56]
	MB: (pH= 6; sorbent dose – 4 g/L; C_0 = (25.6 - 281.6) mg/L) sorption process followed the pseudo-second order kinetic model (k_2= 0.00626 g/ mg·min); ΔG < 0 - thermodynamic spontaneous sorption process	this paper
Cellolignin	MB: (pH= 6; sorbent dose – 4 g/L; C_0 = (25.6 - 281.6) mg/L) sorption process followed the pseudo-second order kinetic model (k_2= 0.0062 g/ mg·min); ΔG < 0 - thermodynamic spontaneous sorption process	this paper

Abbreviations: BRed - Brilliant Red HE-3B; MB – Methylene Blue; RB – Rhodamine B; CV - Crystal Violet; RO – Reactive Orange.

Analysis of thermodynamic parameters must offer supplementary information about the nature of sorption process. Literature data underline that there are two main types of sorption [62, 63]:

- physical, determined by weak forces and characterized by ΔH value no more higher than 4.2 kJ/mol and ΔG values lower than -16 kJ/mol
- chemical, determined by forces much stronger than those implicated in physical process, and characterized by ΔH value higher than 21 kJ/mol and ΔG higher than 20 kJ/mol.

The appreciation of the thermodynamic nature of biosorption processes of studied dyes onto lignocellulosic sorbents were presented in Table 9, together with the main experimental conditions.

In all the studied cases, the negative values of apparent free energy change (ΔG) indicate that the dyes biosorption onto studied materials is spontaneous and feasible thermodynamically on the studied temperature range. The positive values of enthalpy change (ΔH) confirmed the endothermic nature of biosorption process (the case of Orange 16 reactive dye biosorption onto sunflower seed shells and Brilliant Re HE-3B reactive dye sorption onto lignin), and the negative values of enthalpy change (ΔH) underline the exothermic nature of biosorption process (the other cases) (Table 9). The positive value of entropy change (ΔS) suggests the increased randomness at the solid-liquid interface during the biosorption of dyes and some structural changes in the sorbate and the sorbent [55]. It can be suggested that the driving force of biosorption is an entropy effect.

4. AN OVERVIEW OF RECENT LITERATURE ON THE SORPTION OF TEXTILE DYES ONTO LIGNOCELLULOSIC SORBENTS

The importance of introduction of new materials with sorptive properties, preferably ease to be procured, cheap and ease to be conditioned and operated, that will give a high efficiency to the sorption process into favourable economic conditions, is illustrated by the numerous studies from scientific literature [16, 25, 32, 35-37]. It must be underlined especially the natural materials, agro-industrial and industrial by-products, wastes from agro-industrial processing, that have a complex composition, are biodegradable, composed by a complex of lignocellulosic substances. These materials, named also 'biosorbents' were used for removal of pollutants with inorganic origin but also organic (such as dyes) one from different aqueous systems.

The studies were extended also in the field of dyeing effluent treatment. Thus, in Table 10 are selected a series of recent results on sorption/biosorption of some textile dyes onto different lignocellulosic biosorbents.

Table 10. Literature data selection of 'batch biosorption' of different types of dyes onto lignocellulosic biosorbents

Lignocellulosic materials	Dye	Adsorption capacity, mg/g	Ref
Modified spherical sulphonic lignin provided by a bamboo pulp mill	Cationic Red GTL	576.0	62
	Cationic turquoise GB	582.4	
	Cationic Yellow X-5GL	640.8	
Rice husk	Methylene Blue	321.26	63
Rice straw thermochemically modified (with carbonyl groups)	Basic Blue 9	256.4	64
	Basic green 4 (malachite green)	238.1	
Rice straw	Basic green 4 (malachite green)	94.34	65
Rice straw thermochemically modified with citric acid as esterifying agent	Basic green 4 (malachite green)	256.41	
Rice husk	Methylene Blue	40.5833	66
Rice straw – derived char	Basic green 4 (malachite green)	148.74	67
Rice straw modified with phosphoric acid	Methylene Blue	208.33	68
Cedar sawdust	Methylene Blue	142.36	69
Beech wood sawdust	Direct Brown	526.3	70
	Direct Brown 2	416.7	
	Basic Blue 86	136.9	
Beech sawdust	Methylene Blue	9.78	71
	Red Basic 22	20.16	
Prehydrolysed beech sawdust	Methylene Blue	30.5	
	Red Basic 22	24.10	
Salts treated beech sawdust	Methylene Blue		72
$CaCl_2$		12.3	
$ZnCl_2$		11.7	
$MgCl_2$		14.2	
NaCl		9.2	
Poplar sawdust	Metanil Yellow	1.996	73
	Methylene Blue	0.3744	
Tree fern	Basic Red 13	408	74
Wood shavings	Methylene Blue	17.92	75
Treated wood shavings:			
- with HCl 0.5 mol/L	Methylene Blue	1.248	
- with Na_2CO_3	Methylene Blue	268.8	
-with Na_2HPO_4	Methylene Blue	6.08	
Pumpkin seed hull	Methylene Blue	141.92	76
Coffee husks	Methylene Blue	90.1	77
Peanut hull	Methylene Blue	68.03	78
Dehydrated peanut hull	Methylene Blue	108.6	79
Cereal chaff	Methylene Blue	20.3	80
Palm shell powder	Methylene Blue	121.5	81
	Rhodamine 6G	105.0	
Hazelnut shells	Congo Red	13.75	82
Hazelnut shells	Methylene Blue	76.9	
	Acid Blue 25	60.2	
Wood sawdust			
- walnut	Methylene Blue	59.17	
	Acid Blue 25	36.98	83
- cherry	Methylene Blue	39.84	
	Acid Blue 25	31.98	
- oak	Methylene Blue	29.94	
	Acid Blue 25	27.85	

Lignocellulosic materials	Dye	Adsorption capacity, mg/g	Ref
- Pitch-pine	Methylene Blue Acid Blue 25	27.78 26.19	
Sunflower seed shells	Reactive Black 5	0.875	84
Husk of mango seed	Acid Blue 80 Acid Blue 324 Acid green 25 Acid green 27 Acid Orange 7 Acid Orange 8 Acid Orange 10 Acid Red 1	9.2 12.8 8.6 12.3 17.3 15.2 8.3 11.2	85
Waste sugar beet pulp	Gemazol turquoise blue-G	234.8	86
Cashew nut shell	Congo Red	5.184	87
Banana stalk waste	Methylene Blue	243.90	88
Castor seed shell	Methylene Blue	158	89
Wheat straw modified with citric acid	Methylene Blue Crystal Violet	321.50 227.27	90

CONCLUSIONS

The lignocellulosic materials (i.e. lignin, sawdust, sunflower seed shells, corncobs) can be used as good and resistant sorbents for removal of different potential pollutants from wastewaters (e.g., heavy metals, dyes, other organic compounds), due to them fundamental characteristics, such as mechanical resistance, high porosity and specific surface area, hydrophilic character, tolerance to biological adsorbed solid layers, easy functionalization, the possibility of be used in different forms and regimes.

The influence of some operational parameters or factors (i.e. pH, initial pollutant concentration, temperature, contacts time, sorbent concentration etc.) is studied in order to find the best operational condition for high sorption efficiency. These are dependent on the pollutant types, characteristics, chemical structure, initial concentration, and also the initial wastewater flow, working regime, the manner of phase contact (L/S) etc.

There were analyzed some kinetic models, determined the forms of different sorption isotherms, calculated the main sorption constants, sorption kinetic and thermodynamic parameters in order to find the nature of sorption interaction between pollutants and aquatic environment and the optimal conditions in order to apply these laboratory scale results into a real case of a textile chemical finishing effluent.

REFERENCES

[1]   D.Suteu, C.Zaharia, *Removal of textile reactive dye Brilliant Red HE-3B onto materials based on lime and coal ash*, ITC and DC, Book of Proceedings of 4th International Textile, Clothing and Design Conference – Magic World of Textiles, October 5th-8th, 2008, Dubrovnik, Croatia, 2008 , 1118-1123.

[2]   D.Suteu, C.Zaharia, D.Bilba, R.Muresan, A.Popescu, A.Muresan, *Decolourization of textile wastewaters – Chemical and Physical methods*, Textile Industry, 60(5), 2009, 254-263.

[3]   Y.Anjaneyulu, N.Sreedhara Chary, D.Samuel Suman Raj, *Decolourization of industrial effluents – available methods and emerging technologies – a review*, Reviews in Environmental Science and Bio/Technology, 4, 2005, 245-273.

[4]   D.Suteu, C.Zaharia, *Sawdust biosorbent for removal of dyes from wastewaters. Equilibrium study*, Bull.Instit.Politech.Iasi (Romania),tom LIV (LVIII), fasc. 4, 2009, 29 – 38.

[5]   U.Wiesmann, I.S.Choi, E.M.Dombrowski, *Fundamentals of Biological Wastewater Treatment*. Wiley-VCH Verlag GmbHandCo. KgaA, Weinheim, 2007.

[6]   S.J.Allen, B.Koumanova, *Decolourisation of water/wastewater using adsorption* (review), J. of the University of Chemical Technology and Metallurgy, 40(3), 2005, 175- 192.

[7]   A. Bhatnagar, A. K. Jain, *A comparative adsorption study with different industrial wastes as adsorbents for the removal of cationic dyes from water*, Journal of Colloid and Interface Science, 281, 2005, 49-55.

[8]   C.Zaharia, D.Suteu, A.Muresan, R.Muresan, A.Popescu, *Textile wastewater treatment by homogeneous oxidation with hydrogen peroxide*, Environmental Engineering and Management Journal, 8(6), 2009, 1359-1369.

[9]   D. Suteu, D. Bilba, C. Zaharia, *HPAN textile fiber waste for removal of dyes from industrial textile effluents,* Bulletin of the Transilvania University of Brasov (Romania), 2(51), 2009, 217-222.

[10]   D.Suteu, C. Zaharia, A. Muresan, R. Muresan, A. Popescu, *Using of industrial waste materials for textile wastewater treatment*, Environ. Eng. Manage. J., 8(5), 2009, 1097-1102.

[11]   S.Senthilkumaar, P.Kalaamani, K.Porkodi, P.R.Varadarajan, C.V.Subburaam, *Adsorption of dissolved reactive red dye from aqueous phase onto activated carbon prepared from agricultural waste*, Bioresource Technology, 97, 2006, 1618-1625.

[12]   B.Ramesh Babu, A.K.Parande, T.Prem Kumar, *Textile Technology: Cotton Textile Processing: Waste Generation and Effluent Treatment*, The Journal of Cotton Science, 11, 2007, 141 – 153.

[13]   C.Zaharia, *Chemical wastewater treatment,* (in Romanian), Performantica Publishing House, Iasi, Romania, 2006.

[14]   M.Surpateanu, C.Zaharia, *Advanced oxidation processes. Decolourization of some organic dyes with H_2O_2*, Environmental Engineering and Management Journal, 3(4), 2004, 629-640.

[15]   D.Suteu, C.Zaharia, *Sawdust as biosorbent for removal of dyes from wastewaters. Kinetic and thermodynamic study*, Scientific Bulletin of „Politehnica" University from Timisoara, Romania, series: Chemistry and Environment Engineering, in press.

[16]   G.Crini, *Non-conventional low-cost adsorbents for dyes removal: A review*, Bioresource Technology 97, 2006, 1061-1085.

[17]   D.Suteu, C.Zaharia, D.Bilba, M.Surpateanu, *Conventional and unconventional materials for wastewater treatment*, Bulletin of the Transilvania University of Brasov (Romania), IV, 2007, 692 –696.

[18] A.Jayswal, U.Chudasama, *Sorption of water soluble dyes using zirconium-hydroxy ethylidene diphosphonate from aqueous solution, Malaysian Journal of Chemistry* 9(1), 2007, 1-9.

[19] M.Rafatullah, O.Sulaiman, R.Hashim, A. Ahmad, *Adsorption of Methylene Blue on low-cost adsorbent: A review, J. Hazard.Mater.*, 177, 2010, 70-80.

[20] C.Zaharia, D.Suteu, *Orange 16 reactive dye removal from aqueous system using corn cob waste,* Proceeding of the International Scientific Conference UNITECH'09, November 20-21, Gabrovo-Bulgary, Vol.III, 2009, 523- 527.

[21] D.Suteu, C.Zaharia, G.Rusu, E.Muresan, *Sorption of Brilliant Red HE-3B onto modified coal ash. Equilibrium and kinetic studies,* Proceedings of The 14 [th] International Conference "Inventica 2010", June 9-11, Jassy, Romania.

[22] D.Suteu, I.Volf, M.Macoveanu, *Ligno-cellulosic materials for wastewater treatment, Journal of Environmental Engineering and Management*, 5(2), 2006, 119-134.

[23] D.Suteu, T.Malutan, D.Bilba, *Removal of Reactive Dye Brilliant Red HE-3B from aqueous solutions by industrial lignin: Equilibrium and Kinetics Modeling, Desalination,* 255, 2010, 84-90.

[24] A.Demirbas, *Adsorption of lead and cadmium ions in aqueous solutions onto modified lignin from alkali glycerol delignication, J. Hazard.Mater.*, 109(1-3), 2004, 221-226.

[25] W.S.Wan Ngah , M.A.K.M.Hanafiah, *Removal of heavy metal ions from wastewater by chemically modified plant wastes as adsorbents: A review, Bioresource Technology*, 99, 2008, 3935–3948.

[26] Tarioq S.Najim, Nazik J.Elais, Alaya A.Dawood, *Adsorption of copper and iron using low cost material as adsorbent, E-Journal of Chemistry*, 6(1), 2009, 161-168.

[27] R.Malarvizhi, N. Sulochana, *Sorption Isotherm and Kinetic Studies of Methylene Blue Uptake onto Activated Carbon Prepared from Wood Apple Shell, Journal of Environmental Protection Science*, 2, 2008, 40 – 46.

[28] D.Mohan, C.U.Pittman Jr., P.H.Steel, *Single, Binary and Multi-component Adsorption of Copper and Cadmium from Aqueous Solutions on Kraft Lignin-a Biosorbent, J.of Colloid and Interface Science*, 297, 2006, 489-504.

[29] D.Suteu, D.Bilba, C.Zaharia. A.Popescu, *Removal of dyes from textile wastewater by sorption onto ligno-cellulosic materials, Scientific Study and Research* (Romania), IX (3), 2008, 293-302.

[30] D.Suteu, D.Bilba, G.Rusu, M.Macoveanu, *Study of dyes removal from aqueous solution onto Romanian peat,* Annales of Oradea University (Romania), XIV, 2007, 26-31.

[31] M.Akhtar, S.M.Hasany, M.I.Bhanger, S.Iqbal, *Low cost sorbents for the removal of methyl parathion pesticide from aqueous solutions, Chemosphere*, 66, 2007, 1829-1838.

[32] M.Nasiruddin Khan, M.Faeooq Wahab, *Characterization of chemically modified corncobs and its application in the removal of metal ions from aqueous solution, J. Hazard. Mater.*, 141, 2007, 237-244.

[33] M.Jain, V.K.Garg, K.Kadirvelu, *Chromium (VI) removal from aqueous system using Helianthus annuus (sunflower) stem waste, J.Hazard. Mater.*, 162, 2009, 365-372.

[34] B.M.W.P.K.Amarasinghe, R.A.Williams, *Tea waste as a low cost adsorbent for the removal of Cu and Pb from wastewater, Chem.Eng.J.*, 132, 2007, 299-309.

[35] L.Tofan, D.Suteu, O.Toma, L.Bulgariu, *Biosorption in treatment of waste water*, The Scientific Annals of A.I.Cuza University Iasi (Romania), Section: Genetics and Molecular Biology, Tom IV, 2003, 26-35.

[36] M.Gavrilescu, *Removal of heavy metals from the environment by biosorption*, Eng. Life Sci., 4, 2004, 219-232.

[37] G.Crini, P.M.Badot, *Application of chitosan, a natural aminopolysaccharide, for dye removal from aqueous solution by adsorption processes using batch studies: A review of recent literature*, Prog. Polym. Sci., 33, 2008, 399-447.

[38] D.Suteu, M.Harja, L.Rusu, *Sorption of dyes from aqueous solution onto lime-coal ash sorbent*, Proceeding of International Scientific Conference, 23- 24 November 2007, Gabrovo, Bulgaria, 2, 2007, 317–321.

[39] D.Suteu, C.Zaharia, *Studies about sorption equilibrium of Methylene Blue dye onto solid materials based on coal ashes*, Bull.Inst.Polytech.Iasi (Romania), tome LIV (LVIII), 3, 2008, 81-90.

[40] D.Suteu, C.Zaharia, M.Harja, *Residual ash for textile wastewater treatment*, Proceeding of International Scientific Conference, 23 – 24 November 2008, Gabrovo, Bulgaria, 3, 2008, 475–480.

[41] D.Suteu, D.Bilba, C.Zaharia, A.Popescu, *Removal of dyes from textile wastewater by sorption onto ligno-cellulosic materials*, Scientific Study and Research, IX, 2008, 293-302.

[42] D.Suteu, T.Malutan, G.Rusu, *Use of Industrial Lignin for dye removal from aqueous solution by sorption*, Scientific Papers USAMV IASI (Romania), Series Agronomy, 51, 2008, 71-78.

[43] D.Suteu, D.Bilba, R.Muresan, M.Muresan, *Using fibrous materials for textile wastewaters treatment*, Proceeding of 4th International Textile, Clothing and Design Conference – Magic World of Textiles, October 5th - 8th 2008, Dubrovnik, Croatia, 2008, 1112-1117.

[44] E.Voudrias, Fytianos K., Bozani E., *Sorption-desorption isotherms of dyes from aqueous solutions and wastewaters with different sorbent materials*, Global Nest:the Int.J., 4, 2002, 75-83.

[45] W.M.Antumes, A.S.Luna, C.A.Henriques, A.C.A.daCosta, *An evaluation of copper biosorption by a brown seaweed under optimization conditions*, Electronic Journal of Biotechnology, 6, 2003, 174-184.

[46] L.Branza, M.Gavrilescu, *pH Effect on the Biosorption of Cu^{2+} from aqueous solution by Saccharomyces cerevisiae*, Environmental Engineering and Management Journal, 2(3), 2003, 243-254.

[47] G.Carrillo-Morales, M.M.Davila Jimenez, M.P.Elizalde-Gonzalez, A.A.Pelaez-Cid, *Removal of metal ions from aqueous solution by adsorption on the natural adsorbent ACMM2*, J.Chromatogr.A., 938, 2001, 237-242.

[48] A.J.Chaudhary, B.Ganguli, S.M.Grimes, *The use of chromium waste sludge for the adsorption of colour from dye effluent streams*, J. Chem.Technol.Biotechnol., 77, 2002, 767-770.

[49] L.Tofan, D.Suteu, O.Toma, M. Vizitiu, *Cellulosic biomaterials with ecologycal impact*, Analele Şt. ale Univ.A.I.Cuza, Iaşi (Romania), sect.II, tom III, 2002, 88-97.

[50] D.Suteu, D.Bilba, *Equilibrium and kinetic study of reactive dye Brilliant Red HE-3B adsorption by activated charcoal*, Acta Chim.Slov., 52, 2005, 73-79.

[51] L.Tofan, C.Paduraru, *Sorption Studies of Ag (I), Cu (II) and Pb (II) Ions Sulphydryl Hemp Fibers, Croatica Chemica Acta*, 77, 2004, 581-58.

[52] T.Malutan, *Complex valorization of biomass* (in Romanian), Publishing House Performantica, Iasi (Romania), 2008.

[53] X.Guo, X.Shan, S.Zhang, *Adsorption of metal ions on lignin, J.Hazard.Mater.*, 151, 2008, 134-142.

[54] W.H.Cheung, Y.S.Szeto, S.McKay, *Intraparticle diffusion processes during acid dye adsorption onto chitosan, Bioresource Technology*, 98, 2007, 2897-2904.

[55] D.Suteu, C.Zaharia, T.Malutan, *Removal of Orange 16 reactive dye from aqueous solution by wasted sunflower seed shells,* unpublished data.

[56] D.Suteu, T.Malutan, D.Bilba, *Agricultural waste corncob as sorbent for the removal of reactive dye Orange 16: equilibrium and kinetic study, Cell. Chem. Technol.* – in press.

[57] D.Suteu, C.Zaharia, M.Badeanu, *Agriculture wastes used as sorbents for dye removal from aqueous environments*, Scientific Papers USAMV IASI (Romania), Series Agronomy, 53, 2010 – in press.

[58] D.Suteu, T.Malutan, *Removal of Methylene Blue cationic dye onto wasted cellolignin,* unpublished data.

[59] R.Han, J.Zhang, P.Han, Y.Wang, Z.Zhao, M.Tang, *Study of equilibrium, kinetic and thermodynamic parameters about Methylene Blue adsorption onto natural zeolite, Chem.Eng.J.,* 145(3), 2009, 496-504.

[60] M.Wawrzkiewicz, Z.Hubicki, *Equilibrium and kinetic studies on the adsorption of acidic dye by the gel anion exchanger, J. Hazard. Mater.*, 172(2-3), 2009, 868-874.

[61] D.Suteu, C.Zaharia, *Application of lignin materials for dye removal by sorption processes*, unpublished data.

[62] M.H.Liu, J.H.Huang, *Removal and recovery of cationic dyes from aqueous solutions using spherical sulfonic lignin adsorbent, J.Appl.Polym.Sci.*, 101, 2006, 2284-2291.

[63] G.McKay, J.F.Parker, G.R.Prasad, *The removal of dye colours from aqueous solutions by adsorption on low-cost materials, Water Air and Soil Pollut.*, 114, 1999, 423-438.

[64] R.Gong, Y.Jin, J.Sun, K.Zhong, *Preparation and utilization of rice straw bearing carboxyl groups for removal of basic dyes from aqueous solution*, Dyes and Pigments, 76, 2008, 519-524.

[65] R.Gong, Y.Jin, F.Chen, J.Chen., Z.Lin, *Enhanced malachite green removal from aqueous solution by citric acid modified rice straw, J.Haz.Mater.*, B137, 2006, 865-870.

[66] V.Vadivelau, K.Vasanth Kumar, *Equilibrium, kinetics, mechanism, and process design for the sorption of Methylene Blue onto rice husk, J.Colloid.Interface. Sci.*, 286, 2005, 90-100.

[67] B.Hameed, M.I.El-Khaiary, *Kinetics and equilibrium studies of malachite green adsorption on rice straw-derived char, J.Hazard.Mater.*, 153, 2008, 701-708.

[68] R.Gong, Y.Jin, F.Chen, J.Chen., Y. Hu, J. Sun, *Removal of basic dyes from aqueous solution by sorption on phosphoric acid modified rice straw, Dyes and Pigments*, 73, 2007, 332-337.

[69] O.Hamdaoni, *Batch study of liquid-phase adsorption of Methylene Blue using cedar sawdust and crushed brick, J.Hazard.Mater.*, B135, 2006, 264-273.

[70] V.Dulman, S.M.Cucu-Man, *Sorption of some textile dyes by beech wood sawdust, J.Hazard.Mater.*, 162, 2009, 1457-1464.

[71] F.A.Batzias, D.K.Sidiras, *Dye adsorption by prehydrolyzed beech sawdust in batch and fixed-bed systems, Bioresource Technol.*, 98, 2007, 1208-1217.

[72] F.A.Batzias, D.K.Sidiras, *Simulation of methylene blue adsorption by salts-treated beech sawdust in batch and fixed-bed systems, J.Hazard. Mater.*, 149, 2007, 8-17.

[73] H.Pekkkuz, I.Uzun, F.Güzel, *Kinetics and thermodynamics of the adsorption of some dyestuffs from aqueous solution by poplar sawdust, Bioresource Technol.*, 99, 2008, 2009-2017.

[74] Y.S.Ho, T.H.Chiang, Y.M.Hsueh, *Removal of basic dye from aqueous solution using tree fern as a biosorbent, Process Biochemistry*, 40, 2005, 119-124.

[75] P.Janos, S.Coskun, V.Pilarova, J.Rejnek, *Removal of basic(Mthylene blue)and acid (Egacid Orange) dyes from waters by sorption on chemically treated wood shavings, Bioresource Technol.*, 100, 2009, 1450-1453.

[76] B.Hameed, M.I.El-Khaiary, *Removal of basic dye from aqueous medium using a novel agricultural waste material: pumpkin seed hull, J.Hazard. Mater.*, 155, 2008, 601-609.

[77] L.S.Oliveira, A.S.Franca, T.M.Alves, S.D.F.Rocha, *Evaluation of untreated coffee husks as potential biosorbents for treatment of dye contaminanted waters, J.Hazard. Mater.*, 155, 2008, 507-512.

[78] R.Gong, M.Li, C.Yang, Y.Sun, *Removal of cationic dyes from aqueous solution by adsorption on peanut hull, J. Hazard. Mater.*, B.121, 2005, 247-250.

[79] D.Özer, G.Dursum, A.Özer, *Methylene Blue adsorption from aqueous solution by dehydrated peanut hull, J. Hazard. Mater.*, 144, 2007, 171-179.

[80] R.Han, Y.Wang, P.Han, J.Shi, J.Yang, Y.Lu, *Removal of methylene blue from aqueous solution by chaff in batch mode, J. Hazard. Mater.*, B137, 2006, 550-557.

[81] G.Sreelatha, P.Padmaja, *Study of removal of cationic dyes using palm shell powder as adsorbent, J. of Environ. Prot. Sci.*, 2, 2008, 66-71.

[82] R.A.Carletto, F.Chimirri, F.Bosco, F.Ferrero, *Adsorption of Congo Red dye on hazelnut shells and degradation with Phanerochaete chrysosporium, Bioresource*, 3(4), 2008, 1146-1155.

[83] F.Ferrero, *Dye removal by low cost adsorbents. Hazelnuts shells in comparison with wood sawdust, J. Hazard. Mater.*, 142, 2007, 144-152.

[84] J. F.Osma, V.Saravia, J.L.Toca-Herrera, S.Rodriguez Couto, *Sunflower seed shells: A novel an effective low-cost adsorbent for the removal of diazo dye Reactive Black 5 from aqueous solution, J. Hazard. Mater.*, 147, 2007, 900-905.

[85] M.M.Davila-Jimenez, M.P.Elizalde-Gonzalez, V.Hernandez-Montoya, *Performance of mango seed adsorbents in the adsorption of antraquinone and azo dyes in single and binary aqueous solution, Bioresource Technol.*, 100, 2009, 6199-6206.

[86] Z.Aksu, O.A.Isoglu, *Use of agricultural waste sugar beet pulp for the removal of Gemazol turquoise blue-G reactive dye from aqueous solution, J. Hazard. Mater.*, B137, 2006, 418-430.

[87] P.Senthil Kumar, S.Ramalingam, C.Senthamarai, M.Niranjanee, P.Vijayalakshui, S. Sivanesan, *Adsorption of dye from aqueous solution by cashew nut shell: Studies on equilibrium isotherm, kinetics and thermodynamics interaction, Desalination*, 261, 2010, 52-60.

[88] B.H.Hameed, D.K.Mahmond, A.L.Ahmad, *Sorption equilibrium and kinetics of basic dye from aqueous solution using banana stalk waste, J.Hazard. Mater.*, 158, 2008, 499-506.

[89] N.A.Oladoja, C.O.Aboluwoye, Y.B.Oladimeji, A.O.Ashogbou, I.O.Otemuyiwa, *Studies on castor seed shell as a sorbent in basic dye contaminated wastewater remediation, Desalination*, 227, 2008, 190-203.

[90] R.Gong, S.Zhu, D.Zhang, J.Chen, S.Ni, R.Guan, *Adsorption behavior of cationic dyes on citric acid esteriphying wheat straw: kinetic and thermodynamic profile, Desalination*, 230, 2008, 220-228.

In: Lignin
Editor: Ryan J. Paterson

ISBN 978-1-61122-907-3
© 2012 Nova Science Publishers, Inc.

Chapter 9

MICROBIAL DEGRADATION OF LIGNIN FOR ENHANCING THE AGROINDUSTRIAL UTILITY OF PLANT BIOMASS

Smita Rastogi and Ira Chaudhary*

Department of Biotechnology, Integral University, Lucknow, India

ABSTRACT

Lignin is a three dimensional natural plant biopolymer formed by radical coupling of hydroxycinnamyl subunits called monolignols mainly *p*-coumaryl, sinapyl and coniferyl alcohols and creates together with hemicelluloses, a glueing matrix for cellulose microfibrils in tracheary elements and fibers of higher plants. It is the second most abundant component of the cell wall of vascular plants and it protects cellulose towards hydrolytic attack by saprophytic and pathogenic microbes.

Though important for plants, the removal of lignin represents a key step for carbon recycling in land ecosystems. Its removal is also considered as a fundamental issue for the optimal agroindustrial utilization of plant biomass. Lignin polymer is considered to be recalcitrant towards degradation by chemical and biological means. This owes to its molecular architecture in which different non-phenolic phenylpropanoid units form a complex 3D mesh linked by several ether and carbon-carbon bonds. A few microbial species are known for their ligninolytic potential. These ligninolytic microbes exhibit a unique strategy for lignin degradation, which is based on unspecific one-electron oxidation of the benzenic rings catalyzed by synergistic action of extracellular haemperoxidases and peroxide-generating oxidases.

1. INTRODUCTION

Lignin is an amorphous, aromatic, water insoluble, heterogeneous, three-dimensional and cross-linked polymer with low viscosity (Sjöström 1993; Brunow 2001). It is a complex

---

* Mobile: +91 9415759875, Email: smita_rastogi@rediffmail.com.

heteropolymer of three types of methoxylated phenylpropanoids called monolignols which are synthesized by phenylpropanoid pathway starting from *L*-phenylalanine and connected by both ether and carbon-carbon linkages (Howard et al. 2003; Vinardell et al. 2008). It is one of the world's most abundant natural polymers, along with cellulose and chitin. As an integral cell wall constituent, lignin provides plants strength and resistance to plants (Argyropoulos and Menachem 1997). Moreover, lignin participates in water transport in plants and forms a barrier against microbial destruction by protecting the readily assimilable polysaccharides (Monties and Fukushima 2001). From agroindustrial perspective, lignin hinders the optimal utilization of plant biomass in pulp and paper industry, textile industry, forage and biodegradation for carbon recycling. This aggravates the need to devise strategies for the removal of lignin. Lignin is, however, considered to be recalcitrant in nature and no mammalian and animal enzymes are known to degrade lignin. Traditionally, lignin was removed by conventional chemical methods. However, in recent years, looking into the commercial interests of the pulp and paper industry, textile industry, optimal forage utility and biodegradation of lignin for environment restoration, the studies on microbial degradation of lignin have gained momentum (Hatakka et al. 2001; Scott and Akhtar 2001). A few microbes have now been identified to possess ligninolytic enzymes, for example, laccase, lignin peroxidase and manganese peroxidase. The microbial degradation of lignin offers several potential advantages in contrast to chemical methods, which include greater substrate and reaction specificity, lower energy requirements, lower pollution generation and higher yields of desired products (Eriksson et al. 1990; Falcón et al. 1995; Kuhad et al. 1997; Hatakka et al. 2001). The present paper discusses the structure of lignin responsible for its recalcitrant nature and various biolignolytic systems (microbes and enzymes). Further the applicability of biolignolytic systems in various fields is also presented.

2. STRUCTURE OF LIGNIN IMPARTS RECALCITRANCE FOR DEGRADATION

The knowledge about the structure and chemical nature of lignin, bonding patterns, site of deposition and cross-linking with cell wall carbohydrates is important to understand its role in the vitality of plants as well as its recalcitrance to chemical and microbial degradation with a repercussion in the optimal utilization of plant biomass for various agroindustrial purposes or biodegradation of plant remnants. Hence a brief overview of lignification is presented.

Lignin, a biopolymer of monolignols (*p*-coumaryl, coniferyl and sinapyl alcohols), is synthesized from *L*-phenylalanine through phenylpropanoid pathway (Rastogi and Dwivedi 2008). The pathway leading to the formation of monolignols envisages a 'metabolic grid' featuring hydroxylation and methoxylation of the aromatic ring at different levels and conversion of the side chain carboxyl to an alcohol group. After their synthesis in the cytoplasm, the monolignols are converted into glucoside derivatives and transported to the cell wall where these are hydrolyzed before polymerization into lignin. These reactions are catalyzed by UDP-glucosyltransferase and coniferin β-glucosidase, respectively. This is followed by free radical mediated coupling (dehydrogenative polymerization) of monolignols resulting in deposition of a large, heavily cross-linked lignin polymer of undefined dimensions in the cell wall. It has been suggested that the lignin macromolecules themselves act as templates for further lignification once the process has been initiated. The enzymes catalyzing the polymerization step include peroxidase, laccase, phenol oxidase and coniferyl alcohol oxidase.

Lignin formed after polymerization is of two basic types, gymnosperm lignin and angiosperm lignin. Gymnosperm lignin predominantly contains monomethylated guaiacyl units (G units) polymerized from coniferyl alcohol and a small proportion of *p*-hydroxy phenyl units (H units) polymerized from *p*-coumaryl alcohol. Angiosperm lignin contains dimethylated syringyl units (S units) derived from sinapyl alcohol and G units with small proportions of H units. The vessel secondary wall and middle lamella of typical hardwoods contain G units while the secondary walls of fibers and the cell walls of parenchyma cells contain a mixture of G and S units with the latter predominating (Ruel and Ayres 1999).

When the monolignols are polymerized to form lignin, covalent C-C (condensed) and C-O (non-condensed) linkages are formed between them. The main linkages in lignin are β-O-4 (aryl ether), β-5 (phenyl coumaran), 5-5-O-4 (dibenzodioxocin) and β-β (pinoresinol) (Lewis and Davin 1999), but β-O-4 is the most abundant structure in wood lignin, representing over 50% of the linkages.

The nature of these linkages defines the chemical properties of lignin, *i.e.*, 5-5 and β-5 are more difficult to degrade. Lignin is racemic in nature. Even a simple β-O-4-linked dimer that contains two asymmetric carbons exists as four stereoisomers and as the number of isomers increases geometrically with the number of subunits, lignin presents a complex and non-repeating 3D surface.

Besides the common lignin structural linkages, monolignols are also cross-linked with other cell wall biopolymers, *viz.*, polysaccharides, hemicelluloses and proteins. Carbohydrate components of the cell wall exert a mechanical influence on the expanding lignin lamellae during lignification. Transmission electron microscopy has shown that growing lignin particles in the middle lamella and primary wall form roughly spherical structures within the randomly arranged matrix, while in the secondary wall, lignin lamellae form greatly elongated structures, following the orientation of microfibrils in this region of the cell wall (Donaldson 2001).

Lignin also cross-links with hemicellulose, which acts as a bridge or adhesive, ensuring coherence between the matrix phase and the cellulose microfibrils (Hansen and Björkman 1998). Thus, lignin polymerization mechanism based on resonant radical coupling results in a complex three-dimensional network due to both chain branching and inter- or intrachain coupling during polymerization (Gellerstedt and Henriksson 2007).

3. NEGATIVE IMPLICATIONS OF LIGNIN FROM AGROINDUSTRIAL PERSPECTIVE

Lignification of plant cell is one of the major hindrances in the optimal utilization of plant biomass in paper industry, textile industry, forage production and biodegradation of lignocellulosic materials.

3.1. Paper Industry

For paper manufacture, lignin is presently removed from wood by chemical or mechanical means (Sixta 2006). The pulping process is followed by bleaching to remove the residual lignin from the pulp and increase brightness.

These processes, however, require energy, release toxic pollutants into the environment, lead to reduction in pulp yield and decrease the cellulose degree of polymerization due to attack of bleaching agents on cellulose polymer. The presence of residual lignin in paper results in inferior performance characteristics, poor brightness stability or yellowing with age.

3.2. Textile Industry

In the past few years, the popularity of fiber crops had been superseded by synthetic fibers, which are non-renewable and accumulates pollution (Stephens and Halpin 2007). These lignocellulosic fibers, however, contain lignin, which imparts worse elastic properties to the textile. Furthermore, pectin, hemicellulose and lignin content in lignocellulosic fibers interfere with the extractability of the fibers.

3.3. Forage

Lignification also limits digestibility (both *in vivo* and *in vitro*) of plant cell wall polysaccharides by ruminants (Dixon 2004). This is because extensive lignification makes the digestive hydrolytic enzymes inaccessible to the cell wall polysaccharides. The repercussion of reduced digestibility is limited energy yields of forage crops and increased cost of animal production.

3.4. Biodegradation of Plant Remnants

The complex structure and bulky nature of lignin incorporating both ether and C-C linkages between monomers and the abundant chemical cross-linking of lignin makes the lignocellulosic materials less susceptible to microbial degradation and hence their persistence in soil (Monties and Fukushima 2001). Moreover, due to its non-phenolic aromatic nature, lignin units cannot be oxidized by low-redox-potential oxidoreductases, such as the plant peroxidases initiating the polymerization process (Li et al. 2009; Ruiz-Dueñas and Martínez 2009). Consequently plant remnants get accumulated in high amounts in natural environments.

These problems thus signify the need for removal of lignin from plant biomass/remnants in an ecofriendly way.

4. MICROBIAL DEGRADATION OF LIGNIN

In nature, lignin is probably degraded by an array of microorganisms, although abiotic degradation may also occur in special environments, such as those due to alkaline chemical spills (Blanchette et al. 1991) or UV radiation (Vähätalo et al. 1999). In aqueous or other anaerobic environments, polymeric lignin is not degraded and wood may persist in undegraded form for several hundreds or thousands of years (Blanchette 1995). Slow abiotic degradation, favored by high-temperature, acidic, or alkaline environments, releases small fragments. Amongst microorganisms, a few bacteria, actinomycetes and fungi are known to degrade lignin; however, the extent of degradation varies with the type of microorganisms. The degradation of lignin has been studied extensively in recent years to safeguard the commercial interests of the pulp and paper and textile industries, increasing forage digestibility, as well as its biodegradation for environment restoration (Akhtar et al. 1997; Hatakka et al. 2001; Scott and Akhtar 2001).

4.1. Selective and Non-Selective Decay of Lignin

The degradation of lignin is either selective or non-selective (*i.e.* simultaneous decay) (Blanchette 1995; Hatakka et al. 2001). In selective decay, lignin and hemicellulose are degraded significantly more than cellulose, while equal amounts of all components of lignocellulose are degraded in case of non-selective decay. Thus fungi such as *Ceriporiopsis subvermispora*, *Dichomitus squalens*, *Phanerochaete chrysosporium*, *Phlebia radiata* stimulate selective decay, while *Trametes versicolor* and *Fomes fomentarius* catalyze non-selective decay. Selective degradation of lignin occurs under special conditions or at the beginning of decay and is followed by hemicellulose and cellulose degradation (Dix and Webster 1995). Some fungi, such as *Ganoderma applanatum*, *Heterobasidion annosum* and *Phellinus pini*, can execute both types of decay (Blanchette 1995). Nutritional factors also control the type of decay and the fungus may degrade wood both selectively and simultaneously in the very same wood stem (Eaton and Hale 1993).

4.2. Lignin-Degrading Fungi

Fungi are the most rapid and extensive degraders of lignin as compared to any other microbes (Dix and Webster 1995). Lignin-degrading fungi known so far mostly belong to class basidiomycetes and ascomycetes.

The wood-degrading basidiomycetes, for example, brown- and white-rot fungi completely metabolize lignin, exhibit the highest reported rates of degradation and are the most extensively studied. These basidiomycetes are better degraders of lignin in woods, cereals and grasses in contrast to ascomycetes. Thus fungi with ligninolytic activity include *Cariolus pruinosum*, *Cereporiopsis subvermispora*, *Cyathus bulleri*, *C. stercoreus*, *Dichomitus squalens*, *Lentinus edodes*, *Oudemansiella radicata*, *Panus tigrinus*, *Phanerochaete chrysosporium*, *Phlebia radiata*, *P. tremellosus*, *P. ochraceofulva*, *Pleurotus ostreatus*, *P. eryngii*, *P. sajorcaju*, *P. florida*, *Polyporus pletenis*, *P. brumalis*, *Pycnoporus sanguineus, P. cinnabarinus, Rigidosporus lignosus, Sporotrichium thermophile* and *Trametes versicolor,* etc. (Lundell 1993; Martinez et al. 1994; Vasdev and Kuhad 1994; Falcón et al. 1995; Paszczynski and Crawford 1995; Vasdev et al. 1995; Eggert et al. 1996; Hatakka et al. 2001). There are ample evidences to indicate that the growth substrates of these fungi are cellulose and hemicellulose, but lignin degradation occurs at the end of primary growth by secondary metabolism in the deficiency of nutrients, *i.e.,* nitrogen, carbon or sulfur. Fungi catalyze degradation of lignin by an oxidative and non-specific process, which decreases the contents of methoxy, phenolic and aliphatic groups in lignin, cleaves aromatic rings and creates new carbonyl groups. By analysis of wood partially degraded by fungi it has been demonstrated that degradation involves heavy oxidation with formation of numerous carboxyl groups. These changes in the lignin molecule result in depolymerization and carbon dioxide production. Moreover, oxygen stimulates lignin degradation by some fungi and these fungi are able to mineralize up to 75% of lignin. Majority of these fungi grow on hardwoods, but certain species such as *Heterobasidion annosum, Phellinus pini* and *Phlebia radiata* grow on softwoods (Blanchette 1995).

Brown-rot fungi degrade cellulose and hemicellulose from wood very efficiently with a degradation mechanism that is different from all other organisms. It involves non-enzymatic reactions and lacks exoglucanses (Eriksson et al. 1990; Blanchette 1995). The presence of lignin stimulates cellulose degradation by brown-rot fungi, although lignin is degraded to a lesser extent, mainly by demethylating and partly by degrading the lignin-rich middle lamella (Blanchette 1995; Hatakka et al. 2001). Demethylation activity of lignin by brown-rot fungi is stimulated in the presence of wood (Niemenmaa et al. 1992). A brown-rot fungus *P. ostreiformis* produces manganese peroxidase (MnP) and lignin peroxidase (LiP), similar to white-rot fungi, but its lignin degradation capability is significantly lower (Dey et al. 1994). Wood degraded by brown-rot fungi is mainly modified non-degraded lignin (Blanchette 1995). This brown residue of heavily decayed wood may remain in the forest for long periods of time without further degradation, although it is more reactive than native lignin (Blanchette 1995; Hatakka et al. 2001). Brown-rot fungi grow more frequently in softwood than in hardwood, particularly favoring top-layer coniferous forest soils (Blanchette 1995).

The gasteromycete *Cyathus stercoreus,* which is associated with litter decomposition, degrades lignin in wheat straw (Wicklow et al. 1984). The degradation is as extensive as that by various white-rot fungi (Agosin at al 1985). Several other *Cyathus* species degrade grass lignin, but members of the related genera *Nidula* and *Crucibulum* do not (Abbott et al. 1984). Ectomycorrhizal fungi *(Cenococcum, Amanita, Tricholoma* and *Rhizopogon)* has been shown to only slowly mineralize ^{14}C-labeled synthetic lignins and corn lignin (Trojanowski et al. 1984). Among ascomycetes, *Xylaris hypoxylon, X. polymorpha, Ustulina vulgaris* have been reported to facilitate lignin degradation. Laccase from *Monocillium indicum* was the first to be characterized from an ascomycete showing peroxidative activity (Thakker et al. 1992).

In arable soil and in composts, microfungi (soft-rot fungi) are the most important lignin degraders (Daniel and Nilsson 1998). Some examples of soft-rot fungi with lignin degrading ability are *Chaetomium globosum, Daldinia concentrica, Lecythophora cellulolyticum, Fusarium oxysporium, Neurospora crassa, Penicillium pinophilum* and *Pialophora mutabilis* (Eaton and Hale 1993; Daniel and Nilsson 1998). Soft-rots grow only on the surface of wood and require high moisture and nitrogen content. Soft-rot fungi degrade all wood components, but lignin removal is slow and partial with demethylation (Eaton and Hale 1993). Thus, most cellulolytic microfungi are able to degrade grass lignin and wood lignin to a limited extent, nevertheless, relatively little is known about the lignin degradation capability of microfungi (Rodriguez et al. 1996; Daniel and Nilsson 1998). Soft-rot fungi degrade all wood components, but lignin removal is slow and partial with demethylation (Eaton and Hale 1993). *Fusarium oxysporum* (Fakoussa and Frost 1999), *Xylaria* sp. and *Altenaria* sp. produce peroxidases (Hofrichter and Fritsche 1996) and *Botrytis cinerea*, *Myceliophthora thermophila* (Li et al. 1999), *Chaetomium thermophilium* (Chefetz et al. 1998) and *Paecilomyces farinosus* (Fakoussa and Frost 1999) produce laccase. The activity of laccases produced by soft-rot fungi is lower than that of basidiomycetes (Li et al. 1999; Tanaka et al. 2000).

The most important lignin degraders in a compost environment are probably thermophilic fungi such as *Thermoascus aurantiacus* and *Thermomyces lanuginosus* (Kuhad et al. 1997).

4.3. Lignin-Degrading Bacteria

Several bacteria are also capable of degrading lignin (Zimmermann 1990; Rüttimann et al. 1991; Godden et al. 1992; Perestelo et al. 1996). These include a number of bacteria belonging to the genera *Achromobacter, Acinetobacter, Aerobacter, Agrobacterium, Erwinia, Pseudomonas, Nocardia, Serratia and Xanthomonas* (Zimmerman 1990; Perestelo et al. 1994; Morii et al. 1995; Perestelo et al. 1996). The lignin degradation mechanism of bacteria is more specific than that of fungi and one bacterial species is able to cleave only one type of bond in the lignin polymer and bacteria generally cause a low percentage of mineralization of lignin from lignocellulosic materials (Vicuña et al. 1993).

Broadly, bacteria degrade wood slowly and degradation takes place on wood surfaces with high moisture content (Blanchette et al. 1991; Blanchette 1995). These are capable of attacking both softwood and hardwood by first colonizing the parenchyma cell walls.

Wood-degrading bacteria have primarily cellulolytic and pectinolytic activities and pure bacterial cultures are unable to perform efficient lignin degradation. However, bacteria degrade lignocellulose in mixed cultures, either in mixed bacterial cultures, or more commonly, in bacterial and fungal cultures together and bacteria probably consume mainly fungal by-products in mixed cultures (Vicuña et al. 1993; Lang et al. 2000; Vicuña 2000). This is because of the lack of penetrating ability by bacteria alone. Although bacteria can directly attack fibers, vessels and tracheids, few species and strains can degrade all the cell wall components.

Wood-degrading bacteria have a wider tolerance of temperature, pH and oxygen limitations than fungi (Daniel and Nilsson 1998). Thus, bacterial degradation can be observed when the growth of fungi is decreased in acute conditions, such as wood saturated with water, oxygen limitation, high extractive content, high concentration of lignin or wood treated with chemical preservatives (Blanchette 1995; Daniel and Nilsson 1998). This lignin degradation capability of prokaryotes is due to the presence of proteins with typical features of the multi-copper oxidase enzyme family (Claus 2003).

The three major morphological forms of bacterial cell wall degradation are: tunneling, erosion and cavitation (Blanchette 1995). Tunneling bacteria are rare in waterlogged wood because these appear to require a good supply of oxygen. These produce minute tunnels to migrate through the cell wall. Erosion bacteria are responsible for the predominant form of degradation in waterlogged archaeological wood, since these can tolerate near-anaerobic or fully-anoxic environments (Kim et al. 1996; Björdal et al. 1999). Erosion bacteria are typically rod-shaped and these attack the wall from the lumen into the secondary walls singly or in small groups (Holt 1983). Cavitation bacteria, presenting in the wood cell lumen, apparently utilize products derived from the activities of wood degraders (Singh et al. 1990).

4.4. Lignin-Degrading Actinomycetes

Another group of microbes, actinomycetes also live well in environments rich in lignocellulose, such as soil, compost, heaps of hay, straw, or wood chips (Lacey 1988). Actinomycetes have also been reported to mineralize lignin successfully, although not as fast or as comprehensively as fungi (Rüttimann et al. 1991). The degradation of lignin results in the release of lignin rich, water-soluble fragments called acid precipitable polymeric lignin (Pasti et al. 1991; Spiker et al. 1992). Polyphenolic and polymeric lignin fragments have lower molecular mass and lower methoxyl content than native lignin and these are associated with bacterial protein or hemicellulosic carbohydrates (Zimmermann 1990). Although lignin mineralization by actinomycetes is not as efficient as by fungi, it is still more efficient than by unicellular bacteria. Some of the most active lignin degraders among actinomycetes are *Streptomycetes badius*, *S. cyaneus* and *Thermomonospora mesophila*. Actinomycetes have been claimed to produce extracellular peroxidase (Mercer et al. 1996). It is suggested that phenol oxidase produced by *Streptomyces cyaneus* participates in lignin degradation much more than the peroxidase (Berrocal et al. 2000). Laccases are also found in actinomycetes such as *Streptomyces griseus* and *S. lavendulae* (Freeman et al. 1993; Suzuki et al. 2003). In *Streptomyces cyaneus*, a laccase type phenol oxidase is produced during growth under solid substrate fermentation conditions and it is suggested that this enzyme is involved in the solubilization and mineralization of lignin from wheat straw.

4.5. Anaerobic Degradation of Lignin

Substantial amount of lignocellulosic materials is also degraded under anaerobic conditions, for example, in soil and compost microenvironments (Atkinson et al. 1996; Durrant 1996). Lignin degradation under such conditions is significant though slower than that achieved under aerobic conditions (Susmel and Stefanon 1993; McSweeney et al. 1994; Kato et al. 1998).

A few bacterial and fungal species, which are facultative-anaerobes or are capable of growing under microaerophilic conditions, can also degrade lignin. Anaerobic fungi discovered in rumen, such as *Neocallimastix frontalis*, often produce cellulases and xylanases and probably also have some lignin degradation ability (Susmel and Stefanon 1993; McSweeney et al. 1994; Durrant 1996). Under anaerobic conditions, a mixed population isolated from activated sludge has also been reported to cleave the β-O-4 linkage of low molecular mass lignin to produce monoaromatic compounds (Carlile and Watkinson 1994). In rumen, half of the lignin is either solubilized or transformed into a soluble lignin carbohydrate-complex and a variable amount is digested, although the biochemical pathways are unknown (Susmel and Stefanon 1993). Bacteria isolated from compost or soil, namely *Azotobacter*, *Bacillus megaterium* and *Serratia marcescens*, are reported to be capable of decolorizing or solubilizing lignin. Anaerobic lignin degradation also occurs in aquatic environments and termite guts (Blanchette et al. 1991). It has been demonstrated that the microflora of the termite gut demethylates and depolymerizes lignin to a large extent.

5. MECHANISM OF LIGNIN DEGRADATION

5.1. Ligninolytic Enzymes

Lignin degradation requires non-specific and extracellular enzymes because of the random structure and high molecular mass of the lignin molecule (Kirk and Farrell 1987; Ralph 2005). These ligninolytic enzymes cleave various specific bonds in lignin with varying effectiveness and each enzyme is precise to a specific chemical bond. Various chemical bonds are difficult to be cleaved specifically within an enzyme's active site. As a consequence, different enzymes, each with a specific active site are required, for degradation. Three major enzymes implicated in lignin degradation are laccase, lignin peroxidase and manganese peroxidase (Barr and Aust 1994; Hatakka 1994; Vasdev and Kuhad 1994; Vasdev et al. 1995; Pointing 2001). Different wood-decaying fungi and bacteria produce all the three enzymes or either one or two of them.

5.1.1. Laccases (Lac; EC 1.10.3.2)
Laccases are widespread in nature and are implicated in lignin degradation (Freeman et al. 1993; Thurston 1994; Eggert et al. 1996; Alexandre and Zhulin 2000; Suzuki et al. 2003; Sigoillot et al. 2005). Lac [p-benzenediol : oxygen oxidoreductase] is a blue copper-containing oxidase belonging to a family of multicopper oxidases, which also includes ascorbate oxidase, ceruloplasmin and bilirubin oxidase (Hoegger et al. 2006). Lac catalyzes the oxidation of various aromatic compounds (predominantly phenols) with the reduction of oxygen to water (Thurston 1994; Karam and Nicell 1997).

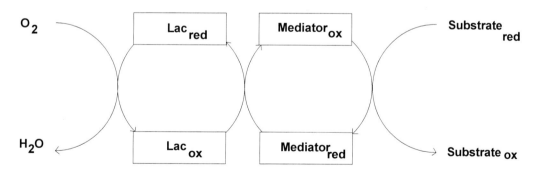

Figure 1. Mechanism of laccase action with reference to mediator.

Copper atoms (four in number) in Lac play important role in its catalytic mechanism. Being large in size (MW 70 kDa), Lac cannot penetrate deep into wood (Bourbonnais et al. 1997). Moreover, due to its 0.5–0.8 V redox potential (low), Lac is unable to oxidize non-phenolic (C4-etherified) lignin units with >1.5 V redox potential (high) (Galli and Gentili 2004). Therefore, an oxidation mediator (i.e., a small molecule that is able to extend the effect of Lac to non-phenolic lignin units and to overcome the accessibility problem) is often used with Lac. Lac-mediated oxidation of non-phenolic lignin units can follow an electron transfer, a radical hydrogen atom transfer, or an ionic mechanism, depending on the mediator (Barreca et al. 2004). Several organic and inorganic compounds, such as thiol and phenol aromatic derivatives, *N*-hydroxy compounds and ferrocyanide are used as mediators (Susana and José 2006). The mechanism of Lac action with respect to mediator is shown in Figure 1.

5.1.2. Lignin Peroxidase (LiP; Ligninase; EC 1.11.1.14)

LiP is another enzyme that catalyzes degradation of lignin. It exists in bacteria, fungi, plants and animals. It is an extracellular heme containing peroxidase, which is dependent on H_2O_2 and has an unusually high redox potential and low optimum pH (4.0 – 8.0), low molecular weight (18 kDa) and typically shows little specificity towards substrates and degrades a variety of lignin-related compounds (Gold and Alic 1993). This heme protein can easily be inhibited by the known heme inhibitor such as potassium cyanide and sodium azide (Howard et al. 2003).

In consideration of the sequence similarity and structural divergence, these are viewed as belonging to a super-family consisting of three major classes (Welinder 1992): mitochondrial yeast cytochrome c peroxidase, chloroplast and cytosol ascorbate peroxidase and gene duplicated bacterial peroxidase (class I), secretory fungal peroxidase (class II); classical, secretory plant peroxidase (class III).

LiP oxidizes the substrate (aromatic nuclei) by two consecutive single electron oxidation steps with intermediate (unstable) cation radical formation (Kirk et al. 1990; Hatakka et al. 2001). These radicals undergo a variety of further reaction, many of which do not involve the enzyme. It preferentially cleaves the Cβ-Cβ bond in the lignin molecule but is also capable of ring opening and other reactions. The mechanism of LiP is shown in Figure 2.

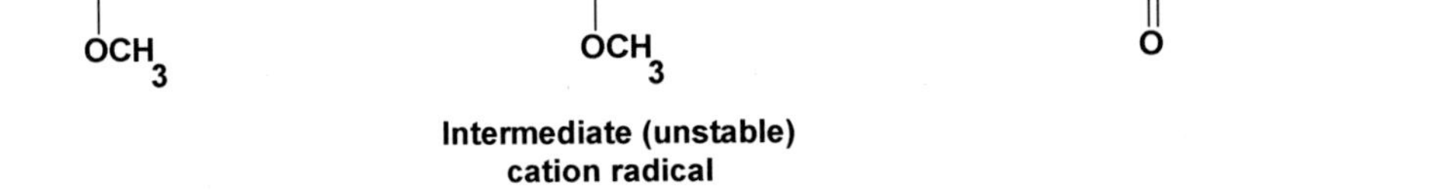

Figure 2. Mechanism of action of lignin peroxidase.

5.1.3. Manganese Peroxidase (MnP; EC 1.11.1.13)

The third major enzyme responsible for lignin degradation is MnP. It is an extracellular heme containing peroxidase with a high molecular weight (32 and 62.5 kDa) and secreted in multiple isoforms (Urzua et al. 1995). It was first isolated from the extracellular medium of ligninolytic cultures of white rot fungus *P. chrysosporium* and it is considered to be a key enzyme in ligninolysis by white-rot fungi.

Native (ferric) peroxidase	+ H_2O_2	$\longrightarrow$	Compound I	+ H_2O
Compound I	+ Mn^{2+}	$\longrightarrow$	Compound II	+ Mn^{3+}
Compound I	+ AH	$\longrightarrow$	Compound II	+ $A^{\cdot}$ + H^{+}
Compound II	+ Mn^{2+}	$\longrightarrow$	Native (ferric) peroxidase	+ Mn^{3+}

Figure 3. Mechanism of action of manganese peroxidase.

Similar to LiP, it oxidizes the substrate by two consecutive single electron oxidation steps with intermediate (unstable) cation radical formation. MnP requires H_2O_2 as a cosubstrate and oxidizes Mn^{2+} found in wood and soil to Mn^{3+} (Hatakka et al. 2001; Hofrichter 2002). Mn^{3+} is highly reactive, which complexed with an organic acid acts as a primary agent in ligninolysis. Mn^{3+} then oxidizes phenolic rings of lignin to unstable free radicals followed by spontaneous decomposition. MnP enzyme intermediate are analogous to other peroxidases (Wariishi et al. 1989). Thus, the native MnP is oxidized by H_2O_2 to Compound I, which can be reduced by Mn^{2+} and phenols to generate Compound II. Compound II is reduced back to a resting state by Mn^{2+}, but not by phenols. Therefore Mn^{2+} is necessary to complete the catalytic cycle and shows saturation kinetics (Pease and Tien 1992). The overall reaction catalyzed by MnP is shown in Figure 3.

The crystal structure of MnP shows similarities with LiP; the active site has a proximal His ligand hydrogen bonded to an Asp and a distal side peroxide-binding pocket consisting of a catalytic His and Arg (Sundaramoorthy et al. 1994). However, there is also a proposed manganese binding site involving Asp-179, Glu-35, Glu-39 and a heme propionate. In MnP there are five disulfide bonds.

5.1.4. Other Lignin-Degrading Enzymes

Other enzymes that participate in the lignin degradation are H_2O_2-producing enzymes and oxidoreductases, which may be compartmentalized either intra- or extracellularly. Bacterial and fungal feruloyl and p-coumaroyl esterases are relatively novel enzymes capable of releasing feruloyl and p-coumaroyl and play an important role in biodegradation of recalcitrant cell walls in grasses. These enzymes act synergistically with xylanases to disrupt the hemicellulose-lignin association, without mineralization of the lignin (Borneman et al. 1990). The accessibility of cell wall polysaccharides from the plant to microbial enzymes is dictated by the extent to which these are associated with phenolic polymers. Another group of peroxidase, versatile peroxidase (VP), which can oxidize not only Mn^{2+} but also phenolic and non-phenolic aromatic compounds including dyes, is regarded as a hybrid between MnP and LiP. VP has been described in species of *Pleurotus* and *Bjerkandera* (Mester and Field 1998).

The above lignin-degrading enzymes are essential for lignin degradation; however, for lignin mineralization these often combine with other processes involving additional enzymes. Such auxillary enzymes, which are by themselves unable to degrade lignin, include glyoxal oxidase and superoxide dismutase for intracellular production of H_2O_2 (Martinez et al. 1994). Lignin has been found to be partly mineralized in cell-free system of lignin-degrading enzyme, with considerably enhanced rates in the presence of co-oxidants produced during the secondary metabolism under the state of nutritional deficiency.

5.2. Lignin-Degradation Pathway

Microbial ligninolytic enzymes, *viz.*, Lac, LiP and MnP oxidize the lignin polymer, thereby generating aromatic radicals. These evolve in different non-enzymatic reactions, including C4-ether breakdown, aromatic ring cleavage, Cα-Cβ breakdown and demethoxylation. The aromatic aldehydes released from the breakdown of Cα-Cβ in lignin, or synthesized *de novo* by microbes are the substrates for aryl-alcohol oxidase (AAO)-catalyzed generation of H_2O_2 in cyclic redox reactions also involving aryl-alcohol dehydrogenase (AAD). If the phenoxy radicals produced after C4-ether breakdown are not first reduced by oxidases to phenolic compounds, may repolymerize on the lignin polymer. The resulting phenolic compounds may again be reoxidized by laccases or peroxidases. Phenoxy radicals can also be subjected to Cα-Cβ breakdown, generating *p*-quinones. Quinones participate in oxygen activation in redox cycling reactions involving quinone reductase (QR), laccases and peroxidases. This results in reduction of the ferric iron present in wood, either by superoxide cation radical or directly by the semiquinone radicals and its reoxidation with concomitant reduction of H_2O_2 to hydroxyl free radical (OH•). Hydroxyl free radical, a very strong oxidizing agent, can initiate the attack on lignin in the initial stages of wood decay, when the small pores in the intact cell wall prevents the penetration of ligninolytic enzymes. Then, lignin degradation proceeds by oxidative attack of the enzymes described above. In the final steps, simple products resulting from lignin degradation enter the microbes and get incorporated into intracellular catabolic routes. The entire mechanism of lignin degradation is depicted in Figure 4.

6. POTENTIAL APPLICATIONS OF LIGNINOLYTIC MICROBES OR PURIFIED LIGNINOLYTIC ENZYMES

Besides the applications in pulp and paper industry, timber industry, textile industry, forage nutritional quality enrichment, bioremediation of contaminating environmental pollutants, a number of other industrial applications for purified bioligninolytic enzymes / ligninolytic microbes include prevention of wine discoloration, oxidation of dye and their precursors, enzymatic conversion of chemical intermediates, production of chemicals from lignin and medical applications, etc.

6.1. Industrial Applications

6.1.1. Paper and Timber Industries

Pretreatment of wood pulp with Lac can provide milder and cleaner strategies of delignification that also respect the integrity of cellulose (Gamelas et al. 2005). Moreover, Lac does not alter pulp brightness, improves auto-adhesion of fibres in medium-density fiberboard, increases tensile strength of sheets derived from mechanical pulp and preserves tensile strength through calendering (Felby et al. 1997; Buchert et al. 1998; Wong and Mansfield 1999).

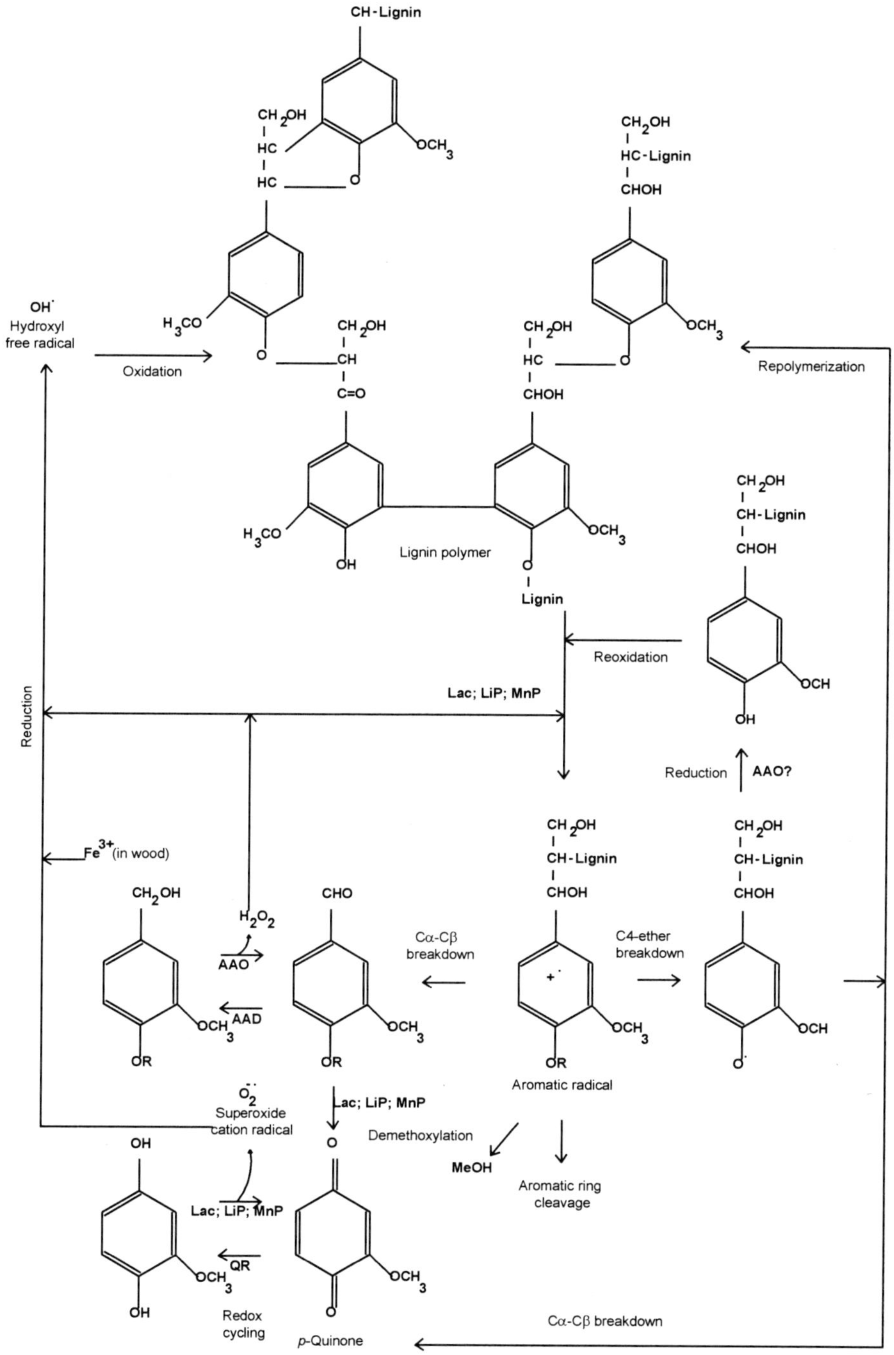

Figure 4. Lignin degradation pathway including enzymatic reactions and oxygen activation.

Lac can also be used for binding fiber-, particle- and paper-boards (Gubitz and Paulo 2003). In the manufacture of particle-boards and liner-boards, Lac has replaced hazardous chemicals such as urea and formaldehyde used to polymerize saccharides (Pederson and Felby 1996).

6.1.2. Organic Synthesis

Bioligninolytic system also acts as a new biocatalyst in organic synthesis which includes ethanol, textile dyes, cosmetic pigments, flavor agents, pesticides and heterocyclic compounds (Mayer and Staples 2002; Riva 2006). For example, to improve the production of fuel ethanol from lignocellulose, the *lac* gene from white-rot fungus *T. versicolor* was expressed in *Saccharomyces cerevisiae* (Larsson et al. 2001). The Lac-producing transformants had the ability to convert coniferyl aldehyde at a very fast rate, which enabled faster growth and ethanol formation.

6.1.3. Treatment of Beverages

Several phenolic compounds (coumaric acids, flavans and anthocyanins) are usually present in beverages (wine, fruit juice and beer) and during their shelf-life, may cause undesirable and deleterious changes such as discoloration, clouding, haze and flavor changes. Positive effects of bioligninolytic action were observed on wine as well as on fruit juice (Servili et al. 2000).

6.2. Analytical Applications

6.2.1. Biosensors

Bioligninolytic system is also used as analytical tools. For example, some lignin-degrading fungi have been employed to estimate the phenolic content of natural juice (Cliffe et al. 1994) or catechol in tea (Ghindilis et al. 1992). The laccases could be immobilized on the cathode of biofuel cells that could provide power, for example, for small transmitter systems (Chen et al. 2001; Barton et al. 2002). The ability of bioligninolytics to catalyze the electroreduction of oxygen via a direct mechanism, without the presence of an electrochemically active mediator was used to develop a gas-phase oxygen biosensor. In the presence of ascorbate the blue chromophore prosthetic group of Lac was reduced and decolorized. When Lac was reoxidized by oxygen, a concomitant return to the blue color occurs, which was recorded both visually and spectrophotometrically at 610 nm. This oxygen biosensor was very active and stable. It has been proposed as a useful tool to measure oxygen levels in products packaged under low-oxygen concentrations for better quality and safety (Gardiol et al. 1996).

6.2.2. Enzyme and Immunological Assays

Bioligninolytic catalysis can be used to assay enzymes including alanine, cystine or leucine-specific aminopeptidases, amylase, alkaline phosphatase, angiotensin I converting enzyme, chymotrypsin, plasmin, thrombin etc (Murao et al. 1985).

6.3. Environmental Applications

The environmental application for the bioligninolytic is the bioremediation of contaminated soils, as lignin-degrading enzymes are able to oxidize toxic organic pollutants, such as polycyclic aromatic hydrocarbons (PAHs) (Collins et al. 1996) and chlorophenols (Gianfreda et al. 1999; Ahn et al. 2002). This involves inoculating the soil with ligninolytic microbes that are efficient Lac-producers, as the use of isolated and purified enzymes is not economically feasible for soil remediation in large-scale.

Lignin-degrading enzymes have also been shown to be useful for the removal of toxic compounds through oxidative enzymatic coupling of the contaminants, leading to formation of insoluble complex structures (Wang et al. 2002). Phenolic compounds are present in wastes from several industrial processes, as coal conversion, petroleum refining, production of organic chemicals and olive oil production among others (Aggelis et al. 2003). Immobilized Lac was found to be useful to remove phenolic and chlorinated phenolic pollutants (Ehlers and Rose 2005).

Lignin-degrading enzymes have also been reported to be responsible for the transformation of 2, 4, 6-trichlorophenol to 2, 6-dichloro-1, 4-hydroquinol and 2, 6-dichloro-1, 4 benzoquinone (Leontievsky et al. 2000). White-rot fungi have also been used to oxidize alkenes, carbazole, N-ethylcarbazole, fluorene and dibenzothiophene in the presence of mediators (Niku-Paavola et al. 2000).

6.4. Medical Applications

Lac can also oxidize iodide to produce iodine, a widely used disinfectant. This iodine generating system may be used in various industrial, medical, domestic applications such as sterilization of drinking water and swimming pool as well as disinfection of minor wounds (Xu et al. 1996).

It can also be used to synthesize several medicinal agents including triazolo (benzo) thiadiazines (a group of anti-inflammatory, analgesic agents etc), vinblastine (cytostatic and antitumor agent), dimerized vindoline (for treatment of neoplastic diseases) etc. Bioligninolytic systems have been also employed to synthesize new cephalosporin antibiotics (Agematu et al. 1993) and to improve the synthesis of actinocin antibiotics (Osiadacz et al. 1999).

CONCLUSION

To sum up, exploring the range of microbial biodiversity is the key to developing effective and environment friendly 'green' technologies for environmental restoration or optimal agroindustrial utilization of plant biomass by mineralization or degradation of lignin. Further, isolation, purification and large-scale production of lignin-degrading enzymes with high activity or strain improvement to get maximum yield of such enzymes is the need of future so as to exploit the plant biomass to its full capacity and in environment-friendly way.

ACKNOWLEDGMENTS

Financial support in the form of research project to Smita Rastogi from CST (UP), India is greatly acknowledged.

REFERENCES

[1]   Abbott TP; Wicklow, DT. Degradation of lignin by *Cyathus* species. *Appl. Environ. Microbiol.,* 1984, 47, 585-587.

[2]   Agematu, H; Kominato, K; Shibamoto, N; Yoshioka, T; Nishida, H; Okamoto, R; Shin, T; Murao, S. Transformation of 7-(4-hydroxyphenylacetamido) cephalosporanic acid into a new cephalosporin antibiotic, 7-[1-oxaspiro(2.5)octa-6-oxo-4,7-diene-2-carboxamido] cephalosporanic acid by laccase. *Biosci. Biotech. Biochem.,* 1993, 57, 1387-1388.

[3]   Aggelis, G; Iconomou, D; Christou, M; Bokas, D; Kotzailias, S; Christou, G; Tsagou, V; Papanikolaou, S. *Water Res.,* 2003, 37, 3897.

[4]   Agosin, E; Odier, E. Solid-state fermentation, lignin degradation and resulting digestibility of wheat straw fermented by selected white-rot fungi. *Appl. Microbiol. Biotechnol.,* 1985, 21, 397-403.

[5]   Ahn, MY; Dec, J; Kim, JE; Bollag, JM. Treatment of 2, 4-dichlorophenol polluted soil with free and immobilized laccase. *J. Environ. Qual.,* 2002, 31, 1509-1515.

[6]   Akhtar, M; Blanchette, RA; Kirk, TK. Fungal delignification and biomechanical pulping of wood. In: Eriksson KEL, editor. Advances in biochemical engineering / biotechnology. Germany: Springer-Verlag; 1997; 159-195.

[7]   Alexandre, G; Zhulin, IB. Laccases are widespread in bacteria. *Trends Biotechnol,* 2000, 18, 41-42.

[8]   Argyropoulos, DS; Menachem, SB. In: Eriksson, KEL, editor. Advances in biochemical engineering biotechnology. Germany: Springer-Verlag; 1997; 127-158.

[9]   Atkinson, CF; Jones, DD; Gauthier, JJ. Putative anaerobic activity in aerated composts. *J. Ind. Microbiol.,* 1996, 16, 182-188.

[10]   Barr, DP ; Aust, SD. Mechanisms that white-rot fungi use to degrade pollutants. *Environ. Sci. Technol.,* 1994, 28, 78-87.

[11]   Barreca, AM; Sjögren, B; Fabbrini, M; Galli, C; Gentili, P. Catalytic efficiency of some mediators in laccase-catalyzed alcohol oxidation. *Biocat. Biotransform.,* 2004, 22, 105-112.

[12]   Barton, SC; Pickard, M; Vazquez-Duhalt, R; Heller, A. Electroreduction of O(2) to water at 0.6 V (SHE) at pH 7 on the "wired" *Pleurotus ostreatus* laccase cathode. *Biosens. Bioelectron.,* 2002, 17, 1071-1074.

[13]   Berrocal, M; Ball, AS; Huerta, S; Barrasa, JM; Hernández, M; Pérez-Leblic, MI; Arias, ME. Biological upgrading of wheat straw through solid-state fermentation with *Streptomyces cyaneus. Appl. Microbiol. Biotechnol.,* 2000, 54, 764-771.

[14]   Björdal, CG; Nilsson, T; Daniel, G. Microbial decay of waterlogged archaeological wood found in Sweden applicable to archaeology and conservation. *Int. Biodeter Biodegrad,* 1999, 43, 63-73.

[15] Blanchette, RA. Degradation of lignocellulose complex in wood. *Can J Bot*, 1995, 73, S999-S1010.

[16] Blanchette, RA; Cease, KR; Abad, AR. An evaluation of different forms of deterioration found in archaeological wood. *Int. Biodeter Biodegrad*, 1991, 28, 3-22.

[17] Borneman, P; Jeffries, T. Mn (II) regulation of lignin peroxidases and manganese-dependent peroxidase from lignin-degrading white-rot fungi. *Appl. Environ. Microbiol.*, 1990, 56, 210-217.

[18] Bourbonnais, R; Paice, MG; Freiermuth, B; Bodie, E; Borneman, S. Reactivities of various mediators and laccases with kraft pulp and lignin model compounds. *Appl. Environ. Microbiol.*, 1997, 63, 4627-4632.

[19] Brunow, G. Methods to reveal the structure of lignin. In: Hofrichter M, Steinbüchel A, editors. Biopolymers: lignin, humic substances and coal. Germany: Wiley-VCH; 2001; 89-116.

[20] Buchert, J; Rättö, M; Mustranta, A; Suurnäkki, A; Ekman, R; Spetz, P; Siikaaho, M; Viikari, L. Enzymes for the improvement of paper machine runnability. In: Proc 7th Int Conf Biotechnol Pulp Paper Ind. 1998, Vancouver, Canada, A225-A228.

[21] Carlile, MJ; Watkinson, SC. The fungi. Great Britain: Acad Press; 1994.

[22] Chefetz, B; Chen, Y; Hadar, Y. Purification and characterization of laccase from *Chaetomium thermophilium* and its role in humification. *Appl. Environ. Microbiol.*, 1998, 64, 3175-3179.

[23] Chen; Barton; Binyamin, G; Gao, Z; Zhang, Y; Kim, H; Heller, A. A miniature Biofuel Cell. *J. Am. Chem. Soc.*, 2001, 123, 8630-8631.

[24] Claus, H. Laccases and their occurrence in prokaryotes. *Arch. Microbiol.*, 2003, 179, 145-150.

[25] Cliffe, S; Fawer, MS; Maier, G; Takata, K; Ritter, G; Enzyme assays for the phenolic content of natural juices. *J. Agric. Food Chem.*, 1994, 42, 1824-1828.

[26] Collins, PJ; Kotterman, MJJ; Field, JA; Dobson, ADW. Oxidation of anthracene and benzo[a]pyrene by laccases from *Trametes versicolor. Appl. Environ. Microbiol.*, 1996, 62, 4563-4567.

[27] Daniel, G; Nilsson, T. Developments in the study of soft rot and bacterial decay. In: Bruce A, Palfreyman JW, editors. Forest products biotechnology. Great Britain: Taylor and Francis; 1998; 37-62.

[28] Dey, S; Maiti, TK; Bhattacharyya, BC. Production of some extracellular enzymes by a lignin peroxidase-producing brown rot fungus, *Polyporus ostreiformis* and its comparative abilities for lignin degradation and dye decolorization. *Appl. Environ. Microbiol.*, 1994, 60, 4216-4218.

[29] Dix, NJ; Webster, J. Fungal ecology. Cambridge, Great Britain: Chapman and Hall, 1995.

[30] Dixon, RA. Molecular improvement of forages: from genomics to GMOs. In: Hopkins A, Wang ZY, Mian R, Sledge M, Barker RE, editors. Molecular breeding of forage and turf. USA: *Kluwer Acad. Publ.*; 2004; 1-19.

[31] Donaldson, L. Lignification and lignin topochemistry an ultrastructural view. *Phytochem.*, 2001, 57, 859-873.

[32] Durrant, LR. Biodegradation of lignocellulosic materials by soil fungi isolated under anaerobic conditions. *Int. Biodeter. Biodegrad.*, 1996, 37, 189-195.

[33] Eaton, RA; Hale, MDC. Wood: decay, pests and protection. Cambridge, Great Britain: Chapman and Hall; 1993.

[34] Eggert, C; Temp, U; Dean, JFD; Eriksson, KEL. Laccase-mediated formation of the phenoxazinone derivative, cinnabarinic acid. *FEBS Lett.*, 1996, 376, 202-206.

[35] Ehlers, GA; Rose, PD. Immobilized white-rot fungal biodegradation of phenol and chlorinated phenol in trickling packed-bed reactors by employing sequencing batch operation. *Bioresource Technol.*, 2005, 96, 1264-1275.

[36] Eriksson, KEL; Blanchette, RA; Ander, P. Microbial and enzymatic degradation of wood and wood components. Berlin, Germany: Springer Verlag; 1990.

[37] Fakoussa, RM; Frost, PJ. *In vivo*-decolorization of coal-derived humic acids by laccase-excreting fungus *Trametes versicolor Appl. Microbiol. Biotechnol.*, 1999, 52, 60-65.

[38] Falcón, MA; Rodríguez, A; Carnicero, A; Regalado, V; Perestelo, F; Milstein, O; de la Fuente, G. Isolation of microorganisms with lignin transformation potential from soil of Tenerife island. *Soil Biol. Biochem.*, 1995, 27, 121-126.

[39] Felby, C; Pedersen, LS; Nielsen, BR. Enhanced autoadhesion of wood fibers using phenol oxidases. *Holzforschung*, 1997, 51, 281-286.

[40] Freeman, JC; Nayar, PG; Begley, TP; Villafranca, *J. Biochem.*, 1993, 24, 4826-4830.

[41] Galli, C; Gentili, P. Chemical messengers: mediated oxidations with the enzyme laccase. *J. Phy. Org. Chem.*, 2004, 17, 973-977.

[42] Gamelas, JAF; Tavares, APM; Evtuguin, DV; Xavier, AMB. Oxygen bleaching of kraft pulp with polyoxometalates and laccase applying a novel multi-stage process. *J. Mol. Cat B: Enzymatic*, 2005, 33, 57.

[43] Gardiol, AE; Hernandez, RJ; Reinhammar, B; Hate, BR. Development of gas-phase oxygen biosensor using a blue copper-containing oxidase. *Enzyme Microbial. Technol.* 1996, 18, 347-352.

[44] Gellerstedt, G; Henriksson, G; Li, J. Lignin depolymerization / repolymerization and its critical role for delignification of aspen wood by steam explosion. *Bioresource Technol.* 2007, 16, 3061-3068.

[45] Ghindilis, AL; Gavrilova, VP; Yaropolov, AI. Laccase-based biosensor for determination of polyphenols: determination of catechols in tea. *Biosens Bioelectron*, 1992, 7, 127-131.

[46] Gianfreda, L; Xu, F; Bollag, JM. Laccases: a useful group of oxidoreductive enzymes. *Bioremediation J.*, 1999, 3, 1-25.

[47] Godden, B; Ball, AS; Helvenstein, P; McCarthy, AJ; Penninckx, MJ. Towards elucidation of the lignin degradation pathway in actinomycetes. *J. Gen. Microbiol.*, 1992, 138, 2441-2448.

[48] Gold, MH; Alic, M. Molecular biology of the lignin-degrading *basidiomycete Phanerochaete chrysosporium. Microbiol. Rev.*, 1993, 57, 605-22.

[49] Gubitz, GM; Paulo, AC. New substrates for reliable enzymes: enzymatic modification of polymers. *Curr. Opin. Biotechnol.*, 2003, 14, 577-582.

[50] Hansen, CM; Björkman, A. The ultrastructure of wood from a solubility parameter point of view. *Holzforschung*, 1998, 52, 335-344.

[51] Hatakka, A. Lignin-modifying enzymes from selected white-rot fungi: production and role in lignin degradation. *FEMS Microbiol. Rev.*, 1994, 13, 125-135.

[52] Hatakka, A; Hofrichter, M; Steinbüchel, A. Biodegradation of lignin in biopolymers: lignin, humic substances and coal. Editors Germany: Wiley-VCH; 2001; 129-180.

[53] Hoegger, PJ; Kilaru, S; James, TY; Thacker, JR; Kuees, U. Phylogenetic comparison and classification of laccase and related multicopper oxidase protein sequences. *FEBS J.*, 2006, 273, 2308-2326.

[54] Hofrichter, M. Review: lignin conversion by manganese peroxidase (MnP). *Enzyme Microbial. Technol.*, 2002, 30, 454-466.

[55] Hofrichter, M; Fritsche, W. Depolymerization of low-rank coal by extracellular fungal enzyme systems: screening for low-rank-coal-depolymerizing activities. *Appl. Microbiol. Biotechnol.*, 1996, 46, 220-225.

[56] Holt, DM. Bacterial degradation of lignified wood cell walls in aerobic aquatic habitats: Decay patterns and mechanisms proposed to account for their formation. *J. Inst. Wood Sci.*, 1983, 9, 212-223.

[57] Howard, RL; Abotsi, E; van Rensburg, JEL; Howard, S. Lignocellulose biotechnology: issues of bioconversion and enzyme production. *Afr. J. Biotechnol.*, 2003, 2, 602-619.

[58] Karam, J; Nicell, JA. Potential applications of enzymes in waste treatment. *J. Chemical Technol. Biotechnol.*, 1997, 69, 141-153.

[59] Kato, K; Kozaki, S; Sakuranaga, M. Degradation of lignin compounds by bacteria from termite guts. *Biotechnol. Lett.*, 1998, 20, 459-462.

[60] Kim, YS; Singh, AP; Nilsson, T. Bacteria as important degraders in water logged archaeological wood. *Holzforschung*, 1996, 50, 389-392.

[61] Kirk, TK; Popp, JL; Kalyanaraman, B. Lignin peroxidase oxidation of Mn^{2+} in the presence of veratryl alcohol, malonic or oxalic acid and oxygen. *Biochem. J.*, 1990, 29, 10475-10480.

[62] Kirk, TK; Farrell, RL. Enzymatic combustion: the microbial degradation of lignin. *Annu. Rev. Microbiol.*, 1987, 41, 465-505.

[63] Kuhad, RC; Singh, A; Eriksson, KEL. Microorganisms and enzymes involved in the degradation of plant fiber cell walls. In: Eriksson KEL, editor. Advances in biochemical engineering / biotechnology. Germany: Springer-Verlag; 1997; 46-125.

[64] Lacey, J. Actinomycetes as biodeteriogens and pollutant of the environment. In: Goodfellow M, Williams ST, Mordarski M, editors. Actinomycetes in biotechnology. Great Britain: Acad Press; 1988; 359-432.

[65] Lang, E; Kleeberg, I; Zadrazil, F. Extractable organic carbon and counts of bacteria near the lignocellulose-soil interface during the interaction of soil microbiota and white-rot fungi. *Bioresource Technol.*, 2000, 75, 57-65.

[66] Larsson, S; Cassland, P; Jönsson, LJ. Development of a *Saccharomyces cerevisiae* strain with enhanced resistance to phenolic fermentation inhibitors in lignocellulose hydrolyzates by heterologous expression of laccase. *Appl. Environ. Microbiol.*, 2001, 67, 1163-1170.

[67] Leontievsky, A; Myasoedova, NM; Baskunov, BP; Evans, CS; Golovleva, LA. Biodegradation, 2000, 11, 331.

[68] Lewis, NG; Davin, LB. Lignans: biosynthesis and function. In: Barton D, Nakanishi K, editors. Comprehensive natural products chemistry. Oxford, UK: Elsevier; 1999.

[69] Li, J; Yuan, H; Yang, J. Bacteria and lignin degradation. *Front Biol. China*, 2009, 4, 29-38.

[70] Li, K; Xu, F; Eriksson, KEL. Comparison of fungal laccases and redox mediators in oxidation of a nonphenolic lignin model compound. *Appl. Environ. Microbiol.*, 1999, 65, 2654-2660.

[71] Lundell, T. Ligninolytic system of the white-rot fungus *Phlebia radiata*. Helsinki, Finland: University of Helsinki; 1993.

[72] Martinez, AT; Camarero, S; Guillen, F; Gutierrez, A; Munoz, C; Varela, E; Martinez, MJ; Barrasa, JM; Ruel, K; Pelayo, JM. Progress in biopulping of non-woody materials: chemical, enzymatic and ultrastructural aspects of wheat straw delignification with ligninolytic fungi from the genus *Pleurotus*. *FEMS Microbiol. Rev.*, 1994, 13, 265-274.

[73] Mayer, AM; Staples, RC. Laccase: new functions for an old enzyme. *Phytochem.*, 2002, 60, 551-565.

[74] McSweeney, CS; Dulieu, A; Katayama, Y; Lowry, JB. Solubilization of lignin by the ruminal anaerobic fungus *Neocallimastix patriciarum*. *Appl. Environ. Microbiol.*, 1994, 60, 2985-2989.

[75] Mercer, DK; Iqbal, M; Miller, PGG; McCarthy, AJ. Screening actinomycetes for extracellular peroxidase activity. *Appl. Environ. Microbiol.*, 1996, 62, 2186-2190.

[76] Mester, T; Field, JA. Characterization of a novel manganese peroxidase-lignin peroxidase hybrid isozyme produced by *Bjerkandera* species strain BOS55 in the absence of manganese. *J. Biol. Chem.*, 1998, 273, 15412-15417.

[77] Monties, B; Fukushima, K. Occurrence, function and biosynthesis of lignins. In: Hofrichter M, Steinbüchel A, editors. Biopolymers: lignin, humic substances and coal. Germany: Wiley-VCH; 2001; 1-64.

[78] Morii, H; Nakamiya, K; Kinoshita, S. Isolation of a lignin-decolorizing bacterium. *J. Ferment. Bioeng.*, 1995, 80, 296-299.

[79] Murao, S; Arai, M; Tanaka, N; Ishikawa, H; Matsumoto, K; Watanabe, S. *Agr. Biol. Chem.*, 1985, 49, 981-985.

[80] Niemenmaa, OV; Uusi-Rauva, AK; Hatakka, AI. Demethoxylation of a [O^{14}CH$_3$]-labelled lignin model compound by white-rot and brown-rot fungi. Kyoto, Japan: Proc 5th Int Conf Biotechnol in the Pulp and Paper Industry (ICBPPI); 1992; 163.

[81] Niku-Paavola, M.L. Viikari, L. Enzymatic oxidation of alkenes. *J. Mol. Cat. B: Enzymatic*, 2000, 10, 435-444.

[82] Osiadacz, J; Al-Adhami, AJH; Bajraszewska, D; Fischer, P; Peczynska-Czoch W. On the use of *Trametes versicolor* laccase for the conversion of 4-methyl-3-hydroxyanthranilic acid to actinocin chromophore. *J. Biotechnol.*, 1999, 72, 141-149.

[83] Pasti, MB; Hagen, SR; Korus, RA; Crawford, DL. The effects of various nutrients on extracellular peroxidases and acid-precipitable polymeric lignin production by *Streptomyces chromofuscus* A2 and *S. viridosporus* T7A. *Appl. Microbiol. Biotechnol.*, 1991, 34, 661-667.

[84] Paszczynski, A; Crawford, RL. Potential for bioremediation of xenobiotic compounds by the white-rot fungus *Phanerochaete chrysosporium*. *Biotechnol. Prog.*, 1995, 11, 368-379.

[85] Pease, EA; Tien, M. Heterogeneity and regulation of manganese peroxidases from *Phanerochaete chrysosporium*. 1992, 174, 3532-3540.

[86] Pedersen, LS; Felby, C. Process for preparing a lignocellulose-based product and product obtainable by the process. *Int. Pat.* WO 96/03596, 1996.

[87] Perestelo, F; Falcón, MA; Carnicero, A; Rodriguez, A; de la Fuente, G. Limited degradation of industrial, synthetic and natural lignins by *Serratia marcescens*. *Biotechnol. Lett.*, 1994, 16, 299-302.

[88] Perestelo, F; Rodriguez, A; Pérez, R; Carnicero, A; de la Fuente, G; Falcón, MA. Isolation of a bacterium capable of limited degradation of industrial and labelled, natural and synthetic lignins. *World J. Microbiol. Biotechnol.*, 1996, 12, 111-112.

[89] Pointing, SB. Feasibility of bioremediation by white-rot fungi. *Appl. Microbiol. Biotechnol.*, 2001, 57, 20-33.

[90] Ralph, K. Actinomycetes and lignin degradation. *Adv. Appl. Microbiol.*, 2005, 58, 125-168.

[91] Rastogi, S; Dwivedi UN. Manipulation of lignin in plants with special reference to *O*-methyltransferase. *Plant Sci.*, 2008, 174, 264-277.

[92] Riva, S. Laccases: blue enzyme for green chemistry. *Trends Biotechnol.*, 2006, 24, 219-226.

[93] Rodriguez, A; Perestelo, F; Carnicero, A; Regalado, V; Perez, R; de la Fuente, G; Falcon, MA. Degradation of natural lignins and lignocellulosic substrates by soil-inhabiting fungi imperfecti. *FEMS Microbiol. Ecol.*, 1996, 21, 213-219.

[94] Ruel, JJ; Ayres, MP. Jensen's inequality predicts effects of environmental variation. *Trends Ecol. Evol.*, 1999, 14, 361-366.

[95] Ruiz-Dueñas, FJ; Martínez, ÁT. Microbial degradation of lignin: how a bulky recalcitrant polymer is efficiently recycled in nature and how we can take advantage of this. *Microbial. Biotechnol.*, 2009, 2, 164-177.

[96] Rüttimann, C; Vicuña, R; Mozuch, MD; Kirk, TK. Limited bacterial mineralization of fungal degradation intermediates from synthetic lignin. *Appl. Environ. Microbiol.*, 1991, 57, 3652-3655.

[97] Scott, GM; Akhtar, M. In: Hofrichter M, Steinbüchel A, editors. Biopolymers: lignin, humic substances and coal. Germany: Wiley-VCH; 2001; 181-207.

[98] Servili, M; G. DeStefano, P; Piacquadio, V; Sciancalepore. A novel method for removing phenols from grape must. *Am. J. Enol. Vitic.*, 2000, 51, 357-361.

[99] Sigoillot, C; Camarero, S; Vidal, T; Record, E; Asther, M; Perez-Boada, M; Martinez, MJ; Sigoillot, JC; Asther, M; Colom, JF; Martinez, AT. Comparison of different fungal enzymes for bleaching high-quality paper pulps. *J. Biotechnol.*, 2005, 23, 333-343.

[100] Singh, AP; Nilsson, J; Daniel, GF. Bacterial attack of *Pinus sylvestris* wood under near anaerobic conditions. *J. Inst. Wood Sci.*, 1990, 12, 143-157.

[101] Sixta, H. Handbook of Pulp. Weinheim, Germany: Wiley-VCH; 2006.

[102] Sjöström, E. Wood chemistry, fundamentals and applications, New York / London: Acad Press; 1993.

[103] Spiker, JK; Crawford, DL; Thiel, EC. Oxidation of phenolic and non-phenolic substrates by the lignin peroxidase of *Streptomycete viridosporus* T7A. *Appl. Microbiol. Biotechnol.*, 1992, 37, 518-523.

[104] Stephens, J; Halpin, C. Lignin manipulation for fiber improvement. In: Ranalli, P, editor. Improvement of crop plants for industrial end uses. The Netherlands: Springer; 2007; 129-153.

[105] Sundaramoorthy, M; Kishi, K; Gold, MH; Poulas, TL. The crystal structure of manganese peroxidase from *Phanerochaete chrysosporium* 2.06Å resolution. *J. Biol. Chem.*, 1994, 269, 32759-32767.

[106] Susana, RC; José, LTH. Industrial and biotechnological applications of laccases: a review. *Biotechnol. Adv.*, 2006, 24, 500-513.

[107] Susmel, P; Stefanon, B. Aspects of lignin degradation by rumen microorganisms. *J. Biotechnol.*, 1993, 30, 141-148.

[108] Suzuki, T; Endo, K; Iro, M; Tsujibo, H; Miyamoto, K; Inamori, Y. A thermostable laccase from *Streptomyces lavendulae* REN-7: purification, characterization, nucleotide sequence, and expression. *Biosci. Biotechnol. Biochem.*, 2003, 67, 2167-2175.

[109] Tanaka, H; Itakura, S; Enoki, A. Phenol oxidase activity and one-electron oxidation activity in wood degradation by soft-rot duteromycetes. *Holzforschung*, 2000, 54, 463-468.

[110] Thakker, GD; Evans, CS; Rao, KK. Purification and characterization of laccase from *Monocillium indicum* Saxena. *Appl. Microbiol. Biotechnol.*, 1992, 37, 321-323.

[111] Thurston, CF. The structure and function of fungal laccases. *Microbiol.*, 1994, 140, 19-26.

[112] Trojanowski, J; Haider, K; Hutterman, A. Decomposition of 14C-labeled lignin, holocellulose and lignocellulose by mycorrhizal fungi. *Arch. Microbiol.*, 1984, 139, 202-206.

[113] Urzúa, U; Fernando, LL; Lobos, S; Larraín, J; Vicuña, R. Oxidation reactions catalyzed by manganese peroxidase isoenzymes from *Ceriporiopsis subvermispora*. *FEBS Lett.*, 1995, 4, 132-136.

[114] Vähätalo, AV; Salonen, K; Salkinoja-Salonen, M; Hatakka, A. Photochemical mineralization of synthetic lignin in lake water indicates rapid turnover of aromatic organic matter under solar radiation. *Biodegrad*, 1999, 10, 415-420.

[115] Vasdev, K; Kuhad, RC. Decolorization of poly R-478 (polyvinylamine sulphonate anthra pyridone) by *Cyathus bulleri*. *Folia Microbiol.*, 1994, 39, 61-64.

[116] Vasdev, K; Kuhad, RC; Saxena, RK. Decolorization of triphenylmethano dyes by the bird's nest fungus *Cyathus bulleri*. *Curr. Microbiol.*, 1995, 30, 269-272.

[117] Vicuña, R. Ligninolysis: a very peculiar microbial process. *Mol. Biotechnol.*, 2000, 14, 173-176.

[118] Vicuña, R; González, B; Seelenfreund, D; Rüttimann, C; Salas, L. Ability of natural bacterial isolates to metabolize high and low molecular weight lignin derived molecules. *J. Biotechnol.*, 1993, 30, 9-13.

[119] Vinardell, M; Ugartondo, V; Mitjans, M. Potential applications of antioxidant lignins from different sources. *Indus Crops Products*, 2008, 27, 220-223.

[120] Wang, CJ; Thiele, S; Bollag, JM. Interaction of 2, 4, 6-trinitrotoluene (TNT) and 4-amino-2, 6-dinitrotoluene with humic monomers in the presence of oxidative enzymes. *Arch. Environ. Contamin. Toxicol.*, 2002, 42, 1-8.

[121] Wariishi, H; Dunford, HB; MacDonald, ID; Gold, MH. Manganese peroxidase from the lignin-degrading basidiomycetes *Phanerochaete chrysosporium*: transient-state kinetics and reaction mechanism. *J. Biol. Chem.*, 1989, 264, 3335-3340.

[122] Welinder, KG. Superfamily of plant, fungal and bacterial peroxidases. *Curr. Opin. Str. Biol.*, 1992, 2, 388-393.

[123] Wicklow, DT; Langie, R; Crabtree, S; Detroy, RW. Degradation of lignocellulose in wheat straw versus hardwood by *Cyathus* and related species (Nidulariaceae). *Can .J. Microbiol.*, 1984, 30, 632-636.

[124] Wong, KKY; Mansfield, SD. Enzymatic processing for pulp and paper manufacture-a review. *Appita J.*, 1999, 52, 409-418.

[125] Xu, F; Shin, W; Brown, SH; Wahleithner, JA; Sundaram, UM; Solomon, EI. A study of a series of recombinant fungal laccases and bilirubin oxidase that exhibit significant differences in redox potential, substrate specificity, and stability. *Biochim. Biophys. Acta*, 1996, 1292, 303-311.

[126] Zimmermann, W. Degradation of lignin by bacteria. *J. Biotechnol.*, 1990, 13, 119-130.

In: Lignin
Editor: Ryan J. Paterson

ISBN: 978-1-61122-907-3
©2012 Nova Science Publishers, Inc.

Chapter 10

ADVANCED APPROACHES TO LIGNIN MODIFICATION

Aicheng Chen[*], Nelson Matyasovszky, Min Tian,*
and Duncan MacDonald

Department of Chemistry, Lakehead University, Thunder Bay, Ontario, Canada

ABSTRACT

Lignocellulosic biomass is present in very large amounts as a result of world-wide pulp & paper processes. Lignin is a major potential renewable, non-fossil source of aromatic and cyclohexyl compounds. It is a feasible raw material for many valuable substances such as activated carbon, vanillin, vanillic acid, dispersing agents, ion-exchange agents, polymer fillers, binding agents for the production of fibre boards, artificial fertilizers and complexing agents. The utilization of lignin can benefit both the green/renewable chemistry and forestry industries. Various methods, including chemical, biological, photochemical and electrochemical techniques have been explored for the modification of lignin to produce value-added products. Each of these advanced approaches are reviewed in this chapter. The chemical treatment of lignin, which leads to the generation of lignin based epoxy resin and activated carbon is also examined. Lignin modification using enzymes in combination with specific chemical mediators is summarized. The photochemical alteration and electrochemical oxidation of lignin for the generation of vanillin, vanillic acid and other value-added products are discussed. In addition, the effectiveness of combining photochemical and electrochemical approaches is evaluated.

1. INTRODUCTION

The global economy drives a continuous search for renewable energy sources to replace our dependence on depleting petroleum reservoirs and to reduce their environmental impact. Interest in biomass as a renewable energy source continues to grow with the belief that it has the potential to be the main provider of clean replenishable energy and chemicals in the

---

[*] Corresponding author. Tel.: 1-807-3438318; Fax: 1-807-346-7775, Email: Email: achen@lakeheadu.ca.

future. For instance, in a 2005 report that was commissioned by the U.S. Department of Energy (DOE) and the U.S. Department of Agriculture (USDA), it was highlighted that the U.S. has the capacity for replacing one-third of its petroleum requirements with biomass technologies [1]. Of the estimated 1 billion tons of biomass required, lignin is a main constituent, which makes research surrounding its applications of ever growing importance [2].

Lignin is the second most abundant renewable organic material on the planet, next to cellulose, as it accounts for close to one-third of plant/tree tissue mass [3, 4]. Its main function is to bind together cellulose fibres, making it an essential component in wood, straw and plant tissues. Lignin constitutes ~30% of the dry weight of softwood and ~20% of hardwood, which supplies the production of more than 50 million tons of technical lignins in Europe each year. All of this available raw material fuels the impetus for finding new applications and developing new lignin based products [5]. Lignin is generally found in large quantities in pulping black liquor and has traditionally been regarded as a cumbersome impediment, because it has lent itself to few commercially viable uses [6]. Normally, the black liquor, which consists of alkaline lignin and inorganic ions, is condensed and burned for fuel and caustic recovery. Overloading of the recovery boilers has become a big problem for many kraft paper mills, making the partial precipitation and removal of lignin prior to recovery boiler entry a viable option, and a possible source of lignin for new commercial applications. As of 2004, only 2% of the lignin produced by the pulp and paper industry was used in commercial applications as low value products in narrow markets [7]. A great magnitude of excess lignin is diverted as effluent into our waterways, where the presence of lignin and its fragments have proven to be difficult to manage, insofar as bioremediation efforts are concerned [8].

Pretreatment of lignin allows for the separation of the desired material from other biomass components and structural modifications, such as fragmentation or sulphur addition. Common pretreatment methods include the kraft lignin process and the lignosulfonate and organosolv processes. In the conventional kraft pulping process, the wood chips and white liquor are digested at a high temperature and pressure to cleave the phenylpropane linkages and increase the water/alkali-solubility of the lignin fragments [9]. The main kraft pulping process involves two stages, namely condensation and degradation. In the condensation process, precipitation may result with an increase in the molecular weight of the fragments, whereas degradation reactions result in fragmentation [10]. Delignification occurs in a series of three steps, with a final bleaching stage required if paper of moderate to high brightness is intended. Several drawbacks of the kraft pulping process include incomplete delignification, relatively high pulping temperatures, exposure of carbohydrates to long pulping periods, and the volatile degradation of some products. Despite these drawbacks, the simplicity, the high quality of the pulp produced and accelerated speed of the kraft pulping process have led to very few changes in over century of use.

Sulphite pulping produces fibres that are more easily bleached than kraft pulps, but they are of diminished strength due to polysaccharide degradation. It yields a sulfonated lignin, lignosulfonate, which has reached production levels of ~1 million tons of solids per year. Its increased functionality allows for its use in some practical applications [11, 12]. However, the higher molecular weight lignosulfonate still exhibits the fragment size variation that is found in kraft lignin samples. The lignosulfonate process is conducted within the pH range of from 2 to 12 and produces a product that is often soluble in water, as well as highly polar organics and amines. Of the lignosulfonates, some success has been achieved with ammonium lignosulfonates (LAS) to generate phenol-formaldehyde (PF) for applications in wood-adhesives and boil-resistant plywood. A recent focus has been to chemically modify the lignosulfonates, using methods such as methylation and phenolation, in an effort to increase their reactivity with formaldehyde [13, 14].

In the organosolv pulping process, wood products are treated with organic solvents instead of the aqueous solutions used in traditional methods. This approach allows for chemical breakdown to occur prior to dissolving the lignin in the organic solvent for the extraction step [15]. The solvent used in the organosolv process can be neutral or acidic with or without the help of an acid catalyst. However, when a neutral organic solvent is used, the acidity of the solution increases throughout the duration of the pulping process due to the release of acetic acid from the wood. Full scale operational organosolv processes were no longer restricted to the Organocell and alkaline sulfite anthraquinone methanol (ASAM) methods, which both used methanol as a solvent. Other organosolv processes using acetic acid and peroxyformic acid as solvents have also been explored. The main chemical modification that occurs during the organosolv process is the cleavage of the α-ether linkages; β-ether cleavage can also take place if the lignin has been separated from the wood. The organosolv pulping process is favoured in small scale operations because of its environmentally friendly nature and ability to produce pulp of comparable quality to kraft pulp [16]. Some advantages of organosolv extracted lignin are its sulphur-free nature and homogenous structure. The organosolv process also allows for the simple extraction of lignin from black liquor though acid precipitation, which isolates a product that can be refined into biofuels and/or other useful chemicals [17].

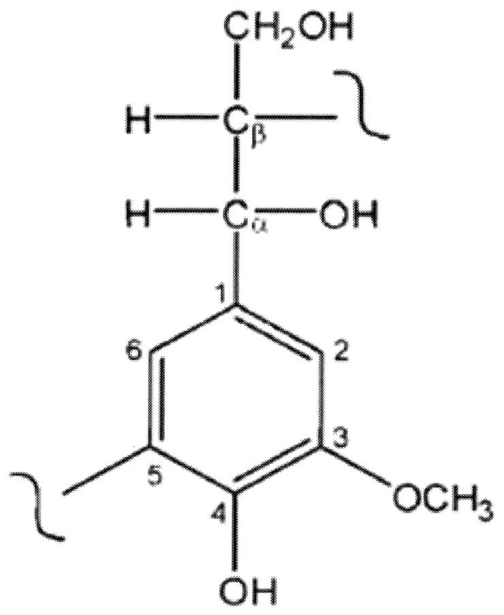

Figure 1. A representation of the basic structural unit found in lignin. Reprinted from Chen, A.; Rogers, E. I.; Compton, R. G. *Electroanalysis* 2010, 22, 1037. Copyright 2010 Elsevier.

Lignin is a complex polymer constructed from phenylpropane monomers that are linked primarily through C-O ether bonds and to a lesser extent through various C-C alkyl-aryl bonds. The dominant C-O ether linkages are β-Aryl (β-O-4), α-Aryl (α-O-4) and diphenyl (4-O-5) types [18, 19]. A representation of the basic lignin structure is shown in Figure 1 [20]. Lignin is enzymatically synthesized in many plant species through the polymerization of precursors: coniferyl alcohol, synapyl alcohol and p-coumaryl alcohol, which form guaiacyl (G), syringyl (S), and p-hydroxyphenyl propane (H) subunits, respectively. According to the composition ratios of the three structural units, lignin can be divided into three general classes, namely softwood, hardwood and grass lignin.

The complex structural nature and low-purity standards currently in place for commercial applications, make lignin an underexploited material, even in the face of dwindling fossil fuel reservoirs. It has been undervalued in its commercial potential for a wide range of applications, including primary chemical reagents, pharmaceutical base agents, resins, coatings, and polymer material additives. The only significant application of lignin in the past has been in biomass fuel production, whereby some of its potential energy may be released. However, initial infrastructure costs may be prohibitively expensive for smaller to medium sized pulp and paper operations. In addition, the products that are generated from the burning process itself contribute adversely to greenhouse gas emissions. A number of different methods have been studied for the purpose of lignin modification. In this chapter, we will focus primarily on the chemical, photochemical, electrochemical, and enzymatic lignin modification methods.

2. LIGNIN MODIFICATION METHODS

2.1. Chemical Modification

Subsequent to mechanical pulping in the kraft pulping process, a significant amount of lignin remains to be removed through bleaching. The chemical structures of native lignin and the lignin isolated from unbleached kraft pulps are much more complex than the lignin that is obtained after bleaching [21]. The secondary condensation and reduction reactions that may occur in the lignin macromolecule result in a substance with decreased reactivity that is harder to degrade. Less than one half of the initial γ-hydroxymethyl groups on the side chains of the lignin molecule remain after the kraft cooking process. This is confirmed by analytical thioacidolysis of wood into non-oligomeric compounds in comparison with the condensed polymers remaining in the kraft pulp. A suggested condensation route results in the formation of quinone methides in the phenolic lignin units via the elimination of an α-hydroxyl or α-ether group. Radical coupling reactions have also been proposed based on studies of model compounds such as syringyl alcohol. Up to 50% of the residual pulping lignin has been removed through oxygen delignification methods, sparing many of the environmental impacts of chemical bleaching and increasing the number of diaryl structures [22].

Onc system for oxidative delignification involves a transition metal polyoxometalate that is comprised of a d^0 metal ion, either V(+5) or Mo(+6), and a transition metal substitute which shows good selectivity for oxidizing lignin. Re-oxidation of the reduced polyoxometalate can involve air, hydrogen peroxide, or other peroxides [23]. It has been suggested that the introduction of a mediator allows for the selective oxidation of specific functional groups on the lignin polymer. The use of a thiol mediator results in lignin oxidation which occurs at the benzylic position. Transition metal mediators may catalyze the production of hydroxyl radicals, which has the undesired effect of degrading cellulosic material. To prevent this degradation various protective agents have been introduced into the pulp. A magnesium sulphate/phenol combination was found to promote the selectivity of oxygen delignification as interpreted by higher recorded viscosity measurements, while preventing phenolic guaiacyl condensation and enriching p-hydroxyphenyl content. In one study, the possible role of surfactants in Fenton's reagent oxidation of kraft black liquor was considered. With the UV/Vis absorbance maximum being measured at 280 and 310 nm, the optimum conditions for the oxidation of black liquor were determined to be 10 mM H_2O_2 and 1 μL of 10% solution of the anionic surfactant [24].

Chemical modification is a potential approach for adding uniform functionality to the lignin macromolecule, which allows for improved compatibility as an additive to other polymers [25, 26, 27]. A great deal of research effort has gone into chemically modifying lignin to achieve more uniform reactivity for improved blending with other polymers. There are generally three different procedures used for the synthesis of epoxy resin from lignin. The first technique involves the blending of lignin derivatives (kraft lignin) directly with the epoxy resin. In the other two methods, lignin derivatives are modified with epoxides or other reagents prior to epoxidization [28]. Numerous papers have investigated the synthesis of epoxy resins with improved biodegradable properties through the modification of lignosulfonates with epichlorohydrine or phenol/epichlorohydrine and incorporation of these epoxies into a synthetic polymer matrix. Zhao et al. investigated the nature of lignin epoxy resin that was generated from calcium lignosulfonate. The calcium lignosulfonate was first modified into phenolated lignosulfonate and then treated with epichlorohydrin in the presence of a sodium hydroxide solution catalyst to generate two types of epoxy resin. A similar procedure has been used with iron and ammonium lignosulfonates to produce both a liquid and a solid resin. Success has been obtained in blending esterified, epoxidized and esterified/epoxidized modified lignin with low density polyethylene [29]. Vasile et al. improved the impact resistance and thermal stability of a polyethylene blend by increasing the esterified flax lignin content.

The intensity of the pulping process is responsible for the conversion of the naturally occurring "native lignins" into the highly condensed "industrial lignins" which are difficult to utilize. The random polymerization, structural variation and solubility issues are characteristics of the "industrial lignins" which have frustrated researchers and led to the current research interest involving lignophenols [30]. Lignophenols have simpler, more readily modifiable structures that are constructed from surface reactions between "native lignins" and phenols under acidic conditions. The acid catalyzes the fragmentation and phenolation of lignin, whereas the addition of a phenol derivative is used as a phenolation agent. Lignophenols are more stable than "native lignins" because of the hydroxyl groups that occupy the benzyl positions, which serve as the active sites for condensation reactions. Lignophenols that were used as a support in a bioreactor system to immobilize an enzyme, while still maintaining their own catalytic activity, have been reported in the literature [31]. Lignocresol of spruce, a lignomonophenol, was determined to have an affinity for bovine serum albumin (BAS) that is 5-10 times that of unmodified lignin, but similar to γ-globulin, haemoglobin and β-glucosidase. Lignophenols were also investigated for their use as sensitizers in dye-sensitive solar cells (DSSC) that were made of porous TiO_2. They showed encouraging voltage-current outputs (although photoactivity in the visible light region needs to be improved), and that hydroxyl groups played an important role in the binding interactions between the lignophenols and the TiO_2 surface.

The modification of lignin with cyclic anhydrides has proved to be a way of introducing new properties into the material. Cyclic anhydrides such maleic, succinic and phthalic anhydride have been used to link lignin through carboxylate groups (-COO-) while also generating a new reactive site for further modification [32, 33]. Interest in the chemical modification of wood and lignin to increase its thermoplasticity may produce a material that is suitable for a wide variety of markets. Timar et al. [34] have chemically modified aspen wood with maleic anhydride in a two step process. In the first step, the esterification of the material occurred with maleic anhydride to produce a monoester and a diester. In the second step, the carboxyl group of the monoester reacted with an epoxide compound and then with the maleic anhydride for a second time to produce the oligoesterified wood. Analysis of the modified lignin material showed reduced hygroscopicity and increased thermoplasticity when compared to the unmodified material.

Activated carbons (ACs) exhibit a porous structure with a large internal surface area that is responsible for its absorption abilities with numerous substances. ACs are commonly prepared from materials like coal, coconut shells, wood and peat,[35]. The high carbon content of lignin makes it a promising precursor for its use as an activated carbon. ACs can be prepared from lignins by either a two-fold carbonization/activation physical process or through a chemical activation method. Chemical activation involves the simultaneous carbonization/activation by treatment with a chemical activator such as alkaline hydroxide or phosphoric acid. The high surface area ACs prepared from kraft lignins by various research groups were shown to be comparable with the best of those that were commercially available. The ideal parameters for the generation of activated carbons produced by the heat treatment of lignin with NaOH or KOH were recorded [36]. KOH had the largest effect on the microporous material with a micropore volume of up to 1.5 cm^3/g, whereas NaOH reached a maximum microporous volume of 1.0 cm^3/g. However, NaOH is much more effective for the activation of carbon materials like anthracites at all experimental parameters. An increase in temperature was seen to increase the activation level, which approached an optimum at 700°C. The mass ratio hydroxide/lignin is the parameter that has the strongest influence over the resulting pore size, with an optimal ratio being 3:1.

2.2. Photochemical Modification

2.2.1. Photodegradation of Lignin at Tio₂

Much of the chemical oxygen demand (COD) in the wastewater from pulp and paper mills is a result of the inefficient decomposition of lignin and its derivatives using conventional treatment processes [37]. One of the strongly developing alternate methods for the removal of organic pollutants from wastewater is hetcrogeneous photocatalysis [38 - 41]. Of the numerous semiconductor photocatalysts that have been investigated, titania (TiO_2) has shown some promising results because of its biological and chemical inertness, cost effectiveness and the strong oxidizing capability of the photogenerated holes [42 - 44]. The photocatalytic oxidation of various pollutants has been studied with a suspension of fine powdered TiO_2 particles, or with TiO_2 particles being supported by a solid surface. A problem that arises with the use of unsupported TiO_2 particles makes efficiency dependant on the surface area, because only a thin layer of the exterior portion of each particle absorbs UV light. The particles in this process tend to form agglomerates, which decrease the surface area, and therefore, reduce catalytic activity. The removal of the suspended particles from the treated solution is a difficult task, which also becomes an issue with this method of wastewater treatment [45].

When the TiO_2 is irradiated with UV light that has an energy greater than the band gap energy ($\lambda < 380$ nm), electrons and holes are generated in the conduction and valence bands, respectively. The electrons and holes generated can either recombine or migrate to the surface to facilitate redox reactions. The holes produced by vacating electrons have a high oxidation potential which can be employed in the oxidative decomposition of targeted organic substances [46]. For photocatalysis to be efficient and productive, the recombination of electrons with holes must be suppressed through additional means. This can be attained: (i) via the use of a suitable electron acceptor, such as O_2, which can undergo transformation to a superoxide radical $O_2^{\bullet-}$ and trap the photogenerated electron, or (ii) by applying an external anodic bias [47].

The photocatalytic degradation of lignin was first studied by Kobayakawa et al. [48]. Using a rutile type TiO_2 powder as a photocatalyst and a 500 W high-pressure mercury lamp to provide the UV-radiation, the absorbance was measured at 280 nm in order to determine the COD value of the test solution. The COD value of the test solution approached zero after seven hours of treatment, indicating that the complete mineralization of lignin had occurred. The importance of the dissolved oxygen was identified, as saturated nitrogen solutions did not exhibit the same colour change or mineralization as the treatment in the presence of oxygen. Oxygen was used as an electron acceptor to prevent the recombination of the photocatalytically generated holes and to promote the productive and efficient decomposition of the lignin. Dehydrodimerization of intermediate phenolics followed by further oxidative degradation was thought to be the reason why vanillin and other aromatics were unrecoverable. The other intermediates detected from the oxidation process were methanol, ethanol, formaldehyde and oxalic acid. Another study involving UV/TiO_2 coupled photodegradation of black liquor based lignin, generated some monomers such as coniferylic alcohol and oxidized monomers in the form of syringaldehyde, vanillic acid and p-coumaric acid. The optimum yields of vanillin and vanillic acid were obtained after 2 hr of UV irradiation ($\lambda > 290$ nm), while the best yields of 4-hydroxy-3,5-dimethoxy benzaldehyde and 3,4,5-trimethoxy benzaldehyde were achieved after 1 hr. The rate of modification was found to be dependent on the initial concentration, as lower lignin concentrations led to effective degradation, while higher concentrations had an inhibitory effect [49].

Dahm et al. [50] investigated the titanium dioxide catalyzed photocatalytic degradation of dissolved lignin. They used a model system with REPAP lignin in water to optimize operational parameters, such as the titanium dioxide loading and the illuminating power for the oxidation. With the use of a 10 g/L rutile TiO_2 suspension a 82% decrease in total lignin concentration was obtained after two hours of photocatalytic treatment at an illumination power of 128W. Above a TiO_2 loading of 10 g/L the degradation decreased due to light refraction by TiO_2 and increased solution turbidity. The higher the power of UV illumination, the faster the initial rate of degradation; hence more lignin could be mineralized throughout the reaction. Doubling the UV lamp output from 64 to 128 Watts resulted in an order of magnitude increase in the initial rates, yet a decrease in overall efficiency. However, the addition of up to 0.05 mM Fe^{2+} ions to the lignin solution increased the photocatalytic oxidation efficiency by up to 25%. A 40% destruction of lignin occurred in five minutes at the highest level of light intensity (128 W), whereas lower levels of light intensity required up to one hour. One UV/TiO_2 coupled study involving the removal of lignin from wastewater, focused on the decolourization ability as measured by ADMI (American Dye Manufacturers Index). The effectiveness of the TiO_2 photocatalyst was shown through the achievement of more than a 99% solution (at pH of 3.0) colour removal with the addition of 10 g/L TiO_2 [51]. The decolourization was quick with 50% occurring in the first 15min. Only 33% of the color and 25% of the DOC (dissolved organic carbon) were removed without the use of the TiO_2 catalyst. The addition of the TiO_2 photocatalyst increased the lignin degradation rate constant by a factor of 2.7. Improved rates of DOC and ADMI were observed during UV exposure with the addition of 1% Pt content into the TiO_2 suspensions. The doping of the TiO_2 particles with Pt increased the rate of the electron-electron hole pair reaction, which enhanced the concentration of hydroxyl radicals on the surface of the catalyst [52, 53]. The addition of Pt enhanced the rate of degradation by as much as 6 times and, as is in previous TiO_2 studies, the reaction rates were found to be highest with an acidic pH.

With benzyl alcohol units accounting for about 30% of the phenylpropane units in the lignin structure, much attention has been given to phenolic and non-phenolic benzyl alcohol lignin model compounds. Antunes et al. reported on the TiO_2-photocatalyzed degradation of four different benzyl alcohol lignin model compounds while studying the role of oxygen (O_2), potassium peroxydisulfide ($K_2S_2O_8$) and nitrous oxide (N_2O) as electron scavengers in this process. Using oxygen as the electron scavenger leads to the formation of a superoxide radical anion $O_2^{\bullet-}$, which leads to the aromatic ring-opening of phenols. When $K_2S_2O_8$ is used as an electron scavenger, a very strong one electron oxidizing agent is formed in $SO_4^{\bullet-}$, which reacts with aromatics to form cation radicals. When N_2O reacts with an electron, a strong oxidizing species is generated in the form of a hydroxyl radical ($^{\bullet}OH$) which reacts with the electron rich aromatic species. The reactivity of the superoxide radical is comparability lower than that of the hydroxyl radical, but a longer lifetime is observed [54]. The photocatalytic degradation of the lignin model compounds, 3,4-deimethoxylbenzyl alcohol (**1**) and 1-(3,4-dimethoxylphenyl)ethanol (**2**), has been investigated in the presence of the previously mentioned electron scavengers at pH levels of 3 and 7. The major products from the photocatalytic treatment of (**1**) and (**2**) were 3,4-dimethoxybenzaldehyde and 3,4-dimethoxyacetophenone, respectively. Based on the results from Antunes et al. it was determined that O_2 was the best electron scavenger for the oxidation of 1-(3,4-dimethoxylphenyl)ethanol (**2**) under experimental conditions. In the presence of O_2, ring opening products and CO_2 were formed and used as an indication of substrate mineralization. Machado et al. observed comparable trends between *Eucalyptus grandis* lignin obtained from peroxyformic acid pulping and two biphenyl lignin model compounds. This relationship confirms the reaction between the benzyl alcohol structures on the lignin and the photocatalytically generated hydroxyl radicals.

2.2.2. Photodegradation of Lignin at Other Semiconductors

Studies investigating other semiconductors that show high photocatalytic ability such as ZnO and CdS have also been conducted [55, 56]. ZnO is a semiconductor oxide that has a band gap energy larger than 3 eV and exhibits very similar electronic properties to TiO_2. One study, involving ZnO photocatalysis in kraft black liquor degradation was most efficient with 0.1 g of ZnO per 5 mL of suspension and a saturated oxygen atmosphere. ZnO impregnated with platinum enhanced the discolourization, which reached 100% after one hour of treatment, but did not achieve the total decomposition of organic matter [57]. Kansal et al. showed that ZnO exhibited a higher rate of lignin degradation than a TiO_2 photocatalyst or any combination of the two. The percentage of lignin degraded ranged from 30% to 84% for different mixture compositions of TiO_2 and ZnO. The maximum percentage of degradation was achieved with a pure ZnO catalyst over a five hour period. The addition of an oxidant such as NaOCl enhanced the rate constant of the lignin degradation, and a similar dependence on the initial lignin concentration was observed. ZnO and CdS did experience some catalyst decomposition in the presence of UV radiation and oxygen. GC analysis of CO_2 evolution was used to study the dependence of the mineralization rate on the reaction parameters. A first-order dependence on dissolved oxygen and a half-order dependence on the black liquor substrate concentration were observed at 25 °C [58]. The majority of the mineralization (more than 95%) was found to occur through heterogeneous ZnO-assisted photocatalysis. The remainder occurred through direct photo-absorption in the reaction where the substrate absorbs light in the presence of oxygen and is directly converted to CO_2.

The combined ZnO photochemical and fungal biodegradative process for the treatment of Kraft E1 effluent is reported by Reyes et al. [59]. Free ZnO catalysts achieved 100% decolourization of *Eucalyptus* and *Pinus* E1 effluents within a two hour period. The ZnO was immobilized on sand and exhibited similar activity to the free ZnO catalysts for the first 30 minutes of treatment. However, poor decolourization (36%) results were observed for the *Pinus* E1 treatment after two hours with a ZnO/sand catalyst. This was accredited to mass transfer and oxygenation problems. When combining the photochemical and biobleaching treatment methods the degradation of biomass increased by 10% and the activation of enzymes such as lignin peroxidase and Mn-peroxidase was observed. In order to improve the efficiency of the photocatalytic process with ZnO, ultrasmall particles have been investigated due to their high surface area. Pal et al. [60] studied the photocatalytic ability of sol-gel produced ZnO thin films on the surfaces of glass plates against a series of phenolic and polyaromatic compounds. The efficiency of the photocatalytic process is enhanced, because the porous films have a higher absorption affinity for the reactant molecules. The FTIR analysis of the reaction products that were generated via the photodegradation of a group of polyaromatics confirmed that hydroxylation and carbonylation were occurring. Phenol degradation produced hydroquinone and benzoquinone as the main intermediates, which were further degraded into CO_2 after a 120-minute irradiation period. To inhibit photocorrosion, a stable small bandgap semiconductor (ZnO) can be coupled with an unstable large bandgap semiconductor (Fe_2O_3). The addition of Fe_2O_3 is thought to aid in the chemisorption of polyphenols as the best performance was achieved with a 0.50 oxide molar ratio [61]. Also promising, are Ag-doped O photocatalysts prepared from silver nitrate photoreduction on a ZnO substrate [62]. Almost total decolourization ($\lambda = 465$ nm) of lignin occurred after 60 min of UV treatment, and up to a 90% reduction in TOC for the Kraft E1 effluent resulted from a 120 min treatment period. Significant modification of the lignin structure was apparent through the almost total suppression of the UV absorbance signal at 290 nm, which indicates the presence of benzenoid compounds. The toxicity of lignin and Kraft E1 effluent towards *E. Coli* was completely eliminated with the use of an Ag-ZnO photocatalyst.

2.3. Electrochemical Modification

2.3.1. Electrochemical Oxidation

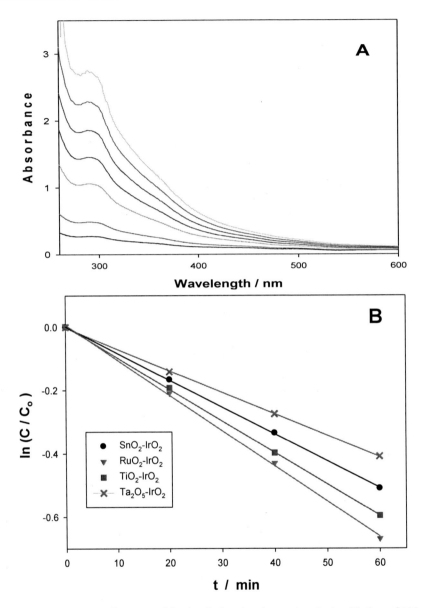

Figure 2. (A) UV/Vis spectra taken every 20 min. during the electrochemical oxidation of 100ppm lignin in 0.5M NaOH at the Ti/RuO$_2$ electrode, with a current density of 300mA/cm^2 and temperature of 60°C. (B) Plots of ln(C/C$_o$) vs. time for the oxidation of 100ppm of lignin in 0.5M NaOH at the four IrO$_2$-based electrodes with 300mA/cm^2 and 60°C. Reprinted from Tolba, R.; Tian, M.; Wen, J.; Jiang, Z. H.; Chen, A. *J. Electroanal. Chem.* 2010, 649, 9-15. Copyright 2010 Elsevier.

Electrochemical processes have been gaining immense attention for the oxidation and modification of organic compounds because of their versatility, compatibility and "green" nature. Their environmentally friendly nature stems from the fact that the primary reactants in electrochemical processes are electrons. Electrochemical oxidation has been studied extensively as an effective method for wastewater treatment, where organic compounds are oxidized by hydroxyl radicals ($^\bullet$OH) and chemisorbed active oxygen species are produced by an electrochemical reaction [63, 64, 65]. Electrochemical oxidation and modification of lignin contaminated pulp and paper effluent has been studied as a possible solution to the pollution problem. Lignin can be oxidatively degraded in aqueous alkaline solutions at anodes of platinum, graphite, nickel, Cu, boron-doped diamond (BDD), dimensionally stable anodes and PbO_2 [66]. Industrial applications have been plagued by the high cost of the process and electrode fouling as the result of polymerization. To improve the applicability of the electrochemical oxidation process, great interest remains in developing a low cost electrocatalyst that has a high activity for lignin oxidization. The nature of the electrode, along with the amount of charge consumed, impacts the yield of the low-molecular weight products produced [67].

One group of dimensionally stable anodes (DSA) encompasses the iridium oxide electrodes, which have been investigated due to their electrocatalytic activity, economical production and stability [68]. Tolba et al. fabricated four different types of IrO_2 based electrodes via the thermal decomposition method and investigated their activity in the electrochemical oxidation of lignin. In an effort to optimize stability and electrocatalytic activity, the electrodes were fabricated on titanium substrates with IrO_2 as the major component. The minor component was used in approximately 30 wt.%, and was either RuO_2, TiO_2, SnO_2 or Ta_2O_5. The oxidizing ability of the different electrodes was investigated by measuring the UV/Vis absorbance for lignin during the electrochemical oxidation process. Figure 2A shows the UV/Vis spectra recorded every 20 minutes for the duration of the electrochemical oxidation of 100 ppm lignin at a Ti/RuO_2-IrO_2 electrode. The peak at 294 nm represents the lignin concentration, which disappeared following two hours of the electrochemical treatment, indicating the complete degradation of lignin. The $\ln(C/C_0)$ vs. time relationship which was used to determine the rate constants for the oxidation of lignin at the different IrO_2 electrodes is shown in Figure 2B. The highest activity for lignin degradation and stability was displayed by the Ti/RuO_2-IrO_2 electrode, which had a rate constant of 9.9 x 10^{-3} min^{-1}. The Ti/Ta_2O_5-IrO_2 electrode displayed the lowest activity towards the oxidation of lignin, with a rate constant of 6 x 10^{-3} min^{-1} as the Ti/Ta_2O_5-IrO_2 electrode exhibited the highest activity toward oxygen evolution. With the Ti/RuO_2-IrO_2 electrode, the most efficient current density was determined to be 500 mA/cm^2 and the activation energy from the Arrhenius plot was 20 kJ/mol. The electrooxidative cleavage of lignin model dimmers at nickel anodes has been studied by Pardini et al. [69]. Structural features such as the β-O-4-aryl ether linkage, the 4-methoxyl or the 4-hydroxyl groups, and the substituents at the α- and γ-positions (vicinal) to the β-O-4-aryl function group, played an important role in the electrooxidative degradation process. The oxygenated Ni(III) anode extracts hydrogen atoms which initiate the oxidation process and the generation of low molecular weight products such as vanillin and guaiacol. Analysis of a preheated, electrochemically oxidized mixture of model dimmers with IR spectroscopy and with ^{1}H and ^{13}C NMR suggested that polymerization and major structural changes had occurred in the products. These structural changes included increased carboxylic functionality, significant vinylic features and a decrease in the presence of methoxyl groups. An illustration of the electrochemical oxidation reaction of a model lignin dimmer is shown below.

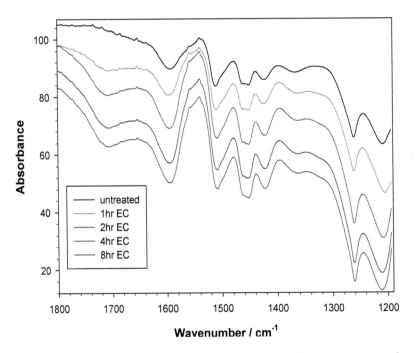

HPLC analysis was used to identify the intermediates produced in the electrochemical oxidation of 500 ppm of lignin at a Ti/RuO$_2$-IrO$_2$ electrode under a 500 mA/cm^2 current and a temperature of 60 °C. A linear increase in the concentrations of vanillic acid and vanillin was observed for the first 45-minutes of oxidation, after which time the production slowed rapidly. The intermediates that are generated in this process may be of great interest to industries involving food, aromas and perfumes. In Figure 3, FTIR spectroscopy was employed to study the effects of electrochemical oxidation on lignin obtained from softwood black liquor. The FTIR peaks centered around 1460, 1510, 1600 and 1712 cm^{-1} are assigned to C-H deformations (asymmetric in methyl, methylene and methoxyl group), the aromatic skeletal vibrations coupled with the C-H plane deformation, the aromatic skeletal vibrations and the carbonyl stretching of unconjugated ketones, respectively. The increase of the peak intensity at 1712 cm^{-1}, subsequent to electrochemical oxidation, is consistent with the expected increase in the number of C=O and carboxylic acid groups (-COOH).

Figure 3. The FTIR spectra of the untreated and electrochemically oxidized lignin from softwood black liquor.

The electrolysis of weak black liquor was investigated and optimized with the intent of reducing the organic load during the caustic recovery process. Lower conductivity and/or different chemical composition was the reason for softwood black liquor requiring more energy (12.5 kWh/kg) per kg of caustic produced in comparison to that of the energy (10 kWh/kg) required for hardwood. Both platinum and iridium-oxide coated titanium were found to be effective corrosion resistant anode materials and favoured the production of oxygen over chlorine gas. However, iridium oxide slightly outperformed platinum with 20% less energy consumption, which was conjectured to result from the decreased resistance at the IrO_2 anode [70]. When taking energy efficiency into account, operation at a higher current density was believed to be more economical. Improved performance was observed along with an increase in temperature, up to 80°C, at which point any further temperature increase made electrode fouling an issue. A variety of methods have been tested to prevent anode fouling, such as a current reversal stage, a high platinum anode current density and increased black liquor flow through the cell. The variety of membranes and the addition of sodium sulphate were found to affect electrolytic performance [71].

The electrochemical oxidation of pulp and paper wastewater, assisted by transition metal (Co, Cu) modified kaolin in a batch reactor with graphite electrodes was investigated by Wang et al. [72]. H_2O_2 produced on the surface of a porous graphite cathode reacted with the catalysts to form a strong oxidant (hydroxyl radicals) that could destroy the pollutants that were adsorbed on the kaolin surface. Soloman et al. [73] studied a series of RuO_2 coated titanium electrode reactor types and determined that the batch recirculation configuration was the most effective, showing 73.3% COD removal. With the continuous flow system, a maximum of 65% COD removal was achieved; however, the energy consumption was only 18% of the batch recirculation process. A COD removal of 80% and colour removal of 90% was achieved using a six plate iron electrode arrangement in a 2 L electrolytic batch reactor, with a current density of 55.56 A/m^3 at a neutral pH. By combining electrochemical treatment with coagulation/flocculation, a COD removal of 91% and a colour removal of 100% were achieved [74]. Gel permeation chromatography (GPC) was used to monitor the changes in lignin molecular weight before and after treatment using a lead dioxide coated titanium anode. The main GPC peak generated by the lignin indicated the presence of a group with a molecular weight of greater than 5000. A minor peak was also present, indicating fragments with molecular weights of less than 1000. The oxidation of lignin was confirmed through the decrease in both of the GPC peaks, which continued with extended electrochemical treatment [75]. The electrochemical oxidation of lignosulfonate at a boron-doped diamond electrode (BDD) was studied through total organic carbon (TOC) oxidation kinetics. The initial chemical oxygen demand, the applied current density and mass transfer coefficient were all related to the observed kinetics, although the identity of the supporting electrolytes did not have an impact on the TOC removal [76].

The most profound modification of lignin occurs when it undergoes chlorination and nitration. Electrochemical sililation, phosphorylation, and fluorination have also been successfully achieved [77, 78]. For electrochemical sililation, a radical cation ($R_2SiCl_2^{+}$) is generated through electrochemical initiation. Successful electrochemical sililation of both the hydroxyl and carbonyl groups of lignin has been achieved on a platinum anode at room temperature with an initiation potential of -0.1V. The maximum silicon content achieved through traditional chemical modification procedures is 10.9%, whereas the electrochemical method can attain 15.5% with no destruction of aromatic structures. The silicon containing lignin has been used in the synthesis of organosilicon polymers, which exhibit improved physical properties with less temperature variation. In the lignin phosphorylation process, electrochemistry can be employed for the pre-modification of lignin, or in electrolysis to generate a phosphoryl radical cation. The pre-modification of lignin through electrochemical oxidation, chlorination and nitration improves lignin reactivity in phosphorylation reactions although only 0.8% of the phosphorus is introduced. When electrochemical oxidation is used to generate a phosphoryl radical cation, it initiates the phosphorylation reaction, whereby 11-17% of the phosphorus has been introduced to both the hydroxyl and carbonyl groups on the lignin macromolecule. The electrochemical fluorination of lignin has been conducted in aqueous alkaline solutions containing sodium or potassium fluorides at a platinum or platinised titanium anode, with a potential of 1.8-2.2 V [79]. Optimization of the electrochemical fluorination of lignin has made it possible for a fluorine content of 20% to be achieved. Fluorinated lignin-based polymers have been studied as antifriction additives whose incorporation into metal-ceramics has been successful at reducing frictional wear by three to four times.

Chlorination and nitration are two very important methods of lignin modification due to the wide range of potential applications for the resulting products. Extensive demethylation of lignin occurs during electrolytic chlorination in aqueous solutions. The demethylation is thought to result from the chlorination of methoxy groups, followed by the hydrolysis of the newly formed C-Cl bonds. The maximum chlorine content achieved though lignin chlorination was 20% and only minimal variation was experienced with a change in the HCl concentration and/or current density. A vigorous reaction rate was seen with a platinum and graphite anode in aqueous solution at the early stages. Most of the chlorination (up to 75%) occurs on the aromatic rings, with the remaining 20-25% occurs on the aliphatic portion of the lignin macromolecule. Chlorolignins have proved to be excellent ingredients for epoxide oligmers and weakly acidic cation exchangers. In electrochemical nitration two types of nitrating species, NO_3^- and NO_2^+, are generated through HNO_3 electrolysis at room temperature. After nitration, all nitrogen present on the lignin macromolecule is found in the form of nitrate groups. Nitrolignins have a diverse range of applications, serving as a regulator for the structural and mechanical properties of clay-based drilling fluids, a plant growth stimulator, filler in polymers and an intermediate in the manufacturing of gun powder. Extensive lignin modification has occurred by combining chlorination and nitration to develop polyfunctional lignins. Polyfunctional lignins have increased reactivity toward monomers and high molecular weight compounds, which allow for the synthesis of highly reactive polymers. Chloronitrolignins have also been used to produce amphoteric ion exchangers, which have adsorption exchange capacities of 4.5-6.5 mg-eq/g and 0.9-1.0 mg-eq/g for cations and anions, respectively.

2.3.2. Electrochemical Modification Couple with Mediators

The use of mediator-based processes for delignification is of great interest as they promise to be environmentally friendly and economic alternatives to currently dominant chlorine-based processes. Various electrochemical heterogeneous catalytic systems have been proposed involving the electrolytic oxidation of a mediator compound, usually a transition metal complex or an organic molecule, followed by the oxidation of lignin by way of the mediator [80, 81]. In this fashion the mediator is constantly recycled while providing a pathway for the oxidation of the insoluble lignin macromolecule. With violuric acid used as a mediator in the electrochemically mediated bleaching process (EMB), approximately 35% delignification was obtained with 4 kg/ton$_{pulp}$ [82]. Violuric acid was selected as a mediator based on the prolonged half-life of its radicals in aqueous solution, its high oxidation potential and low molecular weight, which allows for efficient diffusion throughout the substrate macromolecule. Another beneficial characteristic of violuric acid is its very high selectivity toward lignin oxidation over cellulose. Some evidence has surfaced which shows that violuric acid does not always exhibit the desired mediator recyclability. Kim et al. reported a 30% loss of violuric acid in two hours of bleaching and determined that the degradation products were alloxan, parabanic acid and oxaluric acid. The maximum delignification achieved in a single electrochemical step, when mediated by voiluric acid was recorded at about 40% [83].

Transition metal-based compounds show much promise as electrochemical mediators. With inorganic compounds being used as mediators a temperature and pH dependence was found. However, the inorganic mediators were found not to react using the same radical mechanism as is observed with organic compounds. Due to their stability in either reduced or oxidized states, transition metal mediators can be recycled many times, making them true catalysts unlike organic mediators which are partially consumed in the process. $K_4Mo(CN)_8$ and MoCN have been successful in electrochemical and enzymatic mediated oxidation by showing a high selectivity for lignin over cellulose [84]. Laccase and electrodelignification at a potential of 750 mV for two hours achieved similar delignification results with the aid of a transition metal mediator. A delignification maximum of 60% occurred in a MoCN mediated electrodelignification system with a potential of 750mV over a 45 hour treatment period. Oxygen was determined to have a minimal effect on the MoCN mediated electrochemical process, which suggested that MoCN functions as an electron carrier and not as an oxygen-activating agent. The iron-based complex mediators FeBPY and FeDMBPY have low production costs, low toxicity and show even greater delignification than MoCN, which makes them good mediators.

Cyclic voltammetry, when coupled with a soluble redox mediator for lignin oxidation can be used for rapid Kraft lignin quantification [85]. Generally, in cyclic voltammetry, the peak current (I_p) at the surface of the working electrode is proportional to the concentration of an analyte that comes in direct contact with the electrode. Since lignin has no direct contact with the electrode, the addition of the soluble mediator acts as a coupler. Thus, with a fixed soluble mediator concentration and scan rate, the peak current is proportional to the lignin concentration. The mediator 2,2'-azinobis-(3-ethylbenzthiazoline-6-sulfonate) (ABTS) had two redox couples, one at 520 mV and the other at 920 mV, which both displayed a strong linear correlation between catalytic peak current and kappa number of the pulp samples. It was suggested that the peak current at 520 mV be used to quantify the more readily oxidized lignin functional groups such as phenolic groups, while the peak current at 920 mV should be employed for higher potential functions.

2.4. Combination of Electrochemical and Photochemical Oxidation

By combining a metal oxide coated anode with UV radiation, the observed reaction rate is much higher than the summation of the individual processes. Bertazzoli et al. [86] reported a 70% and 35% reduction in the initial colour and total organic carbon (TOC) after a three hour photoelectrochemical treatment period using a ruthenium and titanium mixed oxide electrode. For a similar TOC removal using the conventional activated sludge method, over 20 days of treatment would be required. The synergistic effects of combined electrochemical and photochemical lignin-sulphonate oxidation processes was observed for DSA® electrodes, comprised of ternary mixtures of *active* oxide (RuO$_2$), *nonactive* oxide (SnO$_2$), and photocatalyst (TiO$_2$) [87]. Two advantages of the photoelectrochemical process using the DSA® electrodes are: the ability of the applied potential to inhibit the recombination of photo-generated electrons with the TiO$_2$ 'holes', and the photoactivation of highly reactive species (O$_3$, OH⁻ and H$_2$O$_2$) on the electrode's surface. SnO$_2$ acts as a "sink" for photo-generated electrons, lowering the band gap-separation to allow for efficient degradation. In order to highlight the enhanced performance of photoelectrochemical oxidation, comparative studies have been done with the individual treatment methods. Electrochemical oxidation of 2000 mg L^{-1} lignin-sulphonate solution with a Ru$_{0.1}$Sn$_{0.6}$Ti$_{0.3}$O$_2$ electrode and an applied potential of 0.6V showed a 19% colour reduction with minimal TOC and phenolic reduction. However, under similar experimental parameters the photoelectrochemical oxidation achieved a 72%, 80% and 50% reduction in colour, TOC and total phenolic content, respectively. An increased degradative capacity was observed for electrodes with higher SnO$_2$ composition. Vanillin and catechol were produced as intermediates at the start of lignin degradation, followed by more basic compounds such as methanol, formaldehyde, ethanol, formic and oxalic acid.

The combined effects of using TiO$_2$ nanotubes (photocatalyst) and a Ta$_2$O$_5$-IrO$_2$ thin film (electrocatalyst) for the photoelectrochemical oxidation of lignin has been reported recently. The UV-Vis spectroscopy absorption band at 295 nm was employed to determine the effective rate of the lignin degradation. Figure 4A shows the UV-Vis absorption spectra for the photochemical oxidation of lignin on the Ti/TiO$_2$ nanotubes under UV irradiation and an applied potential of +600 mV. Applying the potential increased the percentage of lignin modification from 20% to 50% over a two hour treatment period, which indicated the suppression of the electron-hole pair recombination. Figure 4B shows that approximately 66% of the lignin was modified through electrochemical modification with a Ti/Ta$_2$O$_5$-IrO$_2$ electrode and applied potential of +600 mV. Combining the photocatalyst and electocatalyst together under UV radiation with a potential of +600 mV was effective at modifying 92% of the lignin (Figure 4C). Figure 4D presents the plots of ln(C/C$_o$) *vs.* t. The linear relationship shows that the lignin oxidation kinetics was pseudo-first-order, from which the rate constants can be determined. The photoelectrochemical oxidation process had the largest rate constant, which was over twice the size of the electrochemical process. Analysis of the intermediates with HPLC, shown in Figure 5, identified the primary intermediates to be vanillin and vanillic acid.

2.5. Enzymatic Modification

Certain types of fungus, especially the white-rot fungus, have been known to degrade lignin through enzymatic action. The fungus secrets a variety of oxidative enzymes, including laccases, lignin peroxidises and manganese peroxidises, which are responsible for its degrading nature [88]. Most of the enzymatic oxidation methods explored for lignin oxidation employ an additional mediator to catalyze and extend the functional range of the enzyme. The exceptional oxidative properties of laccase have been the driving force toward extensive research involving its application for lignin degradation.

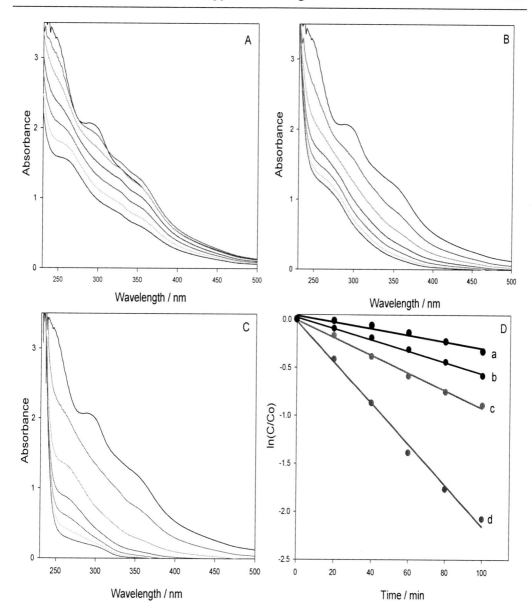

Figure 4. The in situ UV–Vis spectra for the oxidation of 100 ppm of lignin in 0.5 M NaOH on: (A) Ti/TiO$_2$ nanotubes with UV irradiation and an applied potential of +600 mV (vs. Ag/AgCl), (B) Ti/Ta$_2$O$_5$–IrO$_2$ electrode with an applied potential of +600 mV, and (C) TiO$_2$/Ti/Ta$_2$O$_5$–IrO$_2$ electrode with UV irradiation and an applied anodic bias. (D) The ln(C/Co) versus time relationship for lignin oxidation: on Ti/TiO$_2$ nanotubes with UV irradiation without (a), and with the applied anodic bias (b); on the Ti/Ta$_2$O$_5$–IrO$_2$ electrode with the applied potential (c); and on the TiO$_2$/Ti/Ta$_2$O$_5$–IrO$_2$ electrode with UV irradiation and the applied potential (d). Reprinted from Tian, M.; Wen, J.; MacDonald, D.; Asmussen, R.M.; Chen, A. *Electrochem. Commun.* 2010, 12, 527. Copyright 2010 Elsevier.

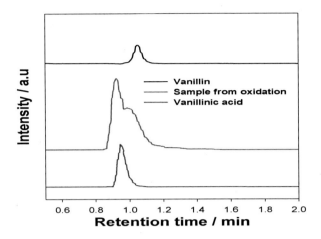

Figure 5. HPLC analysis of the intermediates produced through the photoelectrochemical oxidation of lignin.). Reprinted from Tian, M.; Wen, J.; MacDonald, D.; Asmussen, R.M.; Chen, A. *Electrochem. Commun.* 2010, 12, 527. Copyright 2010 Elsevier

Laccase contains an active site of four copper atoms which oxidizes the substrate through a four electron reduction of O_2 to H_2O. Laccase oxidizes phenolic groups at a reasonable rate while the oxidation of non-phenolic groups occurs at a very slow rate due to a large formal oxidation potential. It is suggested that certain chemical mediators with sufficient formal potentials can improve the oxidation of these non-phenolic groups, as well as improve the overall rate of delignification. A mediator is a small molecule that is used as an electron carrier between the laccase and the lignin macromolecule. The use of a mediator decreases the structural hindrance of the oxidation process, and as a result, a more complete oxidation is achieved. Electrochemical studies of laccase with an ABTS mediator show that the oxidation of non-phenolic veratryl alcohol can occur through the generated $ABTS^{2+}$ dication while the oxidation of phenolic model compounds occurs through the $ABTS^{+\bullet}$ radical cation. With the help of a mediator, the laccase system has been able to achieve delignification values of more than 55% [89]. Other mediator candidates include nitrogen heterocycles as well as transition metal complexes. It is difficult to find a laccase-mediator that effectively oxidizes non-phenolic groups as it is required to have a formula potential close to, or above that of laccase. However, with a formula potential this high, the rate of mediator oxidation by the laccase is thermodynamically unfavourable.

Weihua et al. [90] studied the properties of an extracellular enzyme from an entophytic fungus that exhibited high laccase activity. The fungus isolated enzyme with high laccase activity (MS-Lac) was alkali-stable, with an optimum pH of 10 and a higher affinity for syringaldazine over ABTS. The modification of lignin with MS-Lac resulted in the oxidation of the phenol hydroxyl substituents on the side chains and the carbonyl groups. A shift in the unconjugated to conjugated carbonyl content and the presence of diethoxy was noted after the modification. With the MS-Lac modification, condensation and polymerization of lignin or its degradation products occurred and a decrease in the polydispersity was documented. The modification effect of the laccase-HBT system on lignosulfonates was investigated where the enzymes were isolated from *Trametes villosa* (TvL), *Myceliophthora thermophilia* (MtL), *Trametes hirsuta* (ThL) and *Bacillus subtilis* (BsL) [91]. The fluorescence of lignin compounds was effectively reduced under acidic conditions (pH of 4.0 and 4.5) by ThL and

TvL, while a similar reduction was experienced with MTL and BsI at a pH of above 6. The reduction in lignin fluorescence was attributed to the modification of conjugated carbonyl, biphenyl, phenylcoumarins and stilbene groups. Extensive polymerization due to the laccase modification resulted in an increase in molecular weight by 74% after 17 hours of ThL treatment and by 370% with the TvL-treated samples. However, the FTIR and ^{13}C NMR spectra confirmed that the modification was limited to substituents and that the backbone remained untouched.

Studies have investigated the lignin oxidizing capability of lignin peroxidase (LiP), manganese peroxidase (MnP) and horseradish peroxidase (HRP) [92]. While all three heme peroxidises share comparable formal redox potentials, only LiP is capable of oxidizing non-phenolic substrates with a high redox potential at a low optimum pH. The non-phenolic substrate oxidizing capability of LiP is thought to be a reflection of a different heme environment. After the discovery of heme peroxidises, much attention has been focused on LiP. However, after failing to identify LiP in a few of the important lignin degrading organisms, increased interest in MnP developed [93]. MnP's catalytic cycle for oxidizing phenolic compounds has many similarities to the other heme peroxidises; however, it uses Mn^{2+} as the preferred substrate. A series of MnP and LiP hybrid enzymes that exhibit the capacity to oxidize phenol and non-phenolic aromatic compounds have also been isolated from *Pleurotus eryngii*. A search for an alternate pathway that allows for the oxidation of nonphenolic lignin components has turned to the application of mediators. The oxidation of veratryl alcohol has been achieved with MnP and the aid of reduced glutathione (GSH) as a thiol donating compound. In the thiol mediated oxidation process, a dependence on hydrogen peroxide and Mn (II) was noted along with the accumulation of veratraldehyde. It was thought that the presence of GSH in the veratryl alcohol oxidation pathway favoured side chain oxidation because of the inhibition of aromatic ring oxidation and, as a result, the reaction was stopped at the veratraldehyde stage [94].

Enzymatic lignin modification has been used to produce lignin copolymers, binders for wood complexes, intermediates for the manufacturing of highly reactive reagents, coatings, paints and polymer-template complexes. A great deal of research effort has gone into the synthesis of lignin copolymers for use as a new class of plastics. Liu and co-workers [95] used horseradish peroxidase as a catalyst for the copolymerization of straw-pulp-lignin with cresol in a reversed micellar system. The copolymer displayed a lower glass transition temperature and a higher curing exotherm which opens the door to potential applications in adhesives, bonding agents, laminates and polymeric dispersants. An environmentally friendly formaldehyde-free resin has been developed through a series of laccase catalyzed reactions, which shows promising results for use as a wood binder in the production of wood composites [96]. Laccases, along with an oxidizing agent such as O_2, air or other chemical oxidizing agents, have modified lignins to produce a reactive intermediate that can continue to react with lignin derivatives to produce composite materials. The reactive intermediate product that is generated via this process has the ability to maintain its reactivity through long storage periods and can be easily isolated. Environmentally friendly protective coatings and paints have been synthesized through the enzymatic polymerization of lignin with enzymes such as catechol oxidase, laccase and peroxidase. Lignosulfate was used to form a polymer-template complex as the result of a reaction with a monomer and horseradish peroxidase. The lignosulfate-based complexes show outstanding optical and electrical stability, processibility and water solubility that allow for its use in applications such as rechargeable batteries, electrolytic capacitors, window coatings and biological sensors.

CONCLUSION

The interest and demand for biomass as a renewable energy and chemical source continues to grow with the depletion of, and increased environmental damage caused by, the use of our petroleum reservoirs. Lignin is a main constituent in biomass and will play a major role in the relief of our global fossil fuel energy dependence. A number of advanced modification procedures have been developed to produce a number of fine and bulk chemicals from the lignin macromolecule. The complexity of the native lignin molecule makes the process of obtaining one specific modified product difficult, but does allow for a wide range of products to be produced. Pretreated lignins obtained through sulphate, kraft and organosolv pulping process have been successfully modified through chemical, photochemical, electrochemical, photoelectrochemical and enzymatic methods, to produce a wide range of value-added products.

Considerable research has been devoted to lignin oxidation procedures to generate products with increased functionality. The photochemical oxidation of lignin has been successful through the use of a number of semiconductor oxide photocatalysts, such as TiO_2 and ZnO. Improved photochemical oxidation results were observed with the application of an anodic potential bias, which suppresses electron-hole and electron recombination. Electrochemical studies involving lignin modification have shown improved reaction rates for lignin oxidation and provide effective methods for lignin chlorination, fluoration, nitration, sililation and phosphorylation. Further improvements in the rate of lignin oxidation were observed by combining photochemical and electrochemical processes. Two of the main intermediates produced were vanillic acid and vanillin, which are of great interest to the food, aroma and perfume industries. Laccases, manganese and lignin peroxidases are employed by a wide range of organisms to degrade lignin in nature. The lignin modification capability of these enzymes has been investigated and further improved upon with the addition of a mediator, like ABTS. The mediator is used as an electron carrier that can selectively oxidize various functional groups on the lignin macromolecule.

Over the years of research, advancements in lignin modification have stimulated the creation of a myriad of new applications and have ignited a great interest toward displacing our global energy dependence on fossil fuels with biomass sources. In the short term, biorefinery development is likely to incorporate much of the petroleum refinery infrastructure for cost saving reasons. It has taken considerable effort and many breakthroughs in scientific research over decades to expand the capabilities of the petroleum sector from an industry which produced few products into the highly efficient system that is in existence today. Given a similar developmental period and economic investment, biorefining will be certain to evolve higher levels of efficiency and production, and thus, transition into a major provider of clean energy. For lignin to play a successful role in the biorefining initiative, its utility status must be aggressively transformed from a low-priced waste product into a high-quality/high-value raw material source for the production of the extensive range of bulk and specialty chemicals that society demands.

ACKNOWLEDGMENTS

The work was supported by a Strategic Grant from the Natural Sciences and Engineering Research Council of Canada (NSERC). A. Chen acknowledges NSERC and the Canada Foundation of Innovation (CFI) for the Canada Research Chair Award in Environmental and Materials Chemistry.

REFERENCES

[1]   Perlack, R. D.; Wright, L. L.; Turhollow, A. F.; Graham, R. L.; Stokes, B. J.; Erbach, D. C. *Biomass as Feedstock for a Bioenergy and Bioproducts Industry: The Technical Feasibility of a Billion-Ton Annual Supply*. Oak Ridge National Laboratory: Oak Ridge, TN, 2005; pp 1-78.

[2]   Zakzeski, J.; Bruijnincx, P. C. A.; Jongerius, A. L.; Weckhuysen, B. M. *Chem. Rev.* 2010, 110, 3552-3599.

[3]   Rohella, R. S.; Sahoo, N.; Paul, S. C.; Choudhury, S.; Chakravortty, V. *Thermochim. Acta* 1996, 287, 131-138.

[4]   Forss, K. G.; Fremer, K. E. *The nature and reaction of lignin – a new paradigm*; Nord Print: Helsinki, Finland, 2003; pp 1-558.

[5]   Tian, M.; Wen, J.; MacDonald, D.; Asmussen, R.M.; Chen, A. *Electrochem. Commun.* 2010, 12, 527-530.

[6]   Suhas, P. J. M.; Carrott, M. M. L; Carrott, R. *Bioresour. Technol.* 2007, 98, 2301-2312.

[7]   Gosselink, R. J. A.; de Jong, E.; Guran, B.; Abacherli, A. *Ind. Crops Prod.* 2004, 20, 121-129.

[8]   Dence, C. W.; Reeve, D. W. *Pulp Bleaching: Principles and Practice*, TAPPI PRESS: Atlanta, Gorgia 1996.

[9]   Chakar, F. S.; Ragauskas, A. J. *Ind. Crops Prod.* 2004, 20, 131-134.

[10] Gierer, J. *Wood Sci. Technol.* 1980, 14, 241-266.

[11] Mansouri, N. E.; Salvadó, J. *Ind. Crops Prod.* 2006, 24, 8-16.

[12] Alonso, M. V.; Rodriguez, J. J.; Oliet, M.; Rodriguez, F.; Garcia, J.; Gilarranz, M. A. *J. Appl. Polym. Sci.* 2001, 82, 2661-2668.

[13] Alonso, M. V.; Oliet, M.; Rodriguez, F.; Garcia, J.; Gilarranz, M. A.; Rodriguez, J. J. *Bioresour. Technol.* 2005, 96, 1013-1018.

[14] Cetin, N. S.; Özmen, N. *Int. J. Adhes.* 2002, 22, 477-480.

[15] McDonough. T. J. Tappi J. 1993, 76, 186-193.

[16] Muurinen, E. *University of Oulu*, 2000; 3-262.

[17] Dominguez, J. C.; Oliet, M.; Alonso, M. V.; Gilarranz, M. A.; Rodríguez, F. *Ind. Crops Prod.* 2008, 27, 150-156.

[18] Stolarzewicz, A.; Grobelny, Z.; Pisarski, W.; Losiewicz, B.; Piekarnik, B.; Swinarew, A. *Eur. J. Org. Chem.* 2006, 2485-2497.

[19] Hofrichter, M. *Enzyme Microb. Technol.* 2002, 30, 454-466.

[20] Chen, A.; Rogers, E. I.; Compton, R. G. *Electroanalysis* 2010, 22, 1037-1044.

[21] Gellerstedt, G.; Majtnerova, A.; Zhang, L. C. R. *Biol.* 2004, 327, 817-826.

[22] Fu, S. Y.; Singh, J. M.; Wang, S. F.; Lucia, L. A. J. *Wood Chem. Technol.* 2005, 25, 95-108.

[23] Weinstock, I. A.; Hill, C. L.; Atalla, R. H. *D21C 9/00. Pat.,* 5,695,606, Dec. 9, 1997.

[24] Escalante, M.; Rodriguez-Malaver, A. J.; Araujo, E.; Gonzalez, A. M.; Rojas, O. J.; Penaloza, N.; Bullon, J.; Lara, M. A.; Dmitrieva, N.; Perez-Perez, E. *J. Environ. Biol.* 2005, 26, 709-718.

[25] Vasile, C.; Iwanczuk, A.; Frackoviak, S.; Cazacu, G.; Constantinescu, G.; Kozlowski, M. *Cellul. Chem. Technol.* 2006, 40, 345-351.

[26] Feldman, D; Lacasse, M; Beznaczuk, L. M. Lignin-polymer systems and some applications. *Progress in Polymer Science*, 1986, 12, 271-276.

[27] Gandini, A.; Belgacem, M. N.; Guo, Z. X.; Montanari, S. Lignins as macromonomers for polyesters and polyurethanes; Hu, T. Q. *Chemical Modification properties and Usage of Lignin*. Kluwer Academic Plenum Publishers: New York, NY, 2002; pp 58-80.

[28] Zhao, B.; Chen, G.; Liu, Y.; Hu, K.; Wu, R. *J. Mater. Sci. Lett.* 2001, 20, 859-862.

[29] Sailaja, R. R. N.; Deepthi, M. V. *Mater. Des.* 2010, 31, 4369-4379.

[30] Aoyagi, M.; Funaoka, M. *J. Photochem. Photobiol.* A 164, 53-60.

[31] Funaoka, M. *Polym. Inter.* 1998, 47, 277-290.

[32] Xiao, B.; Sun, X. F.; Sun R. *Polym. Degrad. Stab.* 2001, 71, 223-231.

[33] Hill, C. A. S.; Mallon, S. J. Wood Chem. Technol. 1998, 18, 299-311.

[34] Timar, M. C.; Mihai, M. D.; Maher, K.; Irle, M. *Holzforschung*, 2000, 54, 71-76.

[35] Baklanova, O. N.; Plaksin, G. V.; Drozdov, V. A.; Duplyakin, V. K.; Chesnokov, N. V.; Kuznetsov, B. N. *Carbon* 2003, 41, 1783-1800.

[36] Fierro, V.; Torné-Fernández, V.; Celzard, A. *Microporous Mesoporous Mater.* 2007, 101, 419-431.

[37] Kobayakawa, K; Satio, Y.; Nakamura, S.; Fujishima, A. *Bull. Chem. Soc. Jpn.* 1989, 62, 3433-3436.

[38] Tian, M.; Wu, G. S.; Adams, B.; Wen, J. L.; Chen, A. C. *J. of Phys. Chem.* C 2008, 112, 825-831.

[39] Egerton, T. A.; Christensen, P. A.; Harrison, R. W.; Wang, J. W. *J. Appl. Electrochem.* 2005, 35, 799-813.

[40] Xiong, C. R.; Balkus Jr., K. *J. Phys. Chem.* C 2007, 111, 10359-10367.

[41] Pérez, M.; Torrades, F.; García-Hortal, J. A.; Domènech, X.; Peral, J. *J. Photochem. Photobiol.* A 1997, 109, 281-286.

[42] Zhang, Y. H.; Xu, H. L.; Xu, Y. X.; Zhang, H. X.; Wang, Y. G. *J. Photochem. Photobiol. A* 2005, 170, 279-285.

[43] Leng, W. H.; Liu, H.; Cheng, S.; Zhang, J. Q.; Cao, C.N. *J. Photochem. Photobiol. A* 2000, 131, 125-132.

[44] Legrini, O.; Oliveros, E; Braun, *A. M. Chem. Rev.* 1993, 93, 671-698.

[45] Portjanskaja, E.; Preis, S.; Kallas, *J. Int. J. Photoenergy* 2006, 8, 1-7.

[46] Perez, D. D.; Castellan, A.; Grelier, S.; Terrones, M. G. H.; Machado, A. E. H.; Ruggiero, R.; Vilarinho, A. L. *J. Photochem. Photobiol.* A 1998, 115, 73-80.

[47] Antunes, C. S. A.; Bietti, M.; Salamone, M.; Scione, N. *J. Photochem. Photobiol., A* 2004, 163, 453-462.

[48] Tanaka, K.; Calanag, R. C. R.; Hisanaga, T. *J. Mol. Catal. A: Chem.* 1999, 138, 287-294.

[49] Ksibi, M.; Ben Amor, S.; Cherif, S.; Elaloui, E.; Houas, A.; Elaloui, M. *J. Photochem. Photobiol. A* 2003, 154, 211-218.

[50] Dahm, A.; Lucia, L. A. *Ind. Eng. Chem. Res.* 2004, 43, 7996-8000.

[51] Chang, C.; Ma, Y.; Fang, G.; Chao, A. C.; Tsai, M.; Sung, H. *Chemosphere* 2004, 56, 1011-1017.

[52] Ma, Y. S.; Chang, C. N.; Chiang, Y. P.; Sung, H. F.; Chao, A. C. *Chemosphere* 2008, 71, 998-1004.

[53] Linsebigler, A. L.; Lu, G.; Yates Jr., J. T. *Chem. Rev.* 1995, 95, 735-758.

[54] Machado, A. E. H.; Furuyama A. M.; Falone S. Z.; Ruggiero R.; Perez D. D.; Castellan A. *Chemosphere* 2000, 40 115-124.

[55] Kansal, S. K.; Singh, M.; Sud, D. *J. Hazard. Mater.* 2008, 153, 412-417.

[56] Ohnishi, H.; Matsumura, M.; Tsubomura, H.; Iwaski, M. *Ind. Eng. Chem. Res.* 1989, 28, 719-724.

[57] Mansilla, H. D.; Villasenor, J.; Maturana, G.; Baeza, J.; Freer, J.; Duran, N. *J. Photochem. Photobiol. A* 1994, 78, 267-273.

[58] Villasenor, J.; Mansilla, H. D. *J. Photochem. Photobiol. A* 1996, 93, 205-209.

[59] Reyes, J.; Dezotti, M.; Mansilla, H.; Villasenor, J.; Esposito, E.; Duran, N. *Appl. Catal., B* 1998, 15, 211-219.

[60] Pal, B.; Sharon, M. *Mater. Chem. Phys.* 2002, 76, 82-87.

[61] Villasenor, J.; Duran, N.; Mansilla, H. D. *Environ. Technol.* 2002, 23, 955-959.

[62] Gouvea, C. A. K.; Wypych, F.; Moraes, S. G.; Duran, N.; Peralta-Zamora, P. *Chemosphere* 2000, 40, 427-432.

[63] Tolba, R.; Tian, M.; Wen, J.; Jiang, Z. H.; Chen, A. *J. Electroanal. Chem.* 2010, 649. 9-15.

[64] Matyasovszky, N.; Tian, M.; Chen, A. *J. Phys. Chem. A* 2009, 113, 9348-9353.

[65] Adams, B.; Tian, M.; Chen, A. *Electrochim. Acta* 2009, 52, 1491-1487.

[66] Parpot, P.; Bettencourt, A. P.; Carvalho, A. M.; Belgsir, E. M. *J. Appl. Electrochem.* 2000, 30, 727-731.

[67] Smirnov, V. A.; Kovalenko, E. I. *Sov. Electrochem.* 1992, 28, 485-497.

[68] Trasatti, S. *Electrochim. Acta* 2000, 45, 2377-2385.

[69] Pardini, V. L.; Smith, C. Z.; Utley, J. H. P.; Vargas, R. R.; Viertler, H. *J. Org. Chem.* 1991, 56, 7305-7313.

[70] Blanco, M. A.; Negro, C.; Tijero, J.; deJong A. C. M. P.; Schmal, D. *Sep. Sci. Tech.* 1996, 31, 2705-2712.

[71] Negro, C.; Blanco, A.; Tijero, J.; Villarin, S.; van Erkel, J.; de Jong, M. C. P. *Cellul. Chem. Technol.* 2005, 39, 129-136.

[72] Wang, B.; Gu, L.; Ma, H. *J. Hazard. Mater.* 2007, 143, 198-205.

[73] Soloman, P. A.; Basha, C. A.; Velan, M.; Balasubramanian, N. *J. Chem. Technol. Biotechnol.* 2009, 84, 1303-1313.

[74] Mahesh, S.; Prasad, B.; Mall, I. D.; Mishra, I. M. *Ind. Eng. Chem. Res.* 2006, 45, 2830-2839.

[75] Chiang, L. C.; Chang, J. E.; Tseng, S. C. *Water Sci. Technol.* 1997, 36, 123-130.

[76] Dominguez-Ramos, A.; Aldaco, R.; Irabien, A. *Ind. Eng. Chem. Res.* 2008, 47, 9848-9853.

[77] Kovalenko, E. I.; Kotenko, N.P.; Sherstyukova, N.D. *J. Appl. Chem.* USSR 1988, 61, 768-775.

[78] Kovalenko, E. I.; Popova, O. V.; Aleksandrov, A. A.; Galikyan, T. G. Russ. *J. Electrochem.* 2000, 36, 706-711.

[79] Drakesmith, F.G. *Electrofluorination of Organic Compounds; Topics in Current Chemistry.* Springer Berlin/Heidelberg: New York, NY, 1997; 193, pp 198-242.

[80] Hampp, N. H. *B27K 5/02. Pat.,* US 6,187,170 B1, Feb. 13, 2001.

[81] Rochefort, D.; Leech D.; Bourbonnais, R. *Green Chem.* 2004, 6, 14-24.

[82] Kim, H. C.; Mickel, M.; Bartling S; Hampp, N. *Electrochim. Acta* 2001, 47, 799-805.

[83] Mickel, M.; Kim, H. C.; Noll, S.; Hampp, N. *J. Electrochem. Soc.* 2003, 150, E595-E600.

[84] Rochefort, D.; Bourbonnais, R.; Leech, D.; Renaud, S.; Paice, M. *J. Electrochem. Soc.* 2002, 149, D15-D20.

[85] Bourbonnais, R.; Paice, M. *J. Electrochem.* Soc. 2004, 151, E246-E249.

[86] Bertazzoli, R.; Pelegrini, R. *Quim. Nova* 2002, 25, 477-482.

[87] Pelegrini, R.; Reyes, J.; Duran, N.; Zamora, P. P.; de Andrade, A. R. *J. Appl. Electrochem.* 2000, 30, 953-958.

[88] Bourbonnais, R.; Leech, D.; Paice, M. G. *Biochim. Biophys. Acta* 1998, 1379, 381-390.

[89] Shumakovich, G. P.; Shleev, S. V.; Morozova, O. V.; Khohlov, P. S.; Gazaryan, I. G.; Yaropolov, A. I. *Bioelectrochem.* 2006, 69, 16-24.

[90] Q. Weihua, C. Hongzhang. *Bioresource Technol.* 2008, 99, 5480-5484.

[91] Prasetyo, E. N.; Kudanga, T.; Ostergaard, L.; Rencort, J.; Gutiérrez, A.; del Río, J. C.; Santos, J. I.; Nieto, L.; Jiménez-Barbero, J.; Martínez, A. T.; Li, J.; Gellerstedt, G.; Lepifre, S.; Silva, C.; Kim, S. Y.; Cavaco-Paulo, A.; Klausen, B. S.; Lutnaes, B. F.; Nyanhongo, G. S.; Guebitz, G. M. *Bioresour. Technol.* 2010, 101, 5054-5062.

[92] Oyadomari, M.; Shinohara, H.; Johjima, T.; Wariishi, H. Tanaka, H. *J. Mol. Catal. B: Enzym.* 2003, 21, 291-297.

[93]   Sena-Martins, G.; Almeida-Vara, E.; Duarte, J. C. *Ind. Crops Prod.* 2008, 27, 189-195.

[94] D'Annibale, A.; Crestini, C.; Mattia, E. D.; Scrmanni, G. *J. Biotechnol.* 1996, 48, 231-239.

[95] Liu, J.; Weiping, Y.; Lo, T. *EJB* 1999, 2, 82-87.

[96] Hüttermann, A.; Mai, C.; Kharazipour, A. *Appl. Microbiol. Biotechnol.* 2001, 55, 387-394.

In: Lignin
Editor: Ryan J. Paterson

ISBN 978-1-61122-907-3
© 2012 Nova Science Publishers, Inc.

Chapter 11

PYROLYSIS CHARACTERISTICS OF DIFFERENT KINDS OF LIGNINS

*Pablo R. Bonelli[1,2] and Ana Lea Cukierman[1,2,3]**

[1]Programa de Investigación y Desarrollo de Fuentes Alternativas de Materias Primas y Energía (PINMATE) - Depto. de Industrias, Facultad de Ciencias Exactas y Naturales, Universidad de Buenos Aires. Int. Güiraldes 2620, Ciudad Universitaria. (C1428BGA) Buenos Aires, Argentina
[2]Consejo Nacional de Investigaciones Científicas y Técnicas (CONICET), Av. Rivadavia 1917. (C1033AAJ) Buenos Aires, Argentina
[3]Cátedra de Farmacotecnia II. Depto. de Tecnología Farmacéutica, Facultad de Farmacia y Bioquímica, Universidad de Buenos Aires. Junín 956. (C1113AAD) Buenos Aires, Argentina

ABSTRACT

The present chapter deals with pyrolysis characteristics of different kinds of lignins, focusing on an industrial raw lignin arising from the Kraft pulping process, a commercial alkali lignin, and Klason lignins lab-isolated from two lignocellulosic biomasses with different lignin contents (27 and 57 wt%), emerging from the processing of agro-industrial products. Characterization of the lignins includes determination of ash content, elemental composition, Fourier-transform infrared (FT-IR) spectra, and surface morphological features by scanning electronic microscopy (SEM). Pyrolysis characteristics of the different lignin samples as well as of the whole biomasses from which they are obtained, in the case of Klason lignins, are comparatively examined by non-isothermal thermogravimetric analysis from room temperature up to 1000 °C. In order to investigate possible effects of mineral matter inherently present in the industrial Kraft lignin on its pyrolytic behaviour, pyrolysis characteristics are also determined using samples prior subjected to demineralization by a mild acid treatment. The industrial Kraft lignin possesses the highest contents of ash (16 wt%) and elemental carbon (62.2 wt%) among all the investigated raw lignins. A similar pyrolytic behavior is found for the

---

* E-mail: analea@di.fcen.uba.ar; anacuki@ffyb.uba.ar.

industrial raw Kraft lignin and the commercial alkali one. It is characterized by differentiated thermal degradation domains, as evidenced by three peaks in reaction rate profiles, successively attributable to moisture evolution, primary and secondary pyrolysis, with progressive increase in temperature. Mineral matter reduction of the industrial Kraft lignin induces some structural changes, as suggested by SEM images and FT-IR spectra, and noticeable modifications in its pyrolytic behavior, leading to shift primary pyrolysis to higher temperatures, to increase the maximum primary pyrolysis rate, and to inhibit secondary pyrolysis. Pyrolysis characteristics for binary mixtures composed of equal proportions of the commercial alkali lignin and polyethylene in powder form, as a representative major polymeric waste of massive post-consumed plastics, are also examined following some current research trends towards alternative energy generation based on the advantageously favourable environmental nature of bio-resources and the higher energy content of synthetic polymers. No interactions between the lignin and polyethylene are found, the pyrolytic behaviour of the mixtures arising from independent thermal degradation of the individual constituents. On the other hand, the two Klason lignins separated from sawdust of *Aspidosperma australe* wood and nutshells from *Bertholletia excelsa*, exhibit different ash contents and elemental compositions as well as noticeable differences in their pyrolysis characteristics depending on the biomass source and with respect to the pyrolytic behaviour of the untreated parent biomasses. Compared to the raw alkali lignins, the Klason lignins do not seem to undergo secondary pyrolysis and are more resistant to thermal degradation, likely due to more condensed chemical structures related to the method applied for isolation. Overall, the results highlight the marked influence of both the botanical origin of the bio-resource and extraction method used to obtain the lignins on their physicochemical characteristics and pyrolytic behaviour.

1. INTRODUCTION

Thermo-chemical conversion technology is an interesting route to generate alternative, renewable energy and other useful products from natural bio-resources and industrial by-products (Wongsiriamnuay and Tippayawong, 2010; Jeguirim and Trouvé, 2009; Balat et al., 2009; Di Blasi, 2008; Basso et al., 2005). In particular, lignin represents an attractive option since it is the only renewable aromatic resource in nature and the second most abundant biomass component. Its main function is cementing the cellulose fibres in plants (Nowakowski et al., 2010; Mancera et al., 2010).

Lignin is also the major by-product of second generation bio-ethanol production, and huge quantities of surplus lignin from black liquors emerge from the industrial separation of cellulose from wood for pulp and paper by the Kraft process, attracting growing interest as a potential source of aromatic hydrocarbons for bio-fuels and chemicals (Jiang et al., 2010). Around 63×10^4 metric tons/year of lignin are produced worldwide by pulping (Mohan et al., 2006), mostly employed as in-house low grade fuel. Larger plant capacities and optimization of the pulping process to improve cost effectiveness lead to more by-product lignin than the amount needed for energy requirements. Accordingly, use and/or conversion of surplus lignin from black liquors to added-value products might represent a profitable alternative to incineration (Fierro et al., 2005). In this direction, studies for several different novel applications have been reported in recent years (Mancera et al., 2010; Betancur et al., 2009; El Mansouri et al., 2007; Ramírez et al., 2007; Suhas et al., 2007; El Mansouri and Salvadó,

2006; Baumlin et al., 2006; Braun et al., 2005; Li and Sarkanen, 2005; Velásquez et al., 2003).

Thermo-chemical conversion of lignin by pyrolysis, basically a polymeric structure cracking process in inert atmosphere, constitutes an interesting route since the process enables to obtain several valuable products. From the viewpoint of engineering applications, pyrolysis products are often lumped into three groups: permanent gases, a pyrolytic liquid (bio-oil or tar) and a carbon enriched solid product (char or bio-char) (Di Blasi, 2008). Accordingly, pyrolysis characteristics of lignin are relevant to the efficient design, operation, and/or modeling of full-scale conversion units. Furthermore, interest in development of new processes for co-transformation of lignin with synthetic polymers aimed at obtaining alternative energy, carbon composite materials and/or chemical compounds has increased in recent years (Ohmukai et al., 2008; Sharypov et al., 2003; Sharypov et al., 2002).

Pyrolysis of lignin is complex, and depends on lignin composition and process conditions, such as heating rate, reaction temperature, carrier gas flow rate (Brodin et al., 2010; Yang et al., 2007; Ferdous et al., 2002). Despite several studies exploring different aspects on lignin pyrolysis have been reported, a thorough knowledge on the pyrolytic behaviour of lignin has not been attained and it is still difficult to predict pyrolysis characteristics for a particular lignin since uncertainties about lignin structure and the difficulty in its isolation introduce additional factors of complexity (Wang et al., 2009; Baptista et al., 2006). Besides, studies concerned with the pyrolytic behavior of lignins isolated from lignocellulosic biomasses are still restricted. Low molecular weight model compounds related to structural elements of lignin polymers and/or commercial samples have been employed, but it has been pointed out that they are far from representing the real circumstance (Haykiri-Acma et al., 2010).

Lignins are complex racemic polymers derived mainly from three hydroxycinnamyl alcohol monomers that differ in their degree of methoxylation: p-coumaryl, coniferyl and sinapyl alcohols, which have the amino acid phenylalanine as their bio-synthetical precursor. These species are usually called monolignols, and produce p-hydroxyphenyl, guaiacyl, and syringyl phenylpropanoid units respectively, that are the main building blocks of the lignin polymer. Two main classes of lignin are usually considered: hardwood lignin (Dicotyledonous angiosperm), mainly made up of guaiacyl and syringyl units with traces of p-hydroxyphenil units, and softwood lignin (Gymnosperm), mostly composed of guaiacyl units with low amounts of p-hydroxyphenil. (Faravelli et al., 2010). Besides, the presence of several other phenolics which are frequently used as lignification monomers by certain species of plants makes lignin structure even more complex. It has been extensively examined in the literature (Walker, 2006; Mc Carthy, 2000).

Various methods have been developed to separate the macromolecular bio-polymeric components in lignocellulosic biomass, by employing different chemical reagents. However, the recovery of lignin from lignocellulosic materials is generally difficult due to condensation and oxidation reactions taking place during the isolation processes (Haykiri-Acma et al., 2010). Since each separation method modifies to some degree the chemical structure of the naturally occurring lignin, it is conventionally named after the method of separation. The separation methods can be divided into two categories, mainly including dissolution of lignin into a solution, and hydrolysis of cellulose and hemicellulose by acids leaving lignin as an insoluble residue. In the first group, some important lignins are: alkali lignin, lignosulfonates, Organosolv lignin, and milled wood lignin. All these lignins are co-produced in the pulp and paper industry. Important lignins in the second category are Klason lignin and hydrolytic lignin (Jiang et al., 2010).

Within this context, the present work deals with pyrolysis characteristics of different lignins arising from different separation methods. It focuses on an industrial raw lignin emerging from the Kraft pulping process, and Klason lignins lab-isolated from two lignocellulosic wastes with different contents of lignin, generated in large quantities from the processing of agro-industrial products. The pyrolytic behavior of the lignins is determined by thermogravimetric analysis. Demineralization of the industrial Kraft lignin by applying a mild acid pre-treatment is also carried out in order to examine the effect of the presence of mineral matter on its pyrolysis characteristics. Results for the pyrolysis of the raw and demineralized Kraft lignins are compared with those assessed for a commercially available alkali lignin. Besides, pyrolysis characteristics of binary mixtures composed of equal proportions of the commercial alkali lignin and polyethylene in powder form, the latter representing a major polymeric waste arising from post-consumed plastics, are also investigated and discussed in terms of the behavior and main physico-chemical properties of the individual constituents. Growing importance is given to co-pyrolysis of bio-resources with polymeric wastes arising from massive consumption of plastic products towards sustainable, efficient technologies for clean, alternative energy generation in order to take advantage of the more favorable environmental nature of bio-resources and the higher energy content of polymers. On the other hand, the pyrolysis characteristics of the Klason lignins are examined accounting for the biomass feedstock used for their isolation, and compared with those assessed for the whole biomasses from which they are obtained. Finally, char yields at the highest temperature for the pyrolysis of the alkali and Klason lignins are comparatively analyzed.

2. EXPERIMENTAL SECTION

2.1. Materials and Conditioning

The materials used comprised:

- An industrial raw lignin in powder form (particle diameter between 27 and 70 μm) arising from the Kraft pulping process (RKL). It was supplied by Lignotech Ibérica (Torrelavega, Cantabria, Spain).
- Demineralised Kraft lignin. Demineralisation of the industrial raw lignin was carried out by treatment with sulphuric acid solution (1 wt%) for 1 h, in a proportion of 20 mL of solution per gram of sample, followed by washing extensively with de-ionized water until neutral pH in washing water was attained. The demineralised Kraft lignin is denoted as DKL.
- A commercial alkali lignin (Mw: 28,000; Mn: 5,000) purchased from Sigma Aldrich. This lignin is designated as CAL. Binary mixtures composed of this lignin and polyethylene (PE) in powder form (supplied by Wrigley Fibres, Wellington, Somerset, U.K.) were also prepared. They were obtained by physically mixing the commercial lignin with PE in equal proportions.

– Klason lignins lab-isolated from two lignocellulosic wastes possessing different contents of lignin. The lignocellulosic wastes used and the procedure applied for lignin isolation, are separately depicted below.

2.2. Lignocellulosic Biomasses and Lab-Isolation of Klason Lignins

Lignocellulosic wastes emerging from the processing of two agro-industrial products were employed for lab-isolation of lignin samples: sawdust from *Aspidosperma australe* wood (AWS) and shells from Brazil nuts (BNS). *Aspidosperma australe*, commonly known as Guatambú, is a tree species indigenous to South America, found in Argentine, Brazil, and Paraguay. The wood has many uses in carpentry, owing to its good characteristics, and it is also employed as fuel in some regions. Brazil nut is the fruit of *Bertholletia excelsa* tree. It belongs to the Lecythidaceae family and grows in a vast zone of South America, mostly in the region of Pará (Brazil). The fruit is a large spherical woody capsule or pod. Inside each fruit pod there are several nuts with their own individual shell. The tree can produce around 300 or more of these fruit pods (Bonelli et al., 2001). Brazilian annual production in 2007 was 30,000 tons, generating huge amounts of shells as waste material (de Oliveira Brito et al., 2010).

The Klason method was applied in order to isolate lignin from the lignocellulosic wastes. This procedure has become one of the standard methods for the determination of lignin in wood (Jiang et al., 2010). Briefly, the wastes were subjected to extraction with benzene-ethanol following TAPPI T 204 OM-88. Elimination of fats, resins, photosterols, non-volatile hydrocarbons, waxes, and some water-soluble compounds, i.e. low molecular weight carbohydrates, takes place in the extraction stage. Thus, wastes free from extractives were obtained. Only a small amount of minerals, pectines, and inclusions, such as calcium oxalate, remained.

Table 1. Chemical characteristics of the lignocellulosic biomass employed for isolation of Klason lignins

Characteristic / Sample	AWS	BNS
Biopolymer content (wt%)[a]		
Lignin	27	57
Holocellulose	73	43
Proximate analysis (wt%)[b]		
Volatile matter	72.7	76.1
Ash	1.0	1.7
Fixed carbon[d]	26.3	22.2
Elemental analysis (wt%)[c]		
Carbon	48.2	50.0
Hydrogen	5.6	5.8
Nitrogen	0.9	0.7
Oxygen[d]	45.3	43.5

[a] Dry and extractive-free basis; [b] Dry basis; [c] Dry and ash-free basis;
[d] Estimated by difference.

The extractive-free samples were then subjected to acid hydrolysis by treatment with 72 %volume sulphuric acid at boiling temperature, following TAPPI T 222 OM-88 standard. In this way, the acid hydrolyzes almost all the polysaccharides leaving acid-insoluble solid lignin. Isolation of holocellulose constituting the wastes, namely cellulose and hemicellulose, was performed by acid hydrolysis using $HClO_2$.

Table I reports contents of lignin and holocellulose for both wastes. It also includes ash contents and elemental compositions. The former were determined according to ASTM standards, whereas the latter ones were assessed using a Carlo Erba EA 1108 elemental analyzer. As seen in Table I, content of lignin in the nutshells is remarkably higher than the one for the wood sawdust. It is similar to the one recently reported for hazelnut shells (Haykiri-Acma et al., 2010) on the same basis. Besides, the lignin content of the wood sawdust is comparable to values earlier reported for other species (Cukierman, 2007; Basso et al., 2004).

2.3. Characterization of Lignin Samples

Ash content and elemental composition of the raw and demineralised industrial Kraft lignins, the commercial alkali sample, and the lab-isolated Klason lignins were assessed. The former was determined according to ASTM standard (El Mansouri and Salvadó, 2006), while a Carlo Erba Model EA 1108 instrument (Carlo Erba Strumentazione SPA, Milan, Italy) was used for the latter. Results are reported as average values of triplicate determinations, where the standard error did not exceed 5%.

In addition, Fourier transform infrared (FT-IR) spectroscopy was applied to characterize the lignin samples and the biomasses used for isolation of the Klason lignins. The spectra were recorded using a Nicolet Magna IR 550 spectrometer (Thermo Nicolet Inc., Madison, WI, USA) within the wavenumber range of 400-4000 cm^{-1}. Each sample was mixed with KBr and then ground in an agate mortar at an approximate ratio of 1:100 for preparation of the pellets. The resulting mixture was finally pressed. The background obtained from a scan of pure KBr was automatically subtracted from the sample spectra. The aforementioned procedure was then followed.

Surface morphology of the lignin samples was examined by field emission scanning electron microscopy (SEM), using a FE-SEM-Carl Zeiss, Supra 40 microscope (Carl Ziess AG, Standort, Göttingen, Deutschland). Before the measurements, all the samples were attached to mounting stubs, dried under vacuum, and sputter coated with Au.

2.4. Thermogravimetric Measurements for the Pyrolysis of Lignins

Pyrolysis characteristics of the alkali lignins, lignin-PE mixtures, Klason lignins, and of the lignocellulosic biomasses used for extraction of the latter were examined by non-isothermal thermogravimetric (TG) analysis. For all the samples, average particle diameter lower than 44 μm was employed. Measurements were carried out from ambient temperature up to 1000 °C operating under flowing nitrogen. A TA Instruments SDT Q600 thermogravimetric balance (TA Instruments-Waters LLC, New Castle, DE, USA) equipped with a N_2 mass flow controller and a data acquisition system was used to carry out the measurements.

Preliminary experiments were first conducted in order to assess operating conditions for which diffusional effects were negligible. Accordingly, masses of 10 mg, N_2 flow rates of 400 mL/min, and a heating rate of 10 °C/min were employed. Further details of the different experimental procedures employed may be found in previous own works (Kim et al., 2010; González et al., 2008; Bonelli et al., 2007; Bonelli et al., 2003).

3. RESULTS AND DISCUSSION

3.1. Characteristics of the Lignin Samples

Ash content and elemental composition determined for the industrial raw and demineralised Kraft lignins, the commercial alkali lignin, as well as for the Klason lignins extracted from the wood sawdust and the nutshells, are reported in Table II. As seen, the industrial Kraft lignin has the highest ash content among all the investigated samples. The data in the table show that the acid treatment of the RKL led to an effective reduction in the ash content, in agreement with other reported results (El Mansouri and Salvadó, 2006).

Table 2. Ash content (dry-basis) and elemental composition (dry, ash-free basis) of the lignin samples investigated

Sample	RKL	DKL	CAL[b]	AWL	BNL
Ash (wt %)	16.8	0.2	3.0	0.1	6.0
Carbon (wt %)	62.2	64.6	51.2	58.9	52.7
Hydrogen (wt %)	5.5	5.4	4.8	6.0	4.7
Nitrogen (wt %)	0.3	0.5	-	0.6	0.5
Sulphur (wt %)	1.9	1.4	2.2	0.2	1.4
Oxygen[a] (wt %)	30.1	28.1	41.8	34.3	40.7

[a] Estimated by difference. [b] As provided by the supplier.

Compared with the commercial alkali sample, the industrial Kraft lignins, both the raw and demineralised ones, have higher %C and %H and lower %O. For the industrial raw sample, the elemental contents are similar to values reported in the literature for other alkali lignins (Jiang et al., 2010). Contents of elemental carbon between 57 and 65%, of hydrogen ranging from 5.6 to 6.3%, and of oxygen from 29 to 37%, have been reported for various lignins derived from 28 different species (Faravelli et al., 2010). Accordingly, elemental compositions of the Kraft lignins investigated in the present study lie fairly well within the range of reported values.

In turn, the Klason lignins possess lower contents of elemental carbon than the industrial Kraft lignin, especially the one extracted from the nutshells, but higher than the commercial sample. Major contents of the lignin isolated from the Brazil nutshells are also close to those reported for other samples isolated via acid hydrolysis of softwood (Nowakowski et al., 2010). Overall, elemental compositions point to different chemical structures of the lignin samples.

FT-IR spectra for the industrial raw and demineralised Kraft lignins and for the commercial alkali sample are comparatively illustrated in Figure 1, whereas those for the Klason lignins and the biomasses from which they were extracted are shown in Figure 2. As may be appreciated, all the spectra display a broad band at 3600–3000 cm^{-1} assigned to –OH groups in aliphatic and phenolic structures. Characteristic bands are mainly found at 1500–1600 cm^{-1} and at ca. 1450 cm^{-1} arising from deformations and aromatic ring vibrations. Below 1400 cm^{-1}, spectral analysis becomes more difficult, with the complex bands observed probably arising from contribution of various vibration modes (Boeriu et al., 2004).

The spectrum for the industrial raw lignin is quite similar to the one determined for the commercial lignin, exhibiting certain differences in the relative intensities of the peaks in the fingerprint region, below 1600-1500 cm^{-1}. The raw and demineralised lignins show, in general, characteristic infrared bands in accordance with other reported results (Fierro et al., 2007). A main difference concerns the peak at 1725-1750 cm^{-1} attributable to carboxylic groups in association with the corresponding complementary bands at ca. 1400, 1300 and 900 cm^{-1}; they might be due to the removal of metals composing the minerals present in the raw lignin, related to carbonyl groups (Betancur et al., 2009).

Besides, as seen in Figure 2, the spectrum for the nutshells with a preponderantly high content of lignin (57 wt%), as reported in Table II is not substantially different from that determined for the derived lignin in the region 800-1200 cm^{-1}. On the contrary, differences in the same region are more noticeable between the wood sawdust and the lignin extracted from this biomass.

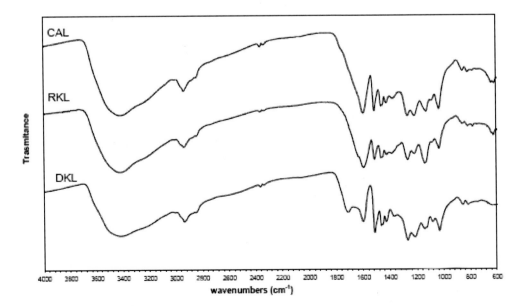

Figure 1. Fourier-transform infrared (FT-IR) spectra for the industrial raw (RKL) and demineralized (DKL) Kraft lignins, and for the commercial alkali lignin (CAL).

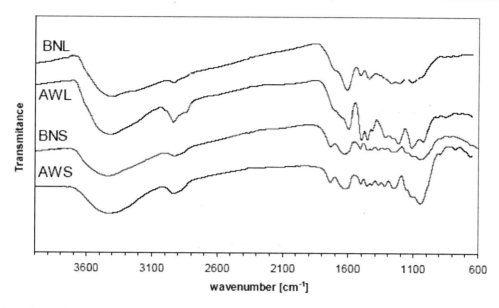

Figure 2. Fourier-transform infrared (FT-IR) spectra for the Brazil nutshells (BNS), sawdust from *Aspidosperma Australe* wood (AWS), and for the Klason lignins extracted from the wood sawdust (AWL) and the nutshells (BNL).

They may be related to C–O, C–C and C–OH bonds, which are characteristic of polysaccharides. As shown in Table II, the content of lignin composing the sawdust is lower than that of holocellulose. Main differences between the lignins extracted from both biomasses may be noticed in the peaks at 1550 and 1450 cm^{-1}. Table III summarizes some characteristic absorption bands of the lignins studied along with their corresponding possible structural units.

Table 3. Characteristic IR absorption bands of the investigated lignins

Wave number (cm^{-1})	Possible Structural Units
3012 – 3042	-OH stretching
3006 – 2840	-CH stretching (aromatic / aliphatic)
1750 – 1720	-C=O stretching carboxylic ester or acid
1628 – 1600	-C=C aromatic ring vibration
1515 – 1504	-C=C aromatic / polyaromatic
1460 – 1450	-CH_2, -CH_3 deformations
1430 – 1425	Aromatic ring vibration (general)
1275 – 1270	Guaiacyl ring breathing
1270 – 1220	C-C, C-O stretching
1100 – 1040	C-H, C-O deformation
900 – 800	C-H out of plane of aromatic rings

SEM images of the lignin samples are shown in Figures 3 to 7, at two different magnifications (2000x and 40000x). Figure 3 (a-b) and Figure 4 (a-b) portray the surfaces corresponding to the industrial raw and commercial alkali lignins, respectively.

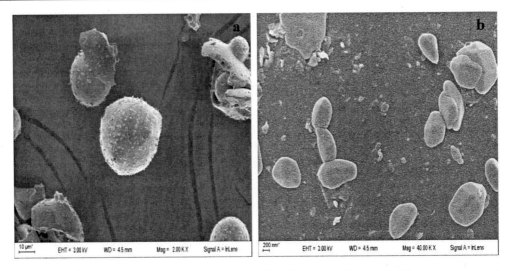

Figure 3. SEM micrographs of the raw industrial Kraft lignin at: a) 2000x; b) 40000x.

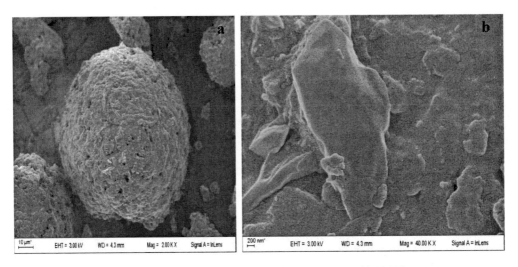

Figure 4. SEM micrographs of the commercial alkali lignin at: a) 2000x; b) 40000x.

As seen from the images at the lower magnification (Figures 3a and 4a), both samples show similar features. They are characterized by round-shaped particles and some oblong ones. Smooth surfaces can be visualized at the higher magnification in both cases (Figures 3b and 4b). Comparison of the micrographs of the industrial raw Kraft lignin (Figure 3 a-b) with those of the sample after demineralization, shown in Figure 5 (a-b), points to some alterations likely occasioned by the acid treatment applied to reduce inorganic matter content. At the higher magnification, the surface looks rougher and certain degree of disintegration is apparent, as inferred from some fragmented particles presumably remaining after minerals solubilisation.

Figures 6 (a-b) and 7 (a-b) illustrate typical representative SEM images for the Klason lignins isolated from the wood sawdust and the nutshells, respectively. The appearance of the surface of these lignins differs considerably from that observed for the raw alkali lignins (Figures 3 and 4).

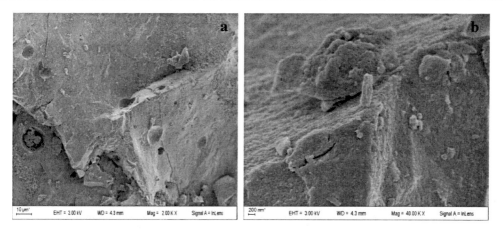

Figure 5. SEM micrographs of the industrial Kraft lignin after demineralization at: a) 2000x; b) 40000x.

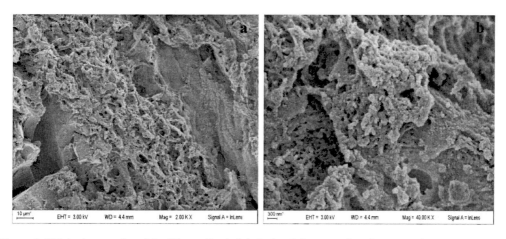

Figure 6. SEM micrographs of the Klason lignin lab-isolated from sawdust from *Aspidosperma Australe* wood at: a) 2000x; b) 40000x.

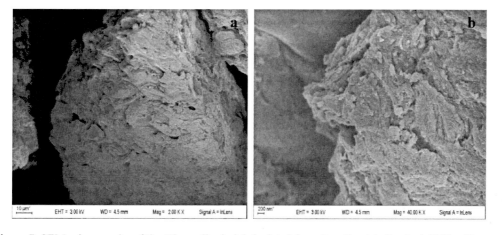

Figure 7. SEM micrographs of the Klason lignin lab-isolated from Brazil nutshells at: a) 2000x; b) 40000x.

It may be attributed to differences in the bio-resources from which they were obtained and in the method applied for their isolation. The latter involving hydrolysis of cellulose and hemicellulose in strong acid conditions leaves lignin as an acid-insoluble residue. As seen, the surfaces look rough and show cavities and/or cracks. Also, differences between the Klason lignins depending on the biomass botanical origin may be noticed.

3.2. Pyrolysis Characteristics of the Alkali Lignins. Effect of Demineralization

Typical representative non-isothermal TG curves for the pyrolysis of the industrial raw and demineralised Kraft lignins, and the commercial sample, are comparatively illustrated in Figure 8.

In this figure, instantaneous weight fractions (w), namely the ratio between the instantaneous weight and the initial weight of the sample, normalized on ash-free basis, are represented as a function of the temperature (T). The reaction rate profiles for the same samples, as obtained by differentiation of the normalized weight losses-time curves, are shown in Figure 9. Main pyrolysis characteristics obtained from TG curves and reaction rate profiles for the lignins are summarized in Table IV.

Table 4. Characteristics of TG curves and reaction rate profiles for the pyrolysis of the industrial raw and demineralized Kraft lignins (RKL, DKL), and the commercial alkali lignin (CAL)

Characteristic	RKL	DKL	CAL
T_{onset1} (°C)	20	20	20
T_{peak1} (°C)	60	60	60
T_{end1} (°C)	130	130	130
T_{onset2} (°C)	160	160	160
T_{peak2} (°C)	360	390	370
T_{end2} (°C)	580	650	580
$(-dw/dt)_{peak2}$ (min^{-1})	0.014	0.028	0.020
T_{onset3} (°C)	600	-	600
T_{peak3} (°C)	730	-	730
T_{end3} (°C)	810	-	810
$(-dw/dt)_{peak3}$ (min^{-1})	0.011	-	0.006
w (T = 950°C)	0.33	0.43	0.37

As seen in Figure 8, lignins decompose gradually in a wide temperature range. Variation of the weight fraction with temperature for the RKL takes place in a rather stepped way, whereas a smoother variation is noticed for the DKL and CAL over all the temperature range.

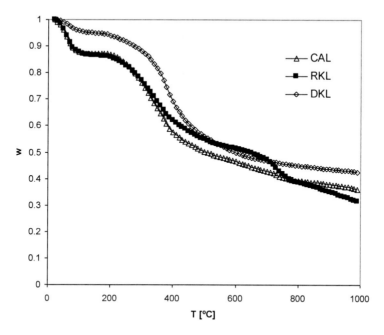

Figure 8. TG curves for the pyrolysis of the industrial raw and demineralized Kraft lignins (RKL, DKL), and for the commercial alkali lignin (CAL).

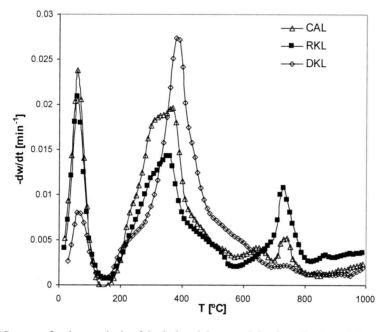

Figure 9. DTG curves for the pyrolysis of the industrial raw and demineralized Kraft lignins (RKL, DKL) and the commercial alkali lignin (CAL).

Small weight losses at the lower temperatures, from room temperature up to 100°C, are due to moisture release. Noticeable changes in weight fractions may be appreciated in the range between 200 °C and ~ 400-500 °C. At higher temperatures, the curves flatten out. Pyrolysis of lignins and related materials generally involve primary and secondary reactions. In primary pyrolysis, products are formed directly from decomposition of the original material. On the other hand, in secondary pyrolysis, volatiles evolved in primary pyrolysis undergo further reactions, such as coking, cracking, and other complex reactions involving free radicals. During thermal decomposition of lignin, relatively weak bonds break at lower temperatures, whereas the cleavage of stronger bonds takes place at higher temperatures (Brodin et al., 2010). As reported in the literature (Faravelli et al., 2010), at temperatures higher than 200 °C lignin pyrolysis takes place by a melt phase radical process in which reactivity is mainly ruled by the competition between initiation, propagation and termination reactions of different propagating radicals. The weakest bonds inside the lignin structure favor radical formation and initiate the radical process. At temperatures higher than 500 °C, and after primary devolatilization, a slow release of the final products takes place before final char formation. The chemistry ruling the system seems to be governed by the progressive formation of poly-cyclic aromatic hydrocarbons, mainly by radical mechanisms. Intermolecular and intramolecular aromatic condensation reactions increase the molecular weight of the macromolecules with continuous release of volatiles and, therefore, with a progressive reduction in O and H contents. The continuous cross-linking and growth of these aromatic structures characterize the formation of the char residue. As temperature increases, the structure further degrades finally releasing hydrogen and carbon monoxide, with the release of CO occurring in the temperature range 400–700 °C. It has also been reported that CO is produced from ether groups, its generation depending on pyrolysis temperature. At lower temperatures, ether bridge joining subunits should be responsible for CO generation, due to their low dissociation energy. At higher temperatures, dissociation of diaryl ethers might additionally cause formation of CO. In turn, CH_4 should be readily produced from weakly bonded methoxy groups ($-OCH_3$). Degradation rate over a range of temperatures seems to arise from cumulative contributions of several independent reactions (Faravelli et al., 2010; Jiang et al., 2010).

DTG curves in Figure 9 show three peaks for the degradation of the raw Kraft lignin and the commercial sample, whereas only two peaks characterize the pyrolysis of the industrial Kraft lignin subjected to demineralization. The first peak corresponds to moisture evolution, whereas the second and third peaks may be attributed to primary and secondary pyrolysis, respectively. Likewise, the temperature at which the reaction rate attains the second maximum value is lower for the raw Kraft lignin (T_{peak2}= 360 °C) than for the commercial lignin (T_{peak2}= 370 °C), whereas it is higher for the demineralized sample (T_{peak2}= 390 °C). The third peak occurs at almost the same temperature (T_{peak3}= 730 °C) for both the raw Kraft lignin and the commercial one.

As inferred from the comparison of the second peak for the raw and demineralized Kraft lignins, mineral matter reduction leads to shift primary pyrolysis to higher temperatures and to duplicate the maximum degradation rate (Table IV, peak 2). In turn, lack of the peak at 730 °C for the demineralized lignin suggests that demineralization inhibits secondary pyrolysis. Similar behaviors have been found by Kumar et al. (2008), who systematically examined the effect of adding alkali salts on the pyrolysis behavior of two different Kraft lignins of different chemical composition, with one of them possessing high ash content (~ 20%). The addition of Na_2CO_3 to both lignins made primary pyrolysis takes place at a lower rate and enhanced the rate for secondary pyrolysis. In particular, these results showed that only the high ash lignin underwent secondary reactions due to the presence of Na_2CO_3 in mineral matter. These reactions were found to be due to interactions of char carbon with alkali metals salts.

In Figure 10, the effect of ash content of the lignins on the maximum degradation rates corresponding to the second and third peaks is illustrated for the three samples investigated. It may be appreciated that the ash content exerts opposite effects. Whereas it leads to decrease maximum reaction rates for the primary pyrolysis (peak 2), the ash content promotes enhancement of those attributed to secondary pyrolysis (peak 3).

Comparison of the results for the pyrolytic behavior between the raw and demineralized Kraft lignins also indicates that mineral matter reduction leads to alter the residual weight fraction at 950 °C. Variations have also been found for the pyrolysis of pristine and previously demineralised whole lignocellulosic biomasses emerging from different sustainable energy plantations (González et al., 2008). Besides, enhancements in the liquid fraction have been reported for other demineralised lignocellulosic biomasses subjected to pyrolysis (Das et al., 2004; Di Blasi et al., 2000; Scott et al., 2000; Raveendran et al., 1995). This fact is particularly relevant to increase bio-oils yields.

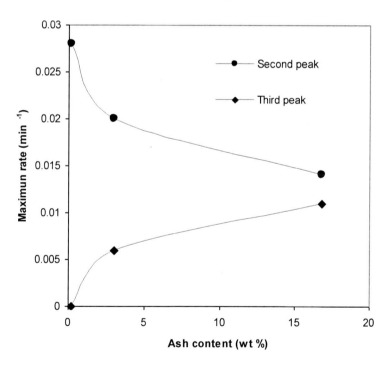

Figure 10. Effect of the ash content of the industrial raw and demineralized Kraft lignins, and of the commercial alkali lignin on the maximum degradation rates corresponding to the second and third peaks.

Furthermore, structural changes occasioned by demineralization, as suggested by SEM images and FT-IR spectra, could also be responsible for the differences in the overall pyrolytic behavior observed between the raw Kraft lignin and the demineralized sample. In this sense, it should be mentioned that Yang et al. (2010) have recently reported that bio-pretreatment of lignin from corn stover by using white – rot fungi, has a remarkable effect on lignin pyrolysis characteristics. It was attributed to strong structural alterations occasioned by the bio-pretreatment applied, which led to destroy the aromatic skeletal carbons and to convert lignin into compounds with relatively simpler structures.

3.3. Pyrolysis Characteristics of the Alkali Lignin-Polymeric Waste Mixtures

Figure 11 shows the TG curves for the pyrolysis of the mixtures composed of equal proportions (50 wt%) of the commercial alkali lignin (CAL) and polyethylene (PE). DTG curves are illustrated in Figure 12. The pyrolysis characteristics of the individual constituents of the mixture are also included in the figures for the sake of comparison. Table V lists characteristic parameters determined from the corresponding TG and DTG curves. In the table, only values corresponding to main peaks have been included.

The results (Figures 11 and 12, Table V) indicate that pyrolysis of PE, which is known to produce olefins of varying chain length (Siddiqui and Redwhi, 2009), begins at a pronouncedly higher temperature than the commercial alkali lignin and proceeds sharply. Despite pyrolysis of lignin takes place even at temperatures for which almost no degradation of PE occurs, lignin is more resistant to degrade than PE. Almost no solid product arises from PE decomposition, whereas pyrolysis of the CAL attains an appreciable residual weight fraction.

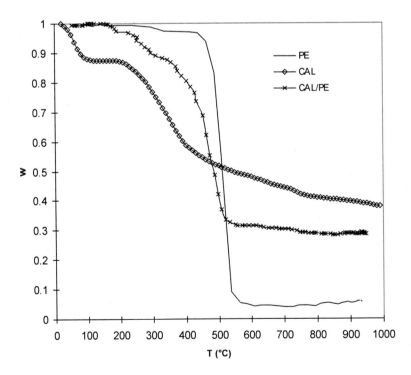

Figure 11. TG curves for the pyrolysis of the commercial alkali lignin (CAL), polyethylene (PE), and for the CAL-PE mixture.

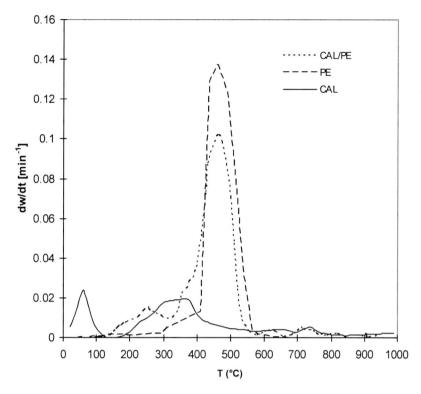

Figure 12. DTG curves for the pyrolysis of the commercial alkali lignin (CAL), polyethylene (PE), and for the CAL-PE mixture.

Table 5. Characteristics of the TG curves and reaction rate profiles for the pyrolysis of the CAL-PE mixture and individual components

Characteristic	CAL	PE	CAL-PE
T_{onset1} (°C)	160	330	160
T_{peak1} (°C)	370	462	254
T_{end1} (°C)	580	578	310
$(-dw/dt)_{peak1}$ (min^{-1})	0.020	0.140	0.014
T_{onset2} (°C)	600		330
T_{peak2} (°C)	730		462
T_{end2} (°C)	810		578
$(-dw/dt)_{peak2}$ (min^{-1})	0.006		0.110
w (T = 950 °C)	0.44	0.04	0.30

Thermal degradation behavior of the commercial lignin and PE is consistent with their inherent chemical nature. DTG curve (Figure 12) shows a sharp, well-defined peak for PE. In comparison, the low degradation rates for lignin make the reaction profile broader, flatter, and more skewed to high temperatures. Pyrolysis of the CAL-PE mixture shows an intermediate behaviour between those of its constituents, with differentiated domains of thermal degradation. Reaction rate profiles show two distinguishable peaks. Comparison of the results for the mixture and individual constituents (Figures 11 and 12) indicates that they are characteristic of the latter ones. The sharp peak at the higher temperature (462 °C) corresponds to pyrolysis of the PE present in the mixture, while the shoulder at the lower temperature (254 °C) is due to the thermal degradation of lignin accompanying PE. Although slight shifts to lower temperatures are detected for the main peaks corresponding to the individual constituents, the presence of lignin in the mixture leads to a noticeable decrease in the maximum reaction rate for the main peak, and to a somewhat lower value for the one related to lignin. In agreement with other reported studies (Sharypov et al., 2002), the results point to no interactions between PE and lignin. Accordingly, the pyrolytic behavior of the mixture seems to result from independent thermal degradation of the individual constituents.

3.4. Pyrolysis Characteristics of the Klason Lignins

TG and DTG curves determined for the pyrolysis of the Klason lignins isolated from the *Aspidosperma australe* wood sawdust and the Brazil nutsells, normalized by the ash content, are illustrated in Figures 13 and 14, respectively. In each figure, TG-DTG curves determined for the pyrolysis of the whole biomasses are also included for the sake of comparison. Characteristic parameters are listed in Table VI.

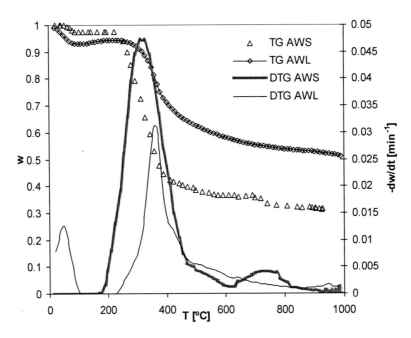

Figure 13. TG-DTG curves for the pyrolysis of the sawdust from Aspidosperma australe wood (AWS) and of the isolated Klason lignin (AWL).

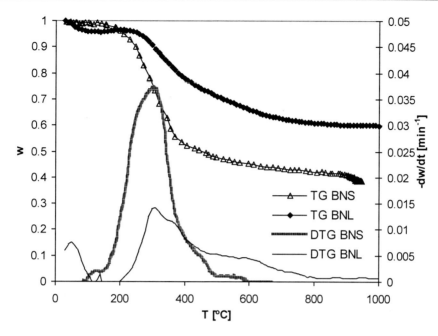

Figure 14. TG-DTG curves for the pyrolysis of the Brazil nutshells (BNS) and of the isolated Klason lignin (BNL).

Table 6. Characteristics of the TG curves and reaction rate profiles for the pyrolysis of the Klason lignins (AWL, BNL) and of the original biomasses (AWS, BNS)

Characteristic/Sample	AWL	AWS	BNL	BNS
T_{onset} (°C)	179	170	160	160
T_{peak} (°C)	360	313	340	309
T_{end} (°C)	600	600	580	580
$(-dw/dt)_{peak}$ (min^{-1})	0.031	0.047	0.012	0.037
w (T =950 °C)	0.52	0.31	0.62	0.38

As may be appreciated in the figures, small changes in weight fractions for all the samples take place at temperatures lower than 100 °C, attributable to moisture evolution. Both biomasses start to decompose at around 160-170 °C, releasing volatile matter. The TG curves show two major weight loss stages, between ~200 and 400 °C and above 400 °C, with the slope of the curves changing between these two temperature intervals. The slope between 200 and 400 °C is higher than that for the higher temperature range.

In agreement with other reported results, distinct weight loss zones may be associated with degradation dynamics of main constituents composing lignocellulosic biomasses. Since the biomasses are mostly composed by cellulose, hemicellulose, and lignin, decrease in weight fractions at around 225–325 °C is generally attributed to the onset of hemicellulose degradation, while contribution of cellulose decomposition to pyrolysis is considered to predominate in the range between 325 and 375 °C. Lignin has a broad decomposition temperature range at temperatures higher than 250 °C. Hence, degradation of cellulose and hemicellulose composing the wood sawdust and the nutshells, possessing a polysaccharide structure relatively easy to breakdown, should take predominantly place at the lower temperatures. At higher temperatures, decomposition of lignin, which is more resistant to degrade than the other two major biopolymers composing the biomasses, due to its cross-linked aromatic structure, appears to become predominant inducing pyrolysis of biomass to slow down (Haykiri-Acma et al., 2010; Wongsiriamnuay and Tippayawong, 2010; González et al., 2008; Basso et al., 2005; Ferdous et al., 2002; Bonelli et al., 2001). Besides, process simulations have also shown that hemicellulose and cellulose decompose independently of one another, the former associated with the shoulder and the latter with the peak of the rate curves, whereas lignin degrades slowly over a very broad range of temperatures (Di Blassi, 2008).

On the other hand, the results (Figures 13 and 14, and Table VI) also show that thermal degradation of the Klason lignins proceeds steadily almost throughout the temperature range in comparison with the pyrolysis of the whole biomasses from which they were isolated. For temperatures above 200 °C, weight fractions for the Klason lignins are pronouncedly higher than those for the whole biomasses. Likewise, the maximum degradation rates are lower for the lignins than for the biomasses and the degradation profiles are shifted to higher temperatures. As expected, higher residual weight fractions are attained for the lignins. Accordingly, comparison of the pyrolytic behaviour for the Klason lignins with that for the biomasses clearly evidences the effect of the other major biopolymers, i.e. cellulose and hemicellulose, and minor constituents on thermal degradation.

Furthermore, noticeable differences in the pyrolytic behaviour between the Klason lignins may be appreciated. The onset of degradation for the lignin extracted from the wood sawdust takes place at a relatively higher temperature (179 °C) than for the sample obtained from the nutshells (160 °C).

The maximum reaction rate and the corresponding peak temperature are also higher for the former, but the residual weight fraction is lower (Table VI). Hence, differences in composition and chemical structures of the lignins related to the different botanical origin of the biomasses from which they were extracted should be responsible for the pyrolytic behaviours observed. As may be noticed, reaction rate profiles for the Klason lignins show a single peak, a main difference in comparison with the pyrolysis of the alkali lignins, suggesting that they do not undergo secondary pyrolysis.

3.5. Comparison of Char Yields for Pyrolysis of the Alkali and Klason Lignins

Figure 15 comparatively illustrates residual weight fractions, namely char yields at 950 °C, attained for the pyrolysis of the alkali and Klason lignins investigated. As may be appreciated in the figure, residual weight fractions are higher for the two Klason lignins than for the alkali samples. Thus, the results indicate that the Klason lignins are more difficult to decompose than the alkali lignins. Among the four samples investigated, the Klason lignin separated from the Brazil nutshells offers the highest resistance to thermal degradation, whereas the industrial alkali lignin shows the lowest one.

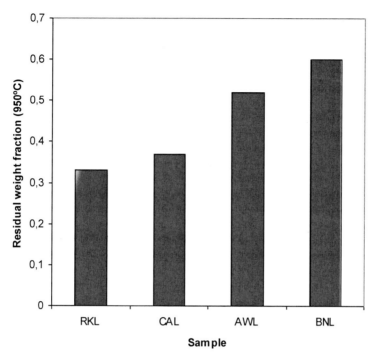

Figure 15. Comparison of the residual weight fractions at T= 950°C for the pyrolysis of the alkali and Klason lignins.

The trend is similar to others reported in the literature and has been related to the methoxy groups contained in lignin reference units (Jiang et al., 2010). Char yields from about 25% up to 40% have also been reported for pyrolysis of several milled wood lignins isolated from grasses, softwoods and hardwoods in inert atmosphere for a heating rate of 20 °C/min (Faravelli et al. 2010; Jakab et al., 1997).

As mentioned earlier, pyrolysis of lignin takes place by a free-radical mechanism. Alkali lignins reportedly possess a larger proportion of ether linkages (β-O-4 and α-O-4), which are liable to break due to their lower bond energy, forming large free radicals. These radicals further undergo decomposition, yielding products. On the other hand, for the Klason lignins possessing less ether bonds due to condensed structures, the free radicals could be predominantly formed by scission of side chains, such as the splitting of methyl radicals (-CH_3) from methoxyl groups on the benzene ring, and of hydroxyl radicals (-OH) from side propanoid chains (Faravelli et al., 2010; Jakab et al., 1997).

Accordingly, the aggressive conditions involved in the acid treatment used to isolate Klason lignins from the wood sawdust and the nutshells should favour condensation reactions, thus leading to more condensed chemical structures for the resulting lignins compared to the alkali lignins. Hence, differences in chemical structures for the alkali and Klason lignins occasioned by the extraction method applied should be responsible for the higher char yields determined for the latter. In turn, differences between the Klason lignins point to dependences on their botanical origin.

CONCLUSION

Pyrolysis characteristics of an industrial raw Kraft lignin, a commercial alkali lignin, and Klason lignins, the latter lab-isolated from two agro-industrial lignocellulosic biomasses with different lignin contents (27 wt% and 57 wt%), have been investigated from room temperature up to 1000 °C. The pyrolytic behaviour of the industrial raw Kraft lignin, possessing the highest contents of ash (16 wt%) and elemental carbon (62.2 wt%) among the investigated samples, and of the commercial alkali lignin shows differentiated domains of thermal degradation with progressive increase of temperature. The reaction rate profiles for both samples exhibit three distinguishable peaks at different temperature ranges, that may be successively related to moisture evolution, primary and secondary pyrolysis.

Demineralization of the industrial Kraft lignin by the mild acid pre-treatment, induces noticeable changes in its pyrolytic behaviour, leading to shift primary pyrolysis to higher temperatures, to increase the maximum degradation rate, and to inhibit secondary pyrolysis. The latter effect is inferred from the absence of the third peak characterizing the alkali lignins at the highest temperature range, likely because the lack of minerals hinders possible interactions with the char carbon. Nevertheless, some structural changes occasioned by demineralization, as suggested by SEM images and FT-IR spectra, could also be responsible for the differences in the overall pyrolytic behavior observed between the raw and demineralized Kraft lignins.

Pyrolysis of the binary mixtures composed by equal proportions of the commercial alkali lignin and polyethylene shows an intermediate behaviour between those of the individual constituents. Reaction rate profiles exhibit two distinguishable peaks, with the one corresponding to polyethylene taking place at higher temperature. Main effect of the presence of lignin in the mixture is to pronouncedly decrease the maximum degradation rate of the polymer. No interactions between lignin and polyethylene appear to occur, the pyrolytic behaviour of the mixture arising from independent thermal degradation of the individual constituents.

On the other hand, the pyrolytic behaviour of the Klason lignins characterized by different ash contents and elemental compositions shows noticeable differences which depend on the biomass from which they were extracted. Thermal degradation of the wood-derived lignin starts at a relatively higher temperature than that of the lignin isolated from the nutshells. The maximum reaction rate and the peak temperature are also higher for the former. Comparison of the pyrolytic behaviour for the Klason lignins with that for the untreated parent biomasses from which they were extracted clearly evidences the favourable effect of cellulose, hemicellulose, and other minor constituents on thermal degradation.

Compared to the alkali raw samples, the Klason lignins do not seem to undergo secondary pyrolysis, as inferred from the lack of the peak detected at high temperatures (730 °C) for the former ones, possibly due to more condensed chemical structures as a consequence of the strong conditions involved in the isolation method. Furthermore, char yields as judged by the residual weight fractions attained at 950 °C, indicate that the Klason lignins are more resistant to degrade than the alkali lignins. The behaviour may be explained accounting for the reportedly larger proportion of methoxy groups characterizing alkali lignins, which are liable to break, in comparison with the Klason lignins.

ACKNOWLEDGMENTS

Grants from Consejo Nacional de Investigaciones Científicas y Técnicas (CONICET) and Universidad de Buenos Aires (UBA) from Argentina are gratefully acknowledged.

REFERENCES

Balat, M., Balat, M., Kırtay, E., Balat, H. 2009. Main routes for the thermo-conversion of biomass into fuels and chemicals. Part 1: Pyrolysis systems. *Energy Conversion and Management* 50, 3147–3157.

Baptista, C., Robert, D., Duarte, A.P., 2006. Effect of pulping conditions on lignin structure from maritime pine kraft pulps. *Chemical Engineering Journal* 121, 153–158.

Basso, M.C., Cerrella, E.G., Cukierman, A.L. 2004. Cadmium uptake by lignocellulosic materials: Effect of lignin content. *Separation Science and Technology* 39, 1163-1175.

Basso, M. C., Cerrella, E. G., Buonomo, E. L., Bonelli, P. R., and Cukierman, A. L. 2005. Thermochemical conversion of *Arundo donax* into useful solid products. *Energy Sources* 27, 1429–1438.

Baumlin, S., Broust, F., Bazer-Bachi, F., Bourdeaux, T., Herbinet, O., Toutie Ndiaye, F., Ferrer, M., Lédé, J. 2006. Production of hydrogen by lignins fast pyrolysis. *International Journal of Hydrogen* 31, 2179–2192.

Betancur, M., Bonelli, P.R., Velásquez, J., Cukierman, A.L. 2009. Potentiality of lignin from the Kraft pulping process for removal of trace nickel from wastewater: Effect of demineralisation. *Bioresource Technology* 100 (3), 1130-1137.

Boeriu, C.G., Bravo, D., Gosselink, R.J.A., van Dam, J.E.G. 2004. Characterisation of structure-dependent functional properties of lignin with infrared spectroscopy. *Industrial Crops and Products* 20 (2), 205-218.

Bonelli, P.R., Della Rocca, P.A., Cerrella, E.G., Cukierman, A.L., 2001. Effect of pyrolysis temperature on composition, surface properties and thermal degradation rates of Brazil nut shells. *Bioresource Technology* 76, 15-22.

Bonelli, P. R., Cerrella, E. G., and Cukierman, A. L. 2003. Slow pyrolysis of nutshells: Characterization of derived chars and of process kinetics. *Energy Sources* 25, 767–778.

Bonelli, P.R., Buonomo, E.L., Cukierman, A.L. 2007. Pyrolysis of sugarcane bagasse and co-pyrolysis with an Argentinean subbituminous coal. Energy Sources. *Part A: Recovery, Utilization, and Environmental Effects* 29(8), 731-740.

Braun, J.L., Holtman, K.M., Kadla, J.F., 2005. Lignin-based carbon fibers: Oxidative thermostabilization of kraft lignin. *Carbon* 43, 385–394.

Brodin, I., Sjohölm, E., Gellerstedt, G., 2010. The behavior of kraft lignin during thermal treatment. *Journal of Analytic and Applied Pyrolysis* 87, 70–77.

Cukierman, A.L. 2007. Metal ion biosorption potential of lignocellulosic biomasses and marine algae for wastewater treatment. *Adsorption Science and Technology* 25 (3-4), 227-244.

Das, P., Ganesh, A., Wangikar, P. 2004. Influence of pretreatment for deashing of sugarcane bagasse on pyrolysis products. *Biomass and Bioenergy* 27, 445-457.

De Oliveira Brito, M.S., Martins Carvalho Andrade, H., Frota Soaresa, L., Pires de Azevedo, R., 2010. Brazil nut shells as a new biosorbent to remove methylene blue and indigo carmine from aqueous solutions. *Journal Hazardous Materials* 174, 84–92.

Di Blasi, C., Branca, C., and D'Errico, G. 2000. Degradation characteristics of straw and washed straw. *Thermochimica Acta* 364:133–142.

Di Blasi, C., 2008. Modeling chemical and physical processes of wood and biomass pyrolysis. *Progress in Energy and Combustion Science* 34, 47–90.

El Mansouri, N.E., Salvadó, J., 2006. Structural characterization of technical lignins for the production of adhesives: Application to lignosulfonate, kraft, soda-anthraquinone, organosolv and ethanol process lignins. *Industrial Crops and Products* 24, 8–16.

El Mansouri, N.E., Pizzi, A., Salvadó, J., 2007. *Lignin-based wood panel adhesives without formaldehyde.* Holz als Roh- und Werkstoff 65, 65–70.

Faravelli, T., Frassoldati, A., Migliavacca, G., Ranzi, E. 2010. Detailed kinetic modeling of the thermal degradation of lignins. *Biomass and Bioenergy* 34, 290 – 301.

Ferdous, D., Dalai, A. K., Bej, S. K., Thring, R. W. 2002. Pyrolysis of lignins: Experimental and kinetics studies. *Energy and Fuels* 16, 1405-1412.

Fierro, V., Torné-Fernández, V., Montané, D., Celzard, A. 2005. Study of the decomposition of kraft lignin impregnated with orthophosphoric acid. *Thermochimica Acta* 433, 142–148.

Fierro, V., Torné-Fernández, V., Celzard, A., Montané, D., 2007. Influence of the demineralisation on the chemical activation of Kraft lignin with orthophosphoric acid. *Journal of Hazardous Materials* 149, 126–133.

González, J.D., Kim, M.R., Buonomo, E.L., Bonelli, P.R., Cukierman, A.L. 2008. Pyrolysis of biomass from sustainable energy plantations: Effect of mineral matter reduction on kinetics and charcoal pore structure. *Energy Sources. Part A: Recovery, Utilization, and Environmental Effects* 30 (9), 809-817.

Haykiri-Acma, H., Yaman, S., Kucukbayrak, S., 2010. Comparison of the thermal reactivities of isolated lignin and holocellulose during pyrolysis. *Fuel Processing Technology,* in press.

Jakab E., Faix O., Till F. 1997. Thermal decomposition of milled wood lignins studied by thermogravimetry/mass spectrometry. *Journal of Analytical and Appllied Pyrolysis* 40–41, 171–186.

Jeguirim, M., Trouvé, G. 2009. Pyrolysis characteristics and kinetics of Arundo donax using thermogravimetric analysis. *Bioresource Technology* 100, 4026–4031.

Jiang, G., Nowakowski, D.J., Bridgwater, A.V., 2010. A systematic study of the kinetics of lignin pyrolysis. *Thermochimica Acta* 498, 61–66.

Kim, M.R., Buonomo, E.L., Bonelli, P.R., Cukierman, A.L., 2010. The thermochemical processing of municipal solid wastes: Thermal events and kinetics of pyrolysis. *Energy Sources. Part A: Recovery, Utilization, and Environmental Effects* 32, 1207–1214.

Kumar, V., Iisa, K., Banerjee, S., Frederick Jr., W. J. 2008. Effect of alkali metals on lignin pyrolysis and gasification. *AIChE Fall and Annual Meeting*, Conference Proceedings, Philadelphia, Pa, USA, Nov. 2008 (CD-ROM).

Li, Y., Sarkanen, S., 2005. Miscible blends of kraft lignin derivatives with low-Tg polymers. *Macromolecules* 38, 2296-2306.

Mancera, A., Fierro, V., Pizzi, A., Dumarçay, S, Gérardin, P., Velásquez, J., Quintana, G., Celzard, A., 2010. Physicochemical characterisation of sugar cane bagasse lignin oxidized by hydrogen peroxide. *Polymer Degradation and Stability* 95, 470-476.

McCarthy JL. In: Glasser WG, Northey RA, Schultz TP, editors. Lignin: historical, biological, and materials perspectives. *American Chemical Society*; 2000. p. 2–100.

Mohan, D., Pittman, Jr., Ch, U, Steele, P.H. 2006. Single, binary and multi-component adsorption of copper and cadmium from aqueous solutions on Kraft lignin—a biosorbent. *Journal of Colloid and Interface Science* 297, 489–504.

Nowakowski, D.J., Bridgwater, A.V., Elliott, D.C., Meier, D., de Wild, P. 2010. Lignin fast pyrolysis: Results from an international collaboration. *Journal of Analytical and Applied Pyrolysis* 88, 53–72.

Ohmukai, Y., Hasegawa, I., Mae, K. 2008. Pyrolysis of the mixture of biomass and plastics in countercurrent flow reactor. Part I: Experimental analysis and modeling of kinetics. *Fuel* 87, 3105–3111.

Ramírez, F., Varela, G., Delgado, E., López-Dellamary, F., Zúñiga, V., González, V., Faix, O., Meier, D. 2007. Reactions, characterization and uptake of ammoxidized kraft lignin labeled with [15]N. *Bioresource Technology* 98, 1494–1500.

Raveendran K, Ganesh AK, Khilar C. 1995. Influence of mineral matter on biomass pyrolysis characteristics. *Fuel* 74, 1812–1822.

Scott, D. S., Paterson, L., Piskorz, J., and Radlein, D. 2000. Pretreatment of poplar wood for fast pyrolysis: Rate of cation removal. *Journal of Analytical and Applied Pyrolysis* 57, 169–176.

Sharypov, V. I., Marin, N., Beregovtsova, N. G., Baryshnikov, S. V., Kuznetsov, B. N., Cebolla, V. L., and Weber, J. V. 2002. Co-pyrolysis of wood biomass and synthetic polymer mixtures. Part I: Influence of experimental conditions on the evolution of solids, liquids and gases. *Journal of Analytical and Applied Pyrolysis* 64, 15–28.

Sharypov, V.I., Beregovtsova, N.G., Kuznetsov, B.N., Baryshnikov, S.V., Marin, N., Weber, J.V., 2003. Light hydrocarbon liquids production by co-pyrolysis of polypropylene and hydrolytic lignin. *Chemistry for Sustainable Development* 11, 427–434.

Siddiqui M. N., Redhwi, H. H., 2009. Pyrolysis of mixed plastics for the recovery of useful products. *Fuel Processing Technology* 90, 545–552.

Suhas, Carrott, PJM, Ribeiro Carrott, MML. 2007. Lignin - from natural adsorbent to activated carbon: a review. *Bioresource Technology* 98(12), 2301-2312.

Velásquez, J.A., Ferrando, F., Salvadó, J., 2003. Effects of Kraft lignin addition in the production of binderless fiberboard from steam exploded *Miscanthus sinensis*. *Ind. Crops Prod.* 18, 17–23.

Walker J.C.F. 2006. *Primary Wood Processing: Principles and Practice*. Springer, Dordrecht, The Netherlands. 2nd Edition.

Wang, S., Wang, K., Liu, Q., Gu, Y., Luo, Z., Cen, K., Fransson, T. 2009. Comparison of the pyrolysis behavior of lignins from different tree species. *Biotechnology Advances* 27, 562–567.

Wongsiriamnuay, T., Tippayawong, N., 2010. Non-isothermal pyrolysis characteristics of giant sensitive plants using thermogravimetric analysis. *Bioresource Technology* 101, 5638–5644.

Yang, H., Yan, R., Chen, F., Lee, D.H., Zheng, Ch., 2007. Characteristics of hemicellulose, cellulose and lignin pyrolysis. *Fuel* 86, 1781–1788.

Yang, X., Zeng, Y., Zhang, X., 2010. Influence of biopretreatment on the character of corn stover lignin as shown by thermogravimetric and chemical structure analyses. *Bioresources* 5(1), 488-498.

In: Lignin
Editor: Ryan J. Paterson

ISBN 978-1-61122-907-3
© 2012 Nova Science Publishers, Inc.

Chapter 12

ORGANOSOLV LIGNINS: EXTRACTION, CHARACTERIZATION AND UTILIZATION FOR THE CONCEPTION OF ENVIRONMENTALLY FRIENDLY MATERIALS

*Nicolas Brosse**

Laboratoire d'Etude et de Recherche sur le MAteriau Bois,
Faculté des Sciences et Technologies, Nancy-Université,
Bld des Aiguillettes, F-54500 Vandoeuvre-lès-Nancy, France

ABSTRACT

Lignin, one of the most abundant natural polymers, is expected to play in the near future an important role as raw material for the world's biobased economy for the production of bioproducts and biofuels. The pulping processes currently used in the paper industry produce a degraded lignin employed in low-added value utilizations and energy production. However, among the various pretreatment methods currently studied for the production of pulp and/or ethanol, the organosolv processes seem to be very promising. Organosolv processes use either low-boiling solvents (e.g., methanol, ethanol, acetone), which can be easily recovered by distillation, or high-boiling solvents (e.g., ethyleneglycol, ethanolamine), which can be used at a low pressure. These procedures not only produce a cellulose-rich pulp but also large amount of high-quality lignins which are relatively pure, primarily unaltered and less condensed than other pretreatment lignins. This review presents the progress of organosolv pretreatment of lignocellulosic biomass for the production of high-quality lignins. Impacts of the process conditions on the chemical structure of the recovered lignin fractions and on delignification mechanisms are exposed. Recent utilizations of organosolv lignins for the production of materials (e.g. biodegradable polymers and adhesives) are given.

---

* E-mail address: Nicolas.Brosse@lermab.uhp-nancy.fr ; tel +333 83 68 48 62 ; fax + 333 83 68 44 98.

1. INTRODUCTION

Organosolv pulping is the process to extract lignin from ligocellulosic feedstocks with organic solvents or their aqueous solutions. Since 1980s, organosolv pulping is an alternative to Kraft and sulfite pulping which have some serious shortcomings such as air and water pollution. It was originally developed as Alcell® pulping process for hardwood. With the recent emerging necessity to develop alternative sustainable transportation fuel, the organosolv process for the production of ethanol is among the pre-treatment strategies currently being studied for the conversion of lignocellulosic feedstock to biofuels and biomaterials. In fact, to improve the overall effectiveness of bioethanol production one strategy is the biorefinery model in which all components of biomass are fully used to produce a wide range of value-added products (Ragauskas et al., 2006). In this context, organosolv processes are very promising because they allow a clean fractionation of lignocellulosic feedstocks into three major components: a cellulose-rich pulp, an organosolv lignin fraction and mono and oligosaccharides (from hemicelluloses) as syrup (Zhao et al., 2009a). In organosolv treatment a mixture water-organic solvent is used as the cooking liquor for the hydrolysis and the solubilization of lignin fragments so produced. It can be performed using a large number of solvents : low boiling points (methanol and ethanol) or high boiling points (ethylene glycol, glycerol) alcohols, carboxylic acids as well as other classes of organic compound. It is generally admitted that if the pre-treatment is conducted at high temperatures (above 185°C), there is no need for addition of catalysts. In these conditions, the cooking liquor becomes acidified due to acetic acid released from the hemicelluloses hydrolysis. For lower temperature, inorganic acid catalysts (HCl or H_2SO_4) are usually added. In these conditions, the rate of delignification is increased and higher yields of xylose are obtained. Most of hemicellulose and lignin are solubilized, but the cellulose remains as solid. The organic solvents used in the process need to be recycled to reduce the cost.

Production of high-quality lignin is one of the unique advantages of the organosolv treatment over alternative processes which generally produce a degraded lignin employed in low added value applications and energy production. Oganosolv lignins are high-purity, low molecular weight and sulfur-free products. Moreover, they are soluble in many organic solvents, possess low Tg's, and are easier to thermally process than kraft lignins. Thus, availability of such high-quality lignin in large quantities should stimulate development in new lignin applications in the fields of fibers, biodegradable polymers, adhesives... From this perspective, it is important to understand how the structure of lignin changes during the organosolv treatment to identify future applications.

The present work deals with reviewing informations about polymeric organosolv lignin : the processes used for its extraction, its chemical structure and its potential for the production of new materials with attempts to find alternatives to petrochemicals and their derivatives.

2. ORGANOSOLV PROCESSES

The organosolv pulping has been the subject of considerable research activity and has generated increasing interest as the pulp and paper industry is moving toward minimization of environmental impact (Zhao et al., 2009). In the 1990's, this technology was developed at the industrial scale in Canada (Alcell® process). However, the development of the process was suspended because of financial difficulties. One of the advantages of the organosolv process is the fractionation of the lignocellulosic materials into three major components: cellulosic fibers, hemicelluloses and lignin. As a result, with the increasing attention devoted to the biorefinery concept, a renewable of interest for the organosolv treatment is currently observed and this technology seems to be promising for the production of ethanol and high value chemicals and materials from lignin, hemicelluloses and extractives (Ragauskas et al., 2006). In this context, utilizations of the high quality lignin produced for the elaboration of bio materials is a key issue and should give an added value to the whole process.

Organosolv pre-treatment of biomass resides on the use of an organic solvent system with enhanced solubilizing properties due to organic components. Effective organosolv pre-treatment technologies need to address several criteria, including: minimization of hemicelluloses degradation, lignin recovery and limiting lignin alteration, minimal energy and capitals and operating costs. Some of the more studied organosolv pretreatments are summarized below.

2.1. Ethanol

Alcohols are the most frequently used solvents in organosolv pre-treatment and due to low cost, ease of recovery and low toxicity, ethanol is the most favoured alcohol for alcohol-based organosolv pretreatment. Biomass is treated with addition of catalyst at low temperature (below 180°C) or without catalyst (auto-catalysis) at higher temperatures (185–210°C). The catalyst generally employed is sulphuric acid. After cooking, the solid residue (pulp) is filtered and washed with an ethanol solution. The resulting black liquor is diluted with water to precipitate the lignin. Recently, this process was found to be very effective for increasing the cellulose enzymatic digestibility and lignin recovery starting from hybrid poplar (Pan et al., 2006), beetle-killed lodgepole pine (Pan et al, 2008), *Miscanthus x giganteus* (Brosse et al., 2009, 2010), *Buddleja davidii* (Hallac et al. 2009), and Loblolly pine (Sannigrahi et al., 2008). The process is schematically presented in Figure 1. In most of these studies, residual organosolv lignins were subjected to a comprehensive physicochemical and structural characterization.

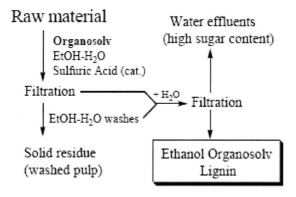

Figure 1. Schematic of the ethanol organosolv treatment.

2.2. High Boiling Point Alcohol

The most obvious advantage for high boiling point alcohol pretreatment is that the process can be performed under atmospheric pressure. High boiling point alcohols enabled a high removal of hemicelluloses and lignin, with good cellulose retention (Rodriguez et al., 2008). Among them, ethyleneglycol, ethanolamine (Jimenez et al., 2007) and glycerol (Sun and Chen, 2008) are the most favored. As an example, it has been demonstrated that aqueous glycerol autocatalytic pre-treatment at 165–225 °C for 8 h removed 50 to 75% of the lignin and retained 95% of the cellulose in wood (Demirbas, 1998). Crude glycerol from oleochemicals industry for pretreatment might be promising for cost reduction (Sun and Chen, 2007, 2008, 2010).

2.3. Organic Acids

The organic acids employed for pre-treatment are mostly acetic and formic acid (Xu et al., 2006 ; Villaverde et al., 2009). Only a few researches were found in the literatures because of several drawbacks:

- corrosive ability of organic acids
- re-precipitation of lignin during the pulping which protected the cellulose of the pulp from enzymatic attack
- acetylation of cellulose during acetic acid pre-treatment which inhibits the interaction between enzymes and cellulose.

However, an industrial process recently developed by CIMV successfully uses a mixture of acetic acid, formic acid and water for the manufacture of paper pulp, sulfur free linear lignin and xylose syrup from annual fiber crops (Figure 2, Delmas, 2008).

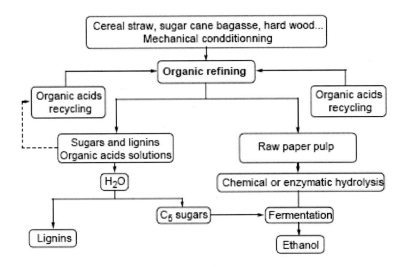

Figure 2. The CIMV bioethanol/lignins biorefinery.

2.4. Organic Peracids

Peracetic acid is the most important organic peracid used for pretreatment of biomass and is prepared by reaction of acetic acid with hydrogen peroxide using sulfuric acid as catalyst. Peracetic acid is a powerful oxidizing agent, used at relatively low temperature under atmospheric pressure (Kham et al., 2005 ; Zhao et al., 2009)). In these conditions, lignin is disrupted through oxidative reactions to generate fragments which are dissolved in the liquid phase.

2.5. Acetone

Acetone is a good solvent for lignin and thus is the most favored ketone used for delignification. The Acetosolv process is generally based on the utilization of HCl-catalyzed acetic acid media) and has proved to be promising process to achieve complete utilization of lignocellulosics (including lignin) under mild conditions. As an example, Arraque et al. (2008) found that the organosolv acetone–water pre-treatment conditions in presence of sulphuric acid as catalyst for *Pinus radiata* shown to be good substrates for ethanol production.

2.6. Two-Steps Processes

During a pre-treatment process, it is desirable to have high yields of glucose and hemicelluloses sugars. However, conditions known to promote lignin depolymerisation also cause degradation of hemicelluloses sugars into furfural, hydroxymethyl furfural (HMF) and carboxylic acid. These degradation products inhibit downstream processing (Thomsen at al., 2009). To circumvent this drawback, two-stage pretreatments are generally considered to be the best option: a 1st step is performed with low severity conditions to hydrolyse the hemicelluloses and a second step, where the solid material from the 1st step is pretreated again with higher severity conditions. This approach permits to obtain higher sugar yields than one-step pre-treatment but also enhanced the dissolution of lignin and has been proposed in the literature several times. Thus, fungal (Munos et al., 2007), dilute sulphuric acid prehydrolysis (Brosse et al, 2009; Patel and Varshney, 1989) or autohydrolysis (Alfaro et al., 2009 ; Caparros et al., 2007) before an organosolv treatment have demonstrated a favourable effect on delignification.

3. LIGNIN STUCTURE

Lignin is an extremely complex three-dimensional, amorphous, cross-linked, and three-dimensional phenolic polymer. It is biosynthesized from three monolignols: p-coumaryl , coniferyl and sinapyl alcohols. These three lignin precursors give rise respectively to the so-called *p*-hydroxyphenyl (H), guaiacyl (G) and syringyl (S) phenylpropanoid units, which show different abundances in lignins from different groups of vascular plants, as well as in different plant tissues and cell-wall layers. During biosynthesis, the monolignols undergo an in situ radical polymerization, yielding several interunit linkages including aryl ether (α-O-4' and β-O-4'), resinol (β-β'), phenylcoumaran (β-5'), biphenyl (5-5'), and 1,2-diaryl propane (β-1') depicted in the figure 3 (Chakar and Ragauskas, 2004).

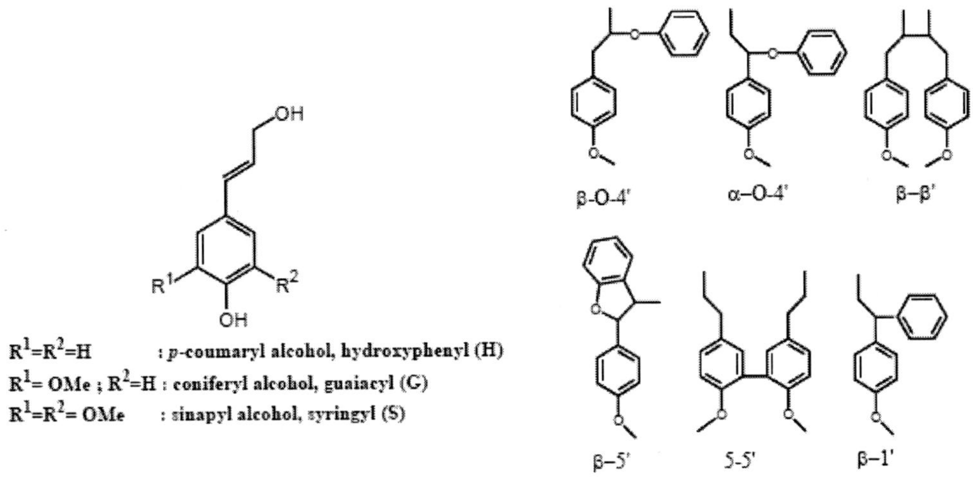

$R^1=R^2=H$: p-coumaryl alcohol, hydroxyphenyl (H)
$R^1=$ OMe ; $R^2=$H : coniferyl alcohol, guaiacyl (G)
$R^1=R^2=$ OMe   : sinapyl alcohol, syringyl (S)

Figure 3. Building blocks of lignin and common linkages between phenyl propane units.

4. CHEMICAL TRANSFORMATION DURING ORGANOSOLV TREATMENT

The organosolv pre-treatment delignifies by a chemical breackdown of the lignin prior to dissolving it. The purpose of this section is to summarize available information regarding lignin organosolv hydrolysis mechanisms.

4.1. Aryl-Ether Cleavage

Numerous studies have led to the conclusion that the cleavage of ether linkage is primarily responsible for lignin hydrolysis and dissolution in organosolv processes. El Hage et al (2009) have examined the structure of two lignin fractions from Miscanthus : milled wood lignin (MWL) which is representative of native lignin and organosolv lignin (EOL) extracted using a sulphuric acid catalysed ethanol organosolv treatment.

As can be seen from comparing the spectra of Figures 4, the ^{13}C NMR spectra of EOL and MWL are not remarkably different: EOL displays a well resolved spectrum with a great similarity with the milled lignin spectrum demonstrating that the organosolv process did not degrade the macromolecule structure to a significant extent. Nevertheless, the spectra display some differences mainly in two regions:

–   in the aliphatic region (60-90 ppm), the signals corresponding to the α, β, γ carbons ;
–   in the aromatic region (145-152 ppm).\

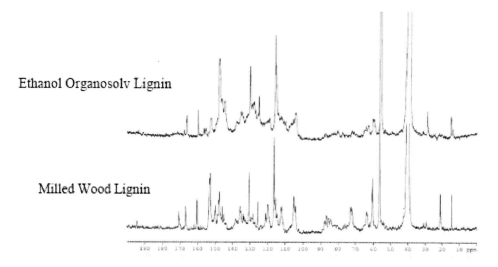

Figure 4. Quantitative C NMR spectrum of Milled Wood Lignin and Ethanol Organosolv Lignin isolated from *Miscanthus*.

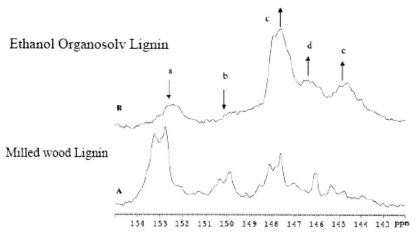

Assignment of the peaks 1-5. 1: C3–C5 S etherified; 2: C3 G etherified; 3: C3–C5 S non etherified; 4: C3 G non etherified; 5: C4 G non etherified.

Figure 5 Expansion of Milled Wood Lignin and Ethanol Organosolv Lignin ^{13}C NMR spectra.

The Figure 5 gives an expansion of this region from which it appears that the intensity of the ArOR carbon signals of the G and S units were lower in EOL spectrum (Figure 5B) and the ArOH higher than in milled lignin.

These modifications, in accordance with a scission of aryl-ether linkages, were confirmed by the ^{31}P NMR study of the phosphitylated lignins performed according a method described by Granata and Argyropoulos (1995). After organosolv treatment, the overall amounts of phenolic OH in the region between 136-144 ppm (syringyl + guaiacyl + *p*-hydroxyphenyl) were found to strongly increase from 0.28 to 0.50 OH group/C9) mainly as a result of β-*O*-4 linkage scission whereas the aliphatic OH were reduced from 0.67 to 0.19, probably because of acid-catalysed elimination reactions (Figure 6).

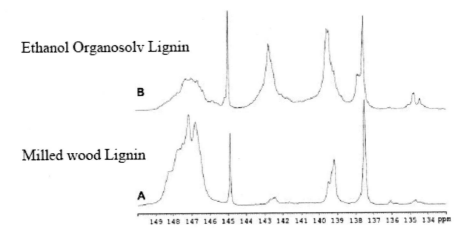

Figure 6. Quantitative ^{31}P NMR spectra of Milled Wood Lignin (A) and Ethanol Organosolv Lignin (B) isolated from *Miscanthus*.

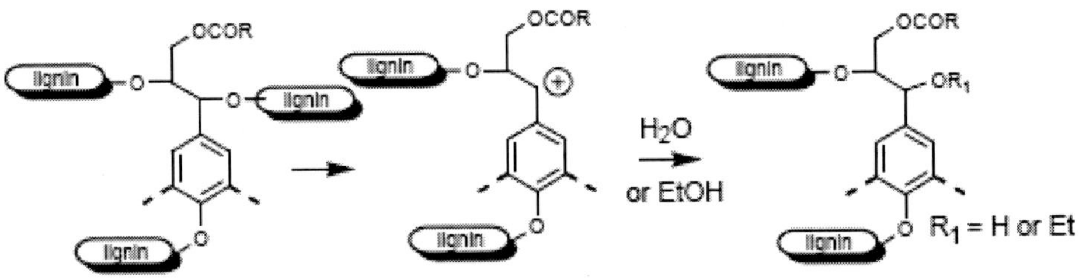

Figure 7. Mechanism of the solvolytic cleavage of α-aryl ether linkage.

According to several authors, the acidic organosolv hydrolysis mainly affected the Cα position of the side chains (McDonough, 1993). In the medium the solvolytic splitting of α-aryl ether leads to the formation of a benzylic carbocation which can react with the solvent (Figure 7). This observation is supported by the fact that model compound studies have shown that α-aryl ether linkages are more easily split than β-aryl ether linkages (Kishimoto and Sano, 2003).

Sarkanen (1980) has pointed out that acidolysis of isolated organosolv lignins substantially increased the content of phenolic hydroxyl groups, indicating the presence of intact β-*O*-4 bonds. More recently, Kubo and Kadla (2004) demonstrated by HMQC 2D NMR that β-*O*-4 structures in Alcell® lignin survive chemical processing. In addition, hydrolysis of α-aryl ethers was found to be the rate controlling reaction step of delignification, with relatively lower activation energy. The activation energies reported by Meshgini and Sarkanen (1989) for the hydrolysis of α-aryl ethers of lignin model compounds were 80–118 kJ/mol, depending on the substituent. Vazquez et al. (1997) reported that, in the case of acetosolv fractionation of pine wood, the activation energies obtained for both auto-catalysed and acid catalysed (HCl) reactions were respectively 78.8 kJ/mol and 69.7 kJ/mol. Since the activation energy reported for β-aryl ether hydrolysis (150 kJ/mol,) is higher, the authors concluded that β aryl ether bond hydrolysis is not the controlling reaction in their delignification processes. In addition, organosolv experiments performed with different levels of severity demonstrated that at the severity conditions generally employed, β-ether cleavage seems not to be the controlling delignification reaction. Only in high severity reaction conditions, extensive aryl-ether bond hydrolysis and β-ether cleavages through the formation of Hibbert's ketones (El Hage et al, 2010) were observed. The same observation was made when model compound (guaiacylglycerol- β- guaiacyl ether) is refluxed with dioaxane-water containing HCl (Adler et al., 1957).

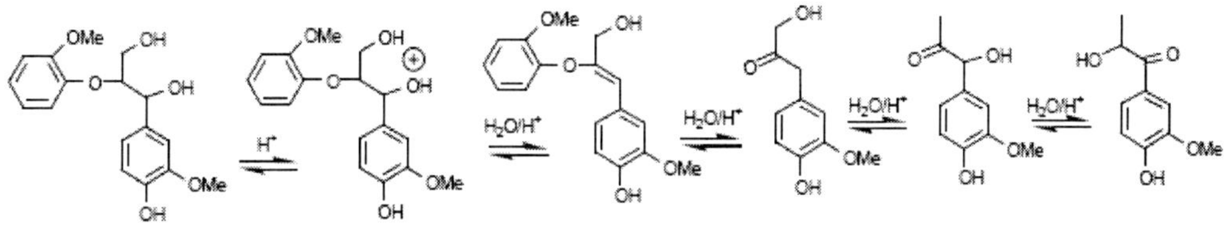

Figure 8. Mechanism of the formation of Hibbert's ketones.

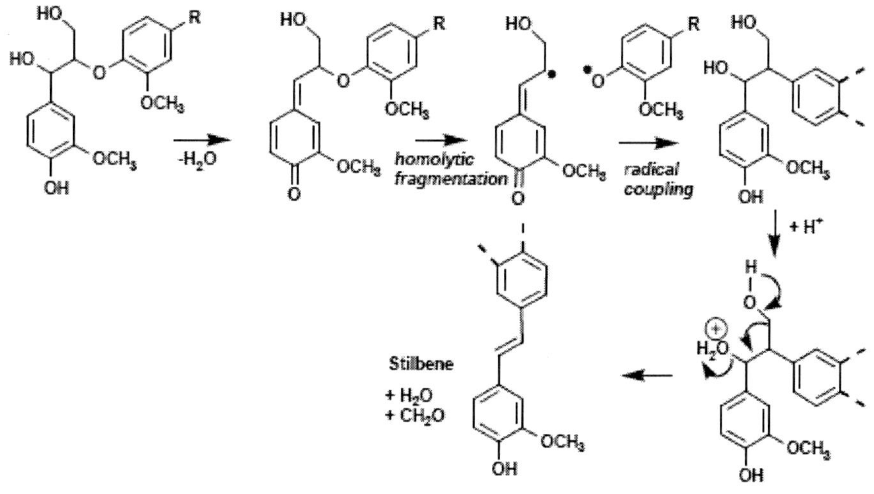

Figure 9. Degradation mechanism of β-aryl ether (β-O-4′) during ethanol organosolv process.

Also, stilbene structures were detected by HSQC 2D NMR during the ethanol organosolv treatment of *Buddleja davidii* (Hallac et al., 2010). Under the harsh conditions used, the homolytic cleavage with an intermediate quinine methide causes the formation of β-1' interlinkage through radical coupling, which then in turn degrades under the acidic medium to give stilbenes through the loss of the γ-methylol group as formaldehyde. Such mechanism and structures were previously identified in Kraft pulping.

4.2. Lignin Condensation

The lignin deconstruction through aryl-ether bonds hydrolysis during organosolv treatment is accompanied by a lignin repolymerization. Sannigrahi et al (2010) studied MWL samples from untreated loblolly pine, post ethanol organosolv treatment, and ethanol organosol lignin by ^{13}C NMR. They showed that EOL was more condensed than the starting lignin and that lignin isolated from the solid residue from the pre-treatment was also condensed and exhibited evidence of repolymerization. Hallac et al (2010) have clearly demonstrated by ^{13}C NMR an increase of the degree of condensation of *B. davidii* EOL with a higher degree of delignification.

In acidic conditions, the repolymerization mainly results from a reaction of the benzylic cation formed by the splitting of α-aryl ether with an electron-rich carbon atom in the aromatic ring of another lignin unit through intra- or inter-molecular reactions (Figure 10).

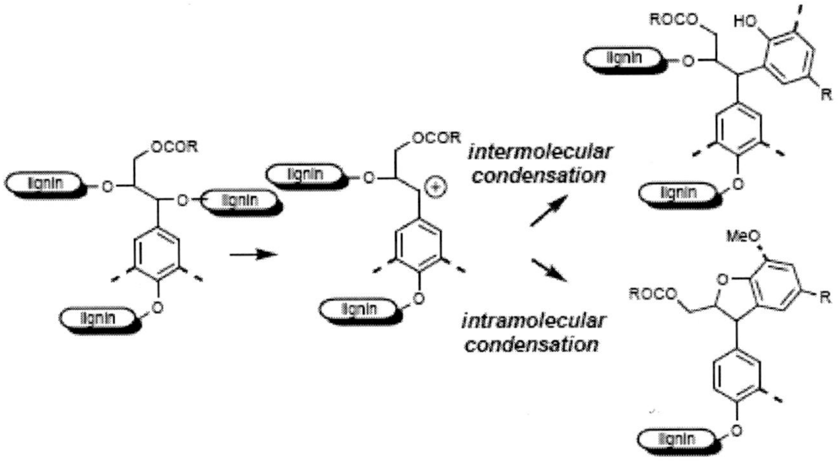

Figure 10. Intra and intermolecular lignin condensation.

These repolymerization reactions, favoured by high temperature and acidity, are counterproductive because they result in producing a network polymer, reducing the delignification rates. Shukry et al. (2008) studied the acetosolv delignification of bagasse and showed that the molecular weighs of extracted lignin were low after 1 h of pulping, increased when pulping time was extended to 3 h and then deceased. They demonstrated by SEC and SEM studies that small lignin fragments came out of small pores in early stages of delignification ; in a second time, with the increase of the pores size, larger fragments formed by condensation reaction diffused out of the pores. This was followed by degradation of the large molecules as the reaction time increased mainly through solvolytic cleavage of ether linkages.

4.3. Ester Hydrolysis

The grass lignins are naturally acylated by *p*-coumaric acid (Buranov and Mazza, 2008). NMR results have confirmed with corn, wheat and other herbaceous crops that *p*-coumarates are exclusively located at the γ position of the lignin side chain (Crestini and Argyropoulos, 1997, Figure 11).

During miscanthus organosolv treatment, extensive hydrolysis of ester bond was observed (El Hage et al., 2009). Furthermore, the organosolv treatment with acetic acid resulted in noticeable amounts of acetyl groups on the lignin side chain through esterification reactions. This was demonstrated by FTIR and NMR techniques (Shukry at al., 2008).

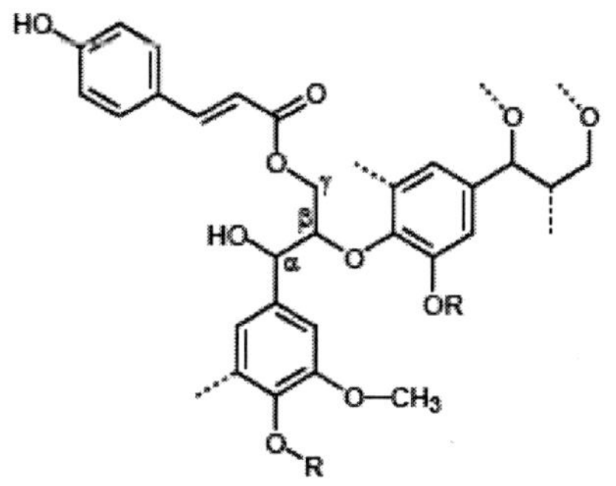

Figure 11. Representation of a coumarylated lignin fragment at the γ position.

4.4. Linear Organosolv Lignins

One of the most promising organosolv processes currently developed at the pilot scale is pulping of wheat straw at atmospheric pressure by a catalyst solvent system of formic acid/acetic acid/water. The industrial process developed by CIMV in France is one of the first biorefinery producing whitened paper pulp, sulphur-free lignin and xylose syrup (Delmas, 2008, see Figure 2). Using Atmospheric pressure photoionization mass spectrometry (APPI-MS), it has been proposed that the organosolv lignins produced are linear polymers (Banoub et al., 2007). Using this technique, it was shown that the native lignin polymer is composed of a mixture of different linear polycondensed coniferyl units (Figure 9). According the authors, the mild experimental conditions used avoid lignin reticulation and repolymerization reactions.

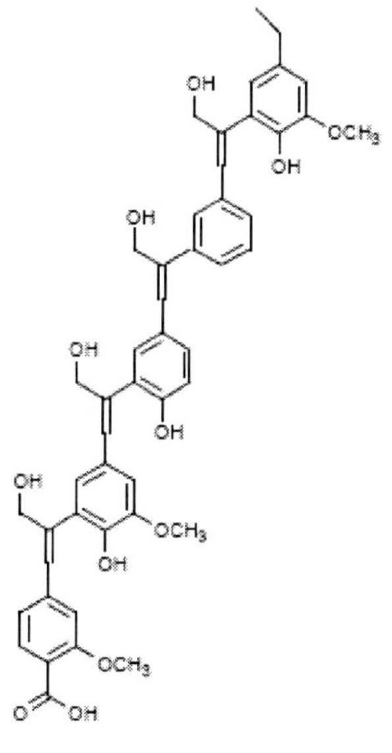

Figure 12. Wheat straw lignins in acetic acid/formic acid/water media.

4.5. Effect of a 2-Steps Process on the Lignin Structure and Organosolv Extractability

Autohydrolysis, a process that treats lignocellulosic materials in a chemical-free, water only media, provides a simple, low-cost and environmental friendly pretreatment technology and has been described by several authors as a prehydrolysis step prior to an organosolv treatment (Alfaro et al., 2009 ; Caparros et al., 2007 ; see subsection 2.5). This step enables the hydrolysis and the dissolution of a great part of the hemicelluloses and the cleavage of lignin-carbohydrate bonds (Garrote et al., 2001, 2002 ; Lee et al, 2009). These changes affect the efficiency of subsequent delignification processes: it has been demonstrated that, after autohydrolysis, kraft pulping of hardwood is accelerated and solubility of lignin in a subsequent extraction with organic solvent is increased (Sixta, 2006; Lora and Wayman, 1978).

Using SEM, the microscopic surfaces of untreated, autohydrolysed and organosolv treated pulp fibers were compared. Examination showed great differences between fibers. Untreated pulp fibers had an intact surface and well-defined folds (Figure 13A) whereas after autohydrolysis, material exhibited fibers with a disrupted surface with small tears and holes (Figure 13B).

This observation could be attributed to a preferential removal of hemicelluloses. Concerning the pulp fibers, as seen in Figure 13C after organosolv treatment, the removal of a large part of the lignin results in a fibrillar material mainly composed of cellulose fibers. These observations confirmed the hypothesis proposed about the removal of a part of the hemicelluloses during the autohydrolysis step (El Hage et al., 2010b) which disrupted enough of the remaining polymers to enhance the hydrolysis of lignin during the organosolv process.

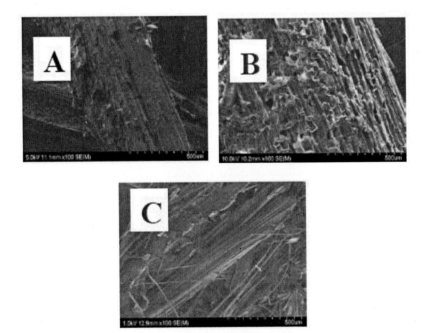

Figure 13. Scanning electron micrographs of (A) Untreated (B) Autohydrolysed (C) Organosolv miscanthus pulp.

It was also found that the maximum of extractability of lignin could be obtain only in a narrow range of reaction severity: when autohydrolysis severity is high (long reaction time and high temperature), the overall delignification rate during kraft pulping and lignin solubility in dioxane/water decreases (Leschinsky et al, 2008a, 2008b, Sixta, 2006). Similar observations were made in 1975 by Kleinert concerning ethanol organosolv delignification of spruce and poplar : delignification rate was lower when wood specimens were dried at high temperature or stored at room temperature for a long time. The explanation proposed for the difficult removal of lignin after intense thermal treatments temperature or long storage time is a lignin repolymerization through formation of a carbonium ion intermediate, which promotes the formation of new linkages of β–β, β–1 and β-5 types (Li and Lundquist, 2000 and as shown on Figure 14). Thus, this hypothesis has been confirmed by an improvement in lignin extractability when autohydrolysis were performed in presence of carbonium ions scavangers like 2-naphthol (Wayman and Lora, 1978, Li and Gellerstedt, 2008 and as shown on Figure 14). Thus, in presence of 2-naphthol the autohydrolysis should be expected to proceed with a depolymerisation of lignin (through both heterolytic acidolysis and homolytic cleavage of β-O-4 linkage) but in the same time, the lignin-lignin condensation should be inhibited by the presence of the highly nucleophilic naphthol through addition at the Cα position (Figure 13). As a result, the presence of 2-naphthol during the autohydrolysis resulted in a better yield of delignification during the organosolv step leaving a pulp with a lower lignin content. It was also recently demonstrated that in presence of 2-naphthol a better sugar recovery was also observed (El Hage et al., 2010b).

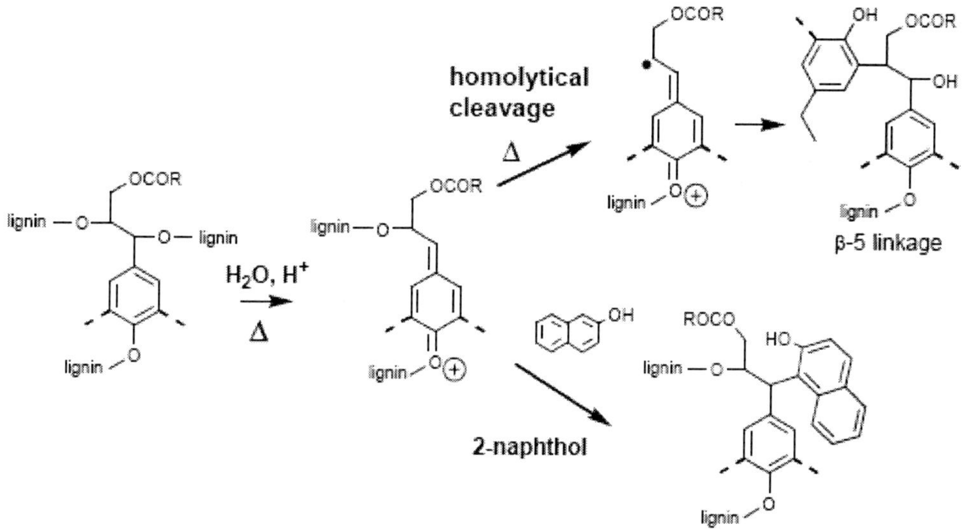

Figure 14. Mechanism of the homolytic cleavage of β-aryl ether linkage and of mechanism of action of 2-naphthol.

5. SOLUBILITY OF ORGANOSOLV LIGNIN

Organosolv lignins are hydrophobic and insoluble in water whereas they are soluble in many organic solvents. The values of solubility of acetosolv lignin (extracted from bagasse) are given in the Table 1 together with the Hildebrand solubility parameters and hydrogen bonding parameters (Schuerch, 1952).

Table 1. δ and δ_h Values of the Solvents and Solubility of the Acetosolv Lignins

Solvent	δ	δ_h	Solubility %
Diethylether	7.4	2.5	Insoluble
Ethylacetate	9.1	3.5	24.3 - 45.1
Chloroform	9.3	2.8	Insoluble
Acetone	9.9	3.4	52 - 88.4
Dioxane	10.0	3.6	83.1- 99.1
Dioxane / water	12.7		82.6 – 97.5
Pyridine	10.7	2.9	88.4 – 98.2
Dimethylsylfoxide	12	5	Soluble
Ethanol	12.7	9.5	20.5-45.1
Methanol	14.5	10.9	23.1 – 48.1

It has been demonstrated that lignin fractions are most soluble in solvents with δ values around eleven, their solubilities decrease fairly regularly as the δ value of the solvents decrease. The best solvents are DMSO, pyridine, dioxane, 80% dioxane-water. The other important parameters to take into account are:

- the capacity of the solvent to form hydrogen bonds. As an example, lignin samples were insoluble in chloroform despite a δ value near that of dioxane because of a poor hydrogen-bonding capacity ($\delta_H = 2.8$);
- the acid-base interaction between basic solvent like pyridine and phenolic group in lignin.

The solubility of organosolv lignin also depends on the pulping conditions : the solubility increases with the severity of the treatment (Shukry et al., 2008). In fact, the organosolv treatment extensively cleaves some inter-unit bonds in lignin to produce low molecular weight lignin fractions soluble in solvents and this degradation is a function of the reaction time, the temperature and the acid concentration.

6. THERMAL BEHAVIOUR OF ORGANOSOLV LIGNIN

The thermal behaviour of organosolv lignin was studied by different authors (Shukry et al., 2008 ; Sun et al., 2001, Uraki et al., 1995). Different lignin preparations are reported to have Tg values between 90 and 180 °C, the higher values corresponding usually to softwood kraft lignins and lower ones to organosolv lignins. This difference of Tg can be rationalized based on differences in chemical structure between the three lignins. Organosolv lignin contains less condensed inter-unit linkages. As a result, its thermal mobility is more important than that of kraft lignin which bears a more complex, network-like structure which hinders the molecular motions. Softwood lignin which is primarily composed of guaiacyl units is the more condensed and contains more condensed inter-unit linkages such as 5-5' and β-5. According Kubo and Kadla (2004), the good thermal properties of the Alcell lignin (low Tg) arise from its unique chemical structure. NMR analysis revealed the presence of alkoxyl chains at the Cα and Cγ positions of the Alcell lignin side chain structure acting as internal plasticizers and enhancing the thermal mobility of the lignin.

The organosolv lignin thermogravimetric curves exhibits that thermal degradation took place after about 190-200°C (Sun et al, 2001; Xu et al., 2006). Nevertheless, Shukry et al (2008) and Pan and Sano (1999) showed that acetosolv lignin underwent two endothermic transitions: the first corresponded to Tg (87°C – 92°C) and the second may be the fusion temperature (155°C – 178°C). According to these authors, fusibility could be a unique property of acetic acid lignins. As a consequence, such lignins could be used as molded materials.

7. APPLICATIONS OF POLYMERIC ORGANOSOLV LIGNIN

The incorporation of lignin into other polymers in the form of blend partners or simply fillers has been investigated for decades, but these efforts have not been translated into any sizable industrial activity most probably because of the limited advantages obtained in terms of gains in specific properties. Sulphur-free organosolv lignins are an emerging class of lignin products which resemble more closely to the structure of native lignins. This opens new utilizations in diverse area, principally as green alternatives to non renewable plastics and resins.

7.1. Synthetic Polymer Blends

Organosolv lignins are hydrophobic polymers, recovered by precipitation in water. Nevertheless, they contain polar groups (phenolic and aliphatic hydroxyl groups, carbonyl groups...) which enable lignin to establish strong interactions with hydrophilic polymers. As a consequence, the spinning properties of the lignin based hydrophobic polymer blends were poor because of phase separation due to a lack of compatibility (Kubo and Kadla, 2005b). It was shown that in lignin–polypropylene fibers, the polypropylene was separated from the lignin. The polypropylene can be seen as fine fibers or "strings" within the bulk of the lignin fiber (Figure 15).

On the other hand, amphiphilic polymers show good affinity toward lignin. Alcell® lignin was thermally blended with poly(ethylene oxide) (PEO) over a range of blend compositions and good spinning properties were observed (Kubo and Kadla, 2004). Compared with kraft lignins, Alcell® lignin displayed better plastic behaviour and the corresponding PEO blends were described to possess good thermoforming properties. As a consequence, fiber spinning was achieved by fusion spinning (Kubo and Kadla, 2003). According these authors, the thermal properties arise because of the chemical structure of the Alcell® lignin : HMQC 2D NMR studies revealed flexible alkoxyl chains introduced at the Cα and Cγ positions of lignin side chain structure during the organosolv process. These moieties seem to act as internal plasticizers enhancing the thermal mobility of the lignin.

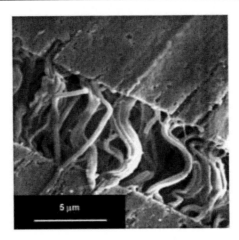

Figure 15. Scanning electron microcraph of a lignin/PP (75/25) fiber. *From Kubo and Kadla, 2005, with permission.*

Alcell® lignin was used as a partial replacement of PVC (Feldman et al., 2002, 1996) for the development of new vinyl flooring formulations with increased resistance to attack by fungi and microorganisms. The influence of the addition of plasticizers for reducing the degree of association existing between lignin molecules was examined (Feldman et al., 2001). The results indicated that lignin could replace up to 20 parts of the copolymer. At this level of replacement, the key mechanical properties of the new composites compared very favorably with those of the control formulations. The results obtained suggested a proton donor/proton acceptor interaction between lignin and PVC chains.

In order to tailor organosolv lignins for specific end uses, many derivatization reactions have been carried out and among them, etherification with alkylene oxides (especially ethylene oxide) has been extensively studied (Lora and Glasser, 2002). This chemical modification is carried out in aqueous alkali at room temperature and produces a unifunctional derivatives with only aliphatic hydroxy groups (Wu and Glasser, 1984 ; Glasser et al., 1984). A drastic reduction of glass transition temperature was observed accompanied by an enhanced solubility in organic solvents. Thus, a decrease in brittleness could be achieved in the resulting material through the introduction of soft molecular segments capable of a plastic response to mechanical deformation.

7.2. Polyurethanes

The utilization of organosolv lignin as macromonomers in polyurethane synthesis was studied by different authors. The high hydroxyl content of organosolv lignin makes possible its direct use without any preliminary chemical modification (Cateto et al., 2008), alone or in combination with other polyols. Studies have demonstrated that different types of polyurethanes having a wide range of mechanical properties can be produced from Alcell® lignin, polyethylene glycol (PEG) and polymeric methyl-diisocyanate (MDI). At low lignin content, flexible but weak polyurethanes were produced but at lignin contents of 30 wt% or above, hard and brittle films were obtained (Thring et al, 1997). Three fractions of distinct molecular weight and chemical functionality from Alcell® lignin were examined: the high molecular weight fraction of Alcell lignin produced stronger polyurethanes, but the materials produced from the medium molecular weight fraction appeared to be tougher and more flexible. Polyurethanes from the low molecular weight fraction were the weakest and could only be prepared with lignin contents below 18 wt%, probably because of the low functionality and/or molecular weight of this fraction (Vanderlaan et al., 1998).

7.3. Carbon Fibers

Carbon fibers produced from organosolv lignins were reported. The process involved thermal spinning at 145°C-165°C followed by carbonization. It was demonstrated that Alcell® lignin exhibited rather thermal processability and continuous fiber spinning was achieved (Kadla et al., 2002). However, lignin fibers and subsequently the carbon fibers exhibited poor mechanical properties and were very brittle (Kubo et al, 2005a).

7.4. Phenolic Resins

The replacement of phenol by lignin and its derivatives has attracted increasing attentions in research and industry. In the literature, the utilisation of organosolv lignin in phenol-formaldehyde resins was achieved through various approaches (Pizzi, 2006). The use of organosolv lignin as a partial replacement for phenolic resins was successfully proposed by Nehez (1997). The use of 20% lignin/80% phenolic resin resulted in competitive advantages relative to controls prepared with 100% phenolic resin. Because of the low chemical reactivity of lignin, utilization of higher lignin content resulted in a decrease in the resin properties : their low reactivity and low level of reactive sites causes that for any percentage of lignin added the cost advantage is lost in the lengthening of the panel press time. This low reactivity has been partially overcome by some pre-treatment methods such as the phenolation and methylolation of lignin before introduction to the phenol–formaldehyde synthesis. Cetin and Ozmen (2002) have demonstrated that phenolated-lignin–formaldehyde resin formulations exhibited satisfactory resin properties. Plywood boards were prepared using resins obtained by copolymerization of phenol, formaldehyde and pre-methylolated acetosolv pine lignin (Vazquez et al. 1999, Figure 16).

Pizzi et al. (1993) developed a promising technology based on pre-methylolated lignin in presence of small amounts of a synthetic phenol-formaldehyde (PF) resin and polymeric 4,4-diphenyl methane diisocyanate (PMDI). The proportion of pre-methylolated lignin used is 65 wt% of the total adhesive, the balance being made up of 10–15% PF resin and of 20– 25% PMDI. This adhesive presses at very fast speed, well within the fastest range used today industrially, contains a high proportion of lignin, and yields exterior grade boards. More recently, mixed wood panel adhesive formulations for interior-grade applications were developed using glyoxalated organosolv lignin. Glyoxal, in opposition to formaldehyde, is a non toxic, non-volatile aldehyde. The ^{13}C-NMR spectrum of lignin after glyoxalation indicated that the conditions used for the reactions modified the lignin fragments in a large extends. Aliphatic region of the ^{13}C NMR spectrum displayed very weak signals arising from the aliphatic C-O on the lignin propyl side chain; this observation indicates that the aryl ether bonds (β-*O*-4 and α-*O*-4) bonds were highly affected by the treatment (El Hage et al., 2010c). However, the glyoxalated lignin by itself gave relatively low wood-joint strength and it cannot be used alone as wood adhesive unless it can be cross-linked furtherly.

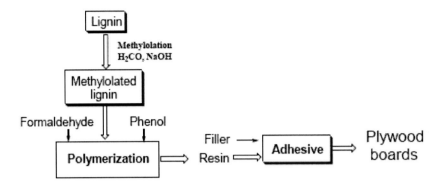

Figure 16. Phenolated-lignin–formaldehyde resin formulations for the formulation of wood adhesive.

Thus, adhesive formulations based on glyoxylated organosolv lignin and mimosa tannin mixtures (50/50) were prepared and the rigidity of bonded wood joint in function of temperature was studied by TMA. Environment-friendly, non-toxic polymeric materials of natural origin constitute as much as 94% of the total panel binder formulation. The wood panel itself is then constituted of 99.5% natural materials. The 0.5% non natural material is composed of glyoxal, a non-toxic and non-volatile aldehyde. The resulting panels present an effective zero formaldehyde emission when tested by the dessicator method (Mansouri et al., 2010).

CONCLUSION

Organosolv processes produce large amount of lignin fractions which are relatively pure, unaltered, sulfur-free and less condensed than other pre-treatment lignins. In addition, their display thermal moldability and good solubility in many organic solvents. These qualities make them suitable for new emerging applications, principally as sustainable alternatives to non renewable products such as polyurethanes, thermoplastic polymers, carbon fibers and epoxy and phenolic resins. However, all the researches currently developed in these fields are carried out at the laboratory scale. Thus, the ability of organosolv lignins to make significant impact as a substitute for polymeric materials depends on its availability in industrial quantities.

REFERENCES

Adler, E., Pepper, J. M., Eriksoo, E. Action of mineral acid on lignin and model substances of guaiacylglycerol-beta-aryl ether type. *Ind. Eng. Chem.* 1957, 49, 1391.

Alfaro, A., Rivera, A., Perez, A., Yanez, R., Garcia, J.C., Lopez, F., 2009. Integral valorization of two legumes by autohydrolysis and organosolv delignification. *Bioresour. Technol.* 100, 440–445.

Araque, E ; Parra, C ; Freer, J ; Contreras, D ; Rodrıguez, J ; Mendonc, R ; Baeza, J. Evaluation of organosolv pretreatment for the conversion of Pinus radiata D. Don to ethanol *Enzyme and Microbial Technology.* 2008 ; 43 : 214–219.

Banoub, J. H.; Benjelloun-Mlayah, B.; Ziarelli, F.; Joly, N.; Delmas, M. *Rapid Commun. Mass Spectrom.* 2007; 21: 2867–2888.

Brosse N, Sannigrahi P, Ragauskas A. Pretreatment of Miscanthus x giganteus using the ethanol organosolv process for ethanol Production. *Ind. Eng. Chem. Res.* 2009; 48: 8328–8334.

Brosse N, El Hage R, Sannigrahi P, Ragauskas A. Dilute Sulphuric Acid and Ethanol Organosolv Pretreatment of *Miscanthus x Giganteus Cellulose Chemistry and Technology* 2010 ; 44 (1-3): 71-78.

Buranov, AU; Mazza, G. Lignin in straw of herbaceous crops. *Ind. Crops Prod.* 2008, 28: 237-259.

Caparros, S., Ariza, J., Garrote, G., Lopez, F., Dıaz, M.J. Optimization of *Paulownia Fortunei L.* Autohydrolysis-Organosolv Pulping as a Source of Xylooligomers and Cellulose Pulp. *Ind. Eng. Chem. Res.* 2007 ; 46: 623-631.

Cateto, CA ; Barreiro, MF ; Rodrigues, AE ; Brochier-Salon, M-C ; Thielemans, W ; Belgacem, MC. Lignins as Macromonomers for Polyurethane Synthesis: A Comparative Study on Hydroxyl Group Determination. *J. Appl. Polym. Sci.* 2008 ; 109: 3008–3017.

Chakar, F S ; Ragauskas, A. Review of current and future softwood kraft lignin process chemistry. *Ind. Crops Prod.* 2004, 20: 131-141.

Crestini, C ; Argyropoulos, DS. Structural Analysis of Wheat Straw Lignin by Quantitative 31P and 2D NMR Spectroscopy. The Occurrence of Ester Bonds and α -O-4 Substructures. *Journal of Agricultural and Food Chemistry.* 1997 ; 45: 1212-1219.

Delmas, M. Vegetal refining and agrichemistry. *Chem. Eng. Technol.* 2008, 31: 792-797.

Demirbas, A. Aqueous glycerol delignification of wood chips and ground wood. *Bioresource Technol.* 1998, 63:179–185.

El Hage, R ; Brosse, N ; Chrusciel, L ; Sanchez, C ; Sannigrahi, P ; Ragauskas, A. Characterization of milled wood lignin and ethanol organosolv lignin from Miscanthus Polymer Degradation and Stability 2009 94(10), 1632-1638.

El Hage, R ; Brosse, N ; Sannigrahi, P ; Ragauskas, A. Effect of the Severity of the Ethanol Organosolv Process on the Chemical Structure of Ethanol Organosolv Miscanthus Lignins. *Polymer Degradation and Stability* 2010a, 95, 997-1003.

El Hage, R ; Chrusciel, L. ; Desharnais, L. ; Brosse, N. Effect of Autohydrolysis of *Miscanthus x giganteus* on Lignin Structure and Organosolv Delignification. Bioressource Technology. 2010b. in press.

El Hage, R ; Brosse, N ; Navarrete, P ; Pizzi, A. Extraction, characterization and utilization of Organosolv Miscanthus Lignin for the conception of Environmentally Friendly Mixed Tannin/Lignin Wood Resins. *Journal of Adhesion Science and Technology* 2010c in press.

Hallac, B ; Sannigrahi, P ; Pu, Y ; Ray, M ; Murphy, RJ ; Ragauskas, A. Biomass Characterization of Buddleja davidii: A Potential Feedstock for Biofuel Production. *J. Agric. Food Chem.* 2009, 57, 1275–1281.

Hallac, B ; Pu, Y; Ragauskas, A. Chemical Transformations of Buddleja davidii lignin during ethanol organosolv pretreatment. *Energy Fuel*. 2010, 24, 2723-2732.

Feldman, D ; Banu, D ; Campanelli, J ; Zhu, H. Blends of Vinylic Copolymer with Plasticized Lignin: Thermal and Mechanical Properties *J. Appl. Polym. Sci*. 2001, 81, 861–874.

Feldman, D ; Banu, D ; Manley, R ; Zhu, R. Highly Filled Blends of a Vinylic Copolymer with Plasticized Lignin: Thermal and Mechanical Properties *J. Appl. Polym. Sci*. 2003, 89, 2000–2010.

Garrote, G., Dominguez, H., Parajo, J.C., 2001. Generation of xylose solutions from eucalyptus globulus wood by autohydrolysis–posthydrolysis processes: posthydrolysis kinetics. *Bioresour. Technol*. 79 (2), 155–164.

Garrote, G., Dominguez, H., Parajo, J.C., 2002. Interpretation of deacetylation and hemicellulose hydrolysis during hydrothermal treatments on the basis of the severity factor. *Process Biochem*. 37, 1067–1073.

Glasser, W ; Barnett, C ; Rials, TG ; Saraf, VP. Engineering plastics from lignin II. Characterization of hydroxyalkyl lignin derivatives. *J. Appl. Polym.Sci*., 1984, 29, 1815-1830.

Granata A, Argyropoulos DS. 2-chloro-4,4,5,5-tetramethyl- 1,3,2-dioxaphospholane, a reagent for the accurate determination of the uncondensed and condensed phenolic moieties in lignins. *J Agric Food Chem*. 1995 ; 43:1538–1544.

Jimenez, L ; Angulo, V ; Caparro, S ; Ariza, J. Comparison of polynomial and neural fuzzy models as applied to the ethanolamine pulping of vine shoots. *Bioresource Technology*. 2007, 98, 3440–3448.

Kadla, JF ; Kubo, S ; Gilbert, RD ; Venditti, RA. Lignin based carbon fibers in Chemical modification, properties and usage of lignin, Hu, TQ, Ed, Kluwer Academic / Plenum Publishers, New York, 2002, pp121-138.

Khama, L ; Le Bigot, Y ; Delmas, M ; Avignon, G. Delignification of wheat straw using a mixture of carboxylic acids and peroxoacids. *Industrial Crops and Products* 21 (2005) 9–15.

Kishimoto, T and Sano, Y. Delignification Mechanism During High-Boiling Solvent Pulping. V. Reaction of Nonphenolic β-O-4 Model Compounds in the Presence and Absence of Glucose. *J. Wood Chem. Technol*. 2003, 23 (3-4), 279–292.

Kleinert, T. N., Pointe C., 1975. Ethanol-water delignification of wood. Rate constants and activation energy. *Can. Tappi*. 58(8), 170-1.

Kubo, S ; Kadla, JF. Poly(Ethylene Oxide)/Organosolv Lignin Blends: Relationship between Thermal Properties, Chemical Structure, and Blend Behavior Macromolecules 2004, 37, 6904-6911.

Kubo, S ; Gilbert, RD ; Kadla, JF. Lignin based polymer blends and biocomposite materials, Mohanty, A ; Misra, M ; Drzal, LT Ed TaylotandFrancis, CRC press, 2005a, pp671-697.

Kubo, S ; Kadla, JF. Lignin-based Carbon Fibers: Effect of Synthetic Polymer Blending on Fiber Properties. *Journal of Polymers and the Environment*, 2005b, 13, (2), 97-105.

Lee, J.M., Shi, J., Venditti, R.A., Jameel, H., 2009. Autohydrolysis pretreatment of Coastal Bermuda grass for increased enzyme hydrolysis. *Bioresource Technology* 100, 6434–6441.

Leschinsky, M., Zuckerstatter, G., Weber, H.K., Patt, R., Sixta, H., 2008a. Effect of autohydrolysis of Eucalyptus globulus wood on lignin structure. Part 1 : Comparison of different lignin fractions formed during water prehydrolysis. *Holzforschung*. 62, 645-652.

Leschinsky, M., Zuckerstatter, G., Weber, H. K., Patt, R., Sixta, H., 2008b. Effect of autohydrolysis of Eucalyptus globulus wood on lignin structure. Part 2 : Influence of autohydrolysis intensity. *Holzforschung,* 62, 653-658.

Li, S., Lundquist K., 2000. Cleavage of arylglycerol β-aryl ethers under neutral and acid conditions. *Nord. Pulp pap. Res. J.* 15, 292-299.

Li, J., Gellesrstedt, G., 2008. Improved lignin properties and reactivity by modifications in autohydrolysis process of aspen wood. *Ind. Crops Prod.* 27, 175-181.

Lora, JH ; Glasser, WG. Recent Industrial Applications of Lignin: A Sustainable Alternative to Nonrenewable Materials *Journal of Polymers and the Environment.* 2002, 10, 39-47.

Lora, J.H., Wayman, L., 1978. Delignification of hardwoods by autohydrolysis and extraction. 61, 47-50.

Mansouri, HR ; Navarrete, P ; Pizzi, A ; Tapin-Lingua, S ; Benjelloun-Mlayah, B ; Pasch, H ; Rigolet, S. Synthetic-resin-free wood panel adhesives from mixed low molecular mass lignin and tannin. *Eur. J. Wood Prod.* 2010, In press.

Meshgini, M ; Sarkanen, KV. Synthesis and kinetics of acid catalysed hydrolysis of some aryl ether lignin model compounds. *Holzforschung.* 1989, 43, 239-243.

McDonough, TJ. The chemistry of organosolv delignification. *Tappi.* 1993, 76, 186-193.

Munoz, C., Mendonc, R., Baeza J., Berlin, A., Saddler, J., Freer, J., 2007. Bioethanol production from bioorganosolv pulps of *Pinus radiata* and *Acacia dealbata J. Chem. Technol. Biotechnol.* 82, 767–774.

Nehez, N. J. "Lignin-based friction material," (1997) *Canadian Patent Application* 2,242,554.

Pan, XJ ; Sano, Y. Atmospheric acetic acid pulping of rice straw. Part 4. Physico-chemical characterization of acetic acid lignins from rice straw and woods. Part 1. Physical characteristics. *Holzforschung.* 1999, 53, 511-518.

Pan X, Kadla J F, Ehara K, Gilkes N, Saddler J N. Organosolv ethanol lignin from hybrid poplar as a radical scavenger: relationship between lignin structure extraction conditions and antioxidant activity. *J. Agric. Food Chem.* 2006; 54: 5806–5813.

Pan X, Xie D, Yu R, W, Saddler J N. The bioconversion of Mountain Pine Beetle-Killed Lodgepole Pine to fuel ethanol using the organosolv process. *Biotechnol. Bioeng.* 2008; 101 (1): 39-47.

Patel, D.P., Varshney, A.K., 1989. The effect of presoaking and prehydrolysis on the organosolv delignification of bagasse. *Indian J. Technol.* 27(6) 285-288.

Sixta, H., 2006. Multistage Kraft pulping. In : Handbook of pulp. Ed Sixta, H. Wiley-VCH. Weinheim. Pp 325-365.

Stephanou, A.; Pizzi, A. Rapid-curing lignin-based exterior wood adhesives. Part II: Esters acceleration mechanism and application to panel products. *Holzforschung.* 1993, 47, 501–506.

Pizzi, A. Recent developments in eco-efficient bio-based adhesives for wood bonding: opportunities and issues. *J. Adhesion Sci. Technol.* 2006, 20 (8), 829–846.

Ragauskas, A.J.; Williams, C.K.; Davison, B.H.; Britovsek, G.; Cairney, J.; Eckert, C.A.; Frederick, W.J., Hallett, J.P.; Leak, D.J.; Liotta, C.L.; Mielenz, J.R.; Murphy, R.; Templer, R.; Tschaplinski, T. The path forward for biofuels and biomaterials. *Science.* 2006 311, 484-489.

Rodriguez, A., L. Serrano, A. Moral, L. Jimenez Pulping of rice straw with high-boiling point organosolv solvents. *Biochemical Engineering Journal* 42 (2008) 243–247.

Sannigrahi, P.; Ragauskas A.J.; Miller. S.J. Effects of high temperature dilute acid pretreatment on the structure and composition of lignin and cellulose in Loblolly pine. *Bioenerg. Res.* 2008 1 (3-4), 205-214.

Sannigrahi, P; Ragauskas, A; Miller, S. Lignin structural modifications resulting from ethanol organosolv treatment of loblolly pine. *Energy Fuel.* 2010, 24 (1), 683–689.

Sarkanen KV. In Progress in Biomass Conversion, Sarkanen KV and Tillman DA, Editors, Academic Press, New York, 1980, vol2, pp127-144.

Schuerch, C. The Solvent Properties of Liquids and Their Relation to the Solubility, Swelling, Isolation and Fractionation of Lignin. *J. Amer. Chem. Soc*, 1952, 74, 5061-5067.

Shukry, N ; Fadel, SM ; Agblevor, FA ; El-Kalyoubi, SF. Some physical properties of acetosolv lignins from bagasse. *J. Appl. Polym. Sci.* 2008, 109, 434-444.

Sun, RC ; Lu, Q ; Sun XF. Physico-chemical and thermal characterization of lignins from Caligonum monogoliacum and Tamarix spp. *Polym. Dgrad. Stab.* 2001, 72, 229-238.

Sun, S ; Chen, H. Evaluation of enzymatic hydrolysis of wheat straw pretreated by atmospheric glycerol autocatalysis *J. Chem. Technol. Biotechnol.* 2007, 82, 1039–1044.

Sun, S ; Chen, H. Enhanced enzymatic hydrolysis of wheat straw by aqueous glycerol pretreatment. *Bioresource Technology.* 2008, 99, 6156–6161.

Sun, F ; Chen, H. Organosolv pretreatment by crude glycerol from oleochemicals industry for enzymatic hydrolysis of wheat straw. *Bioresource Technology.* 2008, 99(13), 5474-5479.

Thomsen, M.H., Thygesen, A., Thomsen, A. B., 2009. Identification and characterization of fermentation inhibitors formed during hydrothermal treatment and following SSF of wheat straw. *Applied Microbiol. Biotechnol.* 83(3), 447-455.

Thring, RW ; Vanderlaanand, MN and Griffin, SL. Polyurethanes from Alcell lignin. *Biomass and Bioenergy.* 1997, 13, (3), 125-132.

Uraki, Y. ; Kubo, S ; Nigo, N ; Sano, Y ; Sasaya, T. Preparation of carbon-fibers from organosolv lignin obtained by aqueous acetic acid pulping. *Holzforschung.* 1995, 49, 343-350.

Vanderlaanand, MN ; Thring, RW. Polyurethanes from Alcell lignin fractions obtained by sequential solvent extraction. *Biomass and Bioenergy.* 1998, 14, 525-531.

Vazquez, G ; Antorrena, G ; Gonzalez, J ; Freire, S ; Lope, S. Acetosolv fractionation of pine wood. Kinetic modelling of lignin solubilization and condensation. *Bioresour. Technol.* 1997; 59, 121–127.

Villaverde, JJ ; Li, J ; Ligero, M ; De Vega A. Native Lignin Structure of Miscanthus x giganteus and Its Changes during Acetic and Formic Acid Fractionation J. *Agric. Food Chem.* 2009, 57, 6262–6270.

Wu, L ; Glasser, WG. Engineering plastics from lignin. I. Synthesis of hydroxypropyl lignin. *J. Appl. Polym.Sci.*, 1984, 29, 1111-1123.

Xu, F ; Sun, JX ; Sun, RC ; Fowler, P ; Baird, MS. Comparative study of organosolv lignins from wheat straw. *Ind. Crops Prod.* 2006, 23, 180-193.

Zhao XB, Cheng KK, Liu DH. Organosolv pretreatment of lignocellulosic biomass for enzymatic hydrolysis *Appl. Microbiol. Biotechnol.* 2009a, 82:815–827.

Zhao XB, Peng F, Cheng KK, Liu DH. Enhancement of the enzymatic digestibility of sugarcane bagasse by alkali–peracetic acid pretreatment. *Enzyme Microb. Technol.* 2009b, 44:156–164.

In: Lignin
Editor: Ryan J. Paterson

ISBN 978-1-61122-907-3
© 2012 Nova Science Publishers, Inc.

Chapter 13

BIOBLEACHING OF PAPER PULP WITH LIGNIN DEGRADING ENZYMES

*Diego Moldes**

Department of Chemical Engineering. University of Vigo.
Isaac Newton Building. Lagoas Marcosende, E-36310 Vigo, Spain

ABSTRACT

Paper pulp bleaching, the most polluting and expensive step on pulp production, consists on removal and decolouration of residual lignin after wood cooking process. Pulp bleaching is performed during several steps using chemical reagents with chlorine species generally involved. Environmental restrictions forced pulp industry to implant ECF (elemental chlorine free) and TCF (totally chlorine free) bleaching processes. Enzymes involved in natural degradation of lignin have been proposed as possible biobleaching agents: peroxidases and laccases can modify the structure of lignin and improve the subsequent chemical steps. The introduction of an enzymatic stage has been studied in several bleaching sequences with different results depending on operational conditions, kind of pulp to be bleached, enzyme used, bleaching sequence employed, etc. A growing interest exists regarding the application of laccase-mediator systems: couples of the enzyme laccase and low molecular weight compounds able to increase the oxidation power of the enzyme and to diffuse through places where the enzyme cannot access due to its large size. Laccases are commercially available in high quantities and their bleaching capability was already demonstrated. The main limitations of laccase application for lignin removal on pulp are the cost of mediator, the effectiveness of some of the tested mediators, and the suitability of enzyme treatments on mill equipment. Lignin modification with these enzymes and their application for pulp biobleaching has been studied for several authors, but more knowledge is needed to get industrial implementation.

---

* E-mail: diego@uvigo.es; diegomoldes@gmail.com.

1. Lignin Degrading Enzymes

Lignin is a complex, three dimensional polymer composed by a random coupling of phenylpropanoid units, namely, syringyl alcohol (S), guaicyl alcohol (G), and p-coumaryl alcohol (H). Up to 20 different types of bonds are presented in this polymer, although the most important one is aryl-aryl ether bond. The main role of lignin is the protection of cellulose and hemicellulose from the plant cell wall. Lignin also provides interesting structural properties to wood in nature like hardness, waterproofness and resistance to microbial attack, as lignin is a very recalcitrant polymer. Only selected organisms can degrade lignin, predominantly fungi (Hammel, 1997). They used an extracellular complex of enzymes that performs different redox reactions leading lignin degradation. There are different groups of enzymes that participate in this process, although each fungi species has its own enzymatic profile (Hatakka, 1994). Basically the attack of lignin is produced by peroxidases and/or laccases (polyphenol oxidases) generating free radicals that perform the oxidation of the polymer by using hydrogen peroxide in the case of peroxidases or molecular oxygen for laccases.

1.1. Peroxidases

There are different peroxidases able to oxidise lignin structure. The first discovered was the lignin peroxidase or LiP (EC. 1.11.1.14) produced by the white rot fungus *Phanerochaete chrysosporium* (Tien and Kirk, 1984). Most white rot fungi produce this heme-containing peroxidase (Hatakka, 1994), a high redox potential enzyme able to oxidize even methoxyl substituents of non-phenolic aromatic rings with a redox potential up to 1.5 V, by a previous generation of cation radicals. The catalytic mechanism of LiP (Figure 1) is a redox cycle where the heme-group is oxidized/reduced in several steps. The LiP oxidation mechanism is activated by hydrogen peroxide, which allows the native enzyme to be converted into LiPI compound. LiPI then produces the generation of free radicals from aromatic compounds performing its own reduction and leading to LiPII compound, which can be also reduced to recover the native enzyme with the necessary oxidation of a new aromatic compound. LiPII could also be inactivated by an irreversible reaction by a moderate excess of hydrogen peroxide.

MnP or manganese peroxidase (EC. 1.11.1.13) was also first discovered in the fungus *Phanerochaete chrysosporium* (Kuwahara et al., 1984). This protein is also a heme-containing enzyme with a catalytic cycle similar to that of LiP. In this case the aromatic compound to be oxidized is substituted by the cation Mn^{2+}, which is oxidized to Mn^{3+}, a strong oxidant. Due to its reactivity and instability, Mn^{3+} should form chelates with organic acids, secreted by fungi in nature, to stabilize and to perform the oxidation of lignin. These chelates may diffuse allowing MnP to attack lignin from the wood cell walls that are inaccessible to enzymes (Archibald and Roy, 1992).

A new type of lignin degrading peroxidases was later discovered in the fungus *Pleurotus eryngii* (Martínez et al., 1996). This third type of ligninolytic peroxidase was called versatile peroxidase or VP (EC. 1.11.1.16), and shares some catalytic properties with LiP and with

MnP, since VP can oxidize Mn^{2+} to Mn^{3+} and also typical LiP substrates: veratryl alcohol, methoxybenzenes and non-phenolic lignin model compounds.

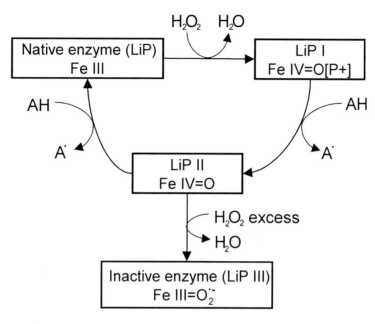

Figure 1. LiP catalytic cycle (Wariishi et al., 1991).

As peroxidases need hydrogen peroxide to be active, they should be accompanied by hydrogen peroxide producing enzymes like glucose oxidase (Eriksson et al., 1986), glyoxal oxidase (Kersten and Kirk, 1987) or aryl alcohol oxidase (Guillén et al. 1990). Another secondary route for hydrogen peroxide production is the oxidation of organic acids by MnP (Urzua et al., 1995).

1.2. Laccases

Laccases (EC. 1.10.3.2) are multicopper oxidases that perform the oxidation of mono-, di- and polyphenols, aminophenols, methoxyphenols, aromatic amines and ascorbate (Galhaup et al., 2002). They are produced by different organisms (e.g. plants, bacteria, insects, etc.) (Mayer and Staples, 2002), being the ones from fungi the most important in order to look for an industrial application. Laccases just need molecular oxygen to carry out the conversion of their phenolic substrates into quinone radicals (Thurston, 1994). The attack of lignin is produced over the phenolic subunits leading to Cα-Cβ cleavage, aryl-alkyl cleavage or Cα oxidation (Archibald et al., 1997). The reducing substrate of laccase loses a single electron and generates a free radical that can undergo further laccase oxidation, usually to form a quinone, or non-enzymatic reaction like hydration or polymerization (Xu, 1999).

Laccases are restricted to phenolic moieties of lignin because of their low redox potential (Reid and Paice, 1994) but the presence of the so called mediator compounds resulted in a higher oxidation capability than the enzyme itself, allowing oxidation of non-phenolic lignin. The mediator is a low molecular weight molecule, substrate of laccase, which generates a reactive radical that can oxidize complex substrates. The mediator allows increasing the range of laccase substrates, and oxidizing non-accessible to laccase molecules due to its lower size. Therefore the mechanism of these laccase mediator systems consists on the oxidation of the mediator by laccase, generating a radical which can undergo non-enzymatic oxidations (Figure 2). The mediator is restored to its original form when the oxidation of the final substrate of reaction is performed. From a practical point of view this simple mechanism is not always produced and some mediators are not restored since their oxidations are not reversible, as it was recently demonstrated by cyclic voltametry (González Arzola et al., 2009). Anyhow, these non-reversible mediator compounds usually improve laccase reactivity and probably the name "laccase enhancer" instead of "laccase mediator" should be used in these particular cases.

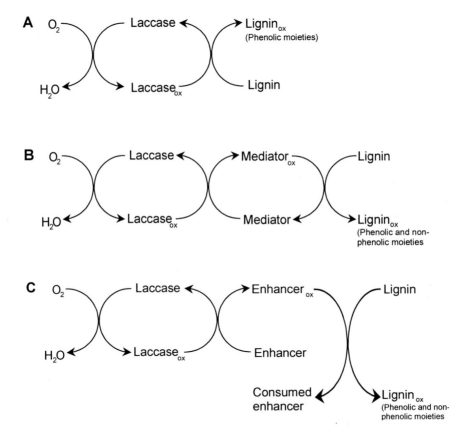

Figure 2. Scheme of the mechanism of lignin oxidation by laccase (A) by laccase-mediator systems (B) and by the new considered laccase-enhancer systems (C).

The complete and detailed oxidation mechanisms of peroxidases and laccases are not explained in this chapter, but excellent reviews are available concerning these aspects for peroxidases (Martínez, 2002; Hofrichter, 2002) for laccases (Madhavi and Lele, 2009; Giardina et al., 2010) and for both (Higuchi, 2004; Wong, 2009).

1.3. Potential Applications of Ligninolytic Enzymes

Ligninolytic enzymes present several and promising potential applications in textile, food, cosmetic and pulp and paper industries. Moreover their capabilities can be also used for detoxification and/or decolourisation of effluents, degradation of xenobiotics for bioremediation, synthesis of organic compounds, production of wood composites (biobased adhesives), biosensors and cathodes for biofuels (Burton, 2003; Riva, 2006; Rodríguez Couto and Toca Herrera, 2006; Widsten and Kandelbauer, 2008; Madhavi and Lele, 2009; Hamid and Rehman, 2009).

Regarding food industry, laccases can be applied in beer and wine stabilization, as indicator of must infection, in fruit juice processing, gelation, biosensor, baking and others (Minussi et al. 2002). In textile industry laccase is used for denim finishing since commercial formulations are already available. Moreover the capability of laccase and peroxidases to degrade textile effluents and dyes has been widely demonstrated (Levin et al., 2004; Husain, 2010; Sasmaz et al., 2010). Some applications of laccase in the field of medical and personal care have also been reported: hair dyeing, deodorant, production of anticancer drugs or even as ingredients in cosmetics (Kunamneni et al., 2008).

There are a lot of references related to the application of ligninolytic enzymes for environmental applications regarding detoxification of effluents and soils, and biodegradation of pollutants (pesticides, polycyclic aromatic hydrocarbons, dyes, drugs, etc.) (Pizzul et al., 2009; Majeau et al., 2010; Haritash and Kaushik, 2009). Peroxidases have high redox potential and so on they can catalyze the degradation of many organic pollutants, and although as a general rule laccases have not this property, laccase mediator systems can be effectively applied for this purpose. The main property of peroxidases and laccases to perform the oxidation of a huge number of pollutants, in addition to the redox potential, is their unspecific nature regarding to substrate.

Pulp and paper industry shows great possibilities for ligninolytic enzymes application, since they could be used in several steps of the production chain. The pretreatment of wood or wood chips before pulping with lignin degrading fungi, usually called biopulping, improves the quality of the final product and reduces the necessary input of energy for pulp production (Singh et al., 2010); the wastewater of pulping process can be successfully degraded by ligninolytic enzymes; pulp and paper properties could be also modified by reactions mediated by laccase and finally, biobleaching is a promising application of these kind of enzymes that will be commented in detail in this chapter.

The wide range of potential applications of peroxidases and laccases led to intense research during last decades. As previously commented, industrial, medical, environmental and other different applications have been reported, but some limitations should be considered. The application of peroxidases and laccases depends on several factors, being the bulk production of these enzymes the most important one. Several efforts for high-scale peroxidase production failed due to the accumulation of peroxidases on insoluble inclusion bodies and the low yield of active protein (Conesa et al., 2000; Gu et al., 2003). However, laccases are already commercially produced in heterologous expression systems, but their low redox potential force the searching of effective, non-expensive and non-toxic mediator compounds allowing and effective enzymatic performance. Peroxidases do not need mediators, therefore their bulk production would be a great advance in biocatalysis with oxidative enzymes.

2. APPLICATION OF LIGNIN DEGRADING ENZYMES IN BIOBLEACHING OF PULP

Wood fibres are the raw material to manufacture pulp and consequently, paper. Basically consist in a mixture of cellulose and hemicellulose where lignin acts as linker, therefore lignin must be removed from the wood fibres in order to obtain single ones easily to organize and to form sheets. Moreover lignin provides undesirable properties to the final product, the paper, so its removing should be as high as possible. Lignin degrading enzymes may contribute to this lignin modification and removal as a new tool to develop cleaner processes.

2.1. Conventional Pulping and Bleaching of Pulp

There are two main methods of pulping: chemical and mechanical. The former consist on the solubilisation of lignin by using the proper chemical treatment, whereas the latter relies on the mechanical treatment of wood to separate the fibres, although no removal of lignin is produced. Even though the mechanical pulping allows a high yield (up to 98% of the starting wood), the quality of pulp is better after chemical processes, although with lower yields (up to 55%). The most important reason is the presence of lignin in mechanical pulps, which provides lower strength to the paper and a marked trend of becoming yellow-brown by the action on sunlight on lignin.

The main method used in pulp mills all around the world is the kraft process because is effective in lignin removal, can be used for many raw materials, high quality of pulp can be obtained and it is non-expensive. The kraft process solubilises lignin of wood chips by using a cooking solution of sodium hydroxide and sodium sulfide at high temperature (160-170°C). Around 90% of lignin is removed, but also 35% of the carbohydrates are solubilised. Although the removal of lignin could be improved by using harder conditions of cooking, it is necessary to assume lower yields and a decreasing quality of pulp. The residual lignin of pulp turns brown during cooking; consequently it should be removed by complementary treatments or bleached. In any case, brown kraft pulp can be used as raw material to manufacture packaging papers, as the appearance of these products is not as important compared with sanitary or printing papers.

Bleaching of kraft pulp is usually performed in a sequence of several stages to arise the desired delignification level and brightening. In essence during the different stages the lignin is oxidized and then solubilised, finally the degradation of residual lignin and chromophore groups is carried out. Some typical bleaching stages are: chlorination (C), treatment with chlorine dioxide (D), alkaline extraction (E), quelation of metals (Q), treatment with hydrogen peroxide (P), etc. The main parameters analysed by standard procedures to determine pulp quality are kappa number as a lignin quantity indicator, brightness as indicator of optical properties of pulp, and viscosity as a measure of cellulose integrity. The objective of bleaching is the reduction of kappa number with a higher increase in brightness while viscosity should be preserved as much as possible.

The chemicals needed in the several stages makes bleaching the higher in cost and the most polluting process of pulp production. Traditionally, bleaching is carried out by elemental chlorine, but the link between this method and chlorinated dioxins emissions and the high content of chlorine compounds on bleaching effluents, forced the implementation of new methods of bleaching. ECF (elemental chlorine free) methods are based on the application of chlorine dioxide and dominate the world market of bleached chemical pulps. However some bleaching methods refuse to use any chlorine chemical and therefore called TCF (totally chlorine free). The bleaching agents of TCF sequences may be oxygen, ozone, hydrogen peroxide and/or peroxygens. The introduction of enzymes in bleaching sequences may contribute to the development of TCF sequences in order to compete with ECF ones and finally remove chlorine from pulp mills.

2.2. Biobleaching of Pulp

The first attempts of pulp biobleaching were performed *in vivo*, by growing fungi on the pulp to be bleached during several days (Kirk and Yang, 1979; Paice et al., 1989; Reid et al., 1990; Fujita et al., 1991). Main results from these innovative experiments show the capability of white-rot fungi and their lignin degrading enzymes to be involved in delignification of different types of pulp and also in reduction of the requirement of reactives to perform the chemical bleaching (Moreira et al., 1997, 1998; Hirai et al, 1994; Fujita et al, 1993). However the direct application of enzymes for pulp biobleaching is preferred, mainly because fungi promote cellulose hydrolysis whereas ligninolytic enzymes oxidize lignin with minor effects on cellulose. Besides, *in vivo* treatment takes long time, and the analysis of data is extremely difficult, since too many uncontrolled factors are involved (growing of fungi, production of different enzymes, nutrient limitation, pulp colonization, metabolites produced, etc.). Biobleaching *in vivo* are still used as a source of interesting scientific data (Machii et al., 2004; Tang et al., 2009) but not as a search of a feasible application.

LiP and MnP were proved to effectively delignify kraft pulp and improve its bleachability (Arbeloa et al., 1992; Kondo et al., 1994; Jimenez et al., 1997; Ehara et al., 1997; Iimori, 1998), but the use of these peroxidases were limited due to their unavailability and stability. As commented in previous sections in this chapter, peroxidases need hydrogen peroxide to be active, but moderate concentrations inactivate the enzyme in an irreversible reaction, consequently a complex system of continuous low level addition of hydrogen peroxide or a generating system of this compound is needed to effectively perform the treatment of pulp with peroxidases. In any case interesting studies were recently performed in an attempt to look for new peroxidases with new properties that could be interesting for application (Kondo, 2002; Li et al., 2002).

Nowadays, biobleaching studies are mainly focused on the application of laccase mediator systems, especially after the description of the Lignozym© process (Call and Mücke, 1997), although the first time that a laccase mediator system was successfully employed in pulp delignification was previously carried out (Bourbonnais and Paice, 1992). The most studied mediators in biobleaching experiments are 1-hydroxybenzotriazole (HBT) and 2,2'-azino-bis(3-ethylbenzthiazoline-6-sulfonic acid) (ABTS). Curiously, the mechanisms of these two laccase mediator systems are different: ABTS forms a cation radical that carry out electronic transfer whereas the radical from HBT oxidizes the lignin by hydrogen atom abstraction (H-abstraction) (Baiocco et al., 2003). More mediators were studied in order to improve the performance of laccase mediator systems: 3-hydroxyanthranilate (3-HAA) (Eggert et al., 1996; Li et al., 2001), violuric acid (Li et al., 1999; Oudia et al., 2007),

2,2',6,6'-tetramethyl-piperidine-N-oxyl (TEMPO) (Barreca et al., 2003), promazine (PZ), chlorpromazine (CPZ), 1-nitroso-2-naphthol-3,6-disulfonic acid (NNDS), 4-hydroxy-3-nitroso-1-naphthalenesulfonic acid (HNNS) (Bourbonnais et al., 1997), phenol red (Li et al., 1999), N-hydroxyacetanilide (NHAA) (Balakshin et al., 2001), N-(4-cyanophenyl)acetohydroxamic acid (NCPA) (Li et al., 2004), etc. A novel approach on laccase mediated system based on the employment of polyoxometalates (POMs) was also recently described (Balakshin et al., 2001; Tavares et al., 2004; Gamelas et al., 2005; 2007).

Therefore a lot of different research experiments are described in the literature performed with several mediators, but also with different pulps to be bleached: hardwood (Ibarra et al., 2006; Oudia et al., 2008a), softwood (Kleen et al., 2003), non-wood materials like wheat straw (Singh et al., 2008), flax (Camarero et al., 2004; Fillat and Roncero, 2009; Fillat et al., 2010), sisal (Aracri et al., 2009), wastepaper pulp (Sadhasivam et al., 2010), etc.; different laccase sources: from *Trametes versicolor*, *Trametes hirsuta*, *Pycnoporus cinnabarinus*, *Trametes villosa*, *Miceliophthora thermophila*, *Fusarium proliferatum*, *Pycnoporus sanguineus*, *Trametes trogii*, *Pleurotus eryngii*, *Pleurotus ostreatus*, etc.; and with different conditions as well (pH, T^a, consistency, treatment time, O_2 pressure, enzyme activity, etc.).

Some of the common aspects of the biobleaching experiments are the frequent use of a reference mediator like HBT or ABTS in order to compare with other mediator compounds or other studies, and the performance of the enzymatic treatment in the first bleaching stages. Then the sequence is completed by a typical chemical bleaching. No specific data of biobleaching are presented in this chapter due to the great variety of experiments, but as a general rule the application of laccase mediator systems on pulp biobleaching improves the delignification and the optical properties of pulp with neither significant cellulose degradation nor important modification of handsheet properties (Wong et al., 1999; Ibarra et al., 2006; Moldes et al., 2010). Results differ depending on the conditions employed (enzyme, mediator, conditions of enzymatic treatment, bleaching sequence employed, etc.), but the properties of biobleached pulp are not the limitation of using lignin degrading enzymes at mill scale.

The modification of residual lignin after kraft cooking by laccase mediator systems was studied by Crestini and Argyropoulos (1998) using lignin model compounds. The data show that many different degradation pathways are used for laccase mediator system like benzylic oxidation, demethylation, hydroxylation, side-chain oxidation and demethoxylation. Sealey and Ragauskas (1998) observed that the laccase–HBT system attacks preferentially the phenolic moieties of lignin allowing its subsequent removal during alkaline extraction. This preference was certificated later by Elegir et al. (2005), although the modification of non-phenolic subunits is also performed (Barreca et al., 2003).

The major limitation of laccase mediator systems is the selection of an effective, non-expensive and non-toxic mediator. During the last years several efforts were made in order to find natural compounds that could link these properties (Camarero et al., 2007; Eugenio et al., 2010; Da Re et al., 2008; Moldes et al., 2008; Aracri et al., 2009). Some phenolic compounds derived from lignin and easy to obtain from the black liquor of the kraft pulping process as by-products or residues (syringaldehyde, acetosyringone, vanillin, etc.), are good candidates for mediators in laccase assisted bleaching, considering the previously commented characteristics. They are apparently non-toxic as a consequence of their natural origin, however this aspect should be carefully treated, since some compounds may increase their toxicity after reaction with laccase (Fillat and Roncero, 2010). These phenolic lignin derivatives react with laccase to produce the corresponding phenoxyl radicals that could move inside the lignin structure and oxidize the non-phenolic moieties of lignin by H-abstraction mechanism (Cañas and Camarero, 2010).

Coupling reactions of mediators on lignin are usually produced, being more or less important depending on the mediator. This reaction is clearly observed when some phenolic mediators are employed, even producing a significant darkening of pulp (Moldes et al., 2008). Although this effect is undesirable for bleaching, the modification of lignin may increase its solubility in subsequent chemical stages. The coupling reactions on lignin have also been proposed as a new method to obtain pulp with different properties (Lund and Felby, 2001; Aracri et al., 2010).

In addition to delignification capability and brightness enhance, the application of laccase mediator systems provides other advantages, as it has been proved to reduce the bleaching chemical requirements for both ECF (Bajpai et al., 2006) and TCF processes (Moldes and Vidal, 2008), to remove lipophilic extractives from pulp avoiding the pitch problem of TCF sequences (Gutiérrez et al., 2007) and to degrade hexenuronic acids (Oudia et al., 2008b; Valls et al., 2010), responsible compounds of paper pulp brightness reversion (yellowing) and bleaching reagents consumers (Vuorinen et al., 1999).

Different configurations from simple enzyme addition can be used for pulp biobleaching. Enzyme immobilization is a very useful approach to improve its stability reducing the quantity of enzyme required as well. The immobilization on solid supports or in membrane systems were applied for several authors with promising results (Sasaki et al., 2001; Gamelas et al., 2005; 2007). Considering that some limitations concerning enzymatic biobleaching are enzyme stability and availability, these immobilized systems could provide significant improvements. Modification of enzymes like fusion of cellulose binding module is also a possibility of improving enzymatic efficiency (Ravalason et al., 2009).

Other enzymes involved in lignocellulosic biodegradation, especially xylanases, can be used in combination with lignin degrading enzymes to boost biobleaching. Bermek et al. (2000) used xylanases in combination with MnP obtaining a synergistic effect on bleaching. Same synergy was observed in the case of laccase mediator system combined by xylanase in the biobleaching of non-wood plant fibres and recycled waste pulp (Kapoor et al., 2007) and for eucalypt kraft pulp (Valls et al., 2010a). Xylanase treatment of eucalypt kraft pulp was proved to help the subsequent laccase mediator stage by making hexenuronic acids more accessible to degradation, allowing a reducing laccase and mediator doses and also the treatment time (Valls and Roncero, 2009; Valls et al., 2010b).

CONCLUSION

Lignin degrading enzymes are a promising alternative to the conventional bleaching methods of paper pulp. Although the fully substitution of the chemical bleaching is not possible, the introduction of enzymes can provide a reducing requirement of bleaching chemicals, better properties to pulp, and in addition a significant contribution to the development of environmentally cleaner processes. In the case of peroxidases, high quantities of not available enzymes are necessary to perform biobleaching at mill scale, therefore the bulk production of peroxidases would be a great advance in order to apply them on pulp bleaching. Concerning laccases, more powerful enzymes and/or more effective and cheaper mediators are needed. Pulp mills impose certain characteristics for enzymatic treatments implementation, like resistant enzymes to high temperatures and aggressive chemical conditions, adaptability to the existing facilities and also economical advantages or higher pulp quality than conventional processes. Some enzymes can provide these characteristics, nevertheless some improvements like new or modified enzymes, combination of different enzymes or their immobilization should be considered to overcome the present limitations for finally applying lignin degrading enzymes in pulp bleaching.

ACKNOWLEDGMENTS

Biobleaching studies performed by the author were supported by the EU project NMP2-CT-2006-026456 BIORENEW. The author thanks to Xunta de Galicia for an Isidro Parga Pondal contract.

REFERENCES

Aracri, E; Colom, JF; Vidal, T. Application of laccase-natural mediator systems to sisal pulp: and effective approach to biobleaching or functionalizing pulp fibres. *Bioresource Technology*, 2009, 100, 5911-5916.

Aracri, E; Fillat, A; Colom, JF; Gutiérrez, A; del Río, JC; Martínez, AT; Vidal, T. Enzymatic grafting of simple phenols on flax and sisal fibres using laccases. *Bioresource Technology*, 2010, 101, 8211-8216.

Arbeloa, M; Leclerc, J; Goma, G; Pommier, JC. An evaluation of the potential of lignin peroxidases to improve pulp. *Tappi Journal*, 1992, 74, 123-127.

Archibald, FS; Bourbonnais, R; Jurasek, L; Paice, MG; Reid, ID. Kraft pulp biobleaching and delignification by *Trametes versicolor*. *Journal of Biotechnology*, 1997, 53, 215-336.

Archibald, FS; Roy, B. Production of manganic chelates by laccase from the lignin degrading fungus *Trametes (Coriolus) versicolor*. *Applied and Environmental Microbiology*, 1992, 58, 1496–1499.

Baiocco, P; Barreca, AM; Fabbrini, M; Galli, C; Gentili, P. Promoting laccase activity towards non-phenolic substrates: a mechanistic investigation with some laccase-mediator systems. *Organic Biomolecular Chemistry*, 2003, 1, 191-197.

Bajpai, P; Anand, A; Sharma, N; Mishra, SP; Bajpai, PK; Lachenal, D. Enzymes improve ECF bleaching of pulp. *Bioresources*, 2006, 1, 34-44.

Balakshin, M; Chen, C-L; Gratzl, JS; Kirkman, AG; Jakob, H. Biobleaching of pulp with dioxygen in laccase-mediator system – effect of variables on the reaction kinetics. *Journal of Molecular Catalysis B: Enzymatic*, 2001, 16, 205–215.

Balakshin, MY; Evtuguin, DV; Pasoal Neto, C; Cavaco-Paulo, A. Polyoxometalates as mediators in the laccase catalyzed delignification. *Journal of Molecular Catalysis B: Enzymatic*, 2001, 16, 131–140.

Barreca, AM; Fabbrini, M; Galli, C; Gentili, P; Ljunggren, S. Laccase/mediated oxidation of a lignin model for improved delignification procedures. *Journal of Molecular Catalysis B: Enzymatic*, 2003, 26, 105-110.

Bermek, H; Li, K; Eriksson, K-EL. Pulp bleaching with manganese peroxidase and xylanase: a synergistic effect. *Tappi Journal*, 2000, 83, 69.

Bourbonnais, R; Paice, MG. Demethylation and delignification of kraft pulp by *Trametes versicolor* laccase in the presence of 2,2'-azinobis-(3-ethylbenzthiazoline-6-sulphonate. *Applied Microbiology and Biotechnology*, 1992, 36, 823–827.

Bourbonnais, R; Paice, MG; Freiermuth, B; Bodie, E; Borneman, S. Reactivities of various mediators and laccase with kraft pulp and lignin model compounds. *Applied and Environmental Microbiology*, 1997, 63, 4627-4632.

Burton, SG. Oxidizing enzymes as biocatalysts. *Trends in Biotechnology*, 2003, 21, 543-549.

Call, HP; Mücke, I. History, overview and applications of mediated lignolytic systems, especially laccase-mediator-systems (Lignozym©-process). *Journal of Biotechnology*, 1997, 53, 163-202.

Camarero, S; García, O; Vidal, T; Colom, J; del Río, JC; Gutiérrez, A; Gras, JM; Monje, R; Martínez, MJ; Martínez, AT. Efficient bleaching of non-wood high-quality paper pulp using laccase-mediator system. *Enzyme and Microbial Technology*, 2004, 35, 113-120.

Camarero, S; Ibarra, D; Martínez, AT; Romero, J; Gutiérrez, A; del Río, JC. Paper pulp delignification using laccase and natural mediators. *Enzyme and Microbial Technology*, 2007, 40, 1264–1271.

Cañas, A.I; Camarero, S. Laccases and their natural mediators: biotechnological tools for sustainable eco-friendly processes. *Biotechnology Advances*, 2010, In press.

Conesa, A; van den Hondel, CAMJJ; Punt, PJ. Studies on the production of fungal peroxidases in *Aspergillus niger*. *Applied and Environmental Microbiology*, 2000, 66, 3016–3023.

Crestini, C; Argyropoulos, DS. The early oxidative degradation steps of residual kraft lignin models with laccase. *Bioorganic and Medical Chemistry*, 1998, 6, 2161-2169.

Da Re, V; Papinutti, L; Villalba, L; Forchiassin, F; Levin, L. Preliminary studies on the biobleaching of loblolly pine kraft pulp with *Trametes trogii* crude extracts. *Enzyme and Microbial Technology*, 2008, 43, 164–168.

Eggert, C; Temp, U; Dean, JFD; Eriksson, KE. A fungal metabolite mediates degradation of non-phenolic lignin structures and synthetic lignin by laccase. *FEBS Letters*, 1996, 391, 144-148.

Ehara, K; Tsutsumi, Y; Nishida, T. Biobleaching of softwood and hardwood kraft pulp with manganese peroxidase. *Mokuzai Gakkaishi/Journal of the Japan Wood Research Society*, 1997, 43, 861-868.

Elegir, G; Daina, S; Zoia, L; Bestetti, G; Orlandi, M. Laccase mediator system: oxidation of recalcitrant lignin model structures present in residual kraft lignin. *Enzyme and Microbial Technology*, 2005, 37, 340-346.

Eriksson, KE; Pettersson, B; Volc, J; Musilek, V. Formation and partial characterisation of glucose-2-oxidase, a hydrogen peroxide producing enzyme in *Phanerochaete chrysosporium*. *Applied Microbiology and Biotechnology*, 1986, 23, 253–257.

Eugenio, ME; Santos, SM; Carbajo, JM; Martín, JA; Martín-Sampedro, R; González, AE; Villar, JC. Kraft pulp biobleaching using and extracellular enzymatic fluid produced by *Pycnoporus sanguineus*. *Bioresource Technology*, 2010, 101, 1866–1870.

Fillat, A; Colom, JF; Vidal, T. A new approach to the biobleaching of flax pulp with laccase using natural mediators. *Bioresource Technology*, 2010, 101, 4104-4110.

Fillat, U; Roncero, MB. Biobleaching of high quality pulps with laccase mediator system: Influence of treatment time and oxygen supply. *Biochemical Engineering Journal*, 2009, 44, 193–198.

Fillat, U; Roncero, MB. Optimization of laccase–mediator system in producing biobleached flax pulp. *Bioresource Technology*, 2010, 101, 181–187.

Fujita, K; Kondo, R; Sakai, K; Kashino, Y; Nishida, T; Takahara, Y. Biobleaching of kraft pulp using white-rot fungus IZU-154. *Tappi Journal*, 1991, 74, 123-127.

Fujita, K; Kondo, R; Sakai, K; Kashino, Y; Nishida, T; Takahara, Y. Biobleaching of softwood kraft pulp using white-rot fungus IZU-154. *Tappi Journal*, 1993, 76, 81-84.

Galhaup, C; Goller, S; Peterbauer, CK; Strauss, J; Haltrich, D. Characterization of the major laccase isoenzyme from *Trametes pubescens* and regulation of its synthesis by metal ions. *Microbiology*, 2002, 148, 2159-2169.

Gamelas, JAF; Pontes, ASN; Evtuguin, DV; Xabier, AMRB; Esculcas, AP. New polyoxometalate–laccase integrated system for kraft pulp delignification. *Biochemical Engineering Journal*, 2007, 33, 141–147.

Gamelas, JAF; Tavares, APM; Evtuguin, DV; Xabier, AMRB. Oxygen bleaching of kraft pulp with polyoxometalates and laccase applying a novel multi-stage process. *Journal of Molecular Catalysis B: Enzymatic*, 2005, 33, 57–64.

Geng, X; Li, K; Xu, F. Investigation of hydroxamic acids as laccase-mediators for pulp bleaching. *Applied Microbiology and Biotechnology*, 2004, 64, 493-496.

Giardina, P; Faraco, V; Pezzella, C; Piscitelli, A; Vanhulle, S; Sannia, G. Laccases: a never-ending story. *Cellular and Mollecular Life Sciences*, 2010, 57, 369-385.

González Arzola, K; Arévalo, MC; Falcón, MA. Catalytic efficiency of natural and synthetic compounds used as laccase-mediators in oxidising veratryl alcohol and a kraft lignin, estimated by electrochemical analysis. *Electrochimica Acta*, 2009, 54, 2621-2629.

Gu, L; Lajoie, C; Kelly, C. Expression of a *Phanerochaete chrysosporium* manganese peroxidase gene in the yeast *Pichia pastoris*. *Biotechnology Progress*, 2003, 19, 1403–1409.

Guillén, F; Martínez, AT; Martinez, MJ. Production of hydrogen peroxide by aryl-alcohol oxidase from the ligninolytic fungus *Pleurotus eryngii*. *Applied Microbiology and Biotechnology*, 1990, 32, 465–469.

Gutiérrez, A; Rencoret, J; Ibarra, D; Molina, S; Camarero, S; Romero, J; del Río, JC; Martínez, AT. Removal of lipophilic extractives from paper pulp by laccase and lignin-derived phenols as natural mediators. *Environmental Science and Technology*, 2001, 41, 4124-4129.

Hamid, M; Rehman, K. Potential applications of peroxidases. *Food Chemistry*, 2009, 115, 1177-1186.

Hammel, KE. Fungal degradation of lignin. In: Cadisch, G and Giller, KE, editors. Driven by Nature: Plant Litter Quality and Decomposition. Wallingford, UK: CAB International; 1997; 33–45.

Haritash, AK; Kaushik, CP. Biodegradation aspects of Polycyclic Aromatic Hydrocarbons (PAHs): a review. *Journal of Hazardous Materials*, 2009, 169, 1-15.

Hatakka, A. Lignin-modifying enzymes from selected white-rot fungi: production and role in lignin degradation. *FEMS Microbiology Reviews*, 1994, 13, 125–135.

Higuchi, T. Microbial degradation of lignin: role of lignin peroxidase, manganese peroxidase, and laccase. Proceedings of the Japan Academy; Ser. B, 2004, 80, 204-214.

Hirai, H; Kondo, R; Sakai, K. Screening of lignin-degrading fungi and their ligninolytic enzyme activities during biological bleaching of kraft pulp. *Mokuzai Gakkaishi*, 1994, 40, 980-986.

Hofrichter, M. Review: lignin conversion by manganese peroxidase (MnP). *Enzyme and Microbial Technology*, 2002, 30, 454-466.

Husain, Q. Peroxidase mediated decolorization and remediation of wastewater containing industrial dyes: a review. *Reviews in Environmental Science and Biotechnology*, 2010, 9, 117-140.

Ibarra, D; Camarero, S; Romero, J; Martínez, MJ; Martínez, AT. Integrating laccase–mediator treatment into an industrial-type sequence for totally chlorine-free bleaching of eucalypt kraft pulp. *Journal of Chemical Technology and Biotechnology*, 2006, 81, 1159-1165.

Ibarra, D; Romero, J; Martínez, MJ; Martínez, AT; Camarero, S. Exploring the enzymatic parameters for optimal delignification of eucalypt pulp by laccase mediator. *Enzyme and Microbial Technology*, 2006, 39, 1319-1327.

Iimori, T. Biobleaching of unbleached and oxygen-bleached hardwood kraft pulp by culture filtrate containing manganese peroxidase and lignin peroxidase from *Phanerochaete chrysosporium*. *Journal of Wood Science*, 1998, 44, 451-456.

Jiménez, L; Martínez, C; Pérez, I; López, F. Biobleaching procedures for pulp from agricultural residues using *Phanerochaete chrysosporium* and enzymes. *Process Biochemistry*, 1997, 32, 297-304.

Kapoor, M; Kapoor, RK; Kuhad, RC. Differential and synergistic effects of xylanase and laccase mediator system (LMS) in bleaching of soda and waste pulps. *Journal of Applied Microbiology*, 2007, 103, 305-317.

Kersten, P J; Kirk, TK. Involvement of a new enzyme, glyoxal oxidase in extracellular hydrogen peroxide production by *Phanerochaete chrysosporium*. *Journal of Bacteriology*, 1987, 169, 2195–2201.

Kirk, TK; Yang, HH. Partial delignification of unbleached kraft pulp with ligninolytic fungi. *Biotechnology Letters*, 1979, 1, 347-352.

Kleen, M; Ohra-aho, T; Tamminen, T. On the interaction of HBT with pulp lignin during mediated laccase delignification – a study using fractionated pyrolysis-GC/MS. *Journal of Analytical and Applied Pyrolysis*, 2003, 70, 589-600.

Kondo, R. Biobleaching and fiber properties improvements by manganese peroxidase (2002) TAPPI Fall technical Conference and Trade Fair, 2002, 875-905.

Kondo, R; Harazono, K; Sakai, K. Bleaching of hardwood kraft pulp with manganese peroxidase secreted from *Phanerochaete sordida* YK-624. *Applied and Environmental Microbiology*, 1994, 60, 4359-4363.

Kunamneni, A; Plou, FJ; Ballesteros, A; Alcalde, M. Laccases and their applications: a patent review. *Recent Patents on Biotechnology*, 2008, 2, 10-24.

Levin, L; Papinutti, L; Forchiassin, F. Evaluation of argentinean white rot fungi for their ability to produce lignin-modifying enzymes and decolorize industrial dyes. *Bioresource Technology*, 2004, 94, 169-176.

Li, K; Horanyi, PS; Collins, R; Phillips, RS; Eriksson, K-EL. Investigation of the role of 3-hydroxyanthranilic acid in the degradation of lignin by white-rot fungus *Pycnoporus cinnabarinus*. *Enzyme and Microbial Technology*, 2001, 28, 301–307.

Li, K; Xu, F; Eriksson, K-EL. Comparison of fungal laccases and redox mediators in oxidation of a nonphenolic lignin model compound. *Applied and Environmental Microbiology*, 1999, 65, 2654-2660.

Li, X; Kondo, R; Sakai, K. Study on hypersaline-tolerant white-rot-fungi (I) Screening of lignin-degrading fungi in hypersaline conditions. *Journal of Wood Science*, 2002, 48, 147-152.

Lund, M; Felby, C. Wet strength improvement of unbleached kraft pulp through laccase catalyzed oxidation. *Enzyme and Microbial Technology*, 2001, 28, 760–765.

Machii, Y; Hirai, H; Nishida, T. Lignin peroxidase is involved in the biobleaching of manganese-less oxygen-delignified hardwood kraft pulp by white-rot fungi in the solid-fermentation system. *FEMS Microbiology Letters*, 2004, 233, 283-287.

Madhavi, V; Lele, SS. Laccase: properties and applications. *Bioresources*, 2009, 4, 1694-1717.

Majeau, JA; Brar, SK; Tyagi RD. Laccases for removal of recalcitrant and emerging pollutants. *Bioresource Technology*, 2010, 101, 2331-2350.

Martínez, AT. Molecular biology and structure-function of lignin-degrading heme peroxidases. *Enzyme and Microbial Technology*, 2002, 30, 425-444.

Martínez, MJ; Ruiz-Dueñas, FJ; Guillén, F.; Martínez, AT. Purification and catalytic properties of two manganese-peroxidase isoenzymes from *Pleurotus eryngii*. *European Journal of Biochemistry*, 1996, 237, 424–432.

Mayer, AM; Staples, RC. Laccase: new functions for an old enzyme. *Phytochemistry*, 2002, 60, 551-565.

Minussi RC; Pastore GM; Durán N. Potential applications of laccase in the food industry. *Trends in Food and Science and Technology*, 2002, 13, 205–216.

Moldes, D; Cadena, EM; Vidal, T. Biobleaching of eucalypt kraft pulp with a two laccase-mediator stages sequence. *Bioresource Technology*, 2010, 101, 6924-6929.

Moldes, D; Díaz, M; Tzanov, T; Vidal, T. Comparative study of the efficiency of synthetic and natural mediators in laccase-assisted bleaching of eucalyptus kraft pulp. *Bioresource Technology*, 2008, 99, 7959-7965.

Moldes, D; Vidal, T. Laccase-HBT bleaching of eucalyptus kraft pulp: influence of the operating conditions. *Bioresource Technology*, 2008, 99, 8565-8570.

Moreira, MT; Feijoo, G; Alvarez, RS; Lema, J; Field, JA. Manganese is not required for biobleaching of oxygen-delignified kraft pulp by the white rot fungus *Bjerkandera* sp. strain BOS55. *Applied and Environmental Microbiology*, 1997, 63, 1749-1755.

Moreira, MT; Feijoo, G; Mester, T; Mayorga, P; Alvarez, RS; Field, JA. Role of organic acids in the manganese-independent biobleaching system of *Bjerkandera* sp. strain BOS55. *Applied and Environmental Microbiology*, 1998, 64, 2409-2417.

Oudia, A; Mészáros, E; Simoes, R; Queiroz, J; Jakab, E. Pyrolysis-GC/MS and TG/MS study of mediated laccase biodelignification of *Eucalyptus globulus* kraft pulp. *Journal of Analytical and Applied Pyrolysis*, 2007, 78, 233-242.

Oudia, A; Queiroz, J; Simoes, R. The influence of operating parameters on the biodelignification of *Eucalyptus globulus* kraft pulps in a laccase–violuric acid system. *Applied Biochemistry and Biotechnology*, 2008a, 149, 23-32.

Oudia, A; Queiroz, J; Simoes, R. Potential and limitation of *Trametes versicolor* laccase on biodegradation of *Eucalyptus globulus* and *Pinus pinaster* kraft pulp. *Enzyme and Microbial Technology*, 2008b, 43, 144-148.

Paice, MG; Jurasek, L; Ho, C; Bourbonnais, R; Archibald, F. Direct biological bleaching of hardwood kraft pulp with the fungus *Coriolus versicolor. Tappi Journal*, 1989, 72, 217-221.

Pizzul, L; Castillo, MdP; Stenström, J. Degradation of glyphosate and other pesticides by ligninolytic enzymes. *Biodegradation*, 2009, 20, 751-759.

Ravalason, H; Herpoël-Gimbert, I; Record, E; Bertaud, F; Grisel, S; de Weert, S; van den Hondel, CAMJJ; Asther, M; Petit-Conil, M; Sigoillot, J-C. Fusion of a family 1 carbohydrate binding module of *Aspergillus niger* to the *Pycnoporus cinnabarinus* laccase for efficient softwood kraft pulp biobleaching. *Journal of Biotechnology*, 2009, 142, 220–226.

Reid, ID; Paice, MG. Biological bleaching of kraft pulps by white-rot fungi and their enzymes. *FEMS Microbiology Reviews*, 1994, 13, 369-375.

Reid, ID; Paice, MG; Ho, C; Jurasek, L. Biological bleaching of softwood kraft pulp with the fungus *Trametes versicolor. Tappi Journal*, 1990, 73, 149-153.

Riva, S. Laccases: blue enzymes for green chemistry. *Trends in Biotechnology*, 2006, 24, 219-226.

Rodríguez Couto, S; Toca Herrera, JL. Industrial and biotechnological applications of laccases: a review. *Biotechnology Advances*, 2006, 24, 500-513.

Sadhasivam, S; Savitha, S; Swaminathan, K. Deployment of *Trichoderma harzianum* WL1 laccase in pulp bleaching and paper industry effluent treatment. *Journal of Cleaner Production*, 2010, 18, 799–806.

Sasaki, T; Kajino, T; Li, B; Sugiyama, H; Takahashi, H. New pulp biobleaching system involving manganese peroxidase immobilized in a silica support with controlled pore sizes. *Applied and Environmental Microbiology*, 2001, 67, 2208-2212.

Sasmaz, S; Gedikli, S; Aytar, P; Güngömedi, G; Cabuk, A; Hür, E; Ünal, A.; Kolankaya, N. Decolorization potential of some reactive dyes with crude laccase and laccase-mediate system. *Applied Biochemistry and Biotechnology*, 2010, In press.

Sealey, J; Ragauskas, AJ. Residual lignin studies of laccase-delignified kraft pulp. *Enzyme and Microbial Technology*, 1998, 23, 422-426.

Singh, G; Ahuja, N; Batish, M; Capalash, N; Sharma, P. Biobleaching of wheat straw-rich soda pulp with alkalophilic laccase from c-proteobacterium JB: optimization of process parameters using response surface methodology. *Bioresource Technology*, 2008, 99, 7472–7479.

Singh, P; Sulaiman, O; Hashim, R; Rupani, PF; Peng, LC. Biopulping of lignocellulosic material using differential fungal species: a review. *Reviews in Environmental Science and Biotechnology*, 2010, 9, 141-151.

Tang, W; Li, X; Zhao, J; Yue, J; Yue, H; Qu, Y. Effect of microbial treatment on brightness and heat-induced brightness reversion of high-yield pulps. *Journal of Chemical Technology and Biotechnology*, 2009, 84, 1631–1641.

Tavares, APM; Gamelas, JAF; Gaspar, AR; Evtuguin, DV; Xavier, AMRB. A novel approach for the oxidative catalysis employing polyoxometalate-laccase system: application to the oxygen bleaching of kraft pulp. *Catalysis Communications*, 2004, 5, 485-489.

Thurston, CF. The structure and function of fungal laccases. *Microbiology*, 1994, 140, 19–26.

Tien, M; Kirk, TK. Lignin-degrading enzyme from *Phanerochaete chrysosporium*: purification, characterization and catalytic properties of a unique hydrogen peroxide-requiring oxygenase. *Proceedings of the National Academy of Sciences USA*, 1984, 81, 2280–2284.

Urzua, U; Larrondo, S; Lobos, S; Larrain, J; Vicuña, R. Oxidation reactions catalysed by manganese peroxidase isoenzymes from *Ceriporiopsis subvermispora*. *FEBS Letters*, 1995, 371, 132–136.

Valls, C; Roncero, MB. Using both xylanase and laccase enzymes for pulp bleaching. *Bioresource Technology*, 2009, 100, 2032–2039.

Valls, C; Vidal, T; Roncero, MB. Boosting the effect of a laccase–mediator system by using a xylanase stage in pulp bleaching. *Journal of Hazardous Materials*, 2010a, 177, 586–592.

Valls, C; Vidal, T; Roncero, MB. The role of xylanases and laccases on hexenuronic acid and lignin removal. *Process Biochemistry*, 2010, 45, 425–430.

Vuorinen, T; Fagerström, P; Buchert, J; Tenkanen, M; Teleman, A. Selective hydrolysis of hexenuronic acid groups and its application in ECF and TCF bleaching of kraft pulps. *Journal of Pulp and Paper Science*, 1999, 25, 155–162.

Wariishi, H; Huang, J; Dunford, HB; Gold, MH. Reactions of lignin peroxidase compounds I and II with veratryl alcohol. *Journal of Biological Chemistry*, 1991, 266, 20694-20699.

Widsten, P; Kandelbauer, A. Laccase applications in the forest products industry: a review. *Enzyme and Microbial Technology*, 2008, 42, 293-307.

Wong, DWS. Structure and action mechanism of ligninolytic enzymes. *Applied Biochemistry and Biotechnology*, 2009, 157, 174-209.

Wong, KKY; Anderson, KB; Kibblewhite, RP. Effects of the laccase-mediator system on the handsheet properties of two high kappa kraft pulps. *Enzyme and Microbial Technology*, 1999, 25, 125-131.

Xu, F. Laccase. In Flickinger, MC. and Drew, SW editors. Encyclopedia of Bioprocess Technology: Fermentation, Biocatalysis, Bioseparation. New York: John Wiley and Sons Inc.; 1999; 1545-1554.

In: Lignin
Editor: Ryan J. Paterson

ISBN 978-1-61122-907-3
© 2012 Nova Science Publishers, Inc.

Chapter 14

LIGNIN: FROM NATURE TO INDUSTRY

Luiz Henrique Saes Zobiole,[1] Wanderley Dantas dos Santos,[2] Edicléia Bonini,[3] Osvaldo Ferrarese-Filho,[3] Robert John Kremer,[4] Rubem Silvério de Oliveira Jr,[1] and Jamil Constantin[1]*

[1]Center for Advanced Studies in Weed Research (NAPD),
Agronomy Department, State University of Maringá (UEM),
5790 Colombo Av., 87020-900, Maringá, Paraná, Brazil
[2]Brazilian Bioethanol Science and Technology Laboratory (CTBE),
Integrate Brazilian Center of Research in Energy and Materials (CNPEM),
Rua Giuseppe Máximo Scolfaro, 10000, Bairro Guará, Distrito de Barão Geraldo,
Campinas, SP, Brazil
[3]Laboratory of Plant Biochemistry, Department of Biochemistry,
University of Maringá, Av. Colombo, 5790, 87020-900,
Maringá, PR, Brazil
[4]United States Department of Agriculture,
Agricultural Research Service, Cropping Systems and
Water Quality Research Unit, Columbia, MO 65211, U. S.

ABSTRACT

Lignin is the second most abundant molecule in nature. The evolution of the lignin pathway allowed plants to conquer dry lands, and as a consequence became essential for the continental success of life as a whole. It is a relevant component of the lignocellulosic biomass which includes, wood, straw, bagasse and other materials that can be used as a renewable resource. Lignin was the first source of bio-energy that men used as firewood and then the main compound as humus in the agricultural revolution. Its biosynthesis results from a free radical polymerization of phenolic compounds. It is composed with various proportions of three main phenylpropanoids: coniferyl, sinapyl and hydroxycoumaryl alcohols as well as other minor compounds. Once completely assembled within cell walls making it impermeable and confers physical and chemical

---

* E-mail address: oferrarese@uem.br.

resistance to plant cells and, ultimately, the entire plant. Lignin has a high caloric content and might be an important compound in new generation biofuels industry, both directly by pyrolysis using energy to produce energy play the mill and indirectly by its thermochemical conversion in bio-oil and other derivatives. The research on lignocellulosic biomass focuses mainly on the processes of conversion of ordinary biomass. On the other hand, breeding, genetic engineering and crop management of plants are also important research areas devoted to improve plant biomass for industrial applications. Both biomass processing and plant production require accurate determination of lignin content and characteristics. In addition, little is known about the changes resulting from genetic modification in glyphosate-resistant (GR) crops in terms of secondary metabolism of plants, in particular lignin biosynthesis. Regarding these effects, some pesticides currently used in weed management, including glyphosate, can directly affect the synthesis of secondary compounds. Even though GR crops are resistant to glyphosate, recent reports suggest that glyphosate, or its main metabolite aminomethylphosphonic acid (AMPA), can decrease lignin content in these crops. Therefore, further research should be conducted to evaluate the different rates of glyphosate in GR crops. Such research is important to evaluate glyphosate effects on GR crop physiology and nutrition, and associated lignin production. In this work, we review: (1) the general properties and chemical composition of lignin; (2) the new uses under study and proposed uses for industry; (3) the potential for processing lignin biomass to provide bioenergy and the contribution to a new generation of biofuels, and (4) the potential threat to its production using current weed management in cropping systems.

1. INTRODUCTION

Lignin is a structural component of cell walls in higher terrestrial plants and, after cellulose, the second most abundant plant polymer. The lignin polymer is composed of phenylpropane units oxidatively coupled through ether and carbon-carbon linkages. This natural polymer can function as a genetically inducible physical barrier in response to microbial attack. In addition to structural support and pathogen defense, lignin acts during the water transport as a hydrophobic constituent of vascular phloem and xylem cells (Zubieta et al. 2002).

Lignified biomass has been used as a crude matter for construction of weapons, tools, sculptures, instruments, furniture, hoses (walls, roofs, floors), cars, ships, shoes, paper, etc. It was the first source of energy that men used out of body (firewood), and as a main compound of humus lignin had a role in the agricultural revolution. Currently, much of the lignin produced by the paper industry is consumed as a fuel. Although there are other marginal applications, such as an adhesive or tanning agent, no major large scale application has so far been found (Gosselink et al., 2004).

The lignin molecule was essential for the evolution of the plants in the land. The resistance of xylem to compressive stresses imposed by water transport and by the mass of the plants is important to growth and development. In addition, the insolubility and complexity of the lignin polymer makes it resistant to degradation by most microorganisms. Therefore, lignin serves an important function in plant defense. Variation in lignin content, composition, and location is likely to affect these essential processes. The constraints on the amount, composition, and localization of lignin for normal xylem function and plant defense are not known (Sarkar et al. 2009).

Lignin composition, quantity, and distribution also affect the agroindustrial uses of plant material. Digestibility and dietary conversion of herbaceous crops are affected by differences in lignin content and its composition. Lignin is an undesirable component in the conversion of wood into pulp and paper; removal of lignin is a major step in the paper making process. Furthermore, the resistance of lignin to microbial degradation enhances its persistence in soils. Lignin is, therefore, a significant component in the global carbon cycle (Campbell et al. 1996).

The integration of new technologies in the molecular genetics and biochemistry of higher plants promises a continuation of the rapid progress that has been made in the study of lignin biosynthesis in recent years and should answer many longstanding questions about this fundamental biochemical pathway in higher plants.

2. LIGNIN COMPOSITION, BIOSYNTHESIS AND PROPERTIES

Lignin is the generic term for a large group of aromatic polymers resulting from the oxidative combinatorial coupling of 4-hydroxyphenylpropanoids. These polymers are deposited predominantly in the walls of secondarily thickened cells, making them rigid and impervious (Vanholme et al. 2010). In addition to developmentally programmed deposition of lignin, its biosynthesis can also be induced upon various biotic and abiotic stress conditions, such as wounding, pathogen infection, metabolic stress, and perturbations in cell wall structure (Tronchet et al., 2010).

Lignin is a complex polymer formed by the oxidative polymerization of hydroxycinnamyl alcohol derivatives termed monolignols. At the start of lignin biosynthesis, monolignols are transported from the cytosol into the cell wall during a specific stage of wall development (Dixon et al. 2001). The three main monolignols, differing only in the substitution pattern on the aromatic ring are coniferil, sinapyl, *p*-coumaryl alcohols (Figure 1). The polymer residues resulting from monolignols incorporated into the lignin are called guaiacyl (G), syringyl (S), and *p*-hydroxyphenyl (H) units. At the start of lignin biosynthesis, monolignols are transported from the cytosol into the cell wall during a specific stage of wall development (Vanholme et al. 2008).

The monolignols themselves are relatively toxic, unstable compounds that do not accumulate to high levels within living plant cells. The monolignols share a conjugated structure (alternation between single and double linkages) which makes them able to stabilize a free radical by resonance. The absorption of an electron radical, make the monolignols actived to crosslink to each other. These reactions are irreversible. Lignin is the combinatorial polymer resulting of these lignin monolignols (Whetten et al. 1995).

Lignin monolignols are secondary metabolites produced only by vascular plants. They are biosynthesized in a sequence of reactions starting from the aromatic amino acid L-phenylalanine. Phenylalanine ammonia-lyase (PAL) is a regulatory enzyme of plant cells which accomplishes the first enzymatic step compromised with production of phenols, the deamination of phenylalanine to produce *trans*-cinnamic acid. Grasses are also capable to convert directly L-tyrosine in *trans*-coumaric acid, a reaction catalyzed by tyrosine ammonia-lyase (TAL) (Boerjan et al. 2003).

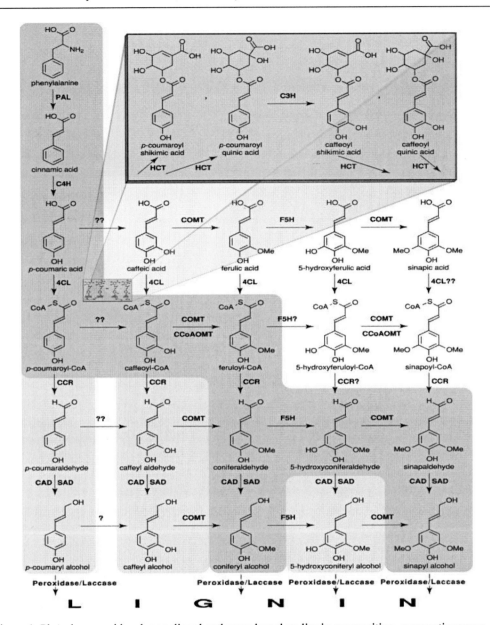

Figure 1. Phenylpropanoid and monolignol pathways based on lignin composition, enzymatic assays and the use of genetically modified plants. The route forming monolignols is most favored in angiosperms. The occurrence of gray routes depends of the plant species and conditions (Boerjan et al. 2003).

The conversion of L-phenylalanine for one of the three lignin precursors, or monolignols, takes place through a series of well-defined reactions. The monomer composition of lignin is known to differ between species, between cell types within a specie, and between developmental stages of a single cell, as does the total lignin content.

Metabolic control over the process is exerted at the level of monolignol synthesis and transport to the wall matrix. In plastids, phenylalanine is derived from the shikimate biosynthetic pathway. Some enzymes of the lignin biosynthetic pathway, namely the cytochrome P450, such as cinnamate 4-hydroxylase (C4H), p-coumarate 3-hydroxylase

(C3H), and ferulate 5-hydroxylase (F5H) are membrane proteins thought to be active at the cytosolic side of the endoplasmic reticulum (Achnine et al. 2004). Although metabolic channeling has been shown between PAL and C4H, it remains unknown whether other enzymatic pathways are also part of metabolic complexes at the endoplasmic reticulum.

Lignin monomers are differentially targeted to discrete regions (lignin initiation sites) of various lignifying cell walls; for example, *p*-coumaryl alcohol is mainly targeted to the middle lamella and coniferyl alcohol to the secondary wall of the xylem elements. By contrast, sinapyl alcohol is targeted to discrete regions in fiber-forming cell walls (Davin et al. 2005).

The physiological significance of this is apparently straightforward: differential targeting permits the construction of lignified cell walls with overall quite distinct biophysical properties. Such differences are, for example, readily apparent in the corresponding wall properties of fiber and xylem elements in plant. For instance, with some notable exceptions, lignins from gymnosperms (softwoods) are composed of G-units with minor amounts of H-units, whereas angiosperm (hardwoods) dicot lignins are composed of G- and S-units, formed from co-polymerization of coniferyl and sinapyl alcohols. Unmethoxylated H-units are elevated in soft compression wood and may be slightly higher in grasses. A variety of less abundant units have been identified from diverse species, and these may be incorporated into the polymer at varying levels (Baucher et al. 1998).

Some residues of lignin are chemically linked to hemicelluloses and pectins of the plant cell wall. Such linkages were confirmed in lignin-carbohydrate complexes isolated from milled pine by 2D NMR. They are benzyl ether, ester and phenol-glycoside linkages. The last one is present in *c*. 8% of pine lignin linkages (Balakshin et al, 2007).

Ferulic acid is a phenylpropanoid present both as a lignin residue as well as an ester-linked branch of polyoses (pectins and arabinoxylan hemicellulose). Ferulic acid is structurally similar to coniferyl alcohol (guaiacyl monolignol), but thanks to its carbonyl group, ferulic acid is able to esterify cell wall carbohydrates. Feruloyl-esters can dimerize cross-linking polyoses themselves and also polymerize with lignin forming bridges that connect lignin with wall polysaccharides, anchoring lignin. Feruloylated polysaccharides have been associated with initiation site of lignification (Boerjan et al, 2003; dos Santos et al. 2008).

2.1. Genetic Regulation

Plants have apparently evolved a particularly elegant process for producing key lignin polymers, rather without direct structural control, but under careful regulation of monolignol supply to lignification sites. The effects of perturbations on the lignin pathway often go beyond alterations in lignin content, composition, and cell wall structure. Indeed, several studies have now demonstrated that perturbing individual steps of the lignin biosynthetic pathway affects the expression level of other lignin pathway genes and also the expression of genes involved in a multitude of other, seemingly unrelated biological processes.

The genetic regulation of the lignin biosynthesis enzymes involves complex and interrelated systems of gene activation. All of the steps in the synthesis of monolignols are likely to be regulated by developmental signals and by environmental stimulation as well. In a few cases, it is apparent that multiple, differentially regulated genes are involved, whereas in other cases, single genes must be regulated by different signal transduction pathways.

The early genes in the phenylpropanoid pathway are expressed in the biosynthesis of many different phenolic and phenylpropanoid products in addition to lignin as flavonoids, coumarines, tannines, etc. The regulatory control of lignin biosynthesis must therefore be part of a complex combinatorial network. It is likely that transcription factors that regulate phenylpropanoid genes in a tissue and pathway specific manner, and thus, the nature of genetic regulatory network of lignin come to be identified soon.

2.2. Properties

Lignin has a relatively amorphous three-dimensional structure. No one has ever reported any evidence of crystalline order in lignin. And the molecule does not appear to be optically active, what is unusual for a biopolymer (Goring, 1989). Lignin has a point of fusion in about 160 °C and it might be blended with plastics to provide biodegradability, mixed with drugs to controlled release and as crude mater for a myriad of other applications.

Lignin is difficult to purify and characterize because of the highly heterogeneous linkages that occur between subunits within lignin and between lignin and polysaccharides. Lignin is highly hydrophobic and is not extracted from tissues by either aqueous or organic solvents. A small proportion (2-3%) of the lignin in wood can be dissolved in ethanol. If the wood is extensively milled, 50% or more of the lignin can be extracted with aqueous dioxane. However, the milling is so severe that it is likely that chemical bonds are broken. Thus, some type of covalent bond rupture seems to be necessary to make the lignin soluble (Goring, 1989).

As lignin have moderately high molecular weights and are highly heterogeneous polymers, deducing their primary structures (i.e. the sequence of H-, G-, and S-units in individual polymers and the bonds between them) is not possible using NMR methods that are typically based on lignin extracted from multiple cell types. In fact, because lignin is combinatorial, sequencing of lignin is meaningless, except for individual molecules. Nevertheless, understanding how monolignols couple (i.e. the propensity to couple a given way) provides insight into the factors that determine lignin properties (Vanholme et al. 2010).

3. METHODS OF QUANTIFICATION

Lignin has a relatively amorphous, three-dimensional structure in contrast to other major abundant polymers such as cellulose or chitin that form repeating, highly ordered linear structures. Lignin is difficult to purify and characterize because of the highly heterogeneous linkages between subunits within lignin and with polysaccharides (Carpita et al. 1993). Characterization of the lignin polymer is made even more difficult because it is not possible to isolate lignin polymer in its intact state. Lignin is highly hydrophobic and is not extracted from plant tissues by either aqueous or organic solvents. Consequently, much of our knowledge of lignin structure comes from the analysis of its chemical degradation products (Goring, 1989).

Lignin composition differs among plant species, tissues in a same plant and even among the different layers of the same cell wall. Lignins are a complex tridimensional polymer with molecular weight of 15,000 or more. No one method is informative enough *in situ* and can not be extracted from cell wall without structural modifications. For instance, phoroglucinol-HCl mix, commonly used in histochemistry to indicate lignin, gives positive only when lignin presents coniferaldehyde end groups what comprises seriously the application of this assay. Together these characteristics make lignin a molecule very difficult to characterize and several methods have been applied to study lignin in a complementary manner (Glasser, 1980).

The quantification of lignin is difficult not only because of its varying monomeric composition but also because lignins are covalently linked with cell wall carbohydrates, proteins, phenolics, or other compounds. These compounds may interfere with determination of lignins leading to over- or underestimation (Brinkmann et al. 2002). Given that lignin deposition and composition can vary with cell wall layer, cell wall type and species, it is necessary to consider the efficacy and limitations of the lignin analytical procedures currently employed (Anterola et al. 2002).

Typically, these include methods of lignin quantification, estimates of lignin monomeric compositions, the isolation of lignin-derived preparations, and their spectroscopic analyses. For determination of lignins, two different principles have commonly been employed: (1) gravimetric methods, which rely on the isolation of lignin or residual cell wall material (Klason lignin, acid detergent fiber lignin (ADF)) and (2) spectrophotometric methods, which are based on the decomposition of lignins into soluble degradation products and the determination of their absorbance in the UV, acetyl bromide method (AB), thioglycolic acid method (TGA). Three of the currently most popular analyses for apparently estimating lignin amounts are Klason, acetyl bromide determinations and thioglycolic acid analysis.

3.1. Klason Method

This method, developed for gymnosperm wood, involves the partial digestion of the plant material with 72% H_2SO_4 thereby leaving an insoluble (Klason lignin) residue. This method also releases smaller amounts of so-called soluble Klason lignin material, the quantities of which are significantly higher (up to 15% or so) when applied to the more acid-susceptible woody angiosperm lignins. The procedure is, however, significantly compromised when used with herbaceous plants. An overestimation caused by co-precipitation of proteinaceous substances and the presence of non-lignin inorganic components (so-called ash constituents). Depending upon the sample and development stage, values can be off by a factor of up to 2 and perhaps even higher. Furthermore, the Klason procedure can also give overestimations when other non-lignin phenolics, such as those in woody bark tissues, are present. Accordingly, overestimations of lignin amounts due to contributions of non-lignin constituents can be sizeable, the extent of which needs to be determined and the lignin amounts revised accordingly. For these reasons, Klason lignin data and their interpretations need to be both carefully collected and analyzed (Lai and Sarkanen, 1971).

3.2. Acetyl Bromide Method (AB)

One popular lignin estimation method is the acetyl bromide determination, which relies upon solubilization of lignin fragments, with the amounts being quantified using an extinction coefficient of 20 L g^{-1} cm^{-1}. However, this method can result in significant lignin overestimations if non-lignin UV-absorbing substances are also solubilized (Brinkmann et al. 2002). Another concern is that it has not been established whether this extinction coefficient can accurately be used for all lignins from all sources and developmental stages.

3.3. Thioglycolic Acid Method (TGA)

Another method spectrophotometric is that of thioglycolic acid extraction. In this procedure, thioglycolic acid derivatization displaces lignin from its normal covalent attachments to the cell wall and enables it to be extracted from cell walls by alkali. Acidification of the alkaline extract precipitates LTGA. After being resolubilized, LTGA can be determined quantitatively by measuring its absorbance at 280 nm (Capelleti et al. 2005). This is method may be used to quantify lignin in seed coats, raises of different cultivates plants. The formation of thiolignoglycolate derivatives is considered a criterion for "genuine" lignin. However, because of this specificity for certain properties of lignins, this method is also likely to underestimate the overall lignin concentration.

In contrast to the TGA method, the AB method is likely to overestimate the lignin concentration slightly because incubation of cell walls with acetylbromide may lead to some oxidative degradation of structural polysaccharides, especially of xylans, resulting in increased absorbance at 280 nm, the wavelength used to determine lignin. One difficulty with the acetyl bromide method, as with the thioglycolate and other methods spectrophotometric, is the need for a well defined lignin standard which one can calibrate the correct absorbance values for quantifying lignin in an unknown sample (Hatfield et al., 1999).

3.4. Other Methods

For routine analysis of large sample numbers high through-put methods are needed. In this respect the use of near-infrared reflectance spectroscopy (NIRS), which takes advantage of the reflectance of chemical bonds in the near infrared, is promising. Near-infrared radiation is absorbed by various chemical linkages such as C----H, O----H, N----H, C==O, C----C, etc., and results in bending, stretching, and twisting of these bonds, leading to characteristic reflectance patterns (Günzler et al. 2002). NIRS can be applied to milled dry samples without further sample preparation and can be used to determine a range of major chemical components. NIRS has obtained a certified status to measure moisture, crude protein, and acid detergent fiber content in forages. It has also been adopted to determine lignin contents, for example, in wood pulp and in leaves or leaf litter (Brinkmann et al. 2002).

Another one method was developed using high-performance size exclusion liquid chromatography (HPSEC) with multi-angle laser light scattering (MALLS), quasi-elastic light scattering (QELS), interferometric refractometry (RI) and UV detection to characterize and monitor lignin. The combination proved very effective at tracking changes in molecular conformation of lignin molecules over time; i.e. changes in molecular weight distribution, radius of gyration, and hydrodynamic radius (Gidh et al. 2006).

As of now, there is no single method that is rapid, noninvasive, handles large sample numbers, and provides accurate measure of cell wall lignin contents. Probably the most important consideration is consistency; one should use the same lignin method for all of samples being analyzed. This will allow a relative comparison of the samples that should avoid the greatest potential problems that might arise from switching from one method to another. Other procedures such as the Klason method can be adapted to obtain nearly a complete cell wall component analysis from one sample. Some procedures, such as the thioglycolate lignin method, may promise of producing a true reflection of lignin, but more research must be undertaken to ensure that all the lignin is precipitated by the acid step.

4. LIGNOCELLULOSE IN THE CONTEXT OF BIOENERGY

With the industrial revolution, civilization has learned the importance of energy availability in technological development and production of richness. On the other hand, we have also learned the bad consequences of large scale extractivism from wood to fossil fuels and the value of renewables. In this sense, while *secure* nuclear energy is not directly available, direct and indirect use of sun energy is our unique option to large scale production of energy. While electricity has worked as revolutionary way to transport energy at long range, the use of electricity to push vehicles with high autonomy is far way from overcome liquid fuels.

Among other technical barrier, the similar low energy per mass ratio of batteries also constrains the use of gaseous fuels in energy per volume. Although hydrogen production driven by sun light or hydroelectricity is a simple, efficient and clean via to produce a renewable fuel, the use H_2 is not so easy. Molecular hydrogen must be transported on pressure demanding improved costs of transportation, limitation in autonomy and risks of accidents.

Liquids fuels are currently more convenient for use in vehicles, considering the volumetric energy density and the easiness of distribution and storage. On the other hand, H_2 it can not be used in substitution to oil derivatives. Among the several forms of energy derived from the sunlight, biomass is the only real candidate able to completely substitute petroleum in the whole chain production as a source of chemical energy to fuels as well as a source raw material to chemicals, from medicines to plastic.

Cellulose and lignin are the most abundant biomolecules in nature and store a lot of free energy. So, to convert lignocellulose biomass in biofuels has been considered a good strategy to produce large scale biofuels. However, despite the high energy content of lignin, its use as raw material to produce biofuel is far from reach commercial viability. Lignins are high molecular weight polymers (15 KD or more) largely insoluble and quite recalcitrant to chemical hydrolysis.

Due the big number of pilot plants in operation and being implanted, biochemical conversion of biomass is already in a pre-commercial stage of development. However, in a biochemical planta lignin is rather an inconvenient. It makes the enzymatic hydrolysis of polysaccharides more difficult, by one side, coating polysaccharides from hydrolytic polysaccharidases, on the other, acting as enzyme inhibitors. Hence, and most of the research in lignocellulosic conversion to biofuels have been driven in removing it from polysaccharides, quit much friendlier to hydrolysis and fermentation.

4.1. Removing Lignin to Produce Lignocellulosic Ethanol

Traditionally, large scale methods for extraction of lignin have been developed by or in close association with the paper and cellulose industry. Such a process named pulping, employ diverse chemical and mechanical methods. Chemical pulping uses chemical reactants to remove 98-99 % lignin and release the fibers of cellulose with yields of *ca.* 40-50 %, high mechanical resistance and stability to whitening. Mechanical pulping yields 75-95%, but the fibers present low mechanical resistance and high susceptibility to yellowing. Although a mechanical pulping planta is very smaller and simpler than a chemical one, it consumes elevated amounts of electric energy.

Alternatively, methods have been developed to make use fungi to degrade the lignocellulose in a process known as biopulping. Brown-rod fungi and white-rod produce polysaccharidases and accessory enzymes able to hydrolyse polysaccharides. Additionally, the second produce also peroxidases, laccases and other low molecular radicals and organometallic complexes able to break also lignin. This biological pretreatment permits reductions from until 40% in the energy consume for mechanical pulping and reduces time, temperature and amount of chemical reactants necessary to chemical pulping.

Such processes have been assayed as pretreatments for biochemical conversion of lignocellulosic biomass in ethanol. However, unlike pulping wood, fast growing cultures as sugarcane (*Saccharum officinarum* x *spontaneum*), switchgrass (*Panicum virgatum*), *Mischantus*, bluestem (*Andropogon gerardii*) among other grass feedstocks wastes have been considered to suit better to bionenergy purposes. In general, such Poales cited above present about 20% of lignin in its lignocellulose biomass and, similarly to what happens in wood pulping reductions in amount of lignin entail significative increase in digestibility of polysaccharides, both for livestock and *in vitro* enzyme cocktails. A point must be made here. Up to a certain point, the more lignin is removed, more linkages takes place between microfibrils. It is good to paper industry cause make stronger paper, but is not so good when one needs to proceed to saccharification.

4.2. Manipulating Lignin *in Planta*

Classic plant breeding programs to reduce lignin content in forage and pulping wood have been successful in improving digestibility. However, while down regulation of phenylpropanoid genes frequently induces increase in cellulose content, probably to offset the lack of lignin, strong reductions in overall biomass production are also observed. A significant contribution to a strict evaluation of such tradeoff was achieved by Chen and Dixon (2007). They showed that at least in one case, an improved saccharification (166 %) of biomass from Arabidopsis mutants silenced for hydroxycinnamoyl:shikimate hydroxycinnamoyl transferase (HCT), more than compensated the reduction in biomass (40 %).

Classical plant breeding also has been successful in revert reduction in the growth of *brown midrib* cell wall mutants with reduced lignin and increased digestibility. A key peace of information in this metabolic puzzle had been produced by the elegant study from Busseau et al. (2007). They revealed that HCT-silenced plants which present reduced lignin content and reduced growth phenotype also accumulate flavonoids and present inhibited auxin transport. Afterwards, they induced simultaneous repression in the synthesis of chalcona synthase, the first enzyme compromised with the synthesis of flavonoids and restored flavonoid content, auxin transport and normal growth.

On the other hand, reducing lignin content in crop trees have been proven to increase the plant susceptibility for biotic and abiotic stresses and occurrence of cavitations, therefore reducing water efficiency and hydraulic conductivity. Maybe these problems may come to be partially overcome by choosing suitable (small) cultivars, modifying rather than simply reducing lignin and/or engineering plants to produce secondary metabolites able to improve resistance against attack from pathogen without conferring recalcitrance. If so, this data offer a promising prospect for genetic manipulation of plant lignin content for biofuel, forage and pulping purposes.

4.3. Lignin as Raw Material for Biofuels

Thermochemical conversion of lignocellulose is considered a promising alternative to biochemical approach to produce biofuels, since it permits avoid coping directly with polymer complexity. The catalytic reform of natural gas and gasified coal is a well established multipurpose route in chemical industry. Heating biomass in complete absence of oxygen (pyrolysis) biomass can be partially converted in a bio-oil suitable for transport and gasification. Syngas can be fermented or catalytically converted in gasoline, diesel, ethanol, methanol, hydrogen and other raw material for the chemical and energy industries.

Another promising technology to manipulate lignin and lignocellulosic biomass is based a small set of low melting point salts dubbed ionic solvents. They have been shown to be able to dissolve lignocellulosic material in mild conditions of temperature and pressure. Such a state seems to be favorable for developing enzymatic or other catalytic processes to convert lignocellulose.

The simplest destination to lignin in a biochemical planta is the burnt to produce thermal and electricity. Indeed, this is the standard destination for lignocellulose in biorrefineries which produces bioethanol from sugarcane, in Brazil. The energy is used to push the mills, heat the distillers, etc, making the planta completely self-sustainable in energy. Usually, the surplus is sold to electricity companies, impacting significantly electric availability in Brazil. Although lignin has a high energy content, burn it is the poorest destination to this high valuable product, both from the thermodynamic and the production point of views.

4.4. Lignin as Raw Material for Chemicals

Besides the energetic destination, other uses to lignin have been studied in the last years. Melting at *ca.* 180 °C, lignin can be used as additive in concrete and asphalt as well as preservative to rubber. Mixed to plastics lignin can improve biodegradability. Water insoluble fractions produced by pirolysis of lignin might be the partial replacement for phenol in phenol-formaldehydes resins. Derivatives of lignin as lignophenol and lignin-aminophenol have been assayed as adsorbents able to immobilize enzymes for industrial biotechnological applications. Completely safe for environment and even to human consume lignin has its use approved for FDA and EPA as food antioxidant, and it is studied as metal scavenger to industrial sewer. Thanks to its safety humans and environment.

Lignin is also being successfully experienced as a polymeric matrix to controlled release of agrochemicals in plantations. Fused with chemical compounds as pesticides or fertilizers, lignin protects the compounds from photodegradation, oxidation, leaching, runoff and biodegradation. On the other hand, the compounds are released by diffusion and slow hydrolysis of eventual linkages among lignin and the compound. As lignin is biodegradable, no one pesticide can persist trapped in the polymer.

5. POTENTIAL THREAT TO LIGNIN PRODUCTION USING CURRENT WEED MANAGEMENT IN CROPPING SYSTEMS

The shikimate pathway leads to synthesis of amino acids, phenolic compounds and a wide range of secondary plant products, including lignin (Boocock and Coggins 1983; Singh et al. 1991; Hernandez et al. 1999). Based on the importance of these metabolic pathways, enzymes involved in the synthesis of essential amino acids are excellent targets for the rational design of herbicide. Consequently, there are numerous potential sites of herbicide action in this highly complex and normally well-coordinated array of enzymes and processes (Devine et al. 1993a).

One of the main herbicides that directly affects the synthesis of secondary compounds is glyphosate (N-(phosphomethyl)glycine) (Devine et al. 1993b), which is a wide-spectrum, foliar-applied herbicide that is translocated throughout the plant to actively growing tissues where it inhibits 5-enolpyruvylshikimate-3-phosphate synthase (EPSPS) in the shikimate pathway of susceptible plant. However, glyphosate resistance (GR) plants carry the gene coding for a glyphosate-insensitive form of this enzyme, obtained from *Agrobacterium* sp. strain CP4. Once incorporated into the plant genome, the gene product, CP4 EPSP synthase, confers crop resistance to glyphosate (Franz et al. 1997).

It is assumed that glyphosate will not interfere with metabolism in the glyphosate resistance plants, for example GR soybeans (*Glycine max* L.), however, with the wide use of glyphosate in current weed control in GR soybeans, many farmers report that some GR soybean varieties are visually injured by glyphosate (Zablotowicz and Reddy 2007). Thus, a recent study was conducted to evaluate the effect of increasing rates of glyphosate on lignin and amino acid content, photosynthetic parameters and dry biomass in the early maturity group cultivar BRS 242 GR soybean (Zobiole et al. 2010).

In that study, plants were grown in half-strength complete nutrient solution and subjected to various rates of glyphosate either as a single or in sequential applications. The treatments under single applications were applied at V4 stage (24 days after emergence, DAE) and for sequential applications at V4 (24 DAE - 50% of dose) and V7 stage (36 DAE - 50% of dose). The authors found that there was a relationship between A and chlorophyll content, which GR soybean receiving glyphosate exhibited reduced A and less chlorophyll, which such effects were more pronounced with increasing glyphosate rates (Figure 2 and 3).

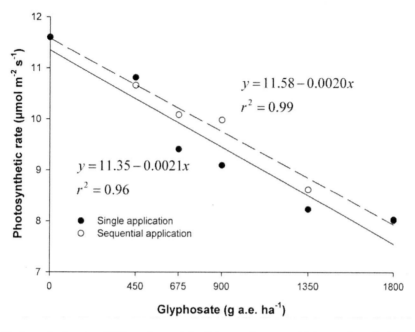

Figure 2. Photosynthetic rate of GR soybean at the R1 growth stage with increasing glyphosate rates applied singly or sequentially (Zobiole et al. 2010).

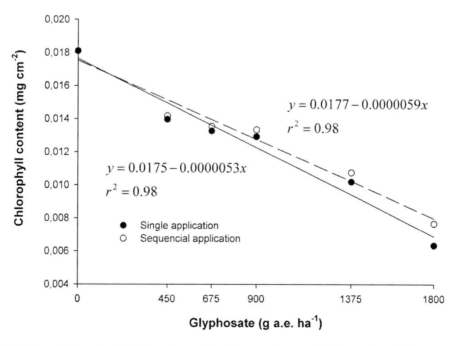

Figure 3. Chlorophyll content of GR soybean at the R1 growth stage with increasing glyphosate rates applied singly or sequentially (Zobiole et al. 2010).

The reduced photosynthetic parameters and chlorophyll content in glyphosate-treated GR plants (Figure 2 and 3) is probably due to the direct damage of the herbicide on chlorophyll production (Reddy et al. 2004) or immobilization of Mg and Mn required for the production and function of chlorophyll (Beale 1978; Taiz and Zeiger 1998). The injuries in GR plants has been attributed, in some cases, by the accumulation of the primary phytotoxic metabolite aminomethylphosphonic acid (AMPA) (Reddy et al., 2004). These results demonstrated that glyphosate or one of its metabolites apparently remained active in soybean through the R1 growth stage or later.

In that experiment the lignin production was also affected, which GR soybean plants treated with glyphosate produced less lignin than non-treated plants, with such reduction was proportional with the glyphosate rate applied. The effects of glyphosate also were greater with a single application at full rate compared with sequential applications with the same total rate applied (Figure 4).

The decreases in photosynthetic rate and chlorophyll contents could explain the reduced lignin content. Otherwise, lignin production is one of several important physiological process controlled by phenylalanine, a key product of the shikimate pathway, which may be impacted through genetic transformation of GR soybeans even though the glyphosate-resistant EPSP synthase is designed to not block the pathway.

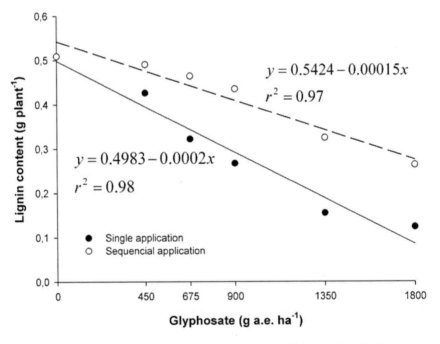

Figure 4. Lignin content of GR soybean at the R1 growth stage with increasing glyphosate rates applied singly or sequentially (Zobiole et al. 2010).

Further studies should be conducted to evaluate the lignin content in different GR crops, in order, to develop a strategy to minimize the use of glyphosate since it can be a potential threat to lignin production using current weed management in cropping systems.

REFERENCES

Achnine, L., Blancaflor, E.B., Rasmussen, S., Dixon, R.A. (2004). Colocalization of L-phenylalanine ammonia-lyase and cinnamate 4-hydroxylase for metabolic channeling in phenylpropanoid biosynthesis. *Plant Cell*, 16: 3098-3109.

Anterola, A.M. and Norman, G.L. (2002) Trends in lignin modification: a comprehensive analysis of the effects of genetic manipulations/mutations on lignification and vascular integrity. *Phytochem.,* 61:221–294.

Baucher, M., Monties, B., Montagu M.V., Boerjan W. (1998) Biosynthesis and Genetic Engineering of Lignin. *Plant Sci,* 17:125 – 197.

Beale, S.I. (1978) δ-Aminolevulinic acid in plants: its biosynthesis, regulation and role in plastid development. *Annu. Rev. Plant Physiol.*, 29:95-120.

Besseau, S., Hoffmann, L., Geoffroy, P., Lapierre, C., Pollet, B., Legrand, M. (2007) Flavonoid Accumulation in *Arabidopsis* Repressed in Lignin Synthesis Affects Auxin Transport and Plant Growth. *The Plant Cell,* 19:148-162.

Biermann, C.J. (1993) Essentials of Pulping and Papermaking, Academic Press, New York.

Boerjan, W., Ralph, J., Baucher, M. (2003) Lignin Biosynthesis. *Annu. Rev. Plant Biol.*, 54: 519-546.

Boocock, M.R. and Coggins, J.R. (1983). Kinetics of 5-enolpyruvylshikimate-3-phosphate synthase inhibition by glyphosate. *FEBS Lett.*, 154:127-133.

Brinkmann, K., Blaschke, L., Polle, A. (2002) Comparison of different methods for lignin Determination as a basis for calibration of Near-infrared reflectance spectroscopy And implications of Lignoproteins. *J. Chem. Ecol.*, 28:2483-2501.

Buckeridge, M.S., Souza, A.P., dos Santos, W.D. (2010) Routes for Cellulosic Ethanol in Brazil. In: Luiz Augusto Barbosa Cortez. (Org.). SUGARCANE BIOETHANOL: RandD for productivity and sustainability. São Paulo: Blucher.

Campbell, M.M. and Sederoff, R.R. (1996) Variation in Lignin Content and Composition - Mechanisms of Control and Implications for the Genetic Improvement of Plants. *Plant Physiol.,* 110: 3-13.

Capeleti, I., Ferrarese, M.L.L., Kryzanowski, F.C., Ferrarese-Filho, O. (2005). A new procedure for quantification of lignin in soybean (Glycine max (L.) Merrill) seed coat and their relationship with resistance to mechanical damage. *Seed Sci. Tech.*, 33:511-515.

Carpita, N.C. and Gibeaut, D.M. (1993). Structural models of primary cell walls in flowering plants: consistency of the molecular structure with the physical properties of the walls during growth. *Plant J.* 3:1–30.

Chen, F., and Dixon, R.A. (2007) Lignin modification improves fermentable sugar yields for biofuel production. *Nature Biotech.,* **25:**759-761.

Davin L.B. and Lewis N.G. (2005) Lignin primary structures and dirigent sites. *Biotech.* 16 407–415.

Devine, M., Duke, S.O., Fedtke, C. (1993a) Oxygen toxicity and herbicidal action; Secondary physiological effects of herbicides. In: Physiology of herbicide action. New Jersey : Prentice-Hall, pp 177-188.

Devine, M., Duke, S.O., Fedtke, C. (1993b) Inhibition of amino acid biosynthesis. In: Physiology of Herbicide Action. New Jersey: Prentice-Hall, pp 251-294.

Dixon, R.A., Chen, F., Guo, D., Parvathi, K. (2001) The biosynthesis of monolignols: a "metabolic grid", or independent pathways to guaiacyl and syringyl units? *Phytochem.*, 57: 1069–1084.

dos Santos, W.D., Ferrarese, M.L.L., Ferrarese-Filho, O. (2008). Ferulic Acid: An Allelochemical Troublemaker. *Funct Plant Science*, 2:47-55.

dos Santos, W.D., Gómez, E.O., Buckeridge, M.S. (2010) Bioenergy and the Sustainable Revolution. In: Buckeridge, M. S., Goldman, G. (Org). *The Routes for Cellulosic Ethanol*. New York: Springer.

Fang, C., Lin, Y., Cheng, X. (2009) Preparation of lignin derivatives and their application as protease adsorbents, *Natural Sci.*, 3:185-190.

Ferraz, A., Souza, J.A., Silva, F.T., Gonçalves, A.R., Bruns, R.E., Cotrim, A.R., Wilkins, R. (1997) Controlled release of 2,4-D from granule matrix formulations based on six lignins. *J. Agric. Food Chem.*, 45:1001-1005.

Ferraz, A., Guerra, A., Mendonça, R., Masarin, F., Vicentin, M.P., Aguiar, A., Pavan, P. C. (2008) Technological advances and mechanistic basis for fungal biopulping. *Enzyme Microb Technol*, 43: 178-185.

Franz, J.E., Mao, M.K., Sikorski, J.A. (1997) Glyphosate: A unique global herbicide. ACS Monograph 189, *American Chemical Society*, pp 521–615.

Gidh, A.V., Decker, S.R., Vinzant, T.B., Himmel, M.E., Williford, C. (2006) Determination of lignin by size exclusion chromatography using multi angle laser light scattering. *J. Chromatogr.*, 1114:102-110.

Goring, D.A.I. The Lignin Paradigm (1989) *ACS Symp. Ser.*, 397:2–10.

Gosselink, R.J.A., DeJong, E., Guran, B., Abächerli, A. (2004) Co-ordination network for lignin—standardisation, production and applications adapted to market requirements. *Ind. Crops Prod.*, 20:121–129.

Günzler, G., Gremlich, H. U. (2002). IR spectroscopy. Wiley-VCH Verlag, Weinheim.

Harris, D. and DeBolt, S. (2010) Synthesis, regulation and utilization of lignocellulosic biomass. *Plant Biotech. J.*, 8:244-262.

Hatfield, R.D., Grabber, J., Ralph, J., Brei, K. (1999) Using the acetylbromide assay to determine lignin concentrations in herbaceous plants: some cautionary notes. *J. Agric. Food Chem.*, 47:628–632.

Hernandez, A., Garcia-Plazaola, J.I., Bacerril, L.M. (1999) Glyphosate effects on phenolic metabolism of nodulated soybean (Glycine max L. Merril). *J. Agric. Food Chem.*, 47, 2920-2925.

Reddy, K.N., Rimando, A.M., Duke, S.O. (2004) Aminomethylphosphonic acid, a metabolite of glyphosate, causes injury in glyphosate-treated, glyphosate-resistant soybean, *J. Agric. Food Chem.*, 52:5139–5143.

Sarkar P., Bosneaga E., Auer, M. (2009) Plant cell walls throughout evolution: towards a molecular understanding of their design principles. *J. Exp. Bot.*, 60: 3615-3635.

Singh, B.K., Siehl, D.L., Connelly, J.A. (1991). Shikimate pathway: why does it mean so much to so many? *Oxf Surv Plant Mol. Cell Biol.*, 7, 143-185.

Taiz, L. and Zeiger, E. (1998) Mineral Nutrition. In: Plant Physiology, Sinauer Associates: Sunderland, pp 111-144.

Tronchet, M., Balague, C., Kroj, T., Jouanin, L., Roby, D. (2010) Cinnamyl alcohol dehydrogenases C and D, key enzymes in lignin biosynthesis, play an essential role in disease resistance in Arabidopsis. *Mol. Plant Pathol.*, 11: 83–92.

Vanholme, R., Demedts, B., Morreel, J., Boerjan, W. (2010) Lignin Biosynthesis and Structure. *Plant Physiol.*, 153:895–905.

Vanholme, R., Morreel, J., Boerjan, W. (2008) Lignin engineering. *Plant Biology* 11:278–285.

Whetten, R. and Sederoff, R. (1995) Lignin Biosynthesis. *The Plant Cell*, 7:1001-1013.

Zablotowicz, R.M. and Reddy, K.N. (2007) Nitrogenase activity, nitrogen content, and yield responses to glyphosate in glyphosate-resistant soybean. *Crop Prot*, 26:370-376.

Zobiole, L.H.S., Bonini, E.A., Oliveira Jr., R.S., Kremer, R.J., Ferrarese-Filho, O. (2010) Glyphosate affects lignin content and amino acid production in glyphosate-resistant soybean. *Acta Physiol Plant*, 32:831-837.

Zubieta C., Kota P., Ferrer J.L., Dixon R.A., Noel J.P. (2002) Structural Basis for the Modulation of Lignin Monomer Methylation by Caffeic Acid/5-Hydroxyferulic Acid 3/5-O-Methyltransferase. *The Plant Cell*, 14:1265-1277.

In: Lignin
Editor: Ryan J. Paterson

ISBN 978-1-61122-907-3
© 2012 Nova Science Publishers, Inc.

Chapter 15

LIGNIN: AN UNTAPPED RESOURCE FOR MULTIPLE INDUSTRIAL APPLICATIONS

Manimaran Ayyachamy, Kevin M. Turner, Vijai Kumar Gupta*
and Maria G. Tuohy
Molecular Glycobiotechnology Group, Biochemistry,
School of Natural Sciences,
National University of Ireland Galway,
Galway City, Ireland

INTRODUCTION

Lignin is the second most abundant natural aromatic polymer after cellulose in terrestrial ecosystems and represents nearly 30% of the organic carbon sequestered in the biosphere (Boudet et al. 2003). Lignins are interlinked with cellulose and hemicellulose conferring structural strength, rigidity and impermeability to the woody cell wall, while providing natural resistance against chemical or microbial attack and environmental stresses (Foster et al. 2010). Additionally, lignin waterproofs the cell wall thus enabling transport of water and solutes through the vascular system.

Lignins are complex racemic aromatic heteropolymers (Wong 2009) derived mainly from *p*-coumaryl, coniferyl, and sinapylalcohols (Figures 1 and 2). The formation of *p*-hydroxyphenyl (H), guaiacyl (G), and syringyl (S)phenylpropanoid units occurs when the respective monolignols are incorporated into the lignin polymer (Karkonen and Koutaniemi 2010; Vogt 2010). The quantity and composition of lignins varies among the plant kingdom and are influenced by developmental and environmental factors. Hardwood lignins consist mainly of Gand Sunits with trace amounts of Hunits, whereas softwood lignins are composed predominantly of Gunits with low levels of Hunits (Bose et al. 2009; Karkonen and Koutaniemi 2010; Vogt 2010).

*Email: manimaran@scientist.com.

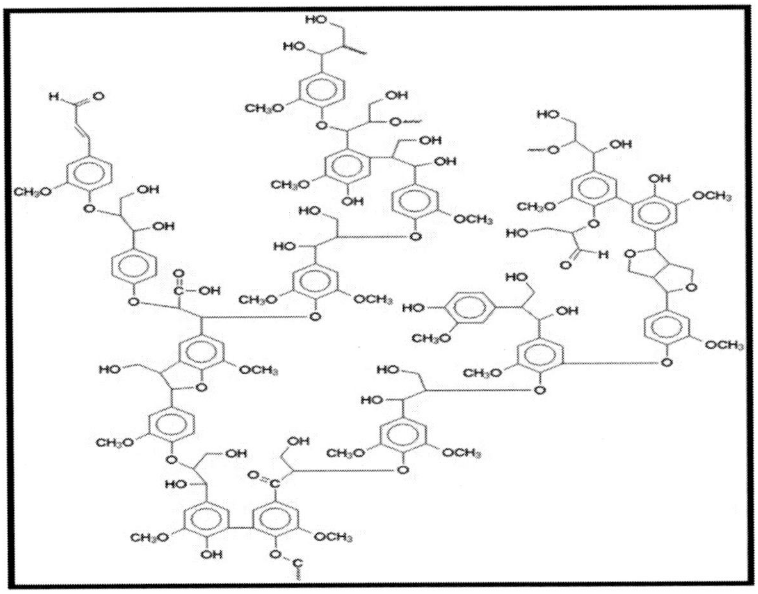

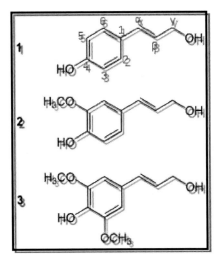

Figure 1. Lignin fragment from softwood. Adapted from www.academic.stanton.edu.

Figure 2. Lignin monomers: *p*-coumaryl alcohol (1); coniferyl alcohol (2) and sinapyl alcohol(3).

Lignin is usually considered as a recalcitrant material that is resistant to microbial degradation; only specialized microbes, predominantly fungi, are capable of secreting extracellular enzymes that cleave these structures into biologically metabolizable forms (Kadam and Drew 1986; Broda et al. 1996; Vicuna 2000). From a functional point of view, lignins impart strength to cell walls, facilitate water transport, and obstruct the degradation of wall polysaccharides, thus acting as a major line of defense against pathogens, insects, and other herbivores (Chafe 1974; Davin and Lewis 2005). Lignin provides the mechanical support for stems and leaves and supplies the strength and rigidity of plant walls. Lignin provides the structural strength needed by large trees to reach heights in excess of 100 m (Radotic et al. 2008). Without lignin these trees would collapse on themselves. Lignin parties a key component in conducting water in plant stems (Miceli et al. 1989; Ouyang et al. 2009). The structure of lignin has not been accurately determined as it usually fragments upon extraction (Avgerinos and Wang 1983; Deschamps et al. 1996; Hintz et al. 1996; Li et al. 2007).

The polysaccharide components of plant cell walls are extremely hydrophilic and thus permeable to water, whereas lignin is more hydrophobic (King et al. 2009). The cross linking of polysaccharides by lignin is an obstacle for water absorption to the cell wall. Thus, lignin makes it feasible for the plant's vascular tissue to conduct water efficiently. Lignin is present in all vascular plants, but not in bryophytes, supporting the suggestion that the original function of lignin was limited to water transport (RosBarcelo 1997; Akin 2007; Del Rio et al. 2007).

Lignin plays a key role in both terrestrial and oceanic carbon cycles. Lignin helps in sequestering atmospheric carbon into the living tissues of woody perennial vegetation (Oglesby et al. 1967; Mishra et al. 1979; Fustec et al. 1989). Lignin is one of the most slowly decomposing components of dead vegetation, contributing a major fraction of the material that becomes humus as it decomposes (Oglesby, Christman et al. 1967). The resulting soil humus generally increases the photosynthetic productivity of plant communities growing on a site as the site transitions from disturbed mineral soil through the stages of ecological succession, by providing increased cation exchange capacity in the soil and expanding the capacity of moisture retention between flood and drought conditions.

Lignin degradation is a rate-limiting step of carbon recycling and its turnover markedly differs from cellulose and hemicellulose (Sorensen 1962; Haider and Domsch 1969; Harayama 1997; Regalado et al. 1997; Grote et al. 2000; Vasil'chenko et al. 2004). It is widely recognized that lignin restricts the accessibility of microbial enzymes to other major cell wall polysaccharides during biogeochemical cycle processes. Lignin and lignin-carbohydrate complexes have both been identified as key factors in determining the rate of organic matter decomposition in forest ecosystems (Song et al. ; Xiang and Lee 2000; Blaschke et al. 2002; Siller et al. 2002).

Lignins can vary in structure according to their method of isolation and their plant source. However, structural differences are not considered to be a limiting factor in the context of prospective uses for lignins in industrial applications. Lignins are non-toxic and extremely versatile in terms of performance. Therefore, several potentially attractive industrial avenues exist for more effective and diverse utilization of lignin. Numerous novel applications have been pin pointed, only some of which have been explored to date.

LIGNOCELLULOSIC BIOMASS

Cellulose is a main component of lignocellulosic biomass together with hemicellulose and lignin. Cellulose and hemicellulose are considered to be macromolecules made of different sugar monomers; while lignin is an aromatic polymer synthesized from phenylpropanoid precursors found in a variety of biomass (Table 1). The chemical composition of these three compounds differs among the plant species (Crawford 1978; Akin 2007; Anderson and Akin 2008).

Table 1. Composition of lignocellulosic biomass

Lignocellulosic residues	Lignin (%)	Hemicellulose (%)	Cellulose (%)	Ash (%)
Hardwood stems	18–25	24–40	40–55	NA
Softwood stems	25–35	25–35	45–50	NA
Nut shells	30–40	25–30	25–30	NA
Corn cobs	15	35	45	1.36
Paper	0–15	0	85–99	1.1–3.9
Rice straw	18	24	32.1	NA
Sorted refuse	20	20	60	NA
Leaves	0	80–85	15–20	NA
Cotton seeds hairs	0	5–20	80–95	NA
Newspaper	18–30	25–40	40–55	8.8–1.8
Waste paper from chemical pulps	5–10	10–20	60–70	NA
Primary wastewater solids	24–29	NA	8–15	NA
Swine waste	NA	28	6	NA
Solid cattle manure	2.7–5.7	1.4–3.3	1.6–4.7	NA
Coastal Bermuda grass	6.4	35.7	25	NA
Switch grass	12.0	31.4	45	NA
S32 rye grass (early leaf)	2.7	15.8	21.3	NA
S32 rye grass (seed setting)	7.3	25.7	26.7	NA
Orchard grass (medium maturity)	4.7	40	32	NA
Grasses (average values for grasses)	10–30	25–50	25–40	1.5
Sugar cane bagasse	19–24	27–32	32–44	4.5–9
Wheat straw	16–21	26–32	29–35	NA
Barley straw	14–15	24–29	31–34	5–7
Oat straw	16–19	27–38	31–37	6–8
Rye straw	16–19	27–30	33–35	2–5
Bamboo	21–31	15–26	26–43	1.7–5
Grass Esparto	17–19	27–32	33–38	6–8
Grass Sabai	22.0	23.9	NA	6.0
Grass Elephant	23.9	24	22	6
Bast fiber Seed flax	23	25	47	5
Bast fiber Kenaf	15–19	22–23	31–39	2–5
Bast fiber Jute	21–26	18–21	45–53	0.5–2
Leaf Fiber Abaca (Manila)	8.8	17.3	60.8	1.1
Leaf Fiber Sisal (agave)	7–9	21–24	43–56	0.6–1.1
Leaf Fiber Henequen	13.1	4–8	77.6	0.6–1
Coffee pulp	18.8	46.3	35	8.2
Banana waste	14	14.8	13.2	11.4
Yuca waste	NA	NA	NA	4.2

Adapted from (Sanchez, 2009).

Lignin is connected to both hemicellulose and cellulose, forming a physical seal that is a tightly packed barrier in the plant cell wall (Figure 3). It provides the structural support, impermeability and resistance against microbial attack and oxidative stress (Sanchez 2009).

Lignins are formed from L-phenylalanine and cinnamic acids via different metabolic routes (Higuchi and Brown 1963). This polymer is synthesized via. peroxidase-mediated dehydrogenation of three different phenyl propionic alcohols such as coniferyl alcohol (guaiacyl propanol), coumaryl alcohol (p-hydroxyphenyl propanol), and sinapyl alcohol (syringyl propanol). These monolignols are derived from phenylalanine via series of metabolic reactions (Figure 4). These heterogeneous alcohols are linked by C–C and aryl-ether linkages, with aryl-glycerol β-aryl ether being the main structures (Boerjan et al. 2003).

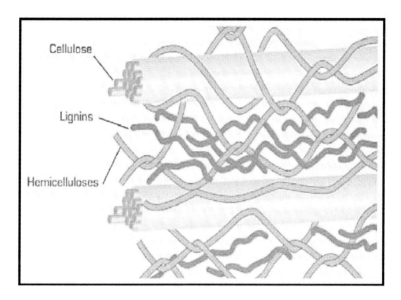

Figure 3.Schematic representation of the lignified secondary wall. Other cell wall constituents are not indicated on the figure (Adapted from Boudet et al, 2003).

The oxidative copolymerization of p-hydroxycinnamyl alcohols (p-coumaryl, coniferyl and sinapyl) contributes the varied degree of polymerization to the macromolecular arrangement depending upon the morphological parts of plants (Huttermann et al. 2001; Grabber and Lu 2007). In addition, factors such as temperature, light intensity, water availability, latitude, harvesting periods and storage conditions also affect lignin content in plant materials (Hatfield and Chaptman 2009)

The composition of lignin in gymnosperm and angiosperm varies in their quantity in terms of V, S, C phenolic units. Gymnosperm wood is composed of mainly V-units (about 80%) coupled with C-units, whereas angiosperm wood is composed of approximately equivalent quantities of V and S units allied with C units (Ritter et al. 1948). Non-woody vascular plants (herbaceous, Gramineae and pine) contain equivalent amounts of V, S and C units as part of the lignin macromolecule (Lam et al. 2001). The comparative contribution of V, S and C units in soils and sediments may mirror the source vegetation. For example, the depiction of the S-to-V vs. C-to-V ratios amassed from numerous studies helps to differentiate organic matter of angiosperm and gymnosperm taxa in soils (Lam et al. 2001).

Lignified plant cell walls are usually considered to be thekey in evolution of terrestrial plants from aquatic ancestors. Lignins are generally found in the secondary cell walls surrounded by xylem tissues, building a dense matrix that binds with cellulose micro fibrils and cross links other cell wall components. Although "lignin-like" complex have been categorized in the primitive green algae, the presence of true lignins in non-vascular aquatic algae has not been explored (Boyce,et al 2004, Peter and Neale 2004). Martone et al (2009) have discovered the secondary walls and lignin within cells of the *Calliarthroncheilosporioides* (intertidal red alga). The abundance of algal lignin may become an important avenue for future biorefinery applications.

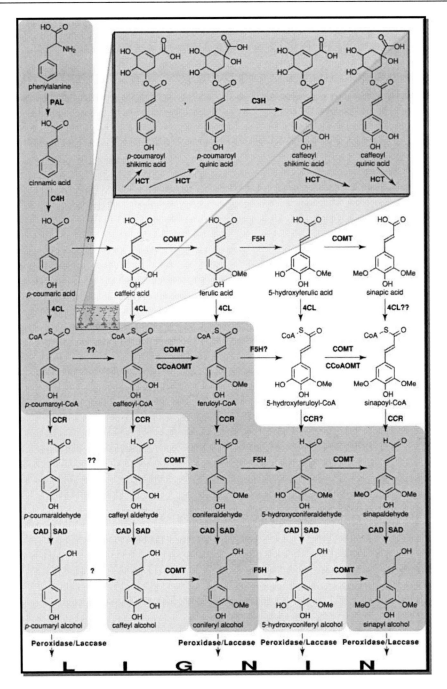

Figure 4. Monolignol biosynthesis pathway (Adapted from Boerjan 2003). (CAD, cinnamylalcoholdehydrogenase; 4CL, 4-coumarate:CoA ligase; C3H, *p*-coumarate 3-hydroxylase; C4H, cinnamate 4-hydroxylase; CCoAOMT, caffeoyl-CoA *O*-methyltransferase; CCR, cinnamoyl-CoA reductase; COMT, caffeic acid *O*-methyltransferase; HCT,*p*-hydroxycinnamoyl-CoA: quinateshikimate*p*-hydroxycinnamoyltransferase; F5H,ferulate 5-hydroxylase; PAL, phenylalanine ammonia-lyase; SAD, sinapylalcoholdehydrogenase).

BIODEGRADATIONOF LIGNIN

Lignin is indigestible by animal enzymes owing to its multifaceted structure; however some fungi and bacteria are able to degrade the polymer (Eggins 1965; Cartwright and Holdom 1973; Shah and Nerud 2002; Romani et al. 2006; Sanchez 2009). The mechanism of the biodegradation is not fully elucidated. Lignin degradation is predominantly carried out by biotic, aerobic and co-metabolic processes (Vicuna 2000). Several studies suggested that lignin degradation occurs via anaerobic as well as abiotic routes. Microbial systems involved in lignolytic activities include bacteria (*Nocardia* sp. *Streptomyces* sp.) and fungi (basidiomycetes - brown-rot and white-rot fungi).

Although most of the lignolytic strainsare capable of altering the lignin structure, only the white-rot basidiomycetes can mineralize the lignin molecule. Biodegradation of lignin takes place due to the catalytic action of non-specific extracellular enzymes, such as lignin peroxidase, manganese peroxidise and laccase (Higuchi, 2004).

It is well known that the decomposition of organic matter in soils is influenced by abiotic factors such as pH, moisture, temperature and nutrients complexity. Several authors have reported the influence of the nitrogen on lignin degradation. The fungal lignin biodegradation is limited in the presence of high nitrogen content. Nitrogen concentration also influences the lignin degradation rate in *Streptomyces.* The N content perhaps regulates the lignin peroxidase (LiP) gene transcription in the microbial systems. However, few studies showed that there were no considerable or negative impacts of N on lignin degradation. In addition, the occurrence of minerals in the soils could also influence the lignin transformation. Therefore the lignin degradation is not only associated to the type of vegetation, but also to the environmental and soil conditions.

Numerous microorganisms are capable of utilizing cellulose and hemicellulose as carbon and energy sources, on the other hand, a smaller group of filamentous white rot fungi possess the exclusive ability to degrade lignin to CO_2 very efficiently. These wood-decay fungi are common inhabitants of forest litter and fallen trees. *Phanerochaetechrysosporium* is extensively studied white rot fungi, belonging to the homobasidiomycetes. This fungus catalyses the depolymerization of lignin via unusual extracellular non-specific enzymes. An example of this is lignin peroxidase is a hemoprotein which exhibits a variety of lignin-degrading reactions, all reliant on hydrogen peroxide to integrate molecular oxygen into reaction products (Wang and Wen 2009; Wen et al. 2009). There are also several other microbial source of enzymes (Levit and Shkrob 1992; Rogalski et al. 2001; Hassett et al. 2009) that are believed to be involved in lignin biodegradation, such as copper oxidases and cellobiose dehydrogenase (Aitken and Irvine 1990; Abdel-Raheem 1997; Henriksson et al. 2000; Cameron and Aust 2001; Bajwa and Arora 2009)

RECOVERY OF LIGNIN

Currently, lignin is extracted from black liquor, a waste discharged from the pulp and paper industries. Due to their very complex structure and amorphous nature, lignin polymers are often used as a fuel source for the pulping process (Bajpai 2004). At present industrial activities are only utilizing 1% of the produced lignin for the development of valuable industrial products. Additionally, large quantities of lignin generated during biomass fractionation processes are under-utilized, which has spurred scientists to seek innovative and economically viable alternative industrial applications. Indeed, current research efforts are investigating lignin as a critically important feedstock to expand the portfolio of products from lignocellulosic biorefineries.The differences in chemical pulping methods give rise to a variety of lignin fragments (Table 2). For example, in sulfite processing, sulphuric acid is being used whereas; in the Kraft process sodium hydroxide and sodium sulphide are utilized. In contrast, the Alcell method employs the use of aqueous ethanol for cooking at 185 and 195°C. Alcell lignins consist of mono-phenolic fragments with low molecular weight and enhanced hydrophobicity (Zhao and Wilkins 2000; Nadif et al. 2002; Baurhoo et al. 2007).

Table 2. Chemical Composition of Biomass Types and Model Compounds

Lignin	Wt (%)					Molar ratio	
	Carbon	Hydrogen	Nitrogen	Sulphur	Calcd O	O/C	H/C
Commercially available standards							
G: Guaiacol	67.7	6.5	0	0	25.8	0.29	1.14
S: Syringol	62.3	6.5	0	0	31.1	0.38	1.25
GGE: Guaiacylglyceryl ether	60.6	7.1	0	0	32.3	0.4	1.4
Cellulose	43.8	6.3	0	0	49.9	0.83	1.67
Industrial lignin sources							
AL: Alkali lignin	48.2	3.4	0	1.1	47.3	0.74	0.84
OS: Organosol lignin	66.3	5.3	0.2	0.3	27.9	0.32	0.95
LS: Lignosulphonate	42	4.6	0	6.3	47.1	0.84	1.31
HL: Hydrolysis lignin	47.6	4.3	0	0.4	47.7	0.75	1.07
Lignin from ethanol plant(Lund SE)	55.2	6	0.1	0.1	38.6	0.52	1.30
Lignin from Ethanol plant(Bergen NO)	63.3	4.7	0	0.6	31.3	0.37	0.88

Modified from Kleinert and Tanja Barth, (2008).

PURIFIED LIGNIN

Purified lignins are manufactured as a co-product of pulp and paper industries and are generally separated from wood pulp via chemical processes. These lignins are characterized as low molecular weight mono-phenolic compounds that have distinct biological properties from native lignin. Using different types of chemicals during the wood-pulping give rise to diverse types of purified lignin.Although, these phenolic fragments may potentially have important applications in agriculture. Purified lignin has been mostly used in industrial and construction purposes, and to date there is no report on its application in animal agriculture (Baurhoo et al. 2007).

APPLICATIONS OF LIGNIN

In the plant biotechnology and pharmaceutical sectors in recent years, numerous studies have focused on the physiological role of various plant bioactive compounds or secondary metabolites. Several R and D projects a barrier highlighting the potential of naturally occurring poly-phenols have been initiated in human medicine and to a lesser extent in animal agriculture. Owing to their unique properties, they have been proven to elicit a number of health benefits *e.g.*, lignins (and lignans) inhibit the oxidation of low-density lipo-proteins (thereby decreasing risks of heart disease), while their potential as an anti-inflammatory, anti-carcinogenic and antioxidants has been suggested. In contrast to native lignin, purified lignin does not represent a barrier to digestion in monogastric or ruminant animals. The purified lignins are low molecular weight mono-phenolic fragments that have different biological characteristics from native lignin; some of these compounds have already been shown to function effectively as prebiotics.

As compared to native lignin, pure lignin does not embody to digestion in monogastric or ruminant animals. Numerous *in vitro* and *in vivo* studies have verified the antimicrobial properties of the phenolic fragments in purified lignin (Baurhoo et al. 2007). In recent times, Alcell lignin has been shown to exhibit prebiotic effects in fowl (Hegde et al. 1982; Baurhoo et al. 2007). These results indicate that purified lignin may possibly exert health benefits in monogastric animals and could potentially be utilized as an alternate natural feed additive (Fardet, 2010). Additional preliminary data suggest that animal responses to the purified lignin are dependent on type and source of the lignin as well as dosage (Baurhoo et al. 2007).

PREBIOTIC EFFECT

Prebiotics are indigestible feed ingredients which selectively trigger the growth or metabolic activities of a number of intestinal microbes in birds and mammals (Gibson and Roberfroid, 1995). *Lactobacilli* and *Bifidobacteria* restrict intestinal colonization of pathogens by competing for nutrients and binding sites (Gibson and Roberfroid, 1995). Baurhoo et al., (2007) have observed that Alcell lignin (12.5 g/kg of DM) improved the cell numbers of *Lactobacilli* and *Bifidobacteria* in broilers intestine. However, the growth of these bacteria was inhibited at higher dose of lignin (25 g/kg of DM).

Bactericidal properties of lignin at higher level have previously been demonstrated in both *in vitro* and *in vivo* studies. However, there was no significant improvement at high lignin dosage suggests that the prebiotic effect of Alcell lignin occurs only within a certain range. Alcell lignin helps in improving the morphological structures of the intestines in monogastric animals. Furthermore, the bactericidal nature of lignin fragments may help in eliminating the intestinal pathogens, thereby ensuring safety of livestock products to humans. However, further R and D is necessary to determine the optimum dosage of lignin, for health benefits, welfare, and safety of animal products.

ANIMAL FEED SUPPLEMENT

Reports on the effect of lignin on animal health benefits are very scarce, and lignin has not been described as a feed additive in livestock production. However, few recent studies have suggested an impact of pure lignin on animal performance. Alcell lignin (12.5 g/kg of DM) enhanced the body weight gain of Holstein calves; but no significant improvement was observed at higher dietary lignin (>25 g/kg) dosages (Phillip et al., 2000). Indulin, a purified Kraft lignin from the paper industry at the dosage of 40 and 80 g/kg of DM, was shown to improve weight gain in broiler chickens.

In geese, supplementation of a pure lignin improved the weight gain, but showed reduced efficiency in feed utilization (Yu et al., 1998). Feed supplementation with Alcell lignin (12.5 g/kg of DM) demonstrated neither body weight gain nor feed efficiency in pigs (Valencia and Chavez, 1997). Similarly, Alcell lignin (12.5 or 25 g/kg of DM) in the absence of antibiotics did not show any improvement in altering body weight or feed efficiency in broiler chickens (Baurhoo et al., 2007). The above findings clearly indicate that differences in form, type and concentration of lignin, as well as animal species, may have contributed to variations in animal responses.

ANTIMICROBIAL PROPERTIES

The antimicrobial properties of the phenolic lignin fragments are well documented. Phenolic compounds such as phenolic acids and flavonoids have long been utilized as one of the main types of food preservatives (Davidson and Branen, 1981). Examples include carvacrol and cinnamaldehyde, which are found to be effective in fresh fruit and vegetables, rice, cheese and meat (Smid et al., 1996; Ultee et al., 2000; Roller and Seedhar, 2002).

The phenolic compounds of lignin have been reported to inhibit the growth of *Escherichia coli, Saccharomyces cerevisiae, Bacillus licheniformis*and *Aspergillusniger* (Zemek et al., 1979). However, the minimum inhibitory concentration of phenolic fragments impacted upon the extent of activity. The chemical moiety of side chain as well as the functional group chemistry of phenolic compounds has significant roles to the antimicrobial properties of lignin. In general, phenolic fragments with oxygen (–OH, –CO, – COOH) in the side chain are less inhibitory to the microbes, whereas isoeugenol has been reported to be the most inhibitory phenolic fragment due to the occurrence of a double bond and methyl group.

Ferulic acid (2.5mM) has been reported to inhibit the growth of *Saccharomyces cerevisiae*. Poly-phenolic fragments of Alcell lignin (100 g/l) diminished the growth rate of *Staphylococcus aureus*and *Pseudomonas* (Nelson et al., 1994). Dosage levels of lignin fragments are also critical to antibacterial efficiency. Inhibition of *E. coli* growth was observed in a medium containing 100 g/l of Alcell lignin vs.50 g/l (Phillip et al., 2000).

Indulin (40 and 100 g/kg of DM) has been reported to decrease the volatile fatty acid levels in ceca and large intestine of chickens (Ricke et al., 1982), suggesting that the pure lignin may have altered the fermentation pattern of intestinal tract. In a mouse model, Alcell lignin did not alter the aerobic bacterial growth in the cecum, but abridged the overall bacterial translocation in the lymph nodes and liver after burn injury (Nelson et al., 1994

Ruminant responses to the dietary lignin fragments are inconsistent. Alcell lignin had no effect in sheep, and did not alter fecal anaerobic, aerobic and coliform loads in calves (Phillip et al., 2000). This could be as a result of degradation of the lignin by rumen microbes. Para-coumaric and ferulic acids are extensively converted into reduced phenolics by ruminal microbial systems (Bourquin et al., 1990). The introduction of phenolic lignin compounds (quinic acid, phenolic benzoic and phenyl acetic acids) into the sheep rumen caused a huge increase in urinary outputs of phenolic acids and phenols. These experimental results put forward that the phenolic monomers are widely metabolized by the rumen microbial system. Therefore, the antimicrobial properties of lignin are dependent on animal species, and are more effective on non-ruminants.

The exact mechanism of antimicrobial action of lignin is still not clear. The poly-phenolic compounds of lignin may cause cell membrane damage to the microbes. Mono-phenolic compounds, iecarvacrol, thymol and cinnamaldehyde, also exhibit anti-bacterial properties (Bozin et al., 2006). Carvacrol and thymol caused the bacterial cell membrane disruption and release of cellular contents (Helander et al., 1998). In contrast, cinnamaldehyde penetrates into the bacterial cell membrane and decrease intracellular pH and thus caused the depletion of ATP (Oussalah et al., 2006).

Natural Antioxidants and Cosmetics

The efficacy of lignins and some lignin related monomeric and dimeric fragments as antioxidants in various materials (rubber, wood, thermo-mechanical pulp), as well as medicines and dietary products have been reported (Ugartondo et al. 2008). Antioxidant effects of lignins are mainly due to the scavenging action of phenolic structures on oxygen containing reactive free radicals. Vinardell et al (2008) have reported the antioxidant potentials and topical applications of lignins obtained from different industrial sources. Studies have confirmed that lignins are not detrimental to the eye and skin. This opens up a novel use of lignin in cosmetic and topical formulations.

Lignin Engineering for Biofuels and Biochemicals Production

Biofuel production from lignocellulosic feedstocks represents a renewable, environmentally friendly alternative to both fossil fuels and fuels derived from corn and sugarcane. In order to achieve efficient bioconversion of lignocellulosic biomass to biofuels, lignin must be separated from the plant tissues prior enzymatic saccharification. The presence of lignin in the plant biomass hinders the conversion of plant polysaccharides to simple fermentable sugars and the subsequent conversion to usable fuels or chemicals via microbial fermentation (Balan et al. 2009; Jegannathan et al. 2009; Carioca 2010).

Although traditional genetic engineering approaches, such as up regulation or down regulation of monolignol biosynthetic genes, have been successful in altering both lignin quantity and chemical composition, however many of them lead to deleterious phenotypes such as dwarfing or vascular collapse. Current advancement in plant genetic engineering strategies, including manipulation of lignin biosynthesis at the regulatory level, controlling monolignol polymerization enzymes, and modification of lignin polymer arrangement, together with the exploration of lignin degradation enzymes from diverse sources should help in developing an efficient energy crops for biofuels production (Gressel 2008).

Pyrolysis

Lignin is one of the main by-products of second generation biofuels production and is also the most valuable impurities obtained during the separation of cellulose from wood pulp. Consequently lignin has attracted extensive attention as a prospective resource of aromatic hydrocarbons for biofuels and bio based chemicals. Biological conversion of lignin into valuable chemicals is one of the key technologies for biomass based industrial sectors. An ever-increasing interest in lignin pyrolysis is furthered by claims of up to 51% yield of phenol and alkyl phenol from lignin (Snell, Huibers, 1984). Most of the R and D on pyrolysis have been focused on use ofproducts with a high yield of phenols. It is vital to understand the kinetics of lignin pyrolysis to provide valuable information for realistic design and scaling-up of pyrolysis reactors. Nevertheless, it is rather difficult to compare the pyrolysis yield data for different lignins, because of the diverse conditions and analytic procedures were employed during the lignin isolation processes (Chen et al. 2003).

Additional Applications

In recent years, a significant amount of research has been devoted for developing lignin-containing polymeric materials. The incorporation of lignin into a polymeric system has been demonstrated by blending large amounts of underivatized or derivatized lignin with a synthetic polymer or incorporating a lesser amount of lignin to stabilize the material against photo and thermal oxidation. Apart from the aforementioned applications, lignin polymers can also be utilized in a number of commercially important materials including in automotive brakes, wood panel products, bio-dispersants, polyurethane foams, epoxy resins, adhesives and tanning agents. Furthermore, lignin has been shown to have a high carbon content and molecular similarity to the bituminous coal, thus opening up the prospect of using this polymer as an ideal alternate precursor for activated carbon production. Althoughresearch to date has identified lignin as a potentially valuable and untapped resource with a variety of commercially viable and economically important applications, advances in understanding the complexity of lignin structure from different sources as well as in harnessing new methods of lignin conversion will play a vital role in the development of biomass based integrated biorefineries.

Economic Significance

Lignified wood is tough and consequently is considered a good raw material for numerous industrial applications(Baucher et al. 2003; Bajpai 2004; Compere and Griffith 2009). It is also an excellent fuel, as lignin yields more energy once burned than cellulose (Grammelis et al. 2003; Chang 2007). Mechanical or high yield pulp utilized in the production of newsprint contains the majority of the lignin originally present in the wood. This lignin causes the characteristic newsprint yellowing with age. Lignin must be separated from the pulp prior to high quality bleached paper manufacturing.

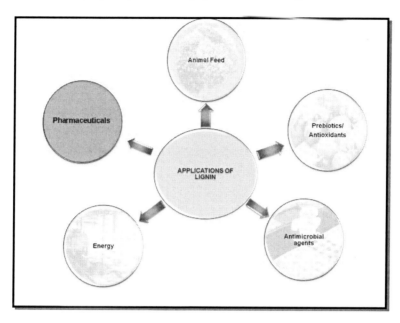

Figure 5. Potential applications of lignin.

In sulfite pulping, lignin is removed from wood pulp as sulphonates (Willauer et al. 2000; Bajpai et al. 2004). These lignosulphonates have numerous applications such as utilization as dispersants in cement applications, water treatment formulations and textile dyes, as additives in specialty oil field applications and chemicals and as raw materials for several chemicals, e.g. vanillin, DMSO, ethanol, xylitol sugar and humic acid (Galletti and Bocchini 1995; Hames 2009; Fitzpatrick et al. 2010).

The first investigations into commercial use of lignin were reported in the 1920's with the first class of products ofpromise being leather tanning agents (Comodo et al. 1970). Lignin removed via the Kraft process (sulfate pulping) is usually burned for its fuel value, providing energy to run the mill and its associated processes (Xiao et al. 2007;Ziaie-Shirkolaee et al. 2008).

Despite the important role of lignin in industrial processes, many uncertainties remain in our perception of lignin's structure, process of formation, analytical recognition, and performance in wood, pulp/man-made materials. These uncertainties need to be addressed with emerging scientific methodologies. This new research combining modern and traditional structural analysis techniques and making use of novel experimental protocols inspired by the fields of biotechnology and materials science, are starting to provide important new insights into the structure of lignin, and its behaviors. The full potential of biomass to biorefinery applications will continue to go unfulfilled until all energy rich components are fully elucidated and understood. Only then can we truly say that bio refining is an economically viable and sustainable alternative to conventional industrial routes.

REFERENCES

Abdel-Raheem, A. M. (1997). "Laccase activity of lignicolous aquatic hyphomycetes isolated from the River Nile in Egypt." *Mycopathologia*139(3): 145-50.

Aitken, M. D. and R. L. Irvine (1990)."Characterization of reactions catalyzed by manganese peroxidase from Phanerochaetechrysosporium."*Arch.Biochem.Biophys.* 276(2): 405-14.

Akin, D. E. (2007)."Grass lignocellulose: strategies to overcome recalcitrance."*Appl.Biochem.Biotechnol.* 137-140(1-12): 3-15.

Anderson, W. F. and D. E. Akin (2008). "Structural and chemical properties of grass lignocelluloses related to conversion for biofuels." *J. Ind.Microbiol.Biotechnol.* 35(5): 355-66.

Avgerinos, G. C. and D. I. Wang (1983)."Selective solvent delignification for fermentation enhancement."*Biotechnol.Bioeng* 25(1): 67-83.

Bajpai, P., (2004). Biological bleaching of chemical pulps.*Crit. Rev.Biotechnol.* 24, 1–58.

Bajwa, P. K. and D. S. Arora (2009)."Comparative production of ligninolytic enzymes by Phanerochaetechrysosporium and Polyporussanguineus."*Can. J.Microbiol.* 55(12): 1397-402.

Balan, V., B. Bals, et al. (2009)."Lignocellulosic biomass pretreatment using AFEX."*Methods Mol. Biol.* 581: 61-77.

Baucher, M., C. Halpin, et al. (2003). "Lignin: genetic engineering and impact on pulping." *Crit. Rev.Biochem Mol. Biol.* 38(4): 305-50.

Baurhoo, B., Letellier, A., Zhao, X., Ruiz-Feria, C.A., (2007) Cecal populations of Lactobacilli and Bifidobacteriaand E. coli populations after in vivo E. coli challenge in birds fed diets with purified lignin or mannanoligosaccharides. Poult.Sci. 86, 2509–2516.farm animals—A review."*Animal Feed Science and Technology*175-184.

Blaschke, L., M. Forstreuter, et al. (2002). "Lignification in beech (Fagussylvatica) grown at elevated CO2 concentrations: interaction with nutrient availability and leaf maturation."*Tree Physiol* 22(7): 469-77.

Boerjan, W., J. Ralph, et al. (2003)."Lignin biosynthesis."*Annu. Rev. Plant Biol.* 54: 519-46.

Bose, S. K., R. C. Francis, et al. (2009). "Lignin content versus syringyl to guaiacyl ratio amongst poplars."*Bioresour. Technol.* 100(4): 1628-33.

Boudet, A. M., S. Kajita, et al. (2003). "Lignins and lignocellulosics: a better control of synthesis for new and improved uses." *Trends Plant Sci.* 8(12): 576-81.

Bourquin, L.D., Garleb, K.A., Merchen, N.R., Fahey Jr., G.C., (1990). Effects of intake and forage level on site and extent of digestion of plant cell wall monomeric compounds by sheep. *J. Anim. Sci.* 68, 2479–2495.

Boyce, C.K., Zwieniecki, M.A., Cody, G.D., Jacobsen, C., Wirick, S.,Knoll, A.H., and Holbrook, N.M. (2004). Evolution of xylem lignification and hydrogel transport regulation. *Proc. Natl. Acad. Sci. USA* 101, 17555–17558.

Bozin, B., Mimica-Dukic, N., Simin, N., Anackov, G., (2006).Characterization of the volatile composition of essential oils of some lamiaceae spices and the antimicrobial and antioxidant activities of the entire oils.*J. Agric. Food Chem.* 54, 1822–1828.

Broda, P., P. R. Birch, et al. (1996). "Lignocellulose degradation by Phanerochaetechrysosporium: gene families and gene expression for a complex process." *Mol.Microbiol.* 19(5): 923-32.

Cameron, M. D. and S. D. Aust (2001)."Cellobiose dehydrogenase-an extracellular fungal flavocytochrome."*Enzyme Microb. Technol.* 28(2-3): 129-138.

Carioca, J. O. B. (2010). "Biofuels: Problems, challenges and perspectives." *BiotechnologyJournal* 5(3): 260-273.

Cartwright, N. J. and K. S. Holdom (1973)."Enzymic lignin, its release and utilization by bacteria."*Microbios* 8(29): 7-14.

Chafe, S. C. (1974). "Cell wall structure in the xylem parenchyma of Cryptomeria."*Protoplasma* 81(1): 63-76.

Chang, M. C. (2007). "Harnessing energy from plant biomass."*Curr.Opin Chem. Biol.* 11(6): 677-84.

Chen, G. Y., M. X. Fang, et al. (2003). "Kinetics study on biomass pyrolysis for fuel gas production."*J. Zhejiang Univ. Sci.* 4(4): 441-7.

Compere, A. L. and W. L. Griffith (2009)."Preparation and analysis of biomass lignins."*Methods Mol. Biol.* 581: 185-212.

Crawford, D. L. (1978)."Lignocellulose decomposition by selected streptomyces strains."*Appl. Environ.Microbiol.* 35(6): 1041-5.

Davidson, P.M., Branen, A.L., (1981). Antimicrobial activity of non-halogenated phenolic compounds.*J. Food Prot.44*, 623–632.

Davin, L. B. and N. G. Lewis (2005)."Lignin primary structures and dirigent sites."*Curr.OpinBiotechnol* 16(4): 407-15.

Del Rio, J. C., G. Marques, et al. (2007). "Occurrence of naturally acetylated lignin units."*J.Agric Food Chem.* 55(14): 5461-8.

Deschamps, F. C., L. P. Ramos, et al. (1996). "Pretreatment of sugar cane bagasse for enhanced ruminal digestion."*Appl.Biochem.Biotechnol.* 57-58: 171-82.

Eggins, H. O. (1965). "A Medium to Demonstrate the Ligninolytic Activity of Some Fungi."*Experientia* 21: 54.

Foster, C. E., T. M. Martin, et al. (2010). "Comprehensive compositional analysis of plant cell walls (Lignocellulosic biomass) part I: lignin." *J. Vis. Exp.* (37).

Fustec, E., E. Chauvet, et al. (1989)."Lignin degradation and humus formation in alluvial soils and sediments."*Appl. Environ.Microbiol.* 55(4): 922-6.

Gibson, G.R., Roberfroid, M.B., (1995). Dietary modulation of the human colonic microbiotica: introducing the concept of prebiotics. *J. Nutr.* 125, 1404–1412.

Grabber, J. H. and F. Lu (2007)."Formation of syringyl-rich lignins in maize as influenced by feruloylated xylans and p-coumaroylatedmonolignols."*Planta* 226(3): 741-51.

Grammelis, P., E. Kakaras, et al. (2003)."Thermal exploitation of wastes with lignite for energy production."*J. Air Waste Manag. Assoc.* 53(11): 1301-11.

Gressel, J. (2008). "Transgenics are imperative for biofuel crops." *Plant Science* 174(3): 246-263.

Grote, M., S. Klinnert, et al. (2000)."Comparison of degradation state and stability of different humic acids by means of chemolysis with tetramethylammonium hydroxide."*J.Environ.Monit* 2(2): 165-9.

Haider, K. and K. H. Domsch (1969)."[Decomposition and transformation of lignified plant material by microscopic soil fungi]."*Arch.Mikrobiol* 64(4): 338-48.

Harayama, S. (1997)."Polycyclic aromatic hydrocarbon bioremediation design."*Curr.Opin.Biotechnol.* 8(3): 268-73.

Hassett, J. E., D. R. Zak, et al. (2009). "Are basidiomycetelaccase gene abundance and composition related to reduced lignolytic activity under elevated atmospheric NO3(-) deposition in a northern hardwood forest?" *Microb. Ecol.* 57(4): 728-39.

Hatfield, R. D. and A. K. Chaptman (2009)."Comparing Corn Types for Differences in Cell Wall Characteristics and p-Coumaroylation of Lignin."*J.Agric Food Chem.*

Hegde, S. N., B. A. Rolls, et al. (1982). "THe effects of the gut microflora and dietary fibre on energy utilization by the chick."*Br. J.Nutr* 48(1): 73-80.

Henriksson, G., G. Johansson, et al. (2000). "A critical review of cellobiose dehydrogenases."*J.Biotechnol* 78(2): 93-113.

Helander, I.M., Alakomi, H.L., Latva-Kala, K., Mattila-Sandhol, T., Pol, I., Smid, E.J., Gorris, L.G.M.,VonWright,A., (1998). Characterization of the action of selected essential oil components on gram-negative bacteria.*J. Agric.Food Chem.* 46, 3590–3595.

Higuchi, T. and S. A. Brown (1963)."Studies of lignin biosynthesis using isotopic carbon.XIII. The phenylpropanoid system in lignification."*Can. J.Biochem. Physiol.* 41: 621-8.

Hintz, R. W., D. R. Mertens, et al. (1996). "Effects of sodium sulfite on recovery and composition of detergent fiber and lignin."*J. AOAC Int* 79(1): 16-22.

Huttermann, A., C. Mai, et al. (2001). "Modification of lignin for the production of new compounded materials."*Appl.Microbiol.Biotechnol.* 55(4): 387-94.

Jegannathan, K. R., E. S. Chan, et al. (2009). "Harnessing biofuels: A global Renaissance in energy production?" *Renewable and Sustainable Energy Reviews* 13(8): 2163-2168.

Kadam, K. L. and S. W. Drew (1986). "Study of lignin biotransformation by Aspergillusfumigatus and white-rot fungi using (14)C-labeled and unlabeled kraft lignins." *Biotechnol.Bioeng.* 28(3): 394-404.

Karkonen, A. and S. Koutaniemi (2010). "Lignin biosynthesis studies in plant tissue cultures." *J.Integr Plant Biol.* 52(2): 176-85.

King, A. W., I. Kilpelainen, et al. (2009). "Hydrophobic interactions determining functionalized lignocellulose solubility in dialkylimidazolium chlorides, as probed by 31P NMR." *Biomacromolecules*10(2): 458-63.

Kajikawa, H., Kudo, H., Kudo, T., Jodai, K., Honda, Y., Kuwahara,Kleinert M and T Ba. (2008). Towards a Lignincellulosic Biorefinery: Direct One-Step Conversion of Lignin to *Hydrogen-Enriched Biofuel Energy and Fuels*, 22, 1371–1379.

Lam, T. B., K. Kadoya, et al. (2001). "Bonding of hydroxycinnamic acids to lignin: ferulic and p-coumaric acids are predominantly linked at the benzyl position of lignin, not the beta-position, in grass cell walls." *Phytochemistry* 57(6): 987-92.

Levit, M. N. and A. M. Shkrob (1992)."[Lignin and ligninase*]."BioorgKhim* 18(3): 309-45.

Li, J., G. Henriksson, et al. (2007). "Lignin depolymerization/repolymerization and its critical role for delignification of aspen wood by steam explosion."*Bioresour. Technol.* 98(16): 3061-8.

Martone P T.,JM. Estevez,F Lu, K Ruel, MW Denny C Somerville John Ralph. (2009). Discovery of Lignin in Seaweed Reveals convergent Evolution of Cell-Wall Architecture.*Current Biology* 19, 169–175.

Miceli, A., D. Traversi, et al. (1989). "[Characterization of the biomass of the stems of sweet sorghum]."*Boll Soc. Ital. Biol.Sper.* 65(12): 1141-7.

Mishra, M. M., C. P. Singh, et al. (1979). "Degradation of lignocellulosic material and humus formation by fungi."*Ann.Microbiol.* (Paris) 130 A(4): 481-6.

Nadif, A., D. Hunkeler, et al. (2002). "Sulfur-free lignins from alkaline pulping tested in mortar for use as mortar additives." *Bioresour. Technol.* 84(1): 49-55.

Nelson, J.L., Alexander, J.W., Gianotti, L., Chalk, C.L., Pyles, T., (1994). Influence of dietary fiber on microbial growth in vitro and bacterial translocation after burn injury in mice. *Nutrition* 10, 32–36.

Oglesby, R. T., R. F. Christman, et al. (1967). "The biotransformation of lignin to humus facts and postulates."*Adv. Appl.Microbiol.* 9: 171-84.

Ouyang, F., Y. Liu, et al. (2009)."[Lignans from stems of Sambucuswilliamsii]."*ZhongguoZhong Yao ZaZhi* 34(10): 1225-7.

Oussalah, M., Caillet, S., Lacroix, M., (2006). Mechanism of action of Spanish oregano, Chinese cinnamon, and savory essential oils against cell membranes and walls of Escherichia coli O157:H7 and Listeria monocytogenes.*J. Food. Prot.* 69, 1046–1055.

Peter, G., and Neale, D. (2004).Molecular basis for the evolution of xylem lignification.*Curr.Opin. Plant Biol.* 7, 737–742.

Phillip, L., Idziak, E.S.,Kubow, S., 2000. The potential use of lignin in animal nutrition, and in modifying microbial ecology of the gut. In: *East.Nutr. Conf. Animal Nutrition* Association of Canada, Montreal, Qu´ebec, Canada,pp. 165–184.

Radotic, K., D. Djikanovic, et al. (2008)."Levels of plant cell wall structural organization revealed by atomic force microscopy."*J.Microsc.* 232(3): 508-10.

Regalado, V., A. Rodriguez, et al. (1997)."Lignin Degradation and Modification by the Soil-Inhibiting Fungus Fusariumproliferatum."*Appl. Environ.Microbiol.* 63(9): 3716-8.

Ritter, D. M., D. E. Pennington, et al. (1948)."Constitution of Gymnosperm Lignin."*Science* 107(2766): 20-2.

Roller, S., Seedhar, P., (2002).Carvacrol and cinnamic acid inhibit microbial growth in fresh-cut melon and kiwifruits at 4 ◦C and 8 ◦C. *Lett. Appl. Microbiol.* 35, 390–394.

Rogalski, J., J. Fiedurek, et al. (2001). "Production of lignolytic and feed-back type enzymes by Phlebiaradiata on different media."*Acta Biol. Hung* 52(1): 149-60.

Romani, A. M., H. Fischer, et al. (2006). "Interactions of bacteria and fungi on decomposing litter: differential extracellular enzyme activities." *Ecology* 87(10): 2559-69.

RosBarcelo, A. (1997). "Lignification in plant cell walls."*Int. Rev.Cytol* 176: 87-132.

Sanchez, C. (2009). "Lignocellulosic residues: biodegradation and bioconversion by fungi." *Biotechnol. Adv.* 27(2): 185-94.

Smid, E.J., Hendriks, I., Boerrigter, H.A.M., Gorris, L.G.M., (1996). Surface disinfection of tomatoes using the natural plant compound trans-cinnamaldehyde.*Postharvest Biol.* Tech. 9, 343–350.

Shah, V. and F. Nerud (2002). "Lignin degrading system of white-rot fungi and its exploitation for dye decolorization." *Can. J.Microbiol* 48(10): 857-70.

Song, F., X. Tian, et al. "Decomposing ability of filamentous fungi on litter is involved in a subtropical mixed forest." *Mycologia*102(1): 20-6.

Sorensen, H. (1962). "Decomposition of lignin by soil bacteria and complex formation between autoxidized lignin and organic nitrogen compounds." *J. Gen.Microbiol* 27: 21-34.

Ugartondo, V., M. Mitjans, et al. (2008)."Comparative antioxidant and cytotoxic effects of lignins from different sources."*Bioresour. Technol.* 99(14): 6683-7.

Ultee, A., Slump, R.A., Steging, G., Smid, E.J., (2000). Antimicrobial activity of carvacrol toward Bacillus cereus on rice.*J. Food Prot.* 63, 620–624.

Vasil'chenko, L. G., K. N. Karapetian, et al. (2004). "[Degradation of lignin-carbohydrate substrate by soil fungi--producers of laccase and cellobiose dehydrogenase]."*PriklBiokhimMikrobiol* 40(1): 51-6.

Vicuna, R. (2000)."Ligninolysis.A very peculiar microbial process."*Mol.Biotechnol* 14(2): 173-6.

Vinardell M.P., V. Ugartondo, M. Mitjans (2008) Potential applications of antioxidant lignins from different sources industrial crops and products 27220–223.

Vogt, T. (2010)."Phenylpropanoid biosynthesis."*Mol. Plant* 3(1): 2-20.

Wang, W. and X. Wen (2009)."Expression of lignin peroxidase H2 from Phanerochaetechrysosporium by multi-copy recombinant Pichia strain."*J. Environ. Sci. (China)* 21(2): 218-22.

Wen, X., Y. Jia, et al. (2009). "Degradation of tetracycline and oxytetracycline by crude lignin peroxidase prepared from Phanerochaetechrysosporium--a white rot fungus." *Chemosphere* 75(8): 1003-7.

Willauer, H. D., J. G. Huddleston, et al. (2000). "Investigation of aqueous biphasic systems for the separation of lignins from cellulose in the paper pulping process."*J.Chromatogr.B Biomed. Sci. Appl.* 743(1-2): 127-35.

Wong, D. W. (2009). "Structure and action mechanism of ligninolytic enzymes."*Appl.Biochem.Biotechnol.* 157(2): 174-209.

Xiang, Q. and Y. Y. Lee (2000)."Oxidative cracking of precipitated hardwood lignin by hydrogen peroxide."*Appl.Biochem.Biotechnol.* 84-86: 153-62.

Yu, B., Tsai, C.C., Hsu, J.C., Chiou, P.W., (1998). Effect of different sources of dietary fibre on growth performance,intestinal morphology and caecalcarbohydrases of domestic geese. *Br. Poult. Sci.* 39, 560–567.

Zemek, J., Kosikova, B., Augustin, J., Joniak, D., (1979). Antibiotic properties of Lignin components.*Folia Microbiol.* 24, 483–486.

Zhao, J. and R. M. Wilkins (2000). "Controlled release of a herbicide from matrix granules based on solvent-fractionated organosolv lignins." *J. Agric. Food Chem.* 48(8): 3651-61.

In: Lignin
Editor: Ryan J. Paterson

ISBN 978-1-61122-907-3
© 2012 Nova Science Publishers, Inc.

Chapter 16

ABIOTIC STRESS RESPONSES IN WOODY PLANTS; MOLECULAR PERSPECTIVE IN ENGINEERING WOODY PLANT TOLERANCE TO ABIOTIC STRESS AND ENHANCE BIOMASS

Yuriko Osakabe,[1] Shinya Kajita[2] and Keishi Osakabe[3]*

[1] Graduate School of Agricultural and Life Sciences,
University of Tokyo, 1-1-1 Yayoi, Bunkyo-ku, Tokyo 113-8657
[2] Graduate School of Bio-Applications and Systems Engineering,
Tokyo University of Agriculture and Technology,
2-24-16 Nakacho, Koganei, Tokyo 184-8588, Japan
[3] Department of Plant Biotechnology,
National Institute of Agrobiological Sciences, 2-1-2 Kannondai,
Tsukuba, Ibaraki 305-8602, Japan

ABSTRACT

Environmental degradation, such as drought and salinity stresses, is a major factor in limiting plant growth and productivity, will become more severe and widespread in the world. To overcome the environmental stress, genetic engineering in woody plants needs to be implemented. The adaptation of plants to environmental stress is controlled by cascades of molecular networks. For woody plant species, the effects of longer periods of stress need to be considered to regard the actual tolerance. This chapter focuses on the molecular mechanism of abiotic stress responses of woody plants. The basis of genetic engineering for enhanced biomass and stress tolerance in woody plants will also summarized in this chapter.

Keywords: abiotic stress; woody plant; reactive oxygen species; transcription factor; lignin.

---

* Corresponding author, e-mail: ayosa@mail.ecc.u-tokyo.ac.jp.

ABBREVIATIONS

ROS, Reactive oxygen species;
HSP, Heat shock proteins;
AP2/EREB, APETALA2 /ethylene responsive element binding;
DRE/CRT, Dehydration-responsive element/C-repeat;
DREB, DRE binding protein.

INTRODUCTION

Abiotic stresses, such as salinity, drought, temperature, chemical toxicity, and oxidative stresses, are major causes for loss of agricultural production and natural vegetation. Plants are exposed to various environmental stresses in their sessile lifestyle. Abiotic stresses cause various physiological, biochemical, and molecular changes that affect plant growth and productivity. Many forest trees, the central component of global ecosystem, suffer from abiotic and biotic stress on a global scale. Tree species need to manage these stresses in their long-term growth. Therefore, plant engineering to improve stress tolerance in tree species is important for preservation of the earth's ecosystem and increase of biomass resources (Figure 1).

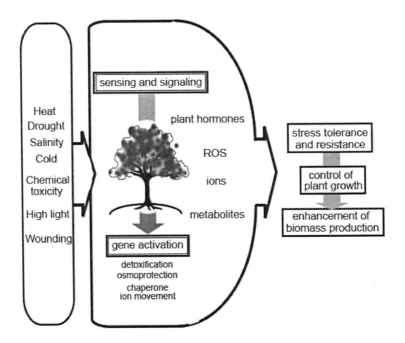

Figure 1. Environmental stress responses of woody plants. Drought, salinity, cold, heat, chemical toxicity, high light or UV, and biotic attacks affect the stress signals that stimulate the downstream pathway and the metabolic reactions, and ionic and plant hormone balances. The signaling pathways control the transcription of stress-responsive genes that include the factors that function in cellular homeostasis and the protection of damaged proteins and membrane structures. Woody plants control the growth and development through the coordinated regulation of stress tolerance and resistance. ROS, Reactive oxygen species.

Stress-responsive mechanisms and acclimation processes to environmental stress are regulated various cellular and molecular events in plants. The expression of a variety of genes that function not only in stress tolerance but also in stress response is upregulated responded to the stress (Hirayama and Shinozaki 2010, Osakabe et al. 2005, Yamaguchi-Shinozaki and Shinozaki 2005). These complex cascades of gene expression in abiotic and biotic stress responses and cross-talk in the stress and plant hormone signaling have been shown in many studies (Baena-Gonzáleza and Sheen 2008, Mitller 2006, Osakabe Y. et al. 2010, Spoel and Dong 2008). These studies focus on the molecular mechanisms of stress responses, and genetic regulatory networks of stress tolerance have been published (Vinocur and Altman 2005, Wang et al. 2003), however, they primarily discuss researches on model herbaceous plants. An understanding of the physiological stress responses and the molecular events in woody plants is now required. Such molecular studies on woody plants would contribute to understanding the molecular mechanism in control of biomass production under stressful conditions.

ABIOTIC STRESS-RESPONSIVE GENES AND TRANSCRIPTION FACTORS

Multiple signaling pathways that activate the expression of various stress-responsive genes regulate plant abiotic stress responses. Transcriptome analyses using *Arabidopsis* and rice have shown that a variety of genes, which are involved in stress response and stress tolerance, are induced by abiotic stresses (Seki et al. 2003, Yamaguchi-Shinozaki and Shinozaki 2006). The general control mechanisms in gene expression under water stress conditions in woody plants and other plant species have shown by transcriptome and proteome analyses. Reactive oxygen species (ROS) production is induced by both abiotic and biotic stress insults, such as high light, osmotic stress, and pathogen attack. ROS is also one of the important factors in osmotic stress responses in poplar, a model species of woody plants (Gu et al. 2004, Xiao et al. 2009, Willins et al. 2009). ROS have now been implicated as important second messengers of abiotic and biotic stress signal transduction (Jammes et al. 2009, Mittler et al. 2004, Torres and Dangl 2005, Osakabe Y. et al. 2010), and therefore ROS might be one of the general important regulatory factors in osmotic stress responses among plant species.

A salt-tolerant popular species, *Populus eupharatica*, has been used in transcriptome analysis to identify the responsive-genes in high salinity condition (Gu et al. 2004), and one gene family of the most highly up-regulated transcripts corresponded to HSPs (heat shock proteins). HSPs function as protein chaperones, which play important roles in the acquisition of not only thermo tolerance but also the adaptation to various environmental stresses including oxidative stress (Wang et al. 2004). HSP genes are also up-regulated by oxidative stress in *Arabidopsis* (Desikan et al. 2001, Kotak et al. 2007, Osakabe Y. et al. 2010). The proteomic analysis, which used two contrasting *Populus cathayana* populations in western China, identified 40 drought-responsive proteins, which include several factors in the regulation of transcription and translation, photosynthesis, ROS, HSPs, the enzymes related to redox homeostasis, and the regulation of secondary metabolism (Xiao et al. 2009). These findings also suggested that the general control mechanisms in gene expression under water stress conditions exist in higher plants including woody plants. Interestingly, the study also suggested that photosynthesis-related genes were down-regulated in response to salt stress (Gu et al. 2004), and the similar down-regulation was also observed in the microarray analysis with *Arabidopsis* (Seki et al. 2002).

The DREB proteins, which have the conserved ERF/AP2 DNA binding domain and belong to the AP2/EREB transcription factor family, specifically recognize the DRE/CRT sequence and activate the transcription of stress-responsive genes in *Arabidopsis* and rice (Sakuma et al. 2006, Yamaguchi-Shinozaki and Shinozaki 2005). *Arabidopsis* DREB2A controls the drought-responsive gene expression (Sakuma et al. 2006). A DREB2-type transcription factor, PeDREB2, was isolated from *Populus eupharatica* (Chen et al. 2009). PeDREB2 binds specifically to DRE sequences and the ectopic expression of PeDREB2 improved salt tolerance in transgenic tobacco. Results showed that the overexpression of stress-responsive transcription factors from woody plants enhanced tolerance to stress in transgenic plants.

STRESS-RESPONSIVE MICRO RNAS

The miRNAs play an important role in abiotic and biotic stress responses in plants, in which stress-responsive miRNAs target negative or positive regulators of stress response. The expression of the specific miRNAs was altered during various abiotic stresses, such as drought, salinity, cold (Sunkar and Zhu, 2004), UV-B (Zhou et al., 2007), phosphate and sulfate starvation (Chiou et al., 2006; Fujii et al., 2005; Jones-Rhoades and Bartel, 2004), oxidative stress (Sunkar et al., 2006), and mechanical stress (Lu et al., 2005). The regulatory mechanism of the gene expression mediated by siRNA during osmotic stress has been shown (Borsani et al., 2005). The nat-siRNA produced from overlapping mRNAs for Δ1-pyrroline-5-carboxylate dehydrogenase (P5CDH) and SIMILAR TO RCD ONE 5 (SRO5) controlled the expression of these genes; the *SRO5* gene is induced by high salinity and siRNAs are generated to down-regulate the *P5CDH* gene expression. The down-regulation of *P5CDH* gene results in the accumulation of osmoprotectant proline (Borsani et al., 2005).

The expressed small RNAs from leaves and buds in *Populus* have been isolated using high throughput pyrosequencing (Barakat et al. 2007), and almost 80,000 small RNAs were identified with 123 novel small RNAs belonging to previously identified miRNA families from other plant species and 48 novel miRNA families that could be *Populus*-specific (Barakat et al. 2007). The putative target genes of *Populus*-specific small RNA were involved in development and resistance to stress. Abiotic stress-responsive small RNAs have also been isolated from *Populus*. Lu *et al.* (2008) have identified abiotic stress-responsive miRNAs from *Populus trichocarpa*, whose expression was altered in response to cold, heat, salt, dehydration, and mechanical stresses. Some of these stress-responsive miRNA families are conserved among various plant species, such as *Arabidopsis*, rice, and *Populus*. The woody-plant specific miRNAs, which might function in the adaptation to long-term growth and survival from stress conditions, have also isolated from *Populus* (Lu et al., 2008). Furthermore, UV-B stress-responsive miRNAs were identified from *Populus* and these miRNAs are highly conserved among various plant species (Jia et al., 2009a). Some of them are also involved in responses to other responses, such as energy metabolism and ROS homeostasis, suggesting a cross-talk among different responses mediated by various miRNAs. These reports provide further understanding of the miRNA-associated regulatory networks, which control the comprehensive responses to abiotic stresses in woody plants. miR398 is one of the conserved miRNAs in *Arabidopsis*, rice, and *Populus*, and targets two closely related copper-zinc superoxide dismutases (Cu/Zn-SODs), cytosolic *CSD1* and plastidic *CSD2* (Bonnet et al. 2004, Jia et al. 2009b, Jones-Rhoades and Bartel, 2004, Sunker et al. 2006). The expression of miR398 was down-regulated by oxidative stress in *Arabidopsis*, and this down-regulation led to increase the expression levels of *CSD1* and *CSD2* and enhance plant adaptation to abiotic stresses (Sunker et al. 2006). It is interesting to note that miR398 was induced by abscisic acid (ABA) and salt stress in *Populus* (Jia et al. 2009b). This study suggested the differential regulation of miR398 in response to stress in *Populus* and *Arabidopsis*, implying that there might be different signaling mechanisms in stress response in woody plants and other plant species.

LIGNIFICATION

In woody plants, a majority of assimilated carbon sources is accumulated in the cell wall, as a phenylpropanoid compound, lignin. Lignin is polymer of hydroxycinnamyl alcohols (*p*-coumaryl, coniferyl and sinapyl alcohols) and provides hydrophobicity and rigidity to the thickening cell wall for water transport and mechanical support in the plant body. The biosynthetic enzymes of various phenylpropanoid isoforms are regulated in response to different plant conditions (Boudet et al. 2003, Boerjan et al. 2003, Osakabe et al. 1999, Raes et al. 2003) and a number of genes involved in phenylpropanoid biosynthesis in woody plants (Hamberger et al. 2007). The spatial and temporal control of lignin biosynthesis is the major focus of study on the regulatory mechanisms of secondary growth and the processes that control the biomass of woody plants. Lignifications and biosynthesis of lignin precursors are also induced by biotic and abiotic stresses, such as wounding, pathogen attack, and drought for prevention of further growth or confine invading pathogens (Vance et al. 1980, Dixon and Paiva 1995, Alvarez et al. 2008, Hu et al. 1999). Chemical structure of the lignin deposited under stress conditions differs from those in normal development. The stress-induced lignins contain elevated amounts of *p*-hydroxyphenyl units, which are derived from *p*-coumaryl alcohol (Lange et al. 1995). Induction of lignin deposition is also observed under mechanical stress. In gymnosperms tree, compression wood, which contains an increased content of lignin, is induced by the compressive strength at the lower part of bending trunks and branches (Du and Yamamoto 2007). Expression of lignin biosynthetic genes is induced by physical stimuli such as bending (Osakabe et al. 2009a, 2009b).

In contrast to the great roles played by lignin in the normal development and stress response in plants, lignin has negative effects on the digestibility of forage crops, pulp productivity of woody biomass in chemical processes, and bioconversion of cellulosic biomass into fermentable sugars for the production of bioethanol (Jung and Vogel 1986, Chen and Dixon 2007, Vanholme et al. 2008). It inhibits enzymatic treatment for the liberation of sugars from cellulosic biomass and adheres to hydrolytic enzymes in the process. Aromatic compounds released from lignin during chemical treatments such as vanillin can inhibit the growth of microorganisms in sugar fermentation for fuel production.

Brown midrib (*bm*) mutants in plants with modified lignin characteristics are often of higher nutritive value than the normal plants because of their increased stem degradability in ruminants (Cherney et al. 1991). The two typical mutants of maize with different alleles, *bm1* and *bm3*, have a mutation in the lignin biosynthetic gene for cinnamyl alcohol dehyrogenase (CAD) and *O*-methyltransferase (OMT), respectively (Vignols et al. 1995, Halpin et al. 1998). In sorghum, *bmr6* and *bmr12,* have the mutation in the same gene as in *bm1* and *bm3* mutants, respectively. In these mutants, lignin and/or cell-wall-bound phenolics were changed in their composition and/or concentration. Lignin content and composition also have great impact in their novel use as feedstock for bioethanol production. The relationship between lignin characteristics and efficiency in saccharification of biomass has lately attracted considerable attention (Sannigrahi et al. 2010). Saballos et al. (2008) reported that stover from four low-lignin mutants of sorghum, *bmr2*, *bmr3*, *bmr6*, and *bmr12* exhibited higher saccharification efficiency (7 to 26% increase) than those of their respective counterparts when they were treated with hydrolyzing enzyme. In contrast to the great benefits of *bm* mutation with increased digestibility, *bm* maize exhibits generally lower grain and stover yields than those of its non-*bm* counterpart. In addition, genetic manipulation of lignin content and/or composition in transgenic plants induces repression of growth in many cases. The negative impact of the lignin modification on plant growth has crucial influence on the industrial and agricultural uses of the transgenic plants, even if their lignocelluloses have the advantages of being easy digestible in ruminants and higher hydrolyzation by enzyme treatment. This inadequacy of the transgenic plants may partially be derived from higher susceptibility of *bm* mutants to various stresses because lack of a function for lignin biosynthesis may be disadvantageous for responses to the stresses and the subsequent growth of the plants. Technology should be developed for the manipulation of lignin for plant growth without negative effects.

GENETIC ENGINEERING IN WOODY PLANTS

In *Populus* and *Eucalyptus*, a number of genetic variations and their genetic information (genetic mapping and quantitative trait locus mapping) for the valuable traits have been identified. The genomic sequencing projects of *Populus trichocarpa* (genotype Nisqually-1) and *Eucalyptus grandis* have been accomplished (for *P. trichocarpa*, Tsukan et al. 2006; for *E. grandis*, the preliminary 8X draft assembly of the *Eucalyptus grandis* genome, http://www.eucagen.org/). These researches would certainly give valuable information to understand the detailed function of genes involved in secondary cell wall biosynthesis and developmental processes of secondary cell walls. "Omics" analyses of woody plants should also provide profound resources for molecular breeding and application. Currently, several online tools for transcriptome analysis have been developed to understand the transcriptome and genome of poplar (The microarray-based expression data (PopGenExpress), Wilkins et al. 2009; The *Populus* Electronic Fluorescent Pictograph (eFP) browser (http://www.bar.utoronto .ca/efppop/cgi-bin/efpWeb.cgi), Sjödin A et al. 2009). The 'omics' analyses and the co-expression analysis of the main factors in secondary wall biosynthesis of woody plants, which have roles on developmental processes and stress responses, will additionally provide an integrative view of the cellular function and information platform.

Although classical breeding method is an important approach to accomplish genetic improvements in woody plants, however, due to the requirement for longer time to obtain progenies, the alternative approach with genetic transformation should be taken to accelerate the production of higher valuable genetic traits. Genetic transformation also allows the production of novel traits, which cannot be found within the natural variations of woody plants. *Populus* is one of the first woody plants to undergo genetic transformation, which was established in early 1970 (see review, Han et al. 1996). In contrast to the case of *Populus*, *Eucalyptus* species are relatively hard to establish the genetic transformation system due to the low efficiency for shoot regeneration in *in vitro* culture. Recently, *Eucalyptus camaldulensis* was identified as an easily regenerable genotype and using this genotype, efficient transformation protocols were established (Tournier et al. 2003, Chen et al. 2006). Currently, the establishment of genetic transformation of other *Eucalyptus* genotypes including commercially important genotypes is on-going. Omics analyses could help to find the genes involved in the regeneration of *Eucalyptus*, and the modification of the superior regenerable genes itself should be an important target to accelerate genome-wide analysis of *Eucalyptus*.

Although several transgenic woody plants were generated for important target genes, more precise genetic modification would be required in woody plants with techniques such as gene targeting and site-directed mutagenesis, which are routinely used in bacteria and yeasts. In general, however, higher plants show very low efficiency of gene targeting. One solution is zinc finger nucleases (ZFNs, see review Weinthal et al. 2010). ZFNs are chimeric proteins composed of a synthetic zinc finger-based DNA binding domain and a DNA cleavage domain. ZFNs can be specifically designed to cleave virtually any long stretch of DNA sequence (Maeder et al. 2009). ZFN-induced double-stranded DNA breaks (DSBs) indeed enhance gene targeting at the target loci in tobacco (Townsend et al. 2009), and maize (Shukla et al. 2009) as high as 10% of transformed cells. Beside of gene targeting, ZFNs can create DSBs, and its DSBs further yield small insertions or deletions (mutations) at a specific genomic location during the DSB repair process (Osakabe et al. 2010, Zhang et al. 2010). The time and effort for the establishment of transgenic plants could be reduced by at least one order of magnitude compared to traditional methods that request long-term molecular or phenotypic screening for the desired characteristics. This technique should be applied to woody plant species, which generally require longer period for the production of engineered plants compared to model plant species, such as *Arabidopsis* or rice.

Environmental stress is the major cause of crop and forest loss worldwide. Future issues such as insufficiency of provisions, environmental conservation, and production increase of biomass will depend on plant biotechnologies. The findings reviewed here would shed light on knowledge of the molecular mechanism in control of biomass production under stressful conditions.

REFERENCES

Alvarez S, Marsh EL, Schroeder SG, Schachtman DP (2008) Metabolomic and proteomic changes in the xylem sap of maize under drought. *Plant Cell Environ.* 31: 325-340.

Barakat A, Wall PK, Diloreto S, Depamphilis CW, Carlson JE. (2007) Conservation and divergence of microRNAs in Populus. *BMC Genomics* 8: 481.

Baena-Gonzáleza E, Sheen J (2008) Convergent energy and stress signaling. Trends in Plant Science 13: 474-482.

Boerjan W, Ralph J, Baucher M (2003) Lignin biosynthesis. *Ann. Rev. Plant Biol.* 54: 519-546.

Bonnet E, Wuyts J, Rouze P, Van de Peer Y (2004) Detection of 91 potential in plant conserved plant microRNAs in *Arabidopsis thaliana* and *Oryza sativa* identifies important target genes *Proc. Natl. Acad. Sci. USA* 101: 11511–11516.

Borsani O, Zhu J, Verslues PE, Zhu JK (2005) Endogenous siRNAs derived from a pair of natural cis-antisense transcripts regulate salt tolerance in *Arabidopsis*. *Cell* 123: 1279-1291.

Boudet AM, Kajita S, Grima-Pettenati J, Goffner D (2003) Lignins and lignocellulosics: a better control of synthesis for new and improved uses. *Trends Plant Sci.* 8: 576-581.

Chen F and Dixon RA (2007) Lignin modification improves fermentable sugar yields for biofuel production. *Nat. Biotech.* 25: 759-761.

Chen J, Xia X, Yin W (2009) Expression profiling and functional characterization of a DREB2-type gene from *Populus euphratica*. *Bioch. Biophy. Res. Comm.* 16: 483-487.

Chen ZZ, Ho CK, Ahn IS, Chiang VL. (2006) Eucalyptus. *Methods Mol. Biol.*, 344: 125-134.

Cherney JH, Cherney DJR, Akin DE, Axtell JD (1991) Potential of brown-midrib, low-lignin mutants for improving forage quality. *Adv. Agron.* 46: 157-198.

Chiou, TJ, Aung K, Lin SI, Wu CC, Chiang SF, Su CL (2006) Regulation of phosphate homeostasis by MicroRNA in Arabidopsis. *Plant Cell* 18: 412-421.

Desikan R, Mackerness SAH, Hancock JT, Neill SJ (2001) Regulation of the Arabidopsis transcriptome by oxidative stress. *Plant Physiol.* 127: 159-172.

Dixon RA, Paiva NL (1995) Stress-Induced Phenylpropanoid Metabolism. *Plant Cell* 7: 1085-1097.

Du S, Yamamoto F (2007) An overview of the biology of reaction wood formation. *J. Integr. Plant Biol.* 49: 131-143.

Fujii H, Chiou TJ, Lin SI, Aung K, Zhu JK. (2005) A miRNA involved in phosphate-starvation response in Arabidopsis. *Curr. Biol.* 15: 2038-2043.

Gu R, Fonseca S, Puskás LG, Hackler L Jr, Zvara A, Dudits D, Pais MS (2004) Transcript identification and profiling during salt stress and recovery of *Populus euphratica*. *Tree Physiol* 24: 265-276.

Halpin C, Holt K, Chojecki J, Oliver D, Chabbert B, Monties B, Edwards K, Barakate A, Foxon GA (1998) *Brown-midrib* maize (*bm1*) - a mutation affecting the cinnamyl alcohol dehydrogenase gene. *Plant J.* 14: 545-553.

Hamberger B, Ellis M, Friedmann M, Souza CDA, Barbazuk B, Douglas CJ (2007) Genome-wide analyses of phenylpropanoid-related genes in *Populus trichocarpa*, *Arabidopsis thaliana*, and *Oryza sativa*: the *Populus* lignin toolbox and conservation and diversification of angiosperm gene families. *Can J. Bot.* 85: 1182-1201.

Han K-H, Gordon MP, Strauss SH (1996) Cellular and molecular biology of Agrobacterium-mediated transformation of plants and its application to genetic transformation of *Populus*. In: Biology of *Populus* and Its Implication for Management and Conservation RF, Stettler HD, Bradshaw Jr. PE Heilman, TM Hinckley (eds), *National Research Council Canada* 201-222.

Hirayama T, Shinozaki K (2010) Research on plant abiotic stress responses in the post-genome era: past, present and future. *Plant J.* 61:1041-1052.

Hu WJ, Harding SA, Lung J, Popko JL, Ralph J, Stokke DD, Tsai CJ, Chiang VL. (1999) Repression of lignin biosynthesis promotes cellulose accumulation and growth in transgenic trees. *Nat. Biotech.* 17: 808-812.

Jammes F, Song C, Shin D, Munemasa S, Takeda K, Gu D, Cho D, Lee S, Giordo R, Sritubtim S, Leonhardt N, Ellis BE, Murata Y, Kwak JM (2009) MAP kinases MPK9 and MPK12 are preferentially expressed in guard cells and positively regulate ROS-mediated ABA signaling. *Proc. Natl. Acad. Sci. USA* 106: 20520-20525.

Jia X, Ren L, Chen QJ, Li R, Tang G (2009a) UV-B-responsive microRNAs in *Populus tremula. J Plant Physiol* 166: 2046-2057.

Jia X, Wang WX, Ren L, Chen QJ, Mendu V, Willcut B, Dinkins R, Tang X, Tang G (2009b) Differential and dynamic regulation of miR398 in response to ABA and salt stress in *Populus tremula* and *Arabidopsis thaliana. Plant Mol. Biol.* 71: 51-59.

Jones-Rhoades MW, Bartel DP (2004) Computational identification of plant microRNAs and their targets, including a stress-induced miRNA. *Mol. Cell* 18: 787-799.

Jung HG, Vogel KP (1986) Influence of lignin on digestibility of forage cell wall material. *J. Anim. Sci.* 62: 1703-1712.

Kajita S, Ishifuji M, Ougiya H, Hara S, Kawabata H, Morohoshi N, Katayama Y (2002) Improvement in pulping and bleaching properties of xylem from transgenic tobacco plants. *J. Sci. Food Agric.* 82: 1216-1223.

Kotak S, Larkindale J, Lee U, von Koskull- Döring P, Vierling E, Scharf KD (2007) Complexity of the heat stress response in plants *Curr. Opin. Plant Biol.* 10: 310-316.

Lange BM, Lapierre C, Sandermann H Jr (1995) Elicitor-induced spruce stress lignin. *Plant Physiol.* 108: 1277–1287.

Lu S, Sun YH, Shi R, Clark C, Li L, Chiang VL (2005) Novel and mechanical stress-responsive MicroRNAs in *Populus trichocarpa* that are absent from Arabidopsis. *Plant Cell* 17: 2186-2203.

Lu S, Sun YH, Chiang VL (2008) Stress-responsive microRNAs in Populus. *Plant J.* 55: 131-151.

Maeder ML, Thibodeau-Beganny S, Sander JD, Voytas DF, Joung JK (2009) Oligomerized pool engineering (OPEN): an 'open-source' protocol for making customized zinc-finger arrays. *Nat. Protoc.* 4: 1471-1501.

Mittler R, Vanderauwera S, Gollery M, Van Breusegem F (2004) Reactive oxygen gene network of plants. *Trends Plant Sci.* 9: 490-498.

Osakabe K, Tsao CC, Li L, Popko JL, Umezawa T, Carraway DT, Smeltzer RH, Joshi CP, Chiang VL. (1999) Coniferyl aldehyde 5-hydroxylation and methylation direct syringyl lignin biosynthesis in angiosperms. *Proc. Natl. Acad. Sci. USA* 96: 8955-8960.

Osakabe K, Osakabe Y, Toki S (2010) Deficiency of end-protection activity altered a profile of error-prone endo-joining after ZFN-mediated cleavage in *Arabidopsis. Proc. Natl. Acad. Sci. USA* 107: 12034-12039.

Osakabe Y, Maruyama K, Seki M, Satou M, Shinozaki K, Yamaguchi-Shinozaki K (2005) Leucine-rich repeat receptor-like kinase1 is a key membrane-bound regulator of abscisic acid early signaling in Arabidopsis. *Plant Cell* 17: 1105-1119.

Osakabe Y, Osakabe K, Chiang VL (2009a) Isolation of 4-coumarate Co-A ligase gene promoter from loblolly pine (*Pinus taeda*) and characterization of tissue-specific activity in transgenic tobacco. *Plant Physiol. Biochem.* 47: 1031-1036.

Osakabe Y, Osakabe K, Chiang VL (2009b) Characterization of the tissue-specific expression of phenylalanine ammonia-lyase gene promoter from loblolly pine (*Pinus taeda*) in *Nicotiana tabacum*. Plant Cell Rep 28: 1309-1317.

Osakabe Y, Mizuno S, Tanaka H, Maruyama K, Osakabe K, Todaka D, Fujita Y, Kobayashi M, Shinozaki K, Yamaguchi-Shinozaki K. (2010) Overproduction of the membrane-bound receptor-like protein kinase1, RPK1, enhances abiotic stress tolerance in Arabidopsis. *J. Biol. Chem.* 285: 9190-9201.

Raes J, Rohde A, Christensen JH, Van de Peer Y, Boerjan W (2003) Genome-wide characterization of the lignification toolbox in Arabidopsis. *Plant Physiology* 133: 1051-1071.

Saballos A, Vermerris W, Rivera L, Ejeta G (2008) Allelic association, chemical characterization and saccharification properties of brown midrib mutants of sorghum (*Sorghum bicolor* (L.) Moench). *Bioenerg. Res.* 1: 193–204.

Sakuma Y, Maruyama K, Osakabe Y, Qin F, Seki M, Shinozaki K, Yamaguchi-Shinozaki K. (2006) Functional analysis of an Arabidopsis transcription factor, DREB2A, involved in drought-responsive gene expression. *Plant Cell* 18: 1292-1309.

Sannigrahi P, Ragauskas AJ, Tuskan GA (2010) Poplar as a feedstock for biofuels: A review of compositional characteristics. *Biofuels Bioprod. Bioref.* 4: 209-226.

Seki M, Narusaka M, Ishida J, Nanjo T, Fujita M, Oono Y, Kamiya A, Nakajima M, Enju A, Sakurai T, Satou M, Akiyama K, Taji T, Yamaguchi-Shinozaki K, Carninci P, Kawai J, Hayashizaki Y, Shinozaki K. (2002) Monitoring the expression profiles of 7000 Arabidopsis genes under drought, cold and high-salinity stresses using a full-length cDNA microarray. *Plant J.* 31: 279-292.

Seki M, Kamei A, Yamaguchi-Shinozaki K, Shinozaki K (2003) Molecular responses to drought, salinity and frost: common and different paths for plant protection. *Curr. Opin. Biotech.* 14: 194-199.

Shukla VK, Doyon Y, Miller JC, DeKelver RC, Moehle EA, Worden SE, Mitchell JC, Arnold NL, Gopalan S, Meng X, Choi VM, Rock JM, Wu YY, Katibah GE, Zhifang G, McCaskill D, Simpson MA, Blakeslee B, Greenwalt SA, Butler HJ, Hinkley SJ, Zhang L, Rebar EJ, Gregory PD, Urnov FD (2009) Precise genome modification in the crop species *Zea mays* using zinc-finger nucleases. *Nature* 459: 437-441.

Spoel SH, Dong X (2008) Making Sense of Hormone Crosstalk during Plant Immune Responses. *Cell Host and Microbe* 6: 348-351.

Sunkar R, Kapoor A, Zhu JK (2006) Posttranscriptional induction of two Cu/Zn superoxide dismutase genes in Arabidopsis is mediated by downregulation of miR398 and important for oxidative stress tolerance. *Plant Cell* 18: 2051-2065.

Sunkar R, Zhu JK (2004) Novel and stress-regulated microRNAs and other small RNAs from Arabidopsis. *Plant Cell* 16: 2001-2019.

Tournier V, Grat S, Marque C, El Kayal W, Penchel R, de Andrade G, Boudet AM, Teulières C (2003) An efficient procedure to stably introduce genes into an economically important pulp tree (Eucalyptus grandis x Eucalyptus urophylla). *Transgenic Res.* 12: 403-411.

Townsend JA, Wright DA, Winfrey RJ, Fu F, Maeder ML, Joung JK, Voytas DF (2009) High-frequency modification of plant genes using engineered zinc-finger nucleases. *Nature* 459: 442-445.

Torres MA, Dangl JL (2005) Functions of the respiratory burst oxidase in biotic interactions, abiotic stress and development. *Curr. Opin. Plant Biol.* 8: 397–403.

Vance CP, Kirk TK, Sherwood RT (1980) Lignification as a mechanism of disease resistance. *Annu Rev Phytopathol* 18: 259-288.

Vanholme R, Morreel K, Ralph J, Boerjan W (2008) Lignin engineering. *Curr. Opin. Plant Biol.* 11: 278-285.

Vignols F, Rigau R, Torres MA, Capellades M, Puigdomenech P (1995) The *brown midrib3* (*bm3*) mutation in maize occurs in the gene encoding caffeic acid *O*-methyltransferase. *Plant Cell* 7: 407-416.

Vinocur B, Altman A (2005) Recent advances in engineering plant tolerance to abiotic stress: achievements and limitations. *Curr. Opin. Plant Biol.* 16: 123-132.

Wang W, Vinocur B, Altman A (2003) Plant responses to drought, salinity and extreme temperatures: towards genetic engineering for stress tolerance. *Planta* 218: 1-14.

Wang W, Vinocur B, Shoseyov O, Altman A (2004) Role of plant heat-shock proteins and molecular chaperones in the abiotic stress response. *Trends Plant Sci.* 9: 244-252.

Weinthal D, Tovkach A, Zeevi V, Tzfira T (2010) Genome editing in plant cells by zinc finger nucleases. *Trends Plant Sci.* 15: 308-321.

Wilkins O, Waldron L, Nahal H, Provart NJ, and Campbell MM (2009) Genotype and time of day shape the *Populus* drought response. *Plant J.* 60: 703-715.

Xiao X, Yang F, Zhang S, Korpelainen H, and Li C. (2009) Physiological and proteomic responses of two contrasting *Populus cathayana* populations to drought stress. *Physiol. Plant* 136: 150-168.

Yamaguchi-Shinozaki K, and Shinozaki K. (2005) Organization of cis-acting regulatory elements in osmotic- and cold-stress-responsive promoters. *Trends Plant Sci.* 10: 88-94.

Yamaguchi-Shinozaki K, Shinozaki K (2006) Transcriptional regulatory networks in cellular responses and tolerance to dehydration and cold stresses. *Annu. Rev. Plant Biol.* 57: 781-803.

Zhang, F, Maeder, ML, Unger-Wallace, E, Hoshaw, JP, Reyon, D, Christian, M, Li, X, Pierick, CJ, Dobbs, D, Peterson, T, Joung, JK, Voytas, DF (2010) High frequency targeted mutagenesis in *Arabidopsis thaliana* using zinc finger nucleases. *Proc. Natl. Acad. Sci. USA* 107: 12028–12033.

Zhou X, Wang G, Zhang W (2007) UV-B responsive microRNA genes in *Arabidopsis thaliana*. *Mol. Syst. Biol* 3: 103.

In: Lignin
Editor: Ryan J. Paterson

ISBN 978-1-61122-907-3
© 2012 Nova Science Publishers, Inc.

Chapter 17

LIGNIN AND FUNGAL PATHOGENESIS

O. I. Kuzmina and I. V. Maksimov

Institute of Biochemistry and Genetics,
Ufa Scientific Center, Russian Academy of Sciences, 450054,
Ufa, pr. Oktyabrya 71, Russia

ABSTRACT

The accumulation of lignin is one of the important plant defense mechanisms against pathogens and wound. In this work the mechanisms of local lignification of pathogen infected zones are discussed. The importance of the structurally functional organization of some peroxidase isoforms, promoting the phenolic compound polymerization with participation of reactive oxygen species (ROS) on a pathogenic fungi mycelium is considered. The analysis of acetylation degree of pathogen polysaccharides in the induced plant defense and its role in the local lignification of infection zones was carried out. The post infectious accumulation analysis of lignin in wheat plants has shown that resistant plant tissues increased activity of lignin synthesis enzymes. We consider this effect is associated with highest activity of pathogen polysaccharide-specific apoplastic peroxidase. For example, an anionic peroxidase in wheat plants is characterized by property to contact with the chitin of pathogenic fungi. Analysis of the amino acid chain of some peroxidases of *Arabidopsis*, wheat, zucchini, corn and rice homologous zone allows allocating isoforms genes from a large amount of genes. In order to prove the polysaccharide-specificity of allocated peroxidases to the site of a gene presumably coding polysaccharide-specific domain of peroxidase have been picked up and designed primers which we have used for an estimation of gene expression of wheat anionic peroxidase gene under infection by fungus disease agents. Sequencing results, received from peroxidase DNA and cDNA of some plants have proved the accuracy of choosing primers. The anionic peroxidase gene expression or repression repeatedly increased or decreased under infection with fungus pathogens and the influence of signal molecules and elicitors. Thus, a comparative estimation of the lignin content, activity of lignification enzymes, ability to sorption on chitin of plant peroxidases, definition of their immunochemical affinity, a homology of molecular structure of polysaccharide-specific zone have allowed to assume, that local lignin accumulation under fungal pathogenesis shows the universality via of activation of polysaccharide-specific peroxidases.

The lignin is a part of the vascular plant cell wall which share at some wood can reach 30%. It, as a rule, is constructed of the residues of hydroxycinnamic alcohols: sinapic, coniferyl and para-coumaric. Surprising characteristic of this biopolymer containing exclusively at the higher plants, unlike many other natural compounds, spatial and structural heterogeneity are necessary to consider of its biochemical and biophysical properties. Besides this biopolymer is badly dissolve and chemically stable. And, composition of hydroxycinnamic alcohols and also physical and chemical properties of a lignin in both different plant organs and ontogenesis phases considerably differ. So, the bark and wood of the same plant can contain lignins of various consistences. Lignins of the basis and of the top part of a plant stem differ from each other. Coniferylic (G) components (75-98%) dominate in a lignin of fern as well as in lignins of gymnospermous plants (50-98%). As a part of a lignin at dicotyledonous plants prevail syringilic (S) components (57-65 %), and oxyphenilic (H) practically are absent. On H components in lignins of monocotyledonous are the share of 10-14% and the share of G compounds (50-56%) as soon as a little exceeds a share S associations (35-45 %). Still there are disagreements concerning presence of a lignin in mosses and algaes. Thus the lignin appearance often connect with emergence of land plants as only they have «the basic device» and the vascular system necessary for transportation of nutrients and water on tissues [Brundrett, 2002]. The lignin is a component of a secondary cell wall and an intercellular blade of practically all plant cells which have finished growth and gives them the strength. This property is necessary for keeping of plant weight and for resistance to negative hydrostatic pressure rising on xylem of water. The cells preserve the ability to lignin formation even in culture *in vitro* over a long time period. This takes place despite lack in calluses the vascular system indicative that a lignin is indispensable component of the higher plant cells [Karkonen, Kautaniemi, 2010].

The lignin accumulation and deposition is one of the important defense mechanisms of plant interaction with pathogens, realized by increase of mechanical strength of plant cell walls, their protection from hydrolysis and destruction by extracellular pathogen enzymes. The lignin does not destruct by fungi and insect hydrolases, with exception ligno-peroxidases of some fungi. The plant tissues, exposing to lignification, become inedible for pathogen and heterotrophs. Probably, the monolignols polymerization in a lignin occurs on free-radical mechanism. There is the question about kind of enzymes which catalyze this process [Hatfield, Vermeris, 2001; Marjamaa et al., 2009]. The urgency of this question is bind with perspective requests of cellulose manufacture as the lignin is one of undesirable biopolymers and difficultly taken from wood [Chen et al., 2001]. According to numerous literary data, in catalysis of polymerization of phenolic monomers in a lignin [Li et al, 2003], a suberin [Bernards et al., 1999] and cutin [Carpin et al., 1999] are involved the apoplastic anionic peroxidase isoforms. Their accumulation after wounding on an internal wall of the periderm by the method of immunofluorescence has found [Quiroga et al., 2000]. A suberin and anionic peroxidase were localized in a periderm of a potato tuber on an internal part of a cell wall. Participation of peroxidases in suberin formation is proved by this fact [Bernards et al., 1999]. Activity of anionic peroxidase and tissue suberinization are elevated in addition to nutrient medium of ions Fe^{3+}. The lignin is synthesized from L-phenylalanine and L-tyrosine through the stage of formation of cinnamic acids. Their condensation leads to formation of dimers, repeatedly exposing to attack peroxidases, forming the dimeric free radicals. These radicals are involved in condensation in more high-oligomerous lignols. The process comes to the end with formation of the disordered three-dimensional polymer named a lignin. These polymers are secreted from cytoplasm in a cell wall. Here they are exposed to oxidation under the influence of connected peroxidase and H_2O_2 in corresponding phenoxylic radicals which are involved in polymerization. Lignin and suberin polymerization occurs not in a random way but by organized mechanism. The lignin phenolic rings in polymer are focused mainly in a plane of a cell wall with participation a polysaccharide. Here the residues of ferulic acid serve lignin synthesis [Ralph et al., 2004]. Earlier the presence of H_2O_2 at a course of lignification tried to reveal in order to confirm the participation of peroxidases in formation of the free radicals of monolignols. Presence H_2O_2 in cell walls of a poplar and its absence in cambial cells is shown. The other important approach to definition of the peroxidase role in lignification is connected with obtaining of transgene plants with the changed peroxidase activity. Group of M. Lagrimini have received some transgenic lines of tobacco with considerably raised expression of anionic peroxidase gene, which presumably participates in lignification, and on the contrary, it's decreased regulation [Dowd, Lagrimini, 2006]. There are data, that the part of plants with the high level peroxidase produces more lignins in comparison with the control, however the reduction of lignin quantity in plants with lower peroxidase activity is not observed.

In connection with role of peroxidase in a plant morphogenetic processes and lignin synthesis special interest represents their ties with processes of plant tissue defense from pathogens and phytophages (wound response). The important feature of a lignin is that only few of a lot of number of parasitic microorganisms (for example, fungus destroying wood) can cleave it. Therefore lignific coates of cells serve as a barrier on a way of distribution of an infection. The induced anionic peroxidase or transgenic plants with constitutively high enzyme activity become toxic for pathogen or phytophage [Behle et al., 2002]. The lignification of plant cell walls creates a physical barrier on a way of distribution a hyphae of parasitic fungus and blocks the diffusion of allocated toxins. It is established, that the lignin accumulation at protective plant reactions occurs as a result of its synthesis de novo, induced by products of enzymatic destructions of cell walls by pathogens. The artificial inhibition of lignification can lead to disorder of the immune response that has been shown on an example of wheat infection by brand [Moldenhauer et al., 2006]. First of all, it is possible to note high stability of vessels of a conductive tissue to pathogens. Besides, initiation of morphogenetic processes in culture of the plant cells also leads to enhancement of their stability to pathogenic fungi [Troshina et al., 2000]. The intensive generation of the active oxygen species and the subsequent lignification of their cell walls with participation of anionic pathogen-induced peroxidase at calluses were found [Troshina et al., 2004, Maksimov et al., 2004]. Similar reaction at not infected calluses has not observed. Fixation of a cell wall in plants occurs through formation in it of intermolecular communications, formation of various polymers, for example, callose to the subsequent and intensive saturation a lignin-like material [Grant, Mansfield, 1999; Verma, Hong, 2001; Hawkins, Boudet, 2003]. There is an opinion, that the specific protective effect of lignification at fungal pathogenesis is connected with creation of a mechanical barrier on a growth way of pathogens, interfering their distribution, protection of a cell wall from hydrolytic enzymes and a cell from the toxins produced by pathogen, limitation of flux to pathogen of host metabolites and water, and toxins in a plant tissues [Cano-Delgado et al., 2003]. As consequence of it packing of infectious structures of a parasite and their destruction is observed. Lignin biosynthesis, as a rule, is observed in places of introduction of infectious structures of a fungus in plant tissues. The formed condensed formations observed at infection have received the name «papilla».

They are not typical neither to a plant, nor a fungi and represent the difficult structural formation between plasmalemma of the host and a cell wall of pathogen. Speed of their formation is important and delay of this process leads to disease development [Heitefuss, 2001]. The pathogen focuses, surrounded by papillae, are thinner and often dendritic. Plant cell wall are saturated by contain the ROS, peroxidase (figure 1a), cellulose, callose, silicon, and phenolic compounds (figure 1b) and, accordingly, a product of their direct interaction – the polymer [Shinogi et al., 2003; Humphrey et al., 2007].

a

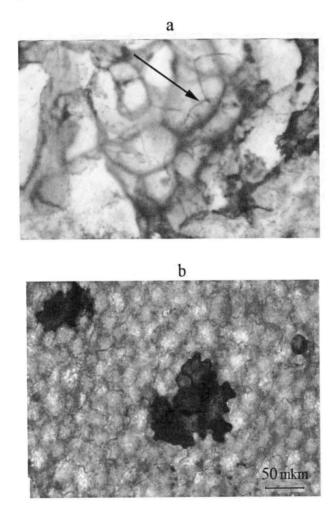

b

Figure 1. Influence of infection by wheat bunt agent on localization peroxidase in wheat calluses [Troshina et al., 2004] (a) and potato phytophthora agent of phenolic compounds in potato leaves [Maksimov et al., 2011] (b). (The arrow show peroxidase activity on pathogen hyphas in plant callus cells, x 800).

Interestingly, that on a surface of cell walls of fungi – *Collelotrichum lagenarium* and *C. cucumerium* – submersed in the reactionary solution containing plant proteins, H_2O_2 and the phenolic substrate, has been found out a material like to the lignin is similar to the postponed on papillae. Formation of these structures in plant cells infected with pathogenic fungus is more intensively in resistant plants, we assume that they perform the protective function. It is connected with interference of papillae, that represent a mechanical barrier, in the active growth of focuses on tissues; they isolate a parasite from plant-host cytoplasm; presence in papillae of the lignin-like compounds incorporating free radicals can lead to direct inhibition of growth of a fungi; the influence of papillae on fungal growth can represent the dynamical and specific character depending on structure of its cell walls [Humphrey et al., 2007]. Unfortunately, the mechanism of local lignification of papillae and concentration of the stress induced peroxidases in this zone as a whole while remains unknown. However, the ability to cooperate with cell wall biopolymers both of a plant and pathogenic fungus, possibly, should play an active role [Maksimov et al., 2005]. Notice that in cytoplasm of infected cells there is a directed polarization of physiological reactions at which Golgi apparatus and components of endoplasmic reticulum are distributed around haustorium. Three stages in lignin synthesis are revealed in plants of the maize infected by *Helmintosporium maydis*. First, there is an accumulation of two ethers of caffeic acid accumulating in a resistant sort through 16 h after fungal penetration. After that ferulic and *n*- coumaric acids alight on host cell walls [Ikegawa et al., 1996]. Last stage is a synthesis of a lignin at both sorts. It is revealed, that the formed lignin contains unusually high level of derivatives of sinapic acid. Also the difference in lignin sensitivity of "intact" and "infected" plants to destruction by various pathogen hydrolases is non-exceptional. Unfortunately, the phenomenon of the directed synthesis of a lignin only in infection places remains unexplained. The participation of a lignin in protective reactions in infection of pines by brands observed in experiments with the conifers. The treatment of suspension culture of pine *Picea abies* L. by preparation evolved from the needle pathogen *Rhizosphaera kalkhoffi*, induced the occurrence in cell walls and in culture medium the lignin-like polymers. It is shown the local formation of a lignin is one of mechanisms of nonspecific resistance to pathogens. This also provides plant protection from specific pathogens. The resistant plants were characterized by hyperactivity of enzyme participating in lignin biosynthesis. Enzymatic activity and lignin deposition were inhibited by cycloheximide treatment. The tissues became susceptible to fungi which does not infect this plant in norm. The lignin accumulates at the supersensitive reaction occurring in reply to infection of wheat by the stem rust activator. The role of lignification in resistance of a cucumber to *C. cucumerianum* was investigated in detail. Lignin formation took place only in plants of the resistant sorts infected by pathogens. The expression of peroxidase genes *AtPrx42*, *AtPrx64* and *AtPrx71* has been found out during formation of a secondary cell wall of conductive elements [Yokoyama, Nishitani, 2006]. Similarly, *AtPrx53* was expressed in vascular bunches at fusion of the regulatory peptide with β – glucuronidase (GUS) [Ostergaard et al., 2000]. Gene *AtPrx66* has shown high degree of homology with a peroxidase gene of zinnia *ZePrx01*, which expresses only at differentiation of tracheal elements, and corresponding protein catalyzes the lignin formation *in vitro* [Sato et al., 2006]. The fact that *AtPrx66* is expressed in root vessels, confirms the involving of this gene in cell wall lignification. The function of protein *AtPrx17* in lignification of pods and, accordingly, in the mechanism of pod dehiscence [Roeder, Yanofsky, 2006] is shown recently. The detailed analysis of powerful regulation of this gene showned that key regulators of its expression are the transcriptional factors *AGL15/18*. Three genes (*AtPrx13*, *AtPrx30* and AtPrx55) were expressed mainly in flowers and were regulated by trance factors *SHP1* and *SHP2*. However, the role of these isoperoxidases is not found out yet. The content of lignin increase through 72 h is revealed at studying of dynamics of lignin accumulation in the rice leaves infected by rice blast. Studying of development *Septoria tritici* on leaves of seedling of susceptible and resistant sorts of wheat has shown, that growth a floccus in intercellular spaces is observed at both sorts.

Resistant sort showed more active cell wall lignification. The results show importance of last stages of lignification where identical accumulation of phenolic compounds in a zone of an infection both at susceptible and at resistant plants is shown. It assumes inhibition of last stages of lignin synthesis by pathogens.

The presented data consider that peroxidases participate in many physiological processes, first of all, in reparation reactions. The enzymatic system of plants represents the important regulatory component of cell. The importances of the structurally functional organization of some isoperoxidases, promoting the concentrated polymerization of the phenolic compounds with participation of ROS on the mycelium surface of pathogenic fungus are discussed. The analysis of importance of polysaccharide acetylation degree both in the induced plant defense and in the local lignification of infection zones was carried out. The lignin post infectious accumulation analysis in wheat plants has shown that plant tissues of resistant sorts differed from susceptible by increased activity of lignin synthesis enzymes and higher polymer concentration in them. Wheat anionic isoperoxidase is characterized by property to contact with the cell wall chitin of pathogenic fungi that explains the presence of the polysaccharide-binding domain in this enzyme. Detection of a homologous zone in the polysaccharide-binding domain of some isoperoxidases in *Arabidopsis*, wheat, zucchini, corn and rice allows to allocate polysaccharide-specific isoforms genes from a large amount of genes. Almost half of century ago in the literature there was a message [Siegel, 1957] in which the author noticed about the increase of lignin level at presence of chitin incubation medium. Then this message has not involved considerable attention from outside researchers. Thus, among diversified plant proteins, we found peroxidases which were sorbed on chitin and preserved their enzymatic activity. An analysis of the isoenzymatic composition and activity of chitin-binding peroxidases revealed the considerable differences between plant species. In particular anionic peroxidases of practically all examined species plants were bind with chitin. In some cases (plants of the *Fabaceae* and *Cucurbitacea* families), the cationic isoforms interacted with chitin along with anionic isoperoxidases. The results obtained suggested that in some cases, the antigenic determinants of chitin-binding anionic isoenzymes were homologous for both monocotyledons and dicotyledons. An increase of plant peroxidase activity is in the presence of chitin and its oligomers [Maksimov et al., 2003]. And its activation can occur both in fraction which to bind on chitin and in the enzyme fraction which is not contacting it. These results confirm the possibility of peroxidase activation at presence of a polysaccharide [Siegel, 1958].

In order to prove the polysaccharide-specificity of allocated peroxidases site of a gene which presumably coding polysaccharide-specific domain of peroxidase have been picked up and designed primers which we have used for an estimation of gene expression of wheat anionic peroxidase gene under infection of fungus disease agents. Sequencing results, received from peroxidase DNA and cDNA of some plants have proved the accuracy of choosing primers. RT-PSR detects that in plant a part of gene transcriptional activity constitutively that might be coordinated with the established opinion about peroxidases as constitutively working enzymes. The anionic peroxidase gene expression repeatedly increased under infection with fungus pathogens and the influence of signal molecules and elicitors

Thus, a comparative estimation of the lignin content, activity of lignification enzymes, ability to sorption on chitin of isoperoxidases from different plant families, definition of their immunochemical affinity, a homology of molecular structure of polysaccharide-specific zone of peroxidases allowed us to assume, that local lignin accumulation under fungal pathogenesis shows the universality via of polysaccharide specific isoperoxidases activation.

ACKNOWLEDGMENTS

The authors are grateful for financial support from the Ministry of Education and Science of Russian Federation P339.

REFERENCES

[1]   Behle R. W., Dowd P. F., Tamez-Guerra P. and Lagrimini L.M. (2002) Effect of Transgenic Plants Expressing High Levels of a Tobacco Anionic Peroxidase on the Toxicity of Anagrapha falcifera Nucleopolyhedrovirus to Helicoverpa zea (Lepidoptera: Noctuidae) // *J. of Economic Entomology.* V.95. P. 81-88.

[2]   Bernards M.A., Fleming W.D., Llewellyn D.B., Priefer R., Yang X., Sabatino A., Plourde G.L. (1999) Biochemical characterization of the suberization-associated anionic peroxidase of potato // *Plant physiol.* V.121. P.135-145

[3]   Brundrett M.C. (2002) Coevolution of roots and mycorrhizas of land plants // *New Phytol. V.* 154. P. 275-304.

[4]   Cano-Delgado A., Penfield S., Smith C. et al. (2003) Reduced cellulose synthesis invokes lignification and defense responses in *Arabidopsis thaliana* // *Plant J. V.* 34. P.351-362.

[5]   Carpin S., Crevecoeur M., Greppin H. and Penel C. (1999) Molecular cloning and tissue-specific expression of an anionic peroxidase in zucchini.// *Plant Physiol. V.* 120. P. 799–810.

[6]   Chen C., Baucher M., Christensen J.H. and Boerjan W. (2001) Biotechnology in trees: towards improved paper pulping by lignin engineering. // *Euphytica. V.* 118. P. 185–195.

[7]   Dowd P.F. and Lagrimini L.M. (2006) Examination of the biological effects of high anionic peroxidase production in tobacco plants grown under field conditions. I. Insect pest damage. // *Transgenic Research.* V. 15. P. 197-204.

[8]   Grant M. and Mansfield J. (1999) Early events in host–pathogen interactions // Cur. *Opin. Plant Biol. N.* 2. P. 312-319.

[9]   Hatfield R. and Vermeris W. (2001). Lignin formation in plants. The dilemma of linkage specifity // *Plant Physiol. V.*126. P.1351-1357.

[10]   Hawkins S. and Boudet A. (2003) Defense lignin and hydroxycinnamyl alcohol dehydrogenase activities in wounded *Eucalyptus gunnii* // *Forest Path. V.* 33. P. 91–104.

[11]   Heitefuss R. (2001) Defense reactions of plants to fungal pathogens: principles and perspectives, using powdery mildew on cereals as an example // *Naturwissenschaften. V.* 88. P. 273–283.

[12]   Humphrey T.V., Boneta D.T. and Goring D.R. (2007). Sentinels at the wall: cell wall receptors and sensors // *New Phytologist. V.*176. P. 7-21.

[13] Ikegawa T., Mayama S., Nakayashiki H. and Kato H. (1996) Accumulation of diferulic acid during the hypersensitive response of oat leaves to *Puccinia coronata* f. sp. *avenae* and its role in the resistance of oat tissues to cell wall degrading enzymes // Physiol. *Mol. Plant Pathol.* V.48. P. 245-255.

[14] Karkonen A. and Kautaniemi S. (2010) Lignin biosynthesis studies in plant tissue cultures // *J. Integr. Plant Biology.* V. 52. P. 176-185.

[15] Li Y., Kajita S., Kawai S., Katayama Y. and Morohoshi N. (2003) Down-regulation of an anionic peroxidase in transgenic aspen and its effect on lignin characteristics // *J. Plant Res.* V. 116. P. 175–182.

[16] Maksimov I.V, Troshina N.B., Yarullina L.G., Surina O.B., Jusupova Z.R. and Cherepanova E.A. (2004) The co-culture of wheat callus and bunt pathogen as a suit test-system for the search of plant resistance inducers // In "Biotechnology and agriculture and the food industry" ed. G.E. Zaikov. NewYork: Nova Science Publ. P. 145-150.

[17] Maksimov I.V., Cherepanova E.A., Yarullina L.G. and Akhmetova I.E. (2005) Isolation of chitin-specific wheat oxidoreductases // *Appl. Biochemistry and microbiology.* V. 41. P. 616-620.

[18] Maksimov I.V., Sorokan A.V., Cherepanova E.A., Surina O.B., Troshina N.B. and Yarullina L. G. (2011) Effect of salicylic and jasmonic acids on the components of pro-/antioxidant system in potato plants infected with late blight // *Russian J. of Plant Physiology.* V. 58. P. 299-306.

[19] Marjamaa K., Kukkola E.M. and Fagerstedt K.V. (2009)The role of xylem class III peroxidases in lignification // *J. of Exp. Botany.* V. 60. P. 367-376.

[20] Moldenhauer J., Moerschbacher B.M. and van der Westhuizen A.J. (2006) Histological investigation of stripe rust (*Puccinia striiformis* f.sp. *tritici*) development in resistant and susceptible wheat cultivars // *Plant Pathology.* V. 55. P. 469–474.

[21] Ostergaard L., Teilum K., Mirza O. et al. (2000) *Arabidopsis* ATP A2 peroxidase. Expression and high-resolution structure of a plant peroxidase with implications for lignification // *Plant Mol. Biology.* V. 44. P. 231–243.

[22] Quiroga M., Guerrero C., Botella M.A., Barcelo A., Amaya I., Medina M.I., Alonso F.J., Milrad de Forchetti S., Tigier H. and Valpuesta V. (2000) A Tomato Peroxidase Involved in the Synthesis of Lignin and Suberin // *Plant Physiol.* V. 122. - P. 1119–1127

[23] Roeder A.H.K. and Yanofsky M.F. (2006) Fruit development in *Arabidopsis*. In: Somerville SC, Myerowitz EM, eds. The Arabidopsis book. Rockville:. *Amer. Soc. of Plant Biologist.*

[24] Sato Y., Demura T., Yamawaki K., Inoue Y., Sato S., Sugiyama M. and Fukuda H. (2006) Isolation and characterization of a novel peroxidase gene ZPO-C whose expression and function are closely associated with lignification during tracheary element differentiation. // *Plant and Cell Physiology.* V.47. P. 493–503.

[25] Shinogi T., Suzuki T., Kurihara T. et al. (2003) Microscopic detection of reactive oxygen species generation in the compatible and incompatible interactions of *Alternaria alternata* Japanese pear pathotype and host plants // *J. Gen. Plant Pathol.* V. 69. P.7–16.

[26] Siegel S.M. (1957) Non - enzymic macromolecules as matrices in biological synthesis. The role of polysaccharides in peroxidase catalyzed lignin polymer formation from eugenol // *J. Amer. Chem. Soc.* V.79. P.1628-1632

[27] Troshina N.B., Maksimov I.V., Surina O.B. and Khairullin R.Ì. (2000) The develop of *Tilletia caries* (D.C.) Tul. in wheat calluses and cell cultures// *Biol. Bullet,*. N. 3. P. 377-381.

[28] Troshina N.B., Maksimov I.V., Yarullina L.G., Surina O.B. and Cherepanova E.A. (2004) Plant resistance inductors and active forms of oxygen. I. The influence of salicylic acid on hydrogen peroxide production in common cultures of wheat callus and bunt pathogen // Cytologia (in Russian). V. 46. P. 1001 – 1005.

[29] Verma D.P.S. and Hong Z. (2001) Plant callose synthase complexes // *Plant Mol. Biol.* V. 47. P. 693–701.

[30] Yokoyama R. and Nishitani K. (2006) Identification and characterization of *Arabidopsis thaliana* genes involved in xylem secondary cell walls. // *J. of Plant Research.* V.119. P. 189–194.

In: Lignin
Editor: Ryan J. Paterson

ISBN 978-1-61122-907-3
© 2012 Nova Science Publishers, Inc.

Chapter 18

APPLICATION OF LIGNIN MATERIALS FOR DYE REMOVAL BY SORPTION PROCESSES

Daniela Suteu[1] and Carmen Zaharia[2]*

[1]Gheorghe Asachi' Technical University of Iasi,
Faculty of Chemical Engineering and Environment Protection,
Department of Organic and Biochemical Engineering,
71A Prof.dr.docent D.Mangeron Blvd, 700050 - Iasi, Romania
[2]Gheorghe Asachi' Technical University of Iasi,
Faculty of Chemical Engineering and Environment Protection,
Department of Environmental Engineering and Management,
71A Prof.dr.docent D.Mangeron Blvd, 700050 - Iasi, Romania

ABSTRACT

Some batch sorption experiments were carried out to remove Methylene Blue cationic dye from aqueous systems using industrial lignin as a low cost sorbent. The solution's pH, amount of industrial lignin, contact time, initial dye concentration and temperature were the studied operating variables.

To establish the most suited type of sorption mechanism to describe the dye retaining onto the solid sorbent, the data were analyzed using the Langmuir, Freundlich and Dubinin-Radushkevich models for the sorption isotherms. The results of this experimental study indicate that the tested solid material has a moderate capacity for dye molecules uptake.

Keywords: lignin, sorption, Methylene Blue, equilibrium.

---

* Corresponding author email: danasuteu67@yahoo.com.

INTRODUCTION

The textile industry is a high water consumer, and the textile finishing process is in the same time a high producer/generator of industrial wastewater with an extremely varied and complex composition.

Over the years, there were implemented different types of textile wastewater treatments based on processes such as: oxidation-reduction, coagulation-flocculation, ionic exchange, ozonation, and adsorption [1- 6].

The recent research studies were focused on emphasizing that adsorption is one of the processes with very good results in term of efficiency vs. treatment costs. The advantages that imposed this process are: easiness of achievement, reduced implementation costs, possibility to use a large variety of adsorptive materials that can be selected also based on identifiable textile dye structure.

The most studied adsorptive materials can be included into the category of: activated coal, chitine, chitosan, ionic exchangeable resins, polyamides, and inorganic polymeric materials. The high costs of synthetic adsorbent preparation re-orientated the researches to the testing of low cost materials (as production wastes) and easily obtainable, included into the category of 'non-conventional' or 'low cost', such as: (i) industrial/agricultural/domestic wastes or industrial/agricultural by-products (ash, sludge, sawdust, textile fibbers, mud, bark, straw, etc.) and (ii) natural materials (peat, seashell, algae, lignite, wood, etc.) [7-12].

From this point of view, the lignin and other lignocellulosic materials, the main by-product of the pulp and agro-industrial industries can be used to remove dyes from textile wastewaters.

The aim of this paper was to evaluate the potential of lignocellulosic sorbents (i.e. industrial lignin) to remove Methylene Blue dye from aqueous media vs. initial dye concentration, temperature and contact time. The sorption equilibrium and kinetic data were analyzed by means of different models in order to understand the best agreed sorption mechanism of the dye molecules onto lignin.

EXPERIMENTAL

Materials

The cationic phenothiazine dye, Methylene Blue (Basic Blue 9) (Figure 1, MW = 319.85 g/mol, absorption maximum at λ_{max}= 660 nm), was used as commercial salt. Working solutions (in concentration of 19-134 mg/L) were prepared by appropriate dilution with bidistilled water of a stock dye solution (320 mg/L).

The experiments were carried out using industrial lignin that represents a macromolecular compound much more reactive than cellulose because of its functional groups (hydroxylic, both phenolic and aliphatic). The lignin used like sorptive materials can be the main by-product of the pulp industry and also can represent a product obtained from renewable resources. The experiments were carried out with alkali lignin (offered by Granit Co.) L_1 (from Wheat straw 100-W-A) with characteristics present in Table 1 [13].

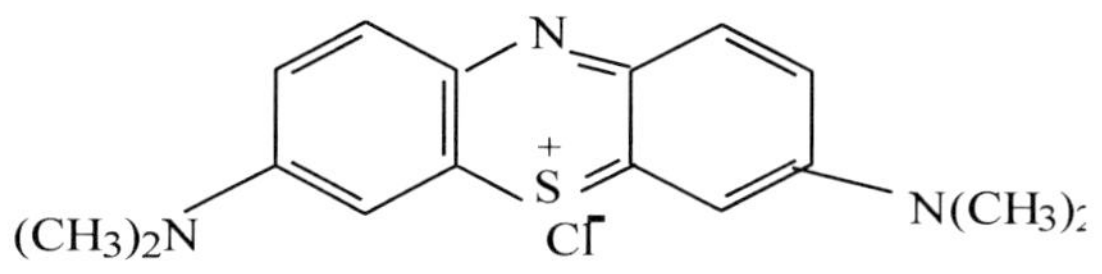

Figure 1. Structure of Methylene Blue cationic phenothiazine dye.

Table 1. The main characteristics of lignin sample

Characteristics	L_1 (100-W-A)	Characteristics	L_1 (100-W-A)
Acid insoluble lignin, %	90	pH (10% dispersion)	2.7
Acid soluble lignin, %	1	MW	3510
COOH, mmole/g	3.8	$S_{oftening}$, ^{0}C	170
Aromatic-OH, mmole/g	1.7-1.8	Solubility in furfuryl alcohol, %	88.5
OH/C9 groups, chemical method	1.02	Solubility in aqueous alkali, pH 12, %	98.5
Ash, %	2.5		

Sorption Working Methodology

The experiments were carried out in batch condition: samples of 0.1 g lignin were contacted with 25 mL of aqueous solution containing various concentrations of dye (25.6 - 281.6 mg/L) into 100 mL conical flasks, under an intermittent stirring. The temperature of aqueous systems was controlled with a thermostatic assembly. After 24 h, the phases were separated by filtration. The concentrations in the filtrates were analyzed by measurement of absorbance at the dye specific λ (660 nm) and calculation from the dye calibration curve, using an UV-VIS SP-830 Plus spectrophotometer, Metertech Inc. Version 1.06.

The sorption capacity of the lignin was evaluated by amount of sorbed dye, q:

$$q = (C_0-C).V/G \ (mg/g) \tag{1}$$

and by percentage of dye removal, R %:

$$R = [(C_0-C)/C_0].100 \ (\%) \tag{2}$$

where: C_0 and C are initial and equilibrium concentration of dye in aqueous system (mg/L), G is amount of sorbent (g), and V is volume of aqueous system (L).

The effect of contact time on the sorption of dyes onto these lignocellulosic materials was determined by the 'limited bath' technique, when sample of sorbent was added under stirring to volumes of dye solution with the necessary pH and knowing the initial dye concentration. Generally, the temperature of dye solutions was maintained constant, at room temperature value. After different contact times, volumes of supernatant were taken for absorbance measurements in order to determine the dye content using the dye calibration curve.

RESULTS AND DISCUSSION

Effect of Solution pH

The behaviour of dyes in the sorption process at different value of solution pH may be correlated with the surface charge of the sorbent in function of the solution pH. The pH_{PZC} (pH of zero charge) of lignin was determined using the method described by Nouri and Haghseresht [14] (Figure 2); the limiting pH (3.5) was considered as value pH_{PZC}, when the sorbent surface is neutral [15].

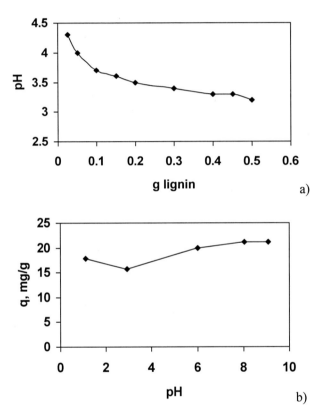

Figure 2. a) The value of pH_{PZC} (pH of zero charge) for lignin; b) The influence of initial pH of solution on sorption of cationic dye onto lignin.

At pH lower than pH_{PZC} the sorbent surface is positive and has affinity for anionic dye and at pH higher than pH_{PZC} the surface of sorbent is negatively charged and is available to electrostatic interactions with cationic dyes. In this context, the effect of pH on the cationic dye sorption onto industrial lignin is insignificant; the Methylene Blue cationic dye is retained from solutions with pH $\geq$ 4. In the following experiments we worked at a pH (6).

Effect of Sorbent Dose

To establish the favourable sorbent dose, different amount of industrial lignin powder were contacted with volumes of solutions containing Methylene Blue cationic dye of a specific concentration (51.2 mg/L), for 24 hours. The results, presented in Figure 3, indicated that the sorption of dye increases with the amount of lignin; the sorbent dose which assures a removal percent of dye higher than 90% is 2 g lignin/L.

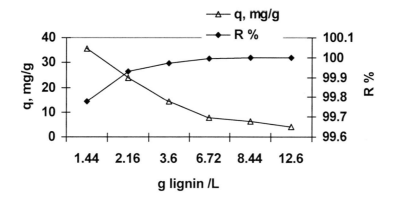

Figure 3. The influence of lignin dose on the dye sorption: 51.2 mg dye /L, 20^0C, pH =6.

Effect of Contact Time

The effect of contact time on the sorption of Methylene Blue cationic dye onto surface of industrial lignin powder is shown in Figure 4. It can be seen that the amount of sorbed dye increases rapidly vs. time up to 150 minutes and after a slow increase can be observed; nearly 5 hours are required to be attained the dye sorption equilibrium.

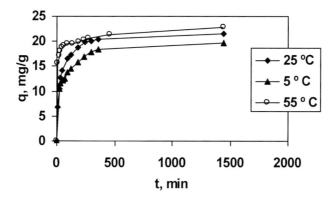

Figure 4. The sorption capacity (q) versus time and temperature for the sorption of Methylene Blue cationic dye onto industrial lignin: pH= 6; 4 g lignin /L, C_0 = 89.6 mg dye/L.

Effect of Initial Dye Concentration

The effect of initial dye concentration on sorption capacity of lignin powder was investigated into dye solutions of pH (6) with a sorbent dose of 4 g lignin/L. The results are presented in Figure 5.

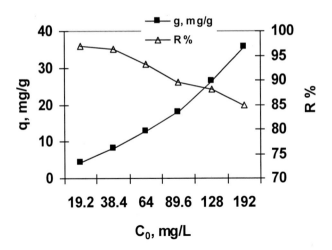

Figure 5. The effect of initial dye concentration on dye sorption onto industrial lignin powder: pH= 6; 4g lignin/L; t = 24 h; T= 25°C.

From this figure it can be seen, that the percentage of cationic dye removal is maximal (96.98 %) in dilute solutions (19.2 mg dye/L) and decreased with increasing of initial dye concentration; however, even at concentrations of 192 mg dye/L the value of dye removal (R, %) is about 84.92 %. At the same time, the sorbed dye amount per mass unit of lignin increased from 4.603 to 56.88 mg/g for an increase in initial dye concentration from 19.2 to 256 mg/L. This is because at higher initial concentrations it is enhanced migration of the large dye molecules in internal macroporous and mesoporous structure of the sorbent.

Effect of Temperature

The influence of temperature on the Methylene Blue sorption onto industrial lignin powder is illustrated in Figures 4 and 6. The data of Figure 6 showed that both percent of dye removal and the amount of dye sorbed onto mass unit of lignin increases with increasing in solution temperature. The effect is more significant at higher temperature in solutions of higher initial concentrations and can be attributed to the endothermic nature of adsorption and also because the fact that the high temperatures favour the large dye molecules diffusion in internal porous structure of sorbent.

Sorption Isotherm

Sorption isotherm gives the dependence of sorption capacity of sorbent and equilibrium concentration of dye in solution (Figure 6). The experimental date of sorption modelling by different sorption model, are used to determine the characteristic quantitative parameters from which it can be obtained some information about sorption mechanism, sorbent surface properties and the affinity of sorbent towards the dye molecule.

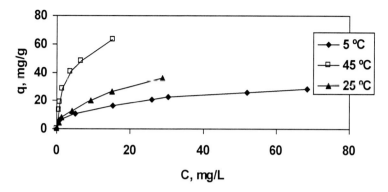

Figure 6. The sorption isotherms of the Methylene Blue cationic dye onto lignin powder: 4 g lignin/L; pH = 6.

The data of sorption equilibrium were processed using three usual models of sorption isotherm: the Freundlich, Langmuir and Dubinin-Radushkevich, expressed by the equations presented in Table 2. In order to establish the best fitting isotherm model for the studied sorption process it is frequently used the correlation coefficient. The constants of isotherms related to each sorption model, calculated from the intercepts and slopes of the corresponding linear plots (Figure 7), together with their correlation coefficients (R^2) are presented in Table 3 [16].

Table 2. Mathematical forms and parameters of the applied isotherm models

Sorption isotherm model-applicability	Equation and \ or their linear form	Isotherms parameters, significance
Freundlich – it is used for non-ideal sorption that involves heterogeneous sorption	$q = K_F \cdot C^{1/n}$ $\lg q = \lg K_F + 1/n \lg C$	K_F – adsorption coefficient $(mg(g)^{-1})(L(mg)^{-1})^{1/n}$ n – adsorption intensity
Langmuir – it is used to estimate the maximum sorption capacity	$q = \dfrac{K_L \cdot C \cdot q_0}{1 + K_L \cdot C}$ $\dfrac{1}{q} = \dfrac{1}{q_0 \cdot K_L \cdot C} + \dfrac{1}{q_0}$	q_0 - saturation capacity $(mg\,(g)^{-1})$ K_L - binding (sorption) energy $(L(mg)^{-1})$
Dubinin-Radushkevich – it is applied to estimate the porosity apparent free energy	$\ln q = \ln q_D - \beta_D \varepsilon^2$ $\varepsilon = RT \ln(1 + \dfrac{1}{C})$ $E = \dfrac{1}{\sqrt{2\beta}}$	q_D - the theoretical saturation capacity $(mg\,(g)^{-1})$ β_D – constant correlated to sorption energy $(mol^2\,(kj^2)^{-1})$ ε - Polanyi potential E – mean free energy of sorption $(kj\,(mol)^{-1})$

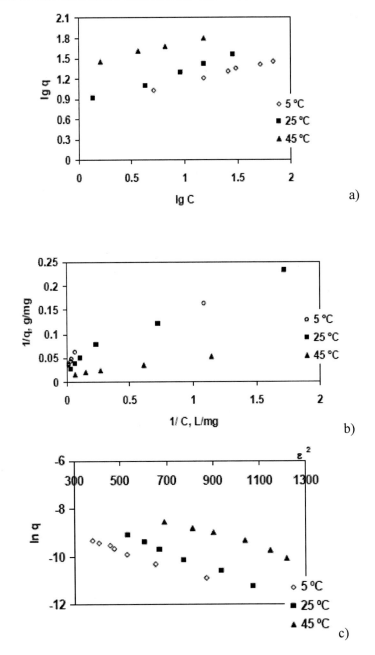

Figure 7. Freundlich (a), Langmuir (b) and Dubinin-Radushkevich (c) plots for the sorption of the Methylene Blue cationic dye onto industrial lignin at three temperatures.

The values from Table 4 indicate that the Freundlich sorption model described better the sorption equilibrium in conditions of low temperatures (5-25 ^{0}C). At higher temperature (> 25^{0}C) the Langmuir isotherm model provides in a better manner the adequacy of this isotherm model with the experimental data for the studied system.

**Table 3. Isotherm parameters for the sorption of Methylene Blue
cationic dye onto industrial lignin**

Type of isotherm	T (K)		
	278	298	318
Freundlich			
K_F (mg/g)(L/mg)$^{1/n}$	6.169	6.179	19.701
n	2.705	1.9	2.149
R^2	0.9955	0.9925	0.9629
Langmuir			
q_0 (mg/g)	23.98	28.01	66.67
K_L (L/mg)	0.37	0.309	0.448
R^2	0.9029	0.9853	0.9955
Dubinin- Radushkevich (DR)			
q_0 (mg/g)	93.88	260.571	502.24
β (mol^2/kJ)	0.0032	0.0038	0.0029
E (kJ/mol)	12.5	11.47	13.13
R^2	0.9898	0.9929	0.9751

The values for *1/n* lower than 1 indicate a normal Langmuir isotherm, and the studied sorbent is favourable for the sorption of the tested cationic dye [17].

The values of the DR isotherm parameters (Table 4) indicate the porous structure of the lignin sorbent (the maximum amounts of dye that can be sorbed under optimal experimental conditions are much higher than the experimental values). The sorption capacity in the D-R equation, which may represent the total specific meso- and macropore volume of the sorbent, was found 11.45 (kJ/ mol) at 25 ^{0}C. The values of mean free sorption energy are between 11.47 and 12.5 kJ/mol, suggesting that in the cationic dye sorption onto the industrial lignin is involved an ion exchange mechanism [18].

Thermodynamic Evaluation of the Process

In order to evaluate the influence of temperature into the sorption process, and also to understand the nature of sorption (physical, chemical or physico-chemical respectively) the apparent thermodynamic parameters including Gibbs free energy change (ΔG), enthalpy (ΔH), and entropy change (ΔS) were calculated using the values of binding Langmuir constant, K_L (L/mol). The values of ΔG were calculated according to the Gibbs equations (3). The ΔH and ΔS changes were determined by the vant`t Hoff equations (Eq. 4) [19]. All obtained values for thermodynamic parameters are presented in Table 5.

$$\Delta G = -RT \ln K_L \qquad (3)$$

$$\ln K_L = -\frac{\Delta H}{RT} + \frac{\Delta S}{R} \qquad (4)$$

where R is the universal gas constant (8.314 kJ/mol K) and T is the temperature (K).

Table 5. The apparent thermodynamic constants of the sorption process of Methylene Blue cationic dye onto industrial lignin

T (K)	ΔG (kJ/mol)	ΔH (kJ/mol)	ΔS (J/mol K)
293	- 26.998		
308	- 28.497	3.7587	109.8
318	- 31.390		

The thermodynamic parameters, ΔH and ΔS, were obtained from the slope and intercept of the plot of linear dependence ΔG vs. T (Figure 8).

The negative values of apparent free energy charge (ΔG) indicate that the cationic dye sorption onto lignin is thermodynamically feasible and spontaneous at any temperature between 5° and 45°C. The literature data [19-21] mentions that values of ΔG smaller than -16 kJ/mol are characteristic to physical sorption by weak interactions (such as van der Waals, dipole-dipole) between adsorption sites of the sorbent and dye molecules and so, the mechanism is not attributed only to ionic exchange [21].

The positive value of ΔH indicates the endothermic nature of dye sorption, favoured by the temperature increases. The literature data [18] indicate that values of ΔH^0 no more than 4.2 kJ/ mol and values of ΔG lower than -16 kJ/mol characterized the physical sorption. According to the calculated data from Table 5, the sorption of Methylene Blue cationic dye onto lignin is rather physical adsorption than chemisorptions.

The positive value of ΔS reflects the electrostatic interactions between opposite charged groups and suggests that the driving force of sorption is an entropic effect (the entropic contribution is even higher than the free energy of sorption, and is induced by the thermal agitation in condition of an apparent quasi-static 'resting' system).

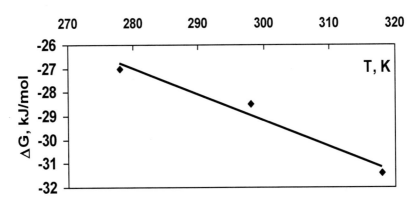

Figure 8. The dependence ΔG vs. T for Methylene Blue cationic dye sorption onto industrial lignin.

In conclusion, it can be said that the mechanism of Methylene Blue cationic dye sorption onto industrial lignin is a combination between adsorption and ionic exchange mechanism. The links and forces involved in dye sorption can range from weak van der Waals forces to electrostatic attractions between the ionized sulfonyl groups of the dye molecule and positively charged surface of sorbent.

CONCLUSIONS

The ability of lignin as sorbent to retain the cationic dyes with relative high molecular weight was investigated for Methylene Blue dye using three sorption isotherm models and thermodynamic experimental parameters.

The results of this study show that the lignin can be used as a potential sorbent for the removal of Methylene Blue cationic dye from aqueous solutions at pH 6. Experimental results provide promising perspective for the utilization of industrial lignin (the main by-product of the pulp industry) as sorbent in reducing pollution of textile effluents.

The equilibrium sorption data were analyzed with the Freundlich, Langmuir, and Dubinin-Radushkevich isotherm models. The Langmuir isotherm represents in a better manner the sorption equilibrium data at lower temperature (> 25 ^{0}C); the monolayer sorption capacity of 28.01 mg/g was obtained at 25 ^{0}C, value in accordance with other literature information about 'low cost' sorbents.

The thermodynamic parameters of the sorption process were also evaluated. The negative value of ΔG confirms the spontaneous nature of sorption process and the positive value of ΔH indicates that the sorption process is endothermic.

The sorption mechanism of Methylene Blue cationic dye onto industrial lignin can be considered more physical by weak interaction than ion exchange.

REFERENCES

[1]   Y.Anjaneyulu, N.Sreedhara Chary, D.Samuel Suman Raj, *Decolourization of industrial effluents – avaible methods and emerging technologies – a review*, Reviews in Environmental Science and Bio/Technology, 4, 2005, 245-273.

[2]   D.Suteu, C.Zaharia, D.Bilba, R.Muresan, A.Popescu, A.Muresan, *Decolourization of textile wastewaters – Chemical and Physical methods*, Textile Industry, 60(5), 2009, 254-263.

[3]   B.Ramesh Babu, A.K.Parande, T.Prem Kumar, *Textile Technology: Cotton Textile Processing: Waste Generation and Effluent Treatment*, The Journal of Cotton Science, 11, 2007, 141 – 153.

[4]   C.Zaharia, D.Suteu, A.Muresan, R.Muresan, A.Popescu, *Textile wastewater treatment by homogeneous oxidation with hydrogen peroxide*, Environmental Engineering and Management Journal, 8(6), 2009, 1359-1369.

[5]   E. Forgacs, T. Cserhati, G.Oros, *Removal of synthetic dyes from wastewaters: a review*, Environ. Internat., 30, 2004, 953-971.

[6]   C. Zaharia, *Chemical wastewater treatment,* (in Romanian), Performantica Publishing House, Iasi, Romania, 2006.

[7]   V.K.Gupta, Suhas, *Application of low cost adsorbents for dye removal – A review*, J.Environ.Manag., 90, 2009, 2313-2342.

[8]   D.Suteu, C.Zaharia, A.Muresan, R.Muresan, A.Popescu, *Using of industrial waste materials for textile wastewater treatment*, Environmental Engineering and Management Journal, 8(5), 2009, 1097-1102.

[9] D.Suteu, C.Zaharia, M.Harja, *Residual ash for textile wastewater treatment*, Proceedings of International Conference – Unitech'08, Gabrovo, Bulgaria, 21-22 November 2008, III, 2008, 475-480.

[10] A. Bhatnagar, A. K. Jain, *A comparative adsorption study with different industrial wastes as adsorbents for the removal of cationic dyes from water*, Journal *of Colloid and Interface Science*, 281, 2005, 49-55.

[11] 11.M. Ahmaruzzaman, *A review on the utilization of fly ash*, Prog. Energy Combust. Sci., 36, 2007, 327-363.

[12] G.Crini, *Non-conventional low-cost adsorbents for dyes removal: A review*, Bioresour. Technol., 60, 2006, 67-75.

[13] T.Malutan, R.Nicu, V.I.Popa, *Contribution to the study of hydroxymetylation reaction of alkali lignin*, Bioresource, 3(1), 2008, 13-20.

[14] S.Nouri, F.Haghseresht, *Adsorption of p-nitrophenol in untreated and treated activated carbon*, Adsorption, 10, 2004, 79-86.

[15] D.Suteu, Malutan T., Rusu G., Use of Industrial Lignin for dye removal from aqueous solution by sorption, *Scientific Papers USAMV Iasi (Romania), Series Agronomy*, 51, 2008, 71-78.

[16] Suteu D., Malutan T., Bilba D., *Agricultural waste corncob as sorbent for the removal of reactive dye Orange 16: Equilibrium and kinetic study*, Cellulose Chemistry and Technology (in press).

[17] S.Thirumalisamy, M.Subbian, *Removal of Methylene blue from aqueous solution by activated carbon prepared from the peel of cucumis sativa fruit by adsorption*, Bioresources, 5(1), 2010, 419-437.

[18] M.M. Dubinin, L.V. Radushkevich, *Equation of the characteristic curve of activated charcoal*, Proc.Acad.Sci USSR, *Phys Chem., Sect.* 55, 1947, 331-333.

[19] G. Crini, P.M.Badot, *Application of chitosan, a natural aminopolysaccharide, for dye removal from aqueous solution by adsorption processes using batch studies: A review of recent literature*, Prog.Polym.Sci., 33, 2008, 399-447.

[20] C.H.Weng, Y.C.Sharma, S.H.Chu, *Adsorption of Cr (VI) from aqueous solution by spend activated clay*, J.Harzard.Mater., 155, 2008, 65-75.

[21] H.B.Senturk, D.Ozdes, C.Duran, *Biosorption of Rhodamine 6G from aqueous solution onto almond shell (Prunus dulcis) as a low cost biosorbent*, Desalination, 252, 2010, 81-87.

In: Lignin
Editor: Ryan J. Paterson

ISBN 978-1-61122-907-3
© 2012 Nova Science Publishers, Inc.

Chapter 19

PLASTIC MOLDABLE LIGNIN

Armando G. McDonald and *Lina Ma*

Forest Products Program, University of Idaho, Moscow,
ID 83844-1132, U. S.

ABSTRACT

The effects of chemical modification of lignin on the moldability of this material were examined. Softwood kraft lignin was chemically modified with benzyl chloride under alkaline conditions at various mole ratios of benzyl chloride to wood hydroxyl groups (ratio = 1-3) at a reaction time of 8 h. The extent of benzylation was assessed by weight gain, and Fourier transform infrared (FTIR) and solid state ^{13}C nuclear magnetic resonance (NMR) spectroscopies. FTIR spectroscopy revealed the reduction of lignin hydroxyl group bands, an increase in aromatic bands and an increase in acryl and alkyl ether bands, which were consistent with etherification. The thermal and flexural properties of the benzylated lignin were assessed by differential scanning calorimetry (DSC), dynamic rheometry, and dynamic mechanical analysis. The results from DSC were consistent with data from rheometry. Results have also shown that the benzylated lignin thermal transition temperature and mechanical properties can be manipulated by the extent of benzylation.

Keywords: benzylation, glass transition temperature, moldable lignin, spectroscopy, viscous modulus

INTRODUCTION

Lignin is a propylphenol-based biopolymer that is a major structural component of all woody plants. Along with cellulose fibrils, lignin acts as a matrix material which forms a composite system capable of imparting the strength and rigidity necessary for plants.

---

[*] Corresponding author: E-mail: armandm@uidaho.edu.

However, lignin presents a problem to industries intent on obtaining cellulosic ethanol and fibers from plants, such as in paper production. The cellulosic polysaccharides must be separated from the rest of the plant, usually by chemical or enzymatic means, leaving behind a lignin-rich material of little value and is typically burned to recover energy. Unfortunately, only 1- 2% of lignin is utilized to make other products [1].

Continuing efforts have been made to try and utilize this surplus of industrial lignin. Limited success has been realized so far in moving lignin utilization from the bench to an industrial scale. This is largely due to the variable nature of lignin structure, as its chemistry is dependent on its biological origin (softwoods, hardwoods, grasses) and processing variables. For example, softwood lignin is predominantly comprised of guaiacyl (G) units with a minor amount of *p*-hydroxyphenyl (H) units while hardwood lignins are comprised of syringyl (S) units and G units [2]. By better understanding the structure of the lignin molecule and approaches for its utilization, lignin's potential as a renewable and widely available raw material can be unlocked [3,4,5].

Lignin has strong inter- and intra-chain hydrogen bonds making it a very stiff and rigid polymer [6]. With substitution of hydroxyl group in the lignin molecule with non-polar groups to inhibit H-bonding, thermoplasticity can be introduced to lignin [5,7,8]. Another benefit of substituting lignin is that protecting the phenolic group can improve the photostability of lignin based materials by impeding free radical formation [9]. There are two approaches for substitution either esterification or alkylation of the hydroxyl groups. Esterification has been shown to be a very effective method to thermoplasticize lignin and the thermal properties of the lignin derivatives can be modified by the ester substituent [5,7,10]. Alkylation is the alternate approach to modifying lignin and work by Li and Sarkanen [8] on fully methylated kraft lignin had given tensile properties similar to that of polystyrene. Despite these promising results, the lignin thermoplastic was brittle. Studies on benzylation of rice straw [11] and wood particles [12,13] have shown that thermoplasticity can be introduced into these substrates. Therefore, by using a bulkier alkylating agent (benzyl) may improve lignin properties over the smaller methyl derivatives.

The aim of this study is to generate lignin moldable materials without the need for synthetic resins. In the present work, kraft lignin was benzylated to produce thermoplastic like material and the resulting materials characterized.

MATERIALS AND METHODS

Softwood kraft lignin (Indulin AT, Mead Westvaco) was dried prior to use. Lignin (5 g) was pre-treated with 10N NaOH (20 mL) for 1h to aid in swelling. The slurry was then transferred into a flask containing benzyl chloride (BC) and the reaction was carried at 110 °C for 8 h with continuous stirring [12,13]. A BC to lignin hydroxyl group mole ratio of 1, 2 and 3 was used for the modifications. The modified lignin was collected by filtration and exhaustively washed with water and then with ethanol to remove any residuals. The final product was vacuum dried at 40°C to constant weight to determine yields. Benzylated lignin was molded into bar (2 mm x 10 mm x 35 mm) and disc (2 mm x 25 mm dia) specimens by injection molding at approximately 180°C using a Dynisco Lab Molding Machine.

Spectroscopy

Fourier transform infrared (FTIR) spectra were obtained on original and benzylated lignin in the attenuated total reflectance (ATR) mode (ZnSe) on an Avatar 370 spectrometer (ThermoNicolet). Benzylation was evidenced by changes of functional group. Solid state cross polarization-magic angle spinning (CP-MAS) ^{13}C nuclear magnetic resonance (NMR) spectra were obtained on a Bruker Avance DRX-400 solid-state NMR spectrometer operating at a frequency of 100.6 MHz equipped with a Chemimagnetics solids probe (Washington State University NMR center). Samples were ground and packed into a 5 mm zirconia rotor which were spun at 5000 Hz. Spectra were recorded using a 3.75 μs proton preparation pulse followed by a 1 ms cross polarization contact time and a acquisition time of 6.6 μs. Spectral analysis was performed using freeware MestRe-C v2.3a.

Thermal Analysis and Mechanical Properties

Differential scanning calorimetry (DSC) was performed on benzylated lignin to determine the glass transition temperature (T_g) according to ASTM D3418 [14]. The benzylated lignins were annealed at 100°C for 4 h prior to DSC analysis (TA Instruments model Q200 DSC, 5mg of sample, 25 to 300°C at 10°C/min). Dynamic mechanical analysis (DMA) was performed on molded specimens (in triplicate) in the 3 point bending mode using a TA Instruments model Q800 DMA. Flexural modulus of elasticity (MOE) at 35°C were obtained from stress-strain curves in the controlled force test using 0.5 N/min ramp force according to ASTM D5934 to maximum strain 0.03% [15]. Data was analyzed by TA Universal Analysis software v4.4A. Dynamic rheometery was performed (in duplicate) using a Bohlin CVO100 rheometer equipped with an extended temperature control module on molded disc specimens in accordance to ASTM D4440 [16]. A negative temperature ramp (250 to 40°C at -2°C /min) was performed at 1 Hz with strain of 0.02% [13]. Data was analyzed using Bohlin rheology software v06.32.

RESULTS AND DISCUSSION

The extent of lignin benzylation was assessed by weight gain. The degree of substitution (DS) was calculated from the percentage weight gain for lignin based on the number of available hydroxyl groups present in lignin (1.08 OH/C_9OCH$_3$ units) [17]. For fully benzylated kraft lignin (DS = 1) a theoretical weight gain of 55% would be obtained. Varying the ratio of BC/OH ratio (1, 2 and 3) resulted in lignin weights gains of 25 (DS 0.45), 57 (DS 1.0) and 137% (DS 2.5), respectively. The DS of benzylated lignin was shown to increase 5-fold with an increase of the BC/OH ratio from 1 to 3 showing that an excess of BC was necessary to obtain high substitution. The high level of substitution above the theoretical level may be due to alkaline induced hydrolysis (cleavage) of lignin ether linkages exposing more available hydroxyl groups that can be substituted.

FTIR Spectroscopic Analysis

FTIR spectra of softwood kraft lignin and benzylated lignin (DS=2.5) are shown in Figure 1. Benzylation of lignin was evidenced [12,18] by: (i) a reduction in the 3450 cm[-1] band (OH stretch); (ii) appearance of bands at 695 cm[-1] (aromatic C-C angular deformation), 736 cm[-1] (aromatic C-H out-of-plane bending, mono-substitution); 1592-1612 cm[-1] and 1496-1454 cm[-1] (aromatic C-C axial deformations), 1810 and 1950 cm[-1] (out of plane deformation vibrations of adjacent hydrogens from mono-substituted aromatic rings), and 3030-3088 cm[-1] (strong multiplet, aromatic C-H stretch) which were attributable to mono substituted benzyl rings; and (iii) an increase of aryl ether band (-C-O-C- stretching) at 1265 cm[-1] and alkyl ether band at 1150-1060 cm[-1] (C-O stretch from alkyl-substituted ether). Positive relationships between DS and (i) intensity of aryl (1265 cm[-1]) and alkyl ether (1150 cm[-1]) peaks and (ii) area of benzyl peak (695 cm[-1]) were observed (data not shown) for benzylated lignin samples. In contrast, a negative relationship was observed between DS and hydroxyl band (3450 cm[-1]) intensity. These results are consistent with the literature [12]. Therefore FTIR spectroscopy can be used to determine the DS of benzylated lignin based materials.

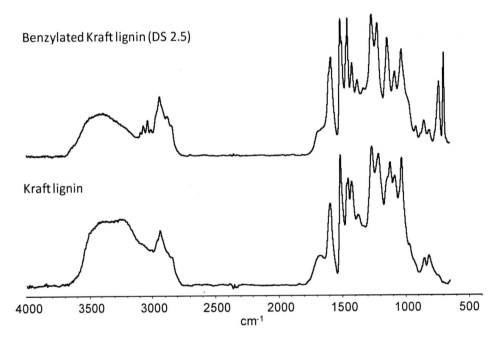

Figure 1. FTIR spectra of (top) benzylated kraft lignin (DS 2.5) and (bottom) kraft lignin.

NMR Spectroscopic Analysis

The ^{13}C NMR spectra of lignin and benzylated lignin (DS 1.0) are shown in Figure 2. Chemical shift assignments for lignin and benzylated lignin are given in Table 1. The ^{13}C spectrum of kraft lignin was partially interpreted due to the fact that lignin is a very heterogeneous polymer, and many signals can overlap [19,5]. Aromatic carbon signals occur in the range of 100 to 155 ppm, while signals from aliphatic carbons occur between 20 and 90 ppm.

The most prominent signal in the lignin ^{13}C spectrum occurs around 56 ppm, attributed to methoxy carbons on G units. Other major signals occur between 145 and 155 ppm. These signals are due to the carbons 3 and 4 on guaiacyl units.

Table 1. C NMR chemical shifts (ppm) of softwood kraft lignin and benzylated lignin

Kraft lignin (ppm)	Benzylated lignin (ppm)	Assignment
-	211.7	Aliphatic ketone
-	196.1	CO
190.7	193.0	α-CHO
173.1	180.9	acetyl
167.3	171.6	Carbonyl
150.5	-	C3,C4 of G phenolic
-	156.0, 152.5	C3 of G
-	140.8	C4 of G or Aromatic C1 of benzyl group
-	131.5	Aromatic C2-C6 on benzyl group
131.3	-	C1 of G
125.6	-	C6 of G
117.4	115.1	C5 of G
110.0	-	C2 of G
	91.0	O-CH$_2$-
83.7	-	OCHCH$_2$OH
76.5	-	OCH$_2$CHOH
-	73.8	-CH$_2$O-R
64.9	-	-CH$_2$OH
57.5	57.5	Methoxyl carbons on G units
-	51.2	Methoxyl aryl ethers
35.6, 32.5	37.2	-CH$_2$-
17.0	17.7, 10.7	-CH$_3$

In the benzylated lignin ^{13}C NMR spectrum (Figure 2, Table 1) an intense signal appeared at 133 ppm and was assigned to aromatic carbons of the benzyl group [20,21]. The signal at 142 ppm was assigned to C4 on benzylated G units [20]. The sharp signal at 171 ppm was assigned to carbonyl groups. A new signal was observed at 51 ppm and assigned to a methoxyl aryl ether group. Carbon signals at 77 and 84 ppm were assigned to etherified aliphatic C's in lignin [22]. These data show that the relative concentration of lignin to benzyl groups decreased upon substitution.

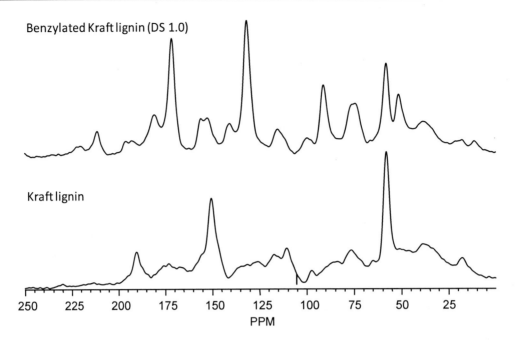

Figure 2. ^{13}C NMR spectra of (top) benzylated kraft lignin (DS 1.0) and (bottom) kraft lignin.

Glass Transition Temperature (T_g)

From the rheological data the temperature at the viscous modulus (G") peak maximum was tentatively assigned to the T_g and was shown to decrease from 183 to 60°C as the extent of lignin benzylation increased (DS = 0.45 to 2.5) (Figure 3, Table 2). These findings are consistent with those reported in the literature for benzylated wood fiber [12,13]. The higher the DS the lower the T_g (start of viscous flow based on G") which indicates an increase in free volume by introducing bulky benzyl substituents and thus increasing molecular mobility [23,24]. It is assumed that the transition was a T_g based on results on benzylated wood fiber [13].

Table 2. Determined T_g and MOE values for benzylated lignin

Benzylated lignin (DS)	T_g (°C) by G"	T_g (°C) by DSC	MOE (GPa)
0.45	184	165	2.82
1.0	104	122	2.69
2.5	60	48	

Thermal transitions of lignin and its derivatives were also analyzed by DSC. No obvious thermal transitions were observed for softwood kraft lignin, however upon benzylation T_g's were readily observed (Figure 4, Table 2). In Figure 4, benzylated lignin (DS 1.0) clearly shows a T_g transition at 122°C. The T_g was shown to decrease from 165 to 48°C as DS increased from 0.45 to 2.5. This trend is consistent with that obtained by dynamic rheology.

However, it is reported that this method reports T_g up to 25°C difference from values determined by DSC [25].

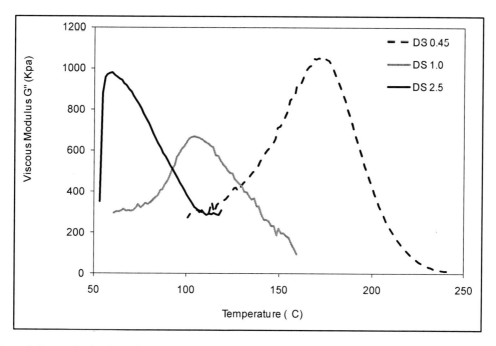

Figure 3. Dynamic rheology thermograms of viscous modulus (G") of benzylated kraft lignin.

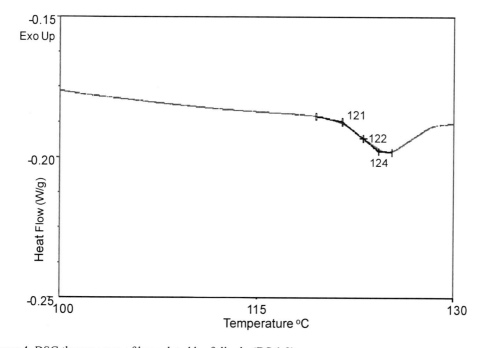

Figure 4. DSC thermogram of benzylated kraft lignin (DS 1.0).

Mechanical Properties

Flexural properties were measured by DMA (3 point bending) to determine the modulus of elasticity (MOE) from the stress-strain curves of molded benzylated lignin samples. MOE values of around 2.7-2.8 GPa were obtained with benzlyated lignin (DS 0.45 to 1.0) (Table 2). However, at higher DS (2.5) it was difficult to test the flexural properties of these molded benzylated lignins since it sagged on the test fixture. The MOE of benzylated lignin, was comparable to values for polypropylene (40%) based WPC at 2.1- 2.8GPa [26], polyethylene-terephthalate at 2.5 GPa, and polystyrene at 3.0 GPa [27].

CONCLUSIONS

Commercial lignin was successfully benzylated in a 2-step process. The extent of benzylation could be governed by the ratio of benzylchloride to wood hydroxyl content. The thermal transition properties of benzylated lignin were successfully plasticized by benzylation.

The modulus of elasticity (MOE) of benzylated lignin (2.7 GPa), at low to moderate levels of substitution, were shown to be comparable to commodity plastics and wood plastic composites.

However, as the benzyl substitution increased the MOE decreased and the material behaved like a thermoplastic rubber. It is possible to produce such a bio-plastic based on lignin benzylation. The outcome of this effort is to develop biobased plastics which will reduce our dependency petroleum based plastics.

ACKNOWLEDGMENTS

The project was supported by USDA-CSREES-Wood Utilization Grant number 2005-34158-14707. The FTIR spectrometer was supported by a USDA-CSREES-NRI grant number 2005-35103-15243.

REFERENCES

[1]   Lora, J. H., and Glasser, W. G. (2002). Recent industrial applications of lignin: A sustainable alternative to nonrenewable materials. *Journal of Polymers and the Environment, 10(1-2)*, 39-48.

[2]   Ralph,J., Brunow, G., Boerjan, W. (2007). Lignins. In Encyclopedia of Life Science. John Wiley and Sons Ltd. (pp 1-10).

[3]   Lindberg, J. J., Kuusela, T. A., and Levon, K. (1989). Specialty polymers from lignin. In W.G. Glasser and S. Sarkanen (Ed.) Lignin: Properties and Materials. (pp 190-204). American Chemical Society, Washington D.C.

[4]   Feldman, D. (2002). Lignin and its polyblends - A review. In T. Q. Hu (Ed.) Chemical
      Modification, Properties, and Usage of Lignin. (pp 81-99). Kluwer Academic/Plenum
      Publishers, New York.

[5]   Fox, S.C., McDonald, A.G. (2010). Chemical and thermal characterizaion of three
      industrial lignins and their corresponding lignin esters. *BioResources J, 5(2)*, 990-1009.

[6]   Kubo, S. and Kadla, J.F. (2005). Hydrogen bonding in lignin: A fourier transform
      infrared model compound study. *Biomacromolecules, 6*, 2815-2821.

[7]   Glasser, W. G. and Jain, R. K. (1993). Lignin derivatives: I. Alkanoates.
      Holzforschung, 47(3), 225-233.

[8]   Li, Y. and Sarkanen, S. (2000). Thermoplastics with very high lignin contents. In W.G.
      Glasser, R.A. Northey, and T. P. Schultz (Ed.) Lignin: Historical, Biological, and
      Materials Perspectives. (pp 351-366). American Chemical Society, Symposium Series
      No. 742. Washington, D.C.

[9]   Chang, H-T., Su, Y-C., Shang-Tzen, C. (2006). Studies on photostability of butyrylated,
      milled wood lignin using spectroscopic analyses. *Polymer Degradation and Stability,
      91*, 816-822.

[10] Steward, D., McDonald, A. G., and Meder, R. (2002). Thermoplastic lignin esters. In
      Proceedings of the 6th Pacific Biobased Composites Symposium. (pp 584-594).
      Portland, Oregon, USA.

[11] Mohammadi-Rovshandeh, J., and Sereshti, H. (2005). The effect of extraction and
      prehydrolysis on the thermoplasticity and thermal stability of chemically modified rice
      straw. *Iranian Polymer Journal, 14(10)*, 855-862.

[12] Hon, D.N.S. and Ou, N-H. (1989). Thermoplastization of wood: Benzylation of Wood.
      Journal of Applied Polymer Science Part A, 27, 2457-2482.

[13] McDonald, A.G., and Ma, L. (2010) Plastic moldable pine fiber by benzylation. In L.F.
      Botannini (Ed.) Wood Types, Properties and Uses. (Chapter 8). Nova Science
      Publishers, Inc.

[14] American Society for Testing and Materials (ASTM D3418-03) (2008) Standard test
      method for transition temperatures and enthalpies of fusion and crystallization of
      polymers by differential scanning calorimetry. ASTM annual book, vol. 08(02). (pp 63-
      69). Conshohocken, PA.

[15] American Society for Testing and Materials (ASTM D5934-02) (2008) Standard test
      method for the determination of modulus of elasticity for rigid and semi-rigid plastic
      specimens by controlled rate of loading using three-point bending. ASTM annual book,
      vol. 08(03). (pp 311-314). Conshohocken, PA.

[16] American Society for Testing and Materials (ASTM D4440-07) (2008) Standard test
      method for plastics: Dynamical mechanical properties melt rheology. ASTM annual
      book, vol. 08(02). (pp 489-492). Conshohocken, PA.

[17] McCarthy, J. L.; Islam, A. (2000). In Glasser, W.G., Northey, R.A.; Schultz, T.P. (Ed.)
      Lignin: Historical, Biological, and Materials Perspectives (pp 2-99) American Chemical
      Society Symp.Ser.742, Washington, DC.

[18] Coates, J. (2000). Interpretation of Infrared Spectra, A Practical Approach. In R.A.
      Meyers (Ed.) Encyclopedia of Analytical Chemistry. (pp. 1-25). John Wiley and Sons
      Ltd, Chichester.

[19] Leary, G. J., and Newman, R. H. (1992). Cross polarization/magic angle spinnin nuclear magnetic resonance (CP/MAS NMR) spectroscopy. In S.Y. Lin and C.W. Dence (Ed.) Methods in Lignin Chemistry. (pp 146-161). Springer-Verlag, Berlin.

[20] Mello, N.C., Ferreira, F.C., Curvelo, A.A.S., Mattoso, L.H.C., and Colnago, L.A. (2000). Study on benzylated sisal fibers by 13C solid state NMR. In Proceedings from the 3rd International Symposium on Natural Polymers and Composites. (pp 32-36). Sao Pedro, Brazil, May 14-17.

[21] Ramos, L.A., Frollini, E., Koschella, A., and Heinze, T. (2005). Benzylation of cellulose in the solvent dimethylsulfoxide/tetrabutylammonium fluoride trihydrate. *Cellulose, 12,* 607–619.

[22] Sipilä, J. and Syrjänen, K. 1995. Synthesis and 13C NMR spectroscopic characterization of six dimeric arylglycerol-β-aryl ether model compounds representative of syringyl and p-hydrosyphenyl structures in lignins. On the aldol reaction in β-ether preparation. *Holzforschung, 49(4),* 325-331.

[23] McCrum, N.G., Read, B.E., and Williams, G. (1991). Anelastic and Dielectric Effects in Polymeric Solids. Dover Publications, New York.

[24] Nielsen, L.E., and Landel, R.F. (1994). Mechanical Properties of Polymers and Composites. Marcel Dekker, Inc., New York.

[25] Menard, K.P. (1999). Dynamic Mechanical Analysis: A Practical Introduction; CRC: Boca Raton, FL:

[26] Wechslera, A., and Hiziroglu, S. (2007). Some of the properties of wood–plastic composites. *Building and Environment, 42(7),* 2637-2644.

[27] Ashby, M.F. and Jones, D.R.H. (2000). Engineering Materials 1. An Introduction to their Properties and Applications, 2nd edition, Butterworth-Heinemann, Burlington, MA.

In: Lignin
Editor: Ryan J. Paterson

ISBN 978-1-61122-907-3
© 2012 Nova Science Publishers, Inc.

Chapter 20

LIGNIN: PROPERTIES AND APPLICATIONS IN BIOTECHNOLOGY AND BIOENERGY

Vishal V. Dawkar, *Umesh U. Jadhav, Ashok D. Chougale and Sanjay P. Govindwar*

Department of Biochemistry, Shivaji University,
Kolhapur-416004, Maharashtra, India
Present Address[1]: Plant molecular Biology Group,
Division of Biochemical Sciences, National Chemical Laboratory,
Pune-411 008, India

ABSTRACT

Lignins are complex phenolic polymers occurring in higher plant tissues and are the second most abundant terrestrial polymer after cellulose. Due to their very complex structure, lignins are amorphous polymers with rather limited industrial use. One of the uses of lignin is production of biofuels. Fuels are generally those made from non-edible lignocellulosic biomass. These biofuels have some clear advantages. Plants can be bred for energy characteristics, and not for food, and a larger fraction of the plant can be converted to fuel. Lignocellulosic crops can be grown on poor quality land, requiring fewer fertilizers. There are substantial energy and environment benefits primarily due to greater biomass usability per unit of land area. Within the bioenergy sector, biotechnology, and in particular genetic engineering, has the potential to be applied to agricultural production - to optimize the productivity of biomass; to raise the ceiling of potential yield per hectare; to modify crops to enhance their conversion to fuels - and to the biomass conversion process, for example by developing more effective enzymes for the downstream processing of biofuels. It has become possible to process lignocellulose at high substrate levels and the enzyme performance has been improved. Also the cost of enzymes has been reduced. Genetic research into dedicated energy crops and manufacturing processes is still at an early stage.

* Email: vvdawkar@gmail.com, Tel: +91-231-2609152, Fax: +91-231-2691533.

INTRODUCTION

Bioenergy

The availability and environmental impact of energy resources will play a critical role in the progress of the world's societies and the physical future of our planet. The majority of human energy needs are currently met using petrochemical sources, coal and natural gases but these fossil fuels are approaching depletion and their continued use has had damaging environmental consequences. Worldwide energy consumption has increased more than twenty-fold in the last century and, with the exception of hydroelectricity and nuclear fusion energy, all current major energy sources are finite. At present usage rates, these sources will soon be exhausted (Srivastava and Prasad, 2000) and this has contributed to soaring fossil fuel prices. As the demand for energy has grown, so have the adverse environmental effects of its production. Emissions of CO_2, SO_2 and NOx from fossil fuel combustion are the primary causes of atmospheric pollution (Ture et al., 1997). The accumulation of carbon dioxide and other greenhouse gases in the atmosphere is thought to be responsible for climate change, which is predicted to have disastrous global consequences for life on this planet (Sheehan et al., 1998).

Over the last two decades there has been a growing interest in using biomass-derived fuels. Initially this interest was driven by concerns for potential shortages of crude oil, concerns about global warming, the soaring cost of gasoline and national security issues have rekindled interest in producing liquid transportation fuels from renewable resources. But in recent years the ecological advantages of biomass fuels have become an even more important factor. Biomass fuels can be considered essentially CO_2 neutral and have a very low sulfur content compared to many fossil fuels. In addition, being a liquid, bio-oil can be easily transported and stored (Czernik and Bridgwater, 2004).

In terms of modern bioenergy, ethanol, biodiesel and biogas are the three major bioenergy products. Ethanol and biodiesel can be used as transportation fuels, and ethanol is also an important raw product in the chemical industry. Therefore, ethanol production has a particularly important role in transforming petroleum-based economies to biomass- based sustainable and environment-friendly economies (Joshua et al., 2008).

Biofuels are at present classified in two categories: "first-generation" and "second generation" fuels. While there are no strict technical definitions for these terms, the main distinction between them is the feedstock used. A first-generation fuel is generally a fuel made from sugars, grains, or seeds; that uses only a specific (often edible) portion of the aboveground biomass produced by a plant; and that is the result of a rather simple manufacturing process. First-generation fuels are already being produced in significant commercial quantities in a number of countries. Second generation fuels are generally those made from non-edible lignocellulosic biomass, either non-edible residues of food crop production (e.g., corn stalks or rice husks) or non-edible whole-plant biomass (e.g., grasses or trees grown specifically for energy). The process to convert the lignocellulosic biomass into a fuel is rather complex (Jegannathan et al., 2009).

Lignin: Properties

Plant cell walls contain lignin, a phenolic polymer that hinders the degradation of cell wall polysaccharides to simple sugars destined for fermentation to ethanol. Lignin is a complex biopolymer based on 4-hydroxyphenylpropanoids and is synthesized by almost all terrestrial plants and also in some algae (Martone,et al., 2009). Lignin plays important roles in mechanical support (Hejnowicz, 2005), stress responses (such as UV radiation), resistance to pests and disease (Bhuiyan, 2009), as well as water and nutrient transport (Boyce et al., 2004). This robust resin is difficult to break down chemically and biochemically. Lignin is one of the most abundant natural raw materials available on earth, second to cellulose if mass is considered, even first if solar energy storage is the criteria (Gargulak and Lebo, 2000).

Biotechnology Solutions for Bioenergy

Lignin is not a polysaccharide and this highly branched polyphenolic macromolecule is strongly resistant to chemical and biological degradation (Larson, 2008). The cellulose and hemicellulose are cemented together by lignin. Lignin is responsible for integrity, structural rigidity, and prevention of swelling of lignocelluloses. Thus, lignin content and distribution constitute the most recognized factor which is responsible for recalcitrance of lignocellulosic materials to enzymatic degradation by limiting the enzyme accessibility; therefore the delignification processes can improve the rate and extent of enzymatic hydrolysis. However, in most delignification methods, part of the hemicellulose is also hydrolyzed, and hence the delignification does not show the sole effect of lignin (Wyman, 1996). Dissolved lignin due to e.g. pretreatment of lignocelluloses is also an inhibitor for cellulase, xylanase, and glucosidase. Various cellulases differ in their inhibition by lignin, while the xylanases and glucosidase are less affected by lignin (Berlin et al., 2006). The composition and distribution of lignin might also be as important as the concentration of lignin. Some softwoods are more recalcitrant than hardwoods. This might be related to the lignin type, since softwoods have mainly guaiacyl lignin while hardwoods have a mix of guaiacyl and syringyl lignin. It has been suggested that guaiacyl lignin restricts fiber swelling and enzyme accessibility more than syringyl lignin (Ramos et al., 1992). Lignin is not fermented to produce liquid biofuels, but instead can be recovered and used as a fuel for heat and electricity at an ethanol production facility (Larson, 2008).

One approach that addresses biological and technological barriers to cost-effective production of biofuels from lignocellusic feedstock would be to reduce the lignin content of the wood or at least make it easier to remove. Reduction of lignin has long been a target of interest in crop and forestry species, because lignin interferes with digestion of plant materials by farm animals and removal of lignin is a costly step in the production of paper. This has motivated researchers to identify and isolate from a variety of species many of the genes coding for enzymes in the lignin pathway. A large body of research involving manipulation of the lignin biosynthetic pathway in transgenic plants has accumulated over the past 15 years; this has been exhaustively reviewed (Anterola and Lewis 2002; Boerjan et al. 2003; Li et al. 2008; Vanholme et al. 2008; Weng et al. 2008). In broad strokes, experiments have shown that lignin content can be significantly reduced in trees and herbaceous plants. For example, several of the transgenic plants with down regulated genes for lignin biosynthetic enzymes in the early steps of the pathway showed that syringyl lignin (S-lignin) content was more strongly affected than guaiacyl lignin (G-lignin) content (Vanholme et al. 2008) Another means of modifying lignin production that may address these negative pleiotropic effects is to alter the expression of transcription factors or other regulators which in turn modify the expression of a whole suite of genes that participate in the lignin biosynthesis pathway. Masaru Ohme-Takagi and colleagues showed that two plant-specific transcription factors, designated NAC secondary wall thickenings promoting factor 1 (NST1) and NST3, regulate the formation of secondary walls in woody tissues of Arabidopsis. (Mitsuda et al. 2007). Interestingly, it has been reported that the composition of lignin does not have a strong effect on biological degradation of cell walls (Grabber et al. 1997).

The future of bioenergy will depend on breakthrough technologies. However, the importance of basic research on pathways and genes involved in cell wall biosynthesis, plant development, and metabolite production should not be ignored. Translational systems biology is needed for biofuel applications. The use of 'omics' Furthermore, bioenergy is not, and should not be, limited to higher plants, although higher plants are likely to provide the most important feedstock for first and second generations of biofuels. Studies of microbes that have the capacity to digest plant cell walls will also be important components of bioenergy research (Yuan et al., 2008).

The white-rot fungi belonging to the basidiomycetes are the most efficient and extensive lignin degraders (Akin et al., 1995; Gold and Alic, 1993) with *P. chrysosporium* being the best-studied lignin-degrading fungus producing copious amounts of a unique set of lignocellulytic enzymes. *P. chrysosporium* has drawn considerable attention as an appropriate host for the production of lignin-degrading enzymes or direct application in lignocellulose bioconversion processes (Ruggeri and Sassi, 2003; Bosco et al., 1999). Less known, white-rot fungi such as *Daedalea flavida*, *Phlebia fascicularia*, *P. floridensis* and *P. radiate* have been found to selectively degrade lignin in wheat straw and hold out prospects for bioconversion biotechnology were the aim is just to remove the lignin leaving the other components almost intact (Arora et al., 2002). Less prolific lignindegraders among bacteria such as those belonging to the genera *Cellulomonas*, *Pseudomonas* and the actinomycetes *Thermomonospora* and *Microbispora* and bacteria with surface-bound cellulase-complexes such as *Clostridium thermocellum* and *Ruminococcus* are beginning to receive attention as representing a gene pool with possible unique lignocellulolytic genes that could be used in lignocellulase engineering (Vicuña, 1988; McCarthy, 1987; Miller (Jr) et al., 1996; Shen et al., 1995; Eveleigh, 1987).

Biofuels are conventionally produced using chemical catalyst processes though recent developments in white biotechnology and green technology have prompted the use of a number of enzymes and microorganisms for the development of products and processes (Villadsen, 2007; Aracil et al., 2006). The use of biotechnology facilitates production of biofuels with efficient use of ligniniocellulosic biomass. The ligniniocellulosic biomass can be first treated to obtain lignin and cellulose. The lignin fraction so obtained can be used as a fuel for heat and electricity. The cellulose fraction can be further treated for production of alcohol or biogas.

Conclusion

Worldwide energy consumption has increased more than twenty-fold in the last century and to cope with the human energy need, biosource is the choice, which will help for ever. Bioenergy is renewable energy can made available from materials derived from biological sources. Biomaterial burning is environmentally friendly compared to other fuels like oil and coal. These sources can also fulfill the need of electrical energy. The use of biotechnology facilitates production of biofuels with efficient use of ligniniocellulosic biomass. Bioenergy will fulfill the needs of humans, devoid of any effects.

References

Akin DE, Rigsby LL, Sethuraman A. 1995. Alterations in the structure, chemistry, and biodegradation of grass lignocelluloses treated with white rot fungi *Ceriporiopsis subvermispora* and *Cyathus stercoreus*. *Appl. Environ. Microbiol.* 61:1591-1598.

Anterola A. M., Lewis N. G. 2002. Trends in lignin modification: a comprehensive analysis of the effects of genetic manipulations/ mutations on lignification and vascular integrity. *Phytochemistry* 61: 221–294.

Aracil J, Vicente M, Martinez M, Poulina M. 2006. Biocatalytic processes for the production of fatty acid esters. *J. Biotechnol.* 124:213–23.

Arora DS, Chander M, Gill PK. 2002. Involvement of lignin peroxidase, manganese peroxidase and laccase in the degradation and selective ligninolysis of wheat straw. *Int. Bioterior. Biodegrad.* 50:115-120.

Berlin, A., Balakshin, M., Gilkes, N., Kadla, J., Maximenko, V., Kubo, S., Saddler, J. 2006. Inhibition of cellulase, xylanase and beta-glucosidase activities by softwood lignin preparations. *J. Biotechnol.* 125, 198-209.

Bhuiyan NH, Selvaraj G, Wei Y, King J. 2009. Role of lignification in **plant** defense. *Plant Signal Behav*, 4:158.

Boerjan W., Ralph J., Baucher M. 2003. Lignin biosynthesis. *Ann. Rev. Plant Biol.* 54: 519–546.

Li X.; Jing-Ke Weng J.-K.; Chapple C. 2008. Improvement of biomass through lignin modification. *Plant J.* 54: 569–581.

Bosco F, Ruggeri B, Sassi G. 1999. Performances of a trickle bed reactor (TBR) for exoenzyme production by *Phanerochaete chrysosporium*: influence of a superficial liquid velocity. *Chem. Eng. Sci.* 54:3163-3169.

Boyce CK, Zwieniecki MA, Cody GD, Jacobsen C, Wirick S, Knoll AH, Holbrook NM. 2004. Evolution of xylem lignification and hydrogel transport regulation. *Proc. Natl. Acad. Sci. USA*. 101:17555.

Eveleigh DE (1987). Cellulase: a perspective. *Phil. Trans. R. Soc. Lond. Ser. A*. 321: 435-447.

Gargulak, J.D., Lebo, S.E., 2000. Commercial use of lignin-based materials. In: Glasser, W.G., Northey, R.A., Schultz, T.P. (Eds.), Lignin: Historical, Biological, and Materials Perspectives. ACS Symposium Series. American Chemical Society, p. 307.

Gold MH, Alic M. 1993. Molecular biology of the lignin-degrading basidiomycetes *Phanerochaete chrysosporium. Microbiol. Rev.* 57(3):605-622.

Grabber J. H.; Ralph J.; Hatfield R. D.; Quideau S. p-hydroxyphenyl, guaiacyl, and syringyl lignins have similar inhibitory effects on cell wall degradation. *J. Agric. Food Chem.* 45: 2530–2532; 1997.

Hejnowicz Z: Unusual Metaxylem Tracheids in Petioles of Amorphophallus (Araceae) Giant Leaves. *Ann. Bot.* 2005, 96:407.

Joshua S. Yuan, David W. Galbraith, Susie Y. Dai, Patrick Griffin and C. Neal Stewart Jr (2008) Plant systems biology comes of age. *Trends Plant Sci.* 13, 165–171.

Joshua S., Kelly H. Tiller, Hani Al-Ahmad, Nathan R. Stewart and C. Neal Stewart Jr, 2008. Plants to power: bioenergy to fuel the future *Trends in Plant Science.* 13, 421-429.

Kenthorai Raman Jegannathan, Eng-Seng Chan, Pogaku Ravindra, Harnessing biofuels: A global Renaissance in energy production? *Renewable and Sustainable Energy Reviews* 13 (2009) 2163–2168.

Larson, E.D. 2008. Biofuel production technologies: status, prospects and implications for trade and development. United Nations Conference on Trade and Development (UNCTAD).

Martone PT, Estevez JM, Lu F, Ruel K, Denny MW, Somerville C, Ralph J: Discovery of lignin in seaweed reveals convergent evolution of cell wall architecture. *Curr. Biol.* 2009, 19:169.

McCarthy AJ (1987). Lignocellulose-degrading actinomycetes. 1987. *FEMS Microbiol. Lett.* 46(2):145-163.

Miller (Jr) RC, Gilkes NR, Johnson P, et al. (1996). Similarities between bacterial and fungal cellulase systems. Proceedings of the 6th International Conference on Biotechnology in the Pulp and Paper Industry: Advances in Applied and Fundamental Research, pp. 531-542.

Mitsuda N.; Iwase, A.; Yamamoto H.; Yoshida, M.; Seki, M.; Shinozaki K.; Ohme-Takagi M. NAC transcription factors, NST1 and NST3, are key regulators of the formation of secondary walls in woody tissues of Arabiodopsis. *Plant Cell* 19: 270–280; 2007.

Ramos, L.P.; Breuil, C.; Saddler, J.N. Comparison of steam pretreatment of eucalyptus, aspen, and spruce wood chips and their enzymic hydrolysis. *Appl. Biochem. Biotechnol.* 1992, 37-48.

Ruggeri B, Sassi G (2003). Experimental sensitivity analysis of a trickle bed bioreactor for lignin peroxidases production by *Phanerochaete chrysosporium. Process Biochem.* 00:1-8.

Czernik S. and A. V. Bridgwater, *Overview of Applications of Biomass Fast Pyrolysis Oil Energy and Fuels* 2004, 18, 590-598.

Sheehan J, Cambreco V, Duffield J, Garboski M, Shapouri H. An overview of biodiesel and petroleum diesel life cycles. A report by US Department of Agriculture and Energy; 1998. p. 1–35.

Shen H, Gilkes NR, Kilburn DG, et al. (1995). Cellobiohydrolases B, a second exo-cellobiohydrolase from the cellulolytic bacterium *Cellulomonas fimi. Biochem. J.* 311:67-74.

Srivastava A, Prasad R. Triglycerides-based diesel fuels. *Renew Sust Energy Rev* 2000;4:111–33.

Ture S, Uzan D, Ture IE. The potential use of sweet sorghum as a non-polluting source of energy. *Energy* 1997;22:17–9.

Vanholme R.; Morreel K.; Ralph J.; Boerjan W. Lignin engineering. *Curr. Opinion Plant Biol.* 11: 278–285; 2008.

Vicuña R (1988). Bacterial degradation of lignin. *Enzyme Microb. Technol.* 10:646-655.

Villadsen J. Innovative technology to meet the demands of the white biotechnology revolution of chemical production. *Chem. Eng. Sci.* 2007;62:6957–68.

Weng J.-K.; Li X.; Bonawitz N.; Chapple C. Emerging strategies of lignin engineering and degradation for cellulosic biofuel production. *Curr. Opinion Biotech.* 19: 166–172; 2008.

Wyman, C.E. *Handbook on bioethanol: production and utilization*; Taylor and Francis: Washington DC, USA, 1996.

In: Lignin
Editor: Ryan J. Paterson
ISBN 978-1-61122-907-3
©2012 Nova Science Publishers, Inc.

Chapter 21

CARBONIZATION KINETICS OF VARIOUS LIGNOCELLULOSIC PRECURSORS

Benoît Cagnon[1] and Xavier Py[2]

[1]Institut de Chimie Organique et Analytique (ICOA-UMR 6005),
Université d'Orléans, IUT, Orléans-cedex 2, France
[2]Laboratoire Procédés, Matériaux et Energie Solaire (CNRS-PROMES UPR 8521),
Université de Perpignan UPVD, Perpignan, France

ABSTRACT

In recent years, considerable effort has been devoted to the thermochemical transformation of agricultural, biomass and industrial residues in order to generate energy, chemicals or activated carbons. Extensive research has been focused on the production of activated carbons from waste materials of agricultural origin, mostly based on lignocellulosic materials. The control of the activated carbon porous texture requires controlling each stage in the activation process. In the case of physical activation, the efficiency of the prior carbonization of the raw material is a key step in reducing the quantity of disorganized carbon inside the char and consequently the extent of subsequent activation required. Similarly, in the case of chemical activation, economic and environmental constraints make it imperative to reduce the amount of activating agent, energy and water needed for subsequent washing as far as possible. This can be carried out by optimizing the corresponding heat treatments. Within the framework of the valorization of agricultural and food industry waste, several recent studies have investigated lignocellulosic precursors for the production of activated carbon. During the carbonization of lignocellulosic precursors, hemicellulose, cellulose and lignin (spatially distributed within the matter) decompose at different rates and within distinct temperature ranges. The differences in reactivity of these three components during carbonization, as well as competition between the reactions and thermal effects accompanying their decompositions make the study of carbonization complex. At this stage of the process, the porosity of the adsorbent is not fully developed. Pyrolysis products such as tars have to be released from the char by means of the subsequent activation step in order to fully open the potential porosity of the material. However, the activation procedure usually induces heterogeneity in the pore size distribution due to the competition between diffusion and chemical phenomena. Moreover, the thermal effect of each reaction induces enhancement or inhibition of other kinetics. Improving the carbonization procedure (in terms of heating rates, plateau,...) can therefore reduce the amount of activation agent needed, lead to narrowing the pore size distribution and reduce the cost of the whole development process. The objective of our study is to optimize the initial carbonization stage of the lignocellulosic precursors. The thermal decomposition of the three major lignocellulosic components (hemicellulose, cellulose and lignin) was first studied separately by thermogravimetry to predict their respective contributions in terms of weight fraction and carbon production in the obtained char. The decomposition kinetics of these lignocellulosic compounds were then studied within both a synthetic blend and natural materials, namely coconut shell and plum stones. A model composed of three independent chemical kinetics was validated for various particle sizes and heat treatment rates. The use of a particular technique is proposed to reduce the number of adjustable parameters needed in the model from 15 to 3.

INTRODUCTION

To address the issue of the valorization of agricultural and food-processing waste, extensive research has focused on the study of lignocellulosic precursors. The most widely-studied precursors are: wood (Solum et al., 1995; Byrne and Nagle 1997), almond shells (Rodriguez-Reinoso et al., 1982, 1984), olive stones (Lopez-Gonzalez et al., 1980; Gonzalez et al., 1994), coconut shells (Banerjee et al., 1976; Mortley et al., 1988), peach stones (Rodriguez-Reinoso et al., 1985; Molina-Sabio et al., 1995), apricot stones (Gergova et al., 1993) and apple pulp (Fernandez et al., 2001). Since the beginning of the XXth century, several authors have studied wood to determine the microstructure, analyze its various chemical components and understand the reactions which take place during combustion (Eickner 1962; Roberts 1971; Shafizadeh 1977). Since the 60s, wood tissue has been described as a heterogeneous system including cells and intercellular material composed of various organic polymers.

Hillis et al., (1989) proposed the following description (Figure 1):

1. Cell wall composed of cellulose molecules (long polysaccharide chains linked together to form elementary fibrils with a width of approximately 3 nm);
2. Intercellular material composed of hemicellulose molecules (polysaccharide chains with a smaller molecular mass, less ordered than the cellulose molecules);
3. Cell wall cemented by lignin molecules (polymers composed of three-dimensional polyphenolic molecules, with a high molecular mass).

The cellulose molecules are ordered in microfibrils embedded in a matrix of hemicellulose and lignin, the lignin acting as cement between the cellulose and hemicellulose fibers.

The cellulose molecule is shown on Figure 2a. This molecule has been fully identified and its formula is well-known. It is composed of elementary linear chains containing approximately 200 monomer molecules bound in 1-4 position. The hemicellulose molecule shown on the Figure 2b, consists of xylan (polyose in which the xylose units are bound by (1,4) glycosidic linkages). Its chemical formula varies depending on the nature of the lignocellulosic material (Haluk 1994; Suhas et al. 2007). The lignin molecule, shown on the

Figure 2c, has a very complex structure. Its monomer components are p.coumarylic, coniferylic and sinapylic alcohols.

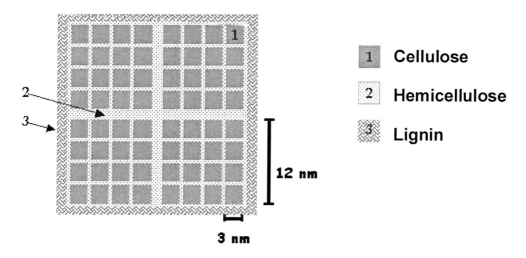

Figure 1. Structural organization of a wood microfibril.

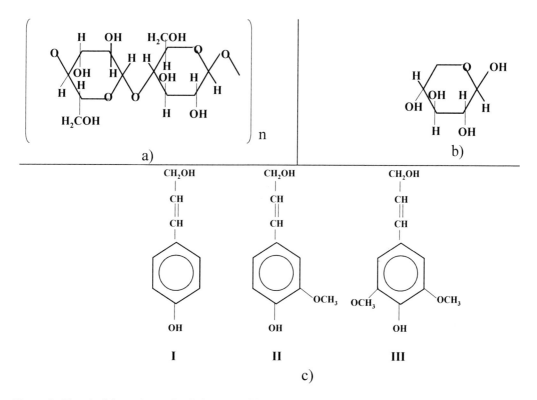

Figure 2. Chemical formula a) of cellulose, b) of ß-D-xylose and c) lignin precursor units composed of I) p-coumarylic, II) coniferylic and III) sinapylic alcohols.

Lignocellulosic materials are currently the focus of much research because of their industrial interest in the field of activated carbons, composite materials, textiles and fibers (Wigmans 1989, Suhas et al., 2007).

As suggested by Haluk (1994), the chemical components of lignocellulosic precursors can be classified in two groups (figure 3):

- Low molecular weight substances (extracts of organic matter and ash);
- Macromolecular substances of the cell walls (hemicellulose, cellulose and lignin).

The low molecular weight substances can be removed by chemical treatment before carbonization. Macromolecular molecules form the carbon mass of the activated carbons. During carbonization, depending on the raw material, hemicellulose produces approximately 20 % of the carbon mass, cellulose between 5 % and 10 %, and lignin approximately 60 % (Shafizadeh 1977).

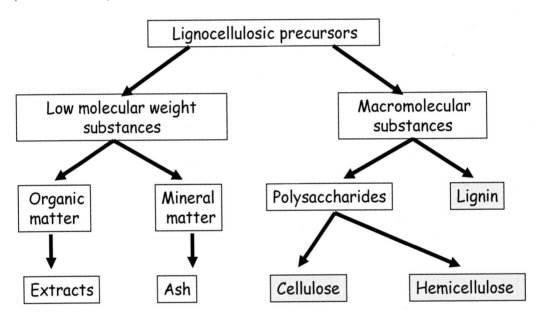

Figure 3. Composition of lignocellulosic precursors.

Macromolecular substances are present according to various amounts (Pollard et al., 1992), depending on both the considered lignocellulosic precursor (Table 1) and the geographical locations (Afrane et al., 2008).

Table 1. Composition of different lignocellulosic precursors (Pollard et al., 1992)

	Hemicellulose (H.)	Cellulose (C.)	Lignin (L.)
Coconut shell	25%	33.6%	38.4%
Pine wood	24.5%	51.4%	19%
Rice straw	24%	30%	12%
Wheat straw	28%	40%	17%

During carbonization of the lignocellulosic precursors, hemicellulose, cellulose, and lignin decompose at different rates and within distinct temperature ranges. Due to the differences in reactivity between these three basic components during pyrolysis step, and to competition between the reactions involved during their decomposition, the study of their carbonization is rather complex. Moreover, the thermal transformation of each component is highly dependent upon the experimental conditions of carbonization (Brunner and Roberts, 1980). The respective proportions of their products at the end of the carbonization step and the structural rearrangement in the chars can also play an important role in the adsorption properties of the final activated carbon (Mackay and Roberts, 1982a, 1982b; Li et al., 2008; Cagnon et al., 2009). It is therefore very important to control the carbonization step and to estimate its influence on the microporosity of activated carbons by studying the involved kinetics and their interactions.

Basically, there are two different methods to produce activated carbons (Figure 4):

- Physical activation composed of two successive heat treatments: carbonization under inert atmosphere followed by activation under oxidizing atmosphere;
- Chemical activation based on a single heat treatment in presence of a chemical agent.

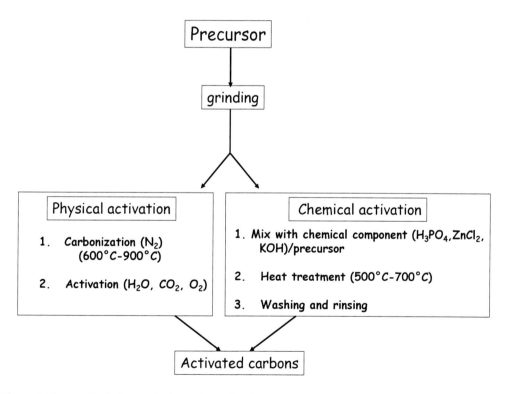

Figure 4. Two methods for producing activated carbons.

1. Carbonization

Traditionally, physical activation is achieved using two successive heat treatments (Bansal et al., 1988):

- Carbonization of the raw precursor;
- Activation of the char i.e. its oxidation, generally carried out under steam or under carbon dioxide atmosphere.

The carbonization step yields to a carbon product by removing heteroatoms (N, H, O, and S) contained in the raw material. Carbonization is typically carried out in the temperature range of 600°C to 900°C, under a flux of inert gas (for example, nitrogen) to avoid combustion. No chemical agent is used for such activation method.

The char obtained during this step is composed of more or less ordered graphitic crystallites (Kaneko et al., 1992). The space between the crystallites forms the rudimentary porous structure (Jankowska et al., 1991). It can be partially blocked by tars deposited at the entrance of the initial microporous system due to volatile matter released during the heat treatment, which is then transformed into disorganized carbon (Figure 5).

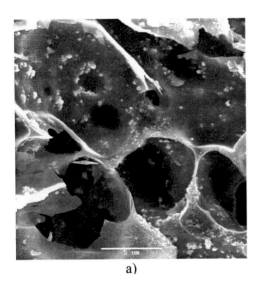 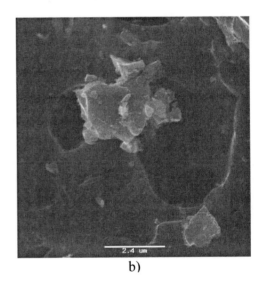

a)   b)

Figure 5. SEM images of chars a) hemicellulose (magnification × 4000) and b) lignin (magnification × 8000).

This rudimentary porous structure composes the first pores network, which allows access to the activation gas during the following step of char activation. The carbonization step is generally quantified in terms of carbon yield defined as the ratio of the weight of the final char to the weight of the initial raw material. The carbon yield depends on the nature (Mackay and Robert 1982a) and the quantity of raw material as well as on the operating parameters of the heat treatment (Cagnon et al., 2009).

Carbonization involves complex mechanisms, resulting from the coupling between the diffusion and chemical phenomena. Figure 6 shows the various phenomena involved during the carbonization of a raw particle.

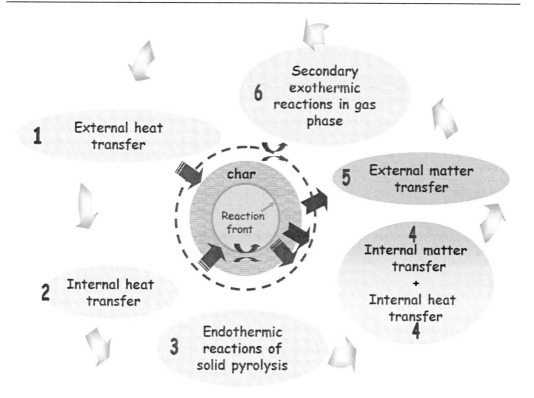

Figure 6. Phenomena involved during carbonization of a raw particle.

The external thermal constraint ① generates simultaneous reactions at different regimes. Complex mechanisms then take place in the following order: diffusional heat transfer generates a reaction front which progresses from the peripheral region towards the particle center ②; then, when the material has reached a required temperature level to allow its degradation, it decomposes into gases and solid residues, leaving behind it a layer of porous char. This decomposition process is composed of the following steps:

- endothermic decomposition reactions ③ (primary reactions);
- heat and matter transfer from the reaction front (cold zone) towards the external gaseous environment ④ (warm zone);
- secondary exothermic reactions of recombination between the products of the primary reactions ⑤ in the surrounding gas;
- secondary reactions between the gases emitted in the reaction front and the obtained char ⑥ (tars which can block the entrance of the micropores).

The products of the secondary reactions can participate again in the reaction mechanism. The internal structure of the particle is modified during carbonization.

The formation of porosity leads to an increasing thermal resistance in the external transfers (① and ⑤) due to the poor thermal conductivity of the char, reducing the temperature of the carbonization front. The carbonization kinetics and the texture of the obtained char cannot thus be radially homogeneous.

The final particle is formed by two types of carbonaceous matters: the carbon of the structure and the carbon of the reaction products (which has to be removed) obtained in various proportions depending on the carbonization protocol (and in particular on the final carbonization temperature (Cagnon et al., 2003)).

The transformation rate of the particle depends on four different limitations occurring in parallel and/or in series:

- chemical reaction;
- mass transfer in the gas layer surrounding the particle;
- external heat transfer ;
- internal heat transfer.

The rate at which the whole transformation takes place is obviously governed by the slowest limitation. As these rates can vary during the transformation process itself, the respective influence of each kinetic evolves with respect to the consumption of the particle. These limitations can be evidenced by thermogravimetric analyses, as a function of the particle size and the heating rate of the carbonization step.

Lignocellulosic materials are still under investigation for the production of activated carbons. It is, however, rather difficult to compare the results obtained by the various studies dealing with physical activation for a given precursor, since the experimental conditions of the carbonization and activation steps, as well as the method of characterisation used, vary considerably from one author to another. Several parameters of the carbonization and activation steps can vary simultaneously in a given study. Furthermore, the porous properties of the chars obtained are not systematically specified. Nevertheless, some studies (Gergova et al., 1994; Cagnon et al., 2003) showed, by comparing two activated carbons produced from two different lignocellulosic precursors, that the porosity, in terms of mean pore size (L_o), potential specific microporous volume (W_o), and pore size distribution, is already partly established at the char level before activation itself. Several studies have also focused on the influence of carbonization on the microporous properties of activated carbons obtained from lignocellulosic materials (Mackay and Roberts 1982b; Reinoso et al., 1984; Gergova et al., 1993; Gonzalez et al., 1994; Attia 1997; Fernandez et al., 2001).

For example, Reinoso et al., (1984) studied the carbonization of almond shells by varying the finale carbonization temperature between 750°C and 900°C. Under those experimental conditions, they concluded that neither the temperature nor the heating rate of the carbonization step had a significant effect on the porous structure of either the char or the corresponding activated carbon. However, Daud et al., (2000) recently showed that the porosity development of activated carbons obtained from palm shells depends strongly on the carbonization temperature.

In the light of these results, and after having determined the minimal final carbonization temperature to avoid partial pyrolysis of lignocellulosic material particles (Cagnon et al., 2003; Cagnon et al., 2009), it seemed worthwhile to study successively the carbonization kinetics (i.e. the thermal decomposition) of hemicellulose, cellulose and lignin separately, within both a synthetic blend and natural materials, such as coconut shell and plum stone.

During a preliminary study, the effects of both the intermediate heat treatment temperatures (IHTT) of carbonization and intermediate cooling (between the carbonization and activation steps) were studied in the particular case of the steam activation process (Cagnon et al., 2003). However, it was shown that IHTT control is not sufficient to manage the microporous properties of the activated carbons properly in this physical activation process. It is also very important to understand the evolution of the kinetics of the different components of the lignocellulosic precursors (hemicellulose, cellulose and lignin) during carbonization. However, the decomposition kinetics of these fractions are still poorly understood owing to the complexity of the reaction and the numerous occurring elementary mechanisms. Among the abundant literature on the subject, there is no kinetic model available which can describe the competition between all these reactions and able to estimate the quantity of char produced during the carbonization step.

Prior studies concerning the kinetics date from the 50s; for example, one can mention the pioneer studies on wood by Stamm (1956) and Eickner (1962), and the work on cellulose by Tang and Bacon (1964). Today, several kinetic models are based on these studies.

As already mentioned, cellulose is one of the three major components of lignocellulosic precursors. It also is the most widely-studied component in the literature. Its decomposition mechanism was established by Tang and Bacon (1964) by combining different analytical techniques: Infra-Red spectra analysis, thermogravimetry and X-ray diffraction. They established that cellulose decomposition proceeds in several steps and within a precise temperature range. Furthermore, dehydratation and depolymerisation reactions also occur. The decomposition kinetic of cellulose can be represented by a first-order power law as: $v=k[A]^n$ (with $n=1$).

Studies concerning hemicellulose and lignin are much less abundant and the obtained results show considerable discrepancies between them. In fact, these kinetic studies were conducted by considering the thermogram of the precursor material as homogeneous (Font et al., 1991). Some authors attempted to break down the study of the different components, for example, by removing hemicellulsoe and cellulose by sulphuric acid washing (Caballero et al., 1997a) or heat treatment (Marcilla et al., 2000a). It is difficult however to isolate lignin without changing its structure and therefore without modifying its kinetic behaviour. In addition, it is difficult to compare the results of these studies since they were carried out for different precursors and using rather different experimental conditions, such as a different atmosphere (vacuum, air or nitrogen) and different temperature ranges.

The data available in the literature cannot be generalized to all lignocellulosic materials. To get a better understanding of the thermal decomposition of hemicellulose, cellulose, lignin, and coconut shell (used currently to produce activated carbons), a kinetic model was therefore developed, taking into account the whole reactions when experimental results were obtained under the same experimental conditions.

In a previous study, hemicellulose, cellulose and lignin were analysed separately in order to estimate the respective contribution of each component to the mass and to the porous properties of the char and the activated carbons obtained from several lignocellulosic precursors (Cagnon et al., 2009).

The present chapter has two main objectives. The first is to study the carbonization kinetics of each constitutive component separately and their mix within both a synthetic blend and natural blends (coconut shell and plum stone), to achieve a better understanding of the interactions between the components and to validate a method for exploiting the thermograms (for different heating rates).

The second aim is to study the carbonization kinetics of coconut shell and plum stone, based on the previous results. The decomposition kinetics of each of the three fractions were found to be well correlated by independent power-law expressions. An empirical global kinetic model was established from three independent reactions, tacking the char formation into account. This model was validated for coconut shell and plum stones.

EXPERIMENTAL PROTOCOL

Thermogravimetry is the standard experimental technique to characterize the kinetics of gas-solid reactions. To study the carbonization kinetics, non-isothermal pyrolysis experiments were conducted using a Setaram 111 thermogravimetric differential scanning calorimeter at atmospheric pressure. Samples of 50 mg were heat-treated under nitrogen flow of 0.50 L min^{-1} in order to remove the pyrolysis gaseous products. Throughout the carbonization process, variations in weight loss of the precursor (representing the volatile matter from char) and the thermal effects due to the different reactions were recorded. A blank experiment was run, under the same conditions, in order to determine the correction required for the base-line. The samples were not dried before carbonization so as to avoid any modification in their initial composition and structure which could change their thermal decomposition within the low temperature range from ambient temperature to 200°C. A series of ground and sieved coconut shell powders (40, 160, 200 and a mix<200µm) were heat-treated from room temperature up to 800°C at different heating rates and kept at the final temperature during one hour. The different heating rate values experimented were: β = 1-5-10-20-25 and 30 °C min^{-1}.

The same experiments were also conducted on the three major fractions known to be present in coconut shell: cellulose (C.), hemicellulose (H.) and lignin (L.). These components were supplied by Aldrich: lignin 8072-93-3, cellulose 9004-34-6 and Xylan 36,355-3. The model was developed with the coconut shell data and was then applied to the plum stone data.

A synthetic mixture (MIX.) of the fractions, corresponding to the composition of a typical coconut shell (35%wt of H., 15%wt of C., 50%wt of L.), was thoroughly homogenized by grinding in a mortar. All the raw materials were crushed and sieved to a particle size of some 200µm.

The coconut shell (CS) and the plum stone (PS) used in this study were composed respectively of 35%wt. and 22%wt in hemicellulose, 15%wt. and 25%wt in cellulose, and 50%wt. and 53%wt in lignin (courtesy G.Chambat CERMAV-CNRS). The technique used consisted in estimating the quantity of glucose in the material. First, total hydrolysis was carried out on the sample according to Seaman's method (Saeman et al., 1954). An internal standard (inositol) was then added. Lastly, gas chromatography was used to obtain the proportions of the different neutral oses present in the initial sample (Sarwardeker et al., 1965). By using this technique, it was possible to estimate the quantity of hemicellulose and cellulose directly. The proportions of lignin and pigments (which accounted for low quantities) were obtained by mass balance.

The elementary compositions of the raw lignocellulosic materials used in this study are given in Table 1 (courtesy Service Central d'Analyse of the C.N.R.S.). Those compositions are different from those reported in the literature (Mortley et al., 1988; Pollard et al., 1992).

Table 1. Composition of the experimented raw lignocellulosic materials

	C (%)	H (%)	O (%)	N (%)	S (%)	Ca	H_2O (%)	Ash (%)
Hemicellulose	38.1	6.0	48.5	<0.1	<0.3	-	6.4	6.7
Cellulose	41.8	6.4	51.2	<0.1	<0.3	-	6.2	<0.3
Lignin	58.6	5.7	30.8	0.7	<0.3	-	5.9	3.9
MIX.	50.1	5.9	38.3	0.5	<0.3	-	6.5	4.9
CS	48.7	5.8	42.5	<0.3	<0.3	380 ppm	5.0	2.7
PS	50.7	6.3	41.1	0.9	<0.3	0.3 %	3.8	0.9

DISCUSSION

1. Thermogravimetry Analyses of the Three Pure Components

1.1. Preliminary Study

The weight fraction (W) can be written as a function of the initial weight fraction (W_o) and the conversion rate (X) following the expression:

$$W = W_o - W_o X \tag{1}$$

Time derivation of the expression yields the following equation:

$$\frac{dW}{dt} = -W_o \times \frac{dX}{dt} \tag{2}$$

The instantaneous conversion rate can also be written as a function of the weight fraction or the conversion rate.

Experimentally, the heat treatments carried out on the materials are based on constant heating rates as expressed by $\beta = dT/dt$ (in °C min^{-1}). Consequently, there is a direct linear relation between time and temperature. Then, the instantaneous conversion rate can be expressed as a function of time or temperature by multiplying by ($1/\beta$) following the equation:

$$-\frac{dX}{dT} = -\frac{dX}{dt} \times \frac{dt}{dT} \tag{3}$$

In this chapter, all the results will be expressed in terms of dX/dT.

Several carbonizations of each component (H., C., L.) and the mix were carried out. These experiments were run at a heating flow of 20°C min^{-1}. Figure 7 shows the rate of decomposition in terms of differential weight loss (DTG profile) against temperature for a heating rate of 20°C min^{-1} (left ordinate: $-dX/dT$, dotted curve) and relative weight loss (right ordinate: mass loss, line curve) as a function of temperature for the three components and the mix.

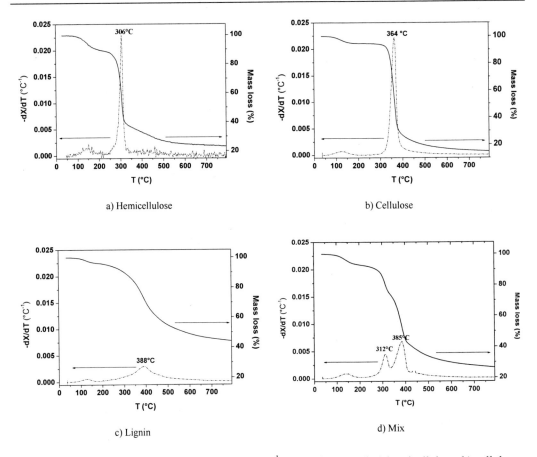

Figure 7. TG-DTG experimental curves at 20°C min^{-1} under nitrogen of: a) hemicellulose, b) cellulose, c) lignin and d) mix.

As shown in Figure 7, the thermal decomposition of the three components occurs at different rates and within distinct temperature ranges.

For each component, the cumulated weight loss observed at 150°C, is rather high, representing approximately 10% of the initial mass. It corresponds to the departure of water which is initially physisorbed in the various raw materials.

The hemicellulose fraction reacts over a narrow temperature range of 250 to 325°C with a high maximum conversion rate at 306°C (-dX/dT=0.023°C^{-1}). The cellulose fraction reacts over a narrow temperature range of 300 to 400°C with a high maximum conversion rate at 364°C (-dX/dT=0.022°C^{-1}) while the lignin fraction decomposes over a wide temperature range of 200 to 800°C with a lower maximum conversion rate at 388°C (-dX/dT=0.003°C^{-1}). These results show that the decomposition rate of hemicellulose and cellulose are similar, and roughly seven times higher than that of lignin. These differences in decomposition rate can be partly explained by the difference in structure evolution of the raw components during the heat treatment. This effect is illustrated in Figures 8, 9 and 10 on which SEM observations of raw and carbonized materials are reported. The particle morphology of hemicellulose and cellulose is not modified after the carbonization step, whereas lignin changes significantly during this step, moving towards a very organized structure. In addition, lignin decomposes over a wide temperature range and begins its degradation before hemicellulose and cellulose.

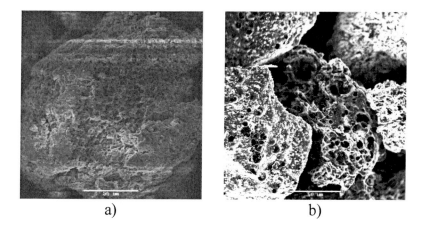

Figure 8. SEM images of a) raw hemicellulose particle (magnification × 450) and b) hemicellulose char (magnification × 350).

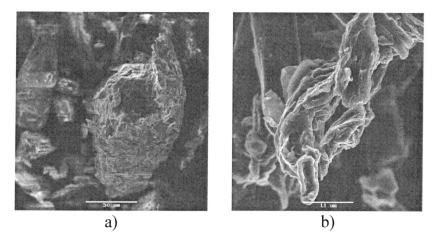

Figure 9. SEM images of a) raw cellulose particle (magnification × 400) and b) cellulose char (magnification × 1800).

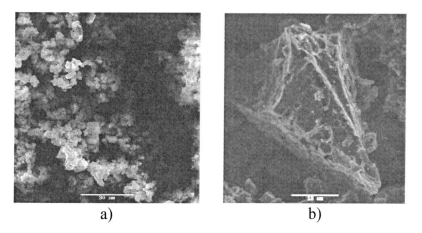

Figure 10. SEM images of a) raw lignin particle (magnification × 700) and b) lignin char (magnification × 1000).

In terms of the temperature range of decomposition and maximum weight-loss rate, the TG–DTG curves on Figure 7 obtained for the H., C., and L. fractions are very similar to those already reported by other authors (Font et al., 1991; Orfao et al., 1999; Haykiri-Acma et al., 2010), but different from those of Kifani-Sahban et al., (1996) who presented DTG curves with two peaks for lignin.

Figure 7 shows that the temperatures corresponding to the maximum in conversion rates of hemicelulose and cellulose (T_{max}) are different when the components are carbonized within the mix or separately. For those particular points, it is important to take into account the weight fraction (%) of the components within the mix. Table 2 gives the maximum in conversion rates (corrected with the weight fraction and the water fraction) of pure hemicellulose and cellulose and within the mix.

Table 2. Comparison of the corrected maximum in conversion rate -(dX/dT)$_{max}$ of the pure components (hemicellulose and cellulose) and within the mix

	Corrected -(dX/dT)$_{max}$ of pure component (°C^{-1})	-(dX/dT)$_{max}$ of the components within the mix (°C^{-1})	Deviation (%)
H.	5.18 10^{-3}	4.52 10^{-3}	12.7
C.	6.93 10^{-3}	6.84 10^{-3}	1.3

The relative deviations of the conversion rates are very low (12.7% for hemicellulose and 1.3% for cellulose). On the other hand, figure 7 shows that there is an increase in the T_{max} of both components (a difference of 6°C for hemicellulose and 21°C for cellulose) which can be attributed to a thermal delay due to the mix. This delay is induced by the competing effects of external thermal constraints, sensible heat, and the thermal effects of each component's reaction. The total heat flow is not available to any of the components under decomposition, since it is partially consumed by the sensible heat of the two other components.

These results clearly show that lignin decomposition is partly hidden by the two much more reactive fractions (hemicellulose and cellulose). This effect explains why it is rather difficult to study the lignin degradation.

Some kinetic models available in the open literature were built using overall kinetic parameters determined from the study of the components within the lignocellulosic precursor (Font et al., 1991 and 1995; Caballero et al., 1997b; Marcilla et al., 2000a; Muller-Hagedorn et al., 2003). This method inevitably leads to inaccuracies in the models. Therefore, it is important to study the kinetics of each component (H., C. and L.) separately in order to establish a kinetic model than can describe more properly the carbonization step of a lignocellulosic precursor such as coconut shell. This study would appear to be mandatory to obtain the carbonization kinetic parameters of each component and thereby describe their carbonization within the precursor.

Figure 11 shows the heat flow variations of the three pure components and of the mix during carbonization as a function of temperature at 20°C min^{-1}. The endothermic effects are negative $(\Delta H=-Q)$, and exothermic effects are positive.

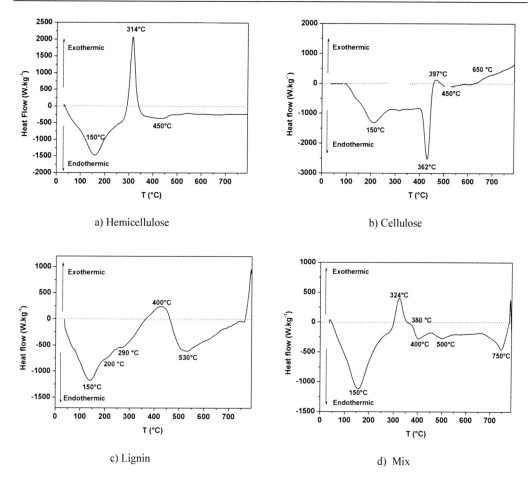

Figure 11. Heat flow variations as a function of temperature during carbonization of a) hemicellulose, b) cellulose, c) lignin and d) mix, at 20°C·min^{-1}.

As can be seen on Figure 11, the three pure components exhibit different thermal behaviours. They have a common peak which is the large endothermic peak at 150°C corresponding to water loss. This effect is very strong with regard to the quantity of desorbed water in the sample but it can hide other thermal effects to be attributed to the decomposition.

Figure 11a shows the heat flow variations of hemicellulose under heat treatment. Hemicellulose presents an exothermic peak between 250°C and 350°C, corresponding to its maximum in weight loss. For temperatures up to 350°C, hemicellulose carbonization is slightly endothermic and there is no significant mass loss.

Figure 11b shows the heat flow variations of cellulose. The strongest endothermic effect is observed at 362°C corresponding to the maximum in weight loss rate of the pure component.

Figure 11c shows the corresponding heat flow variations of lignin. Between ambient temperature and 370°C, lignin presents a large endothermic effect with two bendings respectively at 200°C and 290°C. One can also observe an exothermic effect between 370°C and 470°C, the latter corresponding to the maximum in weight loss. At the end of lignin carbonization, a strong endothermic effect can be observed.

Figure 11d shows the heat flow variations of the synthetic mix of the three basic components H, L, C. This heat flow seems to correspond qualitatively to the sum of the thermal effects obtained separately from the three pure components. However, the resulting peak amplitudes are smaller for the mix than the direct summation of those for the pure components. During the heat treatment of the mix carried out at a constant heating rate, each considered component (hemicellulose, cellulose or lignin) is surrounded by the others and part of its heat of decomposition is exchanged by the sensible heat with the others two components instead of leaving the material. This effect leads to a thermal enhancement or delay for subsequent reactions depending upon the peak has an exothermic or an endothermic character. Consequently, the different peaks amplitudes are lowered and could be also shifted to slightly different temperatures. As a matter of fact, a shift of the T_{max} corresponding to the conversion rate peak is observed. This shift is intensified by the thermicity of the reaction. For an endothermic reaction (for example, cellulose decomposition), the delay will be more pronounced because the decomposition reaction requires a contribution of heat to proceed. On the other hand, the exothermic decomposition reaction of hemicellulose can retrieve a part of the thermal delay. Thus, the shift of the T_{max} corresponding to the conversion rate peak is 21°C for cellulose and only 6°C for hemicellulose. Between 350°C and 800°C, the thermal decomposition effects of hemicellulose, cellulose and lignin are superposed. Some effects are also hidden because several simultaneous complex reactions are thermically antagonist. As an example, this explains why the exothermic effect of lignin at 400°C is hidden by the endothermic peak of cellulose.

According to some authors (Shafizadeh 1977; Antal 1984), the thermal decomposition of lignocellulosic material components is a two-step process. First, primary depolymerisation reactions take place at low temperature. These reactions involve decomposition of the raw precursor and the formation of tars, gas and solid residues. They are characterized by a high weight loss which corresponds to a strong thermal effect. Then, secondary reactions take place with the primary products thus obtained. These reactions are characterized by a thermal effect without any significant weight loss.

Hence, the exothermic peak of hemicellulose between 250°C and 350°C corresponding to its extremum of weight loss is probably due to the primary depolymerisation reactions. The depolymerisation reactions can therefore be considered to be endothermic. This can be explained by the fact that the thermal effects observed with the apparatus are not specific to intraparticle thermal effects only. In fact, it is likely that all the thermal effects of the secondary reactions in the gas phase hide the intraparticle thermal effects. For temperatures up to 350°C, the endothermic peak, with no significant mass loss, corresponds probably to secondary reactions. Nevertheless, the decomposition mechanism of hemicellulose still remains to be correctly established.

For lignin, the exothermic peak between 370°C and 470°C, corresponding to the extremum in weight loss, is probably due to the secondary decomposition reactions of the matter. Due to its complex molecular structure, the thermal decomposition mechanism of lignin is not well-known at this time.

For cellulose carbonization, on the other hand, Tang and Bacon (1964) developed a validated 4-step reaction mechanism in the temperature range of 25°C to 650°C. This mechanism can be written as follows:

- *1st step:* between 25°C and 150°C: physical desorption of water;
- *2nd step:* between 150°C and 240°C: intramolecular dehydratation of cellulose units;
- *3rd step:* between 240°C and 390°C: glycosidic bonds are broken, in particular C-O and C-C bonds, and intermediate component is formed, namely levoglucosan. These depolymerisation reactions lead to the formation of tars and gaseous release of water, CO and CO_2;
- *4th step*: up to 400°C, an aromatization step proceeds leading to the formation of graphitic layers.

According to this mechanism, the strongest endothermic effect observed for cellulose at 364°C with a high weight loss can be attributed to the formation of the intermediate component and depolymerisation reactions.

1.2. Influence of the Heating Rate During the Carbonization Step

The carbonization was performed using a Setaram 111 thermogravimetric differential scanning calorimeter at several heating rates (β=1°C min^{-1}, 5°C min^{-1}, 10°C min^{-1}, 15°C min^{-1}, 20°C min^{-1}, 25°C min^{-1} and 30°C min^{-1}), all the other experimental parameters being kept identical.

Figure 12a shows the extrema evolution of the conversion rate of pure hemicellulose and cellulose as a function of the heating rate. The conversion rate extrema increase for hemicellulose and decrease for cellulose when the heating rate increases. Furthermore, both components present two different domains:

- domain 1 between the values of heating rate 1°C min^{-1} and 15°C min^{-1};
- domain 2 between the values of heating rate 15°C min^{-1} and 30°C min^{-1}.

The Figure 12b shows the respective evolutions of the maximum conversion rate of hemicellulose and cellulose within the synthetic mix as a function of the heating rate. While both components present two different domains, it can be observed that the slopes of the second domain are different, depending on whether the component is pure or within the synthetic mix. The first domain is in the heating rate range of 1°C min^{-1} and 15°C min^{-1}. In this domain, the maximum in conversion rate of hemicellulose increases while those of cellulose decreases. The second domain is in the heating rate range between 15°C min^{-1} and 30°C min^{-1}. In contrast to the first domain, the maximum in conversion rate of hemicellulose slightly decreases while those of cellulose slightly increase.

Figure 12c shows the evolution of the maximum in conversion rates for the lignin as a function of the heating rates. The conversion rates are found to be constant between 1°C min^{-1} and 15°C min^{-1} and increasing between 15°C min^{-1} and 30°C min^{-1}. Compared to the two other compounds, the change in thermal behaviour is also observed at a heating rate of 15°C min^{-1}.

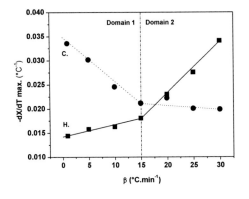

a) Pure hemicellulose and cellulose

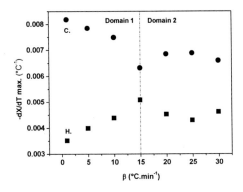

b) Hemicellulose and cellulose within the mix

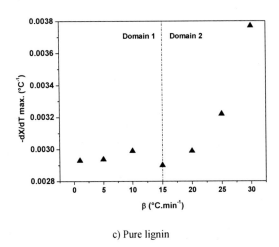

c) Pure lignin

Figure 12. Evolution of the maximum in the conversion rate as a function of the heating rate for a) pure hemicellulose and cellulose, b) hemicellulose (⊠) and cellulose (⊞) within the mix and c) pure lignin (▲).

Table 3 shows the T_{max} corresponding to the maximum in conversion rate for each component.

Table 3. T_{max} of the maximum in conversion rate for each pure component and the synthetic mix

| β | T_{max} (°C) corresponding to the maximum in conversion rate | | | | | | |
	1 °C min^{-1}	5 °C min^{-1}	10 °C min^{-1}	15 °C min^{-1}	20 °C min^{-1}	25 °C min^{-1}	30 °C min^{-1}
H.	260	282	295	302	306	310	316
C.	309	331	348	350	364	367	378
L.	333	365	372	377	388	392	397
Mix	266/330	289/356	297/368	299/369	312/385	307/380	322/394

In Figure 12 and Table 3, one can observe that the heating rate selected for the carbonization step has a considerable influence on the decomposition reactions of the three pure components and the mix. The following conclusion can be written:

When the heating rate increases, one observes:

-    an increase in the conversion rate of hemicellulose and lignin,
-    a decrease in the conversion rate of cellulose,
-    an increase in T_{max}.

Furthermore, all the samples present two domains induced by the competition between the decomposition reactions combined with both the internal heat effects and the external heat effects. For heating rates as low as 15°C min^{-1}, carbonisation takes place in the intraparticle regime (domain 1 on Figure 12a). At low heating rates, the overall kinetics behaviour results from the combination of the decomposition reactions with internal heat transfers. In this domain, the more the temperature increases, the more limiting the internal heat transfer phenomena become. This decreases the peak amplitude of the conversion rate of the cellulose endothermic reaction, but increases the peak amplitude of the conversion rate of the hemicellulose exothermic reaction. The peak amplitude variation in the cellulose conversion rate is greater than that of hemicellulose, since both effects, sensible heat and reaction heat, absorb heat from cellulose. For hemicellulose, in contrast, the two effects compete, which compensates for the heating rate effect. In domain 2 for hemicellulose, the faster the heating rate is, the less the particle reaction heat is evacuated, which accelerates its decomposition. For cellulose, whatever the heating rate, the thermal flow going into the particle is due to external heat transfer, and is therefore constant. Thus, the conversion rate amplitude is constant for heating rates up to 15°C min^{-1}.

1.3. Application of Simple Reaction Kinetics Equation

Initially, the kinetics equation of the thermal decomposition for a simple reaction was used. This equation is not suitable for several simultaneous reactions, but was used merely to compare the experimental results reported in this chapter with available values within the literature. It can be written as in the following equation:

$$\frac{dX}{dt} = k(1-X)^n \tag{4}$$

Where dX/dt is the instantaneous conversion rate,

   k is the global rate constant,
   X is the conversion,
   n is the apparent order of the reaction.

Moreover, the regime was considered to be mainly chemical. Then, the global rate constant (k) can be expressed by the Arrhenius equation:

$$k = A\exp\left(\frac{-E_a}{RT}\right) \qquad (5)$$

Where A is the pre-exponential factor,

E_a is the apparent activation energy,
R is the gas constant ($R = 8.314$ J K^{-1} mol^{-1}),
T is temperature (K).

Inserting equation 5 into equation 4, equation 6 is obtained:

$$\frac{dX}{dt} = \left(A\exp\left(\frac{-E_a}{RT}\right)\right)(1-X)^n \qquad (6)$$

When $n=1$, equation 6 is linear. Based on the experimental thermograms obtained by thermogravimetry for different heating rates, the term $-E_a/R$, which corresponds to the intercept of equation 7, can be identified:

$$-Ln\frac{dX}{dt} = f\left(\frac{-1}{T}\right) \qquad (7)$$

In agreement with the literature, current equation 7 was used here, even though it could not be considered to be truly relevant.

Figure 13 shows the apparent activation energy (E_a) as a function of the conversion rate (X) of each component separately and of the synthetic mix during carbonization. Each point stems from five different thermograms corresponding to five different heating rates, all the other parameters being identical.

As shown in Figure 13, the apparent activation energy evolution as a function of the conversion rate is different for each component:

- the apparent activation energy for hemicellulose increases regularly from 6 kJ mol^{-1} up to 110 kJ mol^{-1};
- the apparent activation energy for cellulose remains constant at 45 kJ mol^{-1}, for a conversion rate between 10% and 60% conversion rate. Beyond 60%, the activation energy increases to 88 kJ mol^{-1};
- the apparent activation energy for lignin increases regularly for a conversion rate from 5% up to 30%. At 30%, there is a plateau at 80 kJ mol^{-1} after which there is an increase up to 100 kJ mol^{-1};
- the apparent activation energy for the synthetic mix corresponds roughly to the sum of the activation energy of the three pure components.

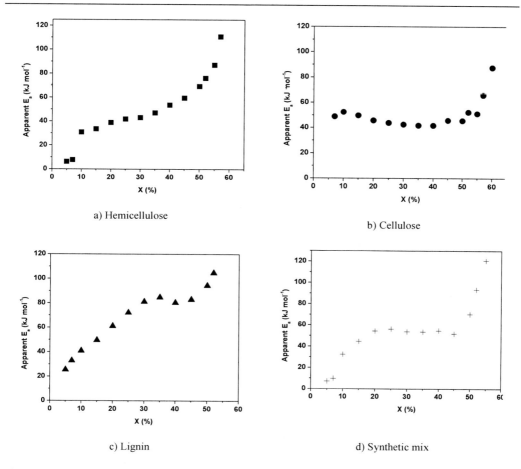

a) Hemicellulose

b) Cellulose

c) Lignin

d) Synthetic mix

Figure 13. Apparent activation energy evolution as a function of the conversion rate during carbonization of: a) hemicellulose (⊠), b) cellulose (⊡), c) lignin (▲) and d) synthetic mix (✎).

Table 4. Comparison between the apparent activation energy calculated from experimental data of pure components obtained in this chapter and literature values

	Temperature range (°C)	Apparent activation energy (kJ mol⁻¹)
Stamm (1956) (Pine wood)		
Hemicellulose	110-220	111.0
Cellulose	110-220	108.0
Lignin	110-220	96.0
Font et al., (1995)		
Pure lignin	450-560	97.5
Kifani-Sahban et al., (1996)		
Pure hemicellulose	195-280	179.5
Pure cellulose	300-375	228.6
Pure lignin	184-400	40.7
Kastanaki et al., (2002) (Olive stones)		
Hemicellulose	ambient-850	92.9
Cellulose	ambient -850	210.6
Lignin	ambient -850	30.6
This chapter		
Pure hemicellulose	ambient-800	40.0
Pure cellulose	ambient-800	45.0
Pure lignin	ambient-800	80.0

In Table 4 are compared the values of the apparent activation energy calculated from experimental data given in this chapter for hemicellulose, cellulose, and lignin to literature values. As shown in the table, these values are close to those obtained by Stamm (1956) and Font et al., (1995), but different from those of Kifani-Sahban et al., (1996) and Kastanaki et al., (2002).

2. THERMOGRAVIMETRY ANALYSES OF NATURAL LIGNICELLULOSIC PRECURSORS

Based on the data obtained in the previous study of carbonization of the three pure components of coconut shell, a model was developed for the analysis of coconut shell thermograms.

2.1. Preliminary Study with Coconut Shell

Figure 14 shows the TG-DTG experimental curves under nitrogen with an instantaneous conversion rate (left ordinate: $-dX/dT$) and relative weight loss (right ordinate: weight loss) of coconut shell carbonization as a function of temperature. This experiment was carried out with a particle size below 200 µm and at 20°C min^{-1}.

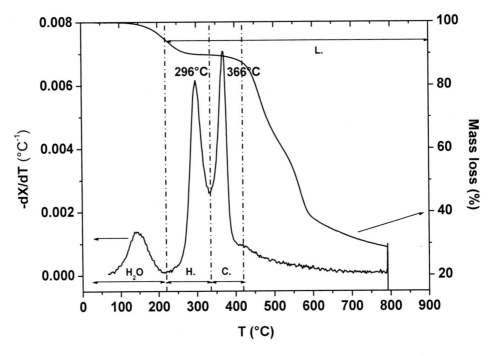

Figure 14. TG-DTG experimental curves of coconut shells under nitrogen at 20°C min^{-1} ($d_p < 200$ µm).

**Table 5. Identification of the different thermogram variation
domains shown on Figure 14**

Temperature	Instantaneous conversion rate -dX/dT	Mass loss (%)	Interpretation and comments
80°C - 200°C	maximum at 2×10^{-3} °C^{-1}	0 to 10%	Water loss of the sample
200°C - 300°C	Fast increase		Simultaneous decompositions of hemicellulose and lignin
300°C - 330°C	strong decrease	55%	Mainly lignin decomposition
330°C - 366°C	Fast increase		Cellulose and lignin decompositions
450°C		70 %	Almost total decomposition of cellulose
Up to 800°C			No significant mass loss: this domain corresponds to secondary reactions with gas release. However, the transformation of lignin remains: mass loss of approximately 73% is obtained at 800°C.

The natural deconvolution of the hemicellulose and cellulose peaks leads to an easier interpretation of the coconut shell thermogram. This is not the case for certain others lignocellulosic materials such as pine wood (Orfao et al., 1999), eucalyptus (Kifani-Sahban et al., 1996), almond shells (Marcilla et al., 2000b), or olive stones (Caballero et al., 1997b). The hemicellulose, cellulose and lignin peaks of these materials overlap, making them more complex and difficult to study. A further advantage is that the global shape of the coconut shell thermogram is similar to the one obtained after carbonization of a mixture of the three components in the same proportions as those of coconut shell (35% hemicellulose, 15% cellulose, 50% lignin) (Figure 7d, synthetic mix). On the DTG profiles of coconut shell, the respective decomposition domains of the three major fractions can be clearly distinguished. Even the lignin fraction, which is known to decompose over a wide temperature range while the other two fractions decompose over some 100°C, is easily identified by the last 680-800°C tail. Table 5 summarizes the identification of the different variation domains of the instantaneous conversion rate and of the mass loss as a function of the temperature (Figure 14).

Table 5 confirms the previous results obtained with pure hemicellulose, cellulose and lignin. Coconut shell carbonization can be interpreted as follows:

−   hemicellulose and cellulose successively and respectively decompose in the temperature ranges 200°C-330°C and 330°C-430°C,
−   lignin decomposes slowly but over a wider temperature range (between 200°C and 800°C).

Furthermore, as seen with the synthetic mix, the decomposition of lignin, which is the major component in coconut shell, is masked by the decomposition of hemicellulose and cellulose. The decomposition rates of hemicellulose and cellulose are similar (though the rate for cellulose is higher than for hemicellulose), and both are faster than that of lignin.

Figure 15 shows the heat flow variation (red) and instantaneous conversion rate (black) during coconut shell carbonization as a function of temperature at 20°C min^{-1} (d$_p$< 200 μm).

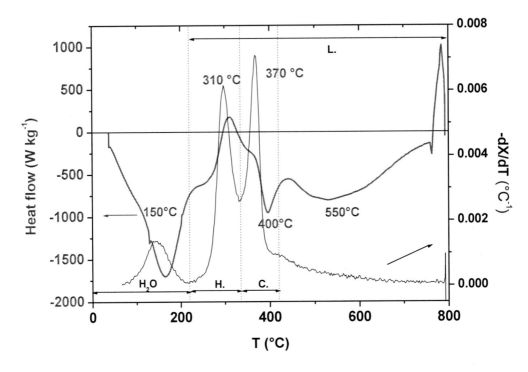

Figure 15. Heat flow variation (red) and instantaneous conversion rate (black) during coconut shell carbonization (d$_p$< 200 μm) at 20°C min^{-1}.

In comparison with the results shown on Figure 11, the observed heat flows of the three components (H., C., and L.) overlap. Nevertheless, each heat flow is well defined and corresponds to the same mass loss domain seen previously.

At 150°C, there is an endothermic peak corresponding to dehydratation of the sample.

At 250°C, this endothermic effect decreases. This temperature corresponds to onset of decomposition for both hemicellulose and lignin.

At 310°C, there is a weak exothermic effect corresponding to hemicellulose decomposition. Between 150°C and 400°C, the decomposition domains of hemicellulose, cellulose and lignin are superposed due to the competition between the numerous complex reactions and simultaneously the influences of the associated thermal effects.

At 400°C, no significant mass loss can be associated to the broad endothermic effect. This phenomenon can be attributed to the secondary reactions of chars with gas release of dihydrogen, both carbon monoxide and dioxide and methane (Shafizadeh 1977). These results are in agreement with the conclusion drawn from previous results. Moreover, it is important to note that the shapes of the heat flow curves for both coconut shell and the synthetic mix (with the same hemicellulose, cellulose, and lignin composition) are similar.

Coconut shell carbonization is globally an endothermic phenomenon. As shown in a previous study (Cagnon et al., 2003), the total carbonization of a lignocellulosic material can only be achieved at temperatures up to 800°C.

As seen in this chapter, coconut shell carbonization is a difficult phenomenon to study. The transformation of the major component in weight (lignin) is hidden by that of hemicellulose and cellulose. These two components are more reactive than lignin. The decomposition temperature ranges depend on the composition of the raw material (Kifani-Sahban et al., 1996; Caballero ct al., 1997b; Orfao et al., 1999) and on the heating rate (Font et al., 1991). Each lignocellulosic material is unique and each one has different carbonization behaviour. To develop a kinetic model, therefore, it is important to study the carbonization behaviour of each component separately. Nevertheless, before any kinetic analysis, it is critical to check if the apparent rate of decomposition observed experimentally is mainly chemically controlled and not limited but a physical or thermal transfer. This particular aspect is developed in the following paragraph.

2.2. Influence of Particle Size and Heating Rate

Figure 16 shows the weight loss curves (TGA curves) of coconut shell for different particle diameters.

It can be clearly seen that particle diameter has no effect on this pyrolysis, as all the curves are superposed. The corresponding data can therefore be considered under chemical limitation without any diffusional effect during the transformation of the different components in coconut shell. Therefore, these results can be directly used for the study of the carbonization kinetics.

Figure 17 shows the evolution of maximum in conversion rate ($-dX/dT$) as a function of the heating rate for hemicellulose and cellulose within coconut shell (dp< 200 μm).

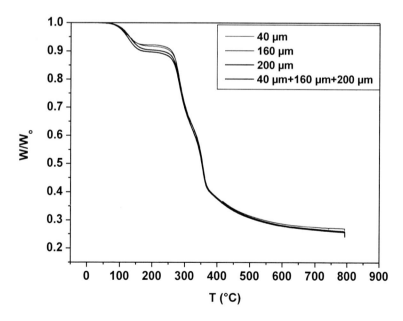

Figure 16. Coconut shell TGA curves during carbonization at 10°C min^{-1} for different particle sizes (40, 160, 200 μm and a mix 40-160-200 μm mix < 200 μm).

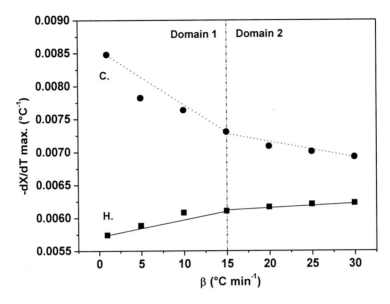

Figure 17. Evolution of the maximum in conversion rate as a function of the heating rate for hemicellulose (⬕) and cellulose (⬓) within coconut shell ($d_p < 200$ μm).

As seen previously in Figure 12, the conversion rate extrema increase for hemicellulose and decrease for cellulose when the heating rate increases. The two components exhibit two different domains:

- domain 1 between 1°C min^{-1} and 15°C min^{-1};
- domain 2 between 15°C min^{-1} and 30°C min^{-1}.

The extremum evolution of hemicellulose and cellulose conversion rates in coconut shell is similar to those of the pure components separately and within the synthetic mix.

Tables 6 and 7 report the T_{max} corresponding to the conversion rate extrema for each component in coconut shell (this chapter) and in almond shell (Font et al., 1991).

Table 6. T_{max} of the conversion rates extrema of hemicellulose and cellulose in coconut shell

	T_{max} (°C) corresponding to the conversion rate extrema						
	1 °C min^{-1}	5 °C min^{-1}	10 °C min^{-1}	15 °C min^{-1}	20 °C min^{-1}	25 °C min^{-1}	30 °C min^{-1}
H.	248	271	282	298	296	307	308
C.	320	344	355	358	366	376	379

Table 7. T_{max} of the conversion rates extrema of hemicellulose and cellulose in almond shell (Font et al., 1991)

	T_{max} (°C) corresponding to the conversion rate extrema				
	5 °C min^{-1}	10 °C min^{-1}	15 °C min^{-1}	20 °C min^{-1}	30 °C min^{-1}
H.	293	310	313	314	316
C.	360	368	373	378	379

Tables 6 and 7 show that while the T_{max} values corresponding to the hemicellulose and cellulose peaks differ depending on the raw lignocellulosic precursor, the evolution of the conversion rate extrema as a function of the heating rate is similar: there is an increase for both precursors (coconut shell and almond shell) but in different temperature ranges. This may be due to the different compositions of the raw materials and to more or less strong bonds between the three components (H., C., and L.) depending on the structure of the raw precursor.

3. CARBONIZATION KINETIC MODEL

3.1. Kinetic Equation of a Simple Reaction

To calculate the global activation energy of the coconut shell decomposition reaction, equation 4 (kinetic equation of a simple reaction) was applied to coconut shell. The chemical composition of coconut shell was assumed to be homogeneous. The obtained results were compared to those of the synthetic mix. It is important to note that this equation applied to coconut shell is significant only from a qualitative point of view.

Figure 18 shows the apparent activation energy (E_a) evolution as a function of the conversion rate (X) during coconut shell carbonization for different particle sizes. Each point stems from five different thermograms corresponding to five different heating rates, and each curve corresponds to a particle size between 40 µm and 200 µm (all the other parameters being identical).

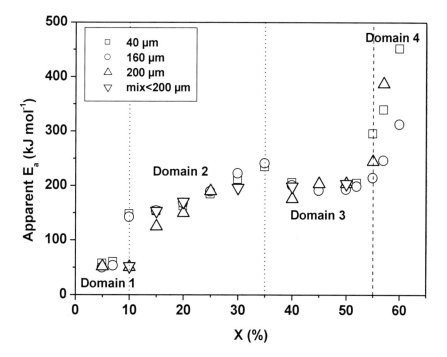

Figure 18. Apparent activation energy evolution of coconut shell during its carbonization as a function of the conversion rate and the particle size.

Different domains of coconut shell apparent activation energy evolution can be distinguished:

- *domain 1:* the value is constant, approximately 50 kJ mol^{-1} between 5% and 10% of conversion rate,
- *domain 2:* the apparent activation energy increases from 150 kJ mol^{-1} to 240 kJ mol^{-1} between 10% and 35% of conversion rate,
- *domain 3:* the apparent activation energy value is approximately 200 kJ mol^{-1} between 40% and 55% of conversion rate,
- *domain 4:* the apparent activation energy increases from 200 kJ mol^{-1} to 450 kJ mol^{-1} for conversion rate values up to 55%.

On the four domains, the different data, corresponding to different particle sizes, are superposed. These results confirm that particle diameter has no effect on pyrolysis. It is important to note that the overall shape of the apparent activation energy evolution of coconut shell as a function of the conversion rate is similar to that of the synthetic mix (Figure 13). However, the values of coconut shell global activation energy are three times higher than those obtained with the synthetic mix. The interactions between the three components influence the intrinsic behaviour of each component. For example, cellulose depolymerisation may be different if lignin is present, probably due to the fact that lignin acts as cement between the cellulose and hemicellulose fibers.

Thus, the differences in the behaviour of coconut shell and the synthetic mix (which have the same composition) can be attributed without doubt to breaking of the bonds between the different components in coconut shell.

There is, however, a serious drawback in using the simple kinetic equation in the case of a lignocellulosic precursor: the precursor composition is assumed to be homogeneous. The abscissa of Figure 18 corresponds to the global conversion rate of coconut shell, whereas coconut shell is in fact composed of hemicellulose, cellulose, and lignin which have different carbonization behaviours within the lignocellulosic material. It is very important therefore to take the heterogeneity of coconut shell composition into account in order to lay down a kinetic model, by considering the intrinsic effect of each component.

3.2. Kinetic Model with Three Independent Reactions Applied to Coconut Shell

The non-isothermal decomposition kinetics of a lignocellulosic material is difficult to study. The thermal decomposition of the major component (lignin in terms of weight fraction) is partly masked by the two much more reactive fractions (cellulose and hemicellulose) (Figure 15). To allow any kinetic analysis, deconvolution of the DTG curve (Figure 15) into three independent contributions is therefore needed.

The decomposition kinetics of each of the three fractions is easily and efficiently correlated by independent power-law expressions (Marcilla et al., 2000b). The kinetic expressions of lignocellulosic precursor decomposition are generally written in terms of variation in weight fraction of each component as a function of temperature, taking the initial mass of the anhydrous component as the reference mass:

$$\frac{dW_i}{dT} = K_i (1 - W_i)^{ni} \tag{8}$$

Where n_i is the global order of component i of the reaction (i=H., C. or L.),

$K_i = k_i \times \beta \times w_{TOT}^{(n_i - 1)}$ is the kinetic constant of the global rate with k_i the real kinetic constant rate,

$\beta = \dfrac{dT}{dt}$ the heating rate,

w_{TOT} the initial total mass of the anhydrous precursor.

Following equation 8, the overall pyrolysis rate (dW/dT) is equal to the sum of the decomposition kinetics of each fraction (i=1,3) balanced by their respective weight fractions (W_i) and the corresponding chars weight fraction (Wci) balanced by their stoechiometric coefficients (v_i).

$$\frac{dW}{dT} = \sum_i \left(\frac{dW_i}{dT} + v_i \frac{dW_{ci}}{dT} \right) \tag{9}$$

Where W is the weight fraction of the precursor,
W_i is the weight fraction of each component,
v_i is the stoechiometric coefficients of the char formed by each component,
W_{ci} is the weight fraction of the char formed by each component i=H., C. or L.

The stoechiometric coefficients of the chars inherited from each of the three compounds were identified from the yield measured on the decomposition of the isolated hemicellulose, cellulose, and lignin. These coefficients linked to the respective weight of each fraction have to agree with the overall yield of the lignocellulosic material or mix.

$$W_\infty = \sum_i W_{ci} \tag{10}$$

$$W_{ci} = v_i \left(W_{io} - W_i \right) \tag{11}$$

In most of the published studies, the contribution of the char to the overall kinetics is neglected. This is due to the failure to assess the initial composition of the lignocellulosic material and stoechiometric coefficients of the char. As a result, the decomposition rates are usually under-estimated and the corresponding identified kinetics parameters are erroneous. In the present model, the decomposition rates are written in terms of weight fraction variations taking the initial anhydrous lignocellulosic material weight as a reference and char weight production is also considered. Applying equation 9 to the three components gives a system with 3 equations and 9 unknown parameters:

$$\frac{dW_H}{dT} = K_H \left(1 - W_H\right)^{n_H} \tag{12}$$

$$\frac{dW_C}{dT} = K_C \left(1 - W_C\right)^{n_C} \tag{13}$$

$$\frac{dW_L}{dT} = K_L \left(1 - W_L\right)^{n_L} \tag{14}$$

This kinetic model is based on 15 adjustable parameters the values of which have to be identified using the available experimental data (equation 15).

$$\frac{dW}{dT} = \frac{k_{cio}}{\beta} \times W_o^{n_i-1} \times W_i^{n_i} \times \exp\left(-\frac{E_i}{R \times T}\right) \tag{15}$$

Mathematically, this number of parameters is too high to lead to a unique set of solutions. For each component, the real constant rate (k_i), the global reaction order (n_i) and the apparent activation energy (E_{ai}) have to be identified (corresponding to 9 adjustable parameters). The initial proportions of each component and the char yield have also to be determined. Six parameters (two for each component) were identified using the experimental carbonization results of the isolated pure components. Therefore, the initial system to be solved contains 15 parameters.

In order to reduce this number of parameters, the observation of the deconvoluated thermogram led to the identification of additional constraints which have to be fulfilled by the model. The integration of those constraints led to new relationships between several parameters and then reduced the number of possible solutions.

The first additional constraint is defined at the top of each peak where the second derivative of the weight function has to be zero and the first derivative equal to the experimental decomposition rate at the observed temperature. By applying this principle to each fraction, the model was reduced to only 3 adjustable parameters. Given that the activation energy values (E_{aH}, E_{aC} and E_{aL}) reported in the literature are rather homogeneous, these parameters were selected as the remaining adjustable ones and all the others calculated applying the additional criteria.

Figure 19 shows the experimental coconut shell (CS) TGA curve and model, during carbonization at 10°C min^{-1} taking into account or not the char formation of the three components (H., C. and L.) in the calculations.

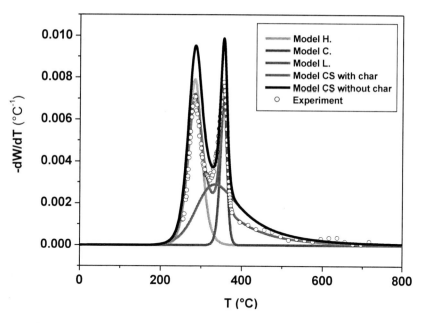

Figure 19. Experimental coconut shell TGA curve and model during carbonization at 10°C min^{-1} with and without char formation of each component.

The reduced model taking into account the formed char of the three components (pink curve) illustrated in Figure 19 is in very good agreement with both the experimental data (circle) and the corresponding kinetic parameters of the literature. In comparison, if the weight of the formed char is not taken into account by the reduced model (black curve), a significant shift between the model curve and the experimental data (circle) is observed; the parameters of the last model are thus under estimated.

It i important to note that the global mass loss rate observed is not only the summation of the carbonization rates of the three components. As shown on Figure 19, the char production of the three compounds contributes significantly to the global kinetic of carbonization.

The coconut shell carbonization global kinetic parameters (10°C min^{-1}) calculated with the model with or without the char production of the three components are presented in Table 8.

Table 8. Coconut shell carbonization global kinetic parameters obtained with the model with or without char production of the three components at 10°C min^{-1}

	Model with char production	Model without char production
E_{aH} (kJ mol^{-1})	220	115
E_{aC} (kJ mol^{-1})	550	400
E_{aL} (kJ mol^{-1})	100	45
n_H	1.9	1.3
n_C	1.7	1.4
n_L	3.8	2.8
k_H (s^{-1})	$1.75 \ 10^{23}$	$6.90 \ 10^{9}$
k_C (s^{-1})	$4.58 \ 10^{47}$	$4.43 \ 10^{33}$
k_L (s^{-1})	$1.13 \ 10^{19}$	$4.35 \ 10^{9}$

These results show that the kinetic parameter values are different when the char production of each component is taken into account or is neglected. These values are in agreement with literature values (Stamm 1956; Font et al., 1995).

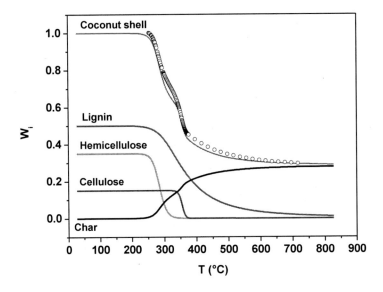

Figure 20. TGA experimental (circles) and model (lines) curves of coconut shell, hemicellulose, cellulose, lignin and char production at 10°C min⁻¹.

Integrating each equation (equations 12 to 14), the theoretical mass loss curve (TGA curves) obtained by the model can be plotted as a function of the temperature and can be compared to the experimental data.

Figure 20 shows the theoretical mass loss curve (TGA curves) of coconut shell, hemicellulose, cellulose, and lignin, and the theoretical curve of the char production of the three components, plotted with the carbonization yields obtained for each isolated component. The coconut shell TGA curve corresponding to the theoretical mass loss is obtained by adding the individual contributions of each compound and the carbonaceous weight fraction; it is representative of the distinct total mass loss. The empirical model is in perfect agreement with the experimental data, as can be seen in Figure 20.

3.3. Kinetic Model with Three Independent Reactions Applied to Plum Stones

Figure 21 shows the decomposition rate of plum stones in terms of differential weight loss (DTG profile) against temperature for a heating rate of 10°C min⁻¹. On this DTG profile, the respective decomposition domains of the three major fractions can also be very clearly distinguished and are in perfect agreement with the observed DTG profiles of the isolated compounds. As seen previously with coconut shell, the heat flow of each component (H., C., and L.) overlaps. Nevertheless, each heat flow is well defined and corresponds to the same mass loss domain seen previously. In terms of kinetics, the hemicellulose and cellulose decomposition rates in coconut and plum stone are different.

Despite these differences, Figure 22 shows that the experimental results obtained on plum stone are also very well predicted by the model. In Figures 20 and 22 one can clearly see the different weight loss curves of the three fractions in both materials with a highly similar composition.

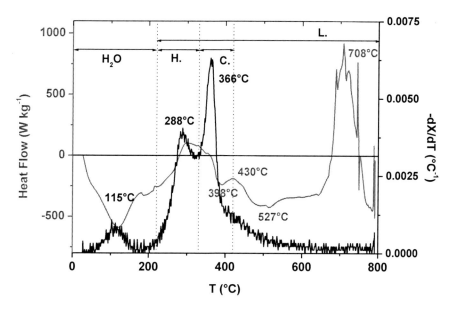

Figure 21. TG-DTG experimental curves for plum stones under nitrogen at 10°C min⁻¹.

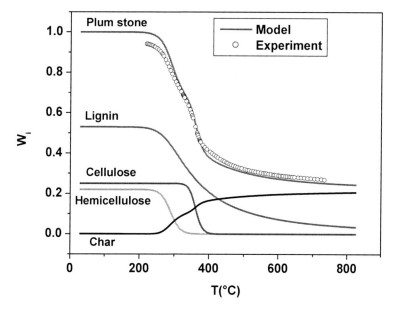

Figure 22. TGA experimental (circles) and model (lines) curves of plum stones, hemicellulose, cellulose, lignin and char production at 10°C min⁻¹.

The plum stone carbonization global kinetic parameters (10°C min⁻¹) calculated with the model with the char production of the three components are presented in Table 9.

Table 9. Plum stones carbonization global kinetic parameters obtained with the model with char production of the three components at 10°C min[-1]

	Model with char production
E_{aH} (kJ mol^{-1})	200
E_{aC} (kJ mol^{-1})	300
E_{aL} (kJ mol^{-1})	100
n_H	2.1
n_C	1.8
n_L	5.4
k_H (s^{-1})	$1.65 \ 10^{22}$
k_C (s^{-1})	$4.16 \ 10^{26}$
k_L (s^{-1})	$1.29 \ 10^{27}$

In term of global kinetic parameters, the cellulose and lignin decomposition rates in coconut shell and plum stone are different while both precursors have a nearly similar composition (Coconut shell : 35%wt in H., 15%wt in C., 50%wt in L., Plum stone : 22%wt in H., 25%wt in C., 53%wt in L.).

CONCLUSION

The respective decomposition temperature ranges of the three major components in lignocellulosic materials have been determined from carbonization thermogravimetric analyses. The coconut shell results show that hemicellulose and cellulose decomposition conversion rates are different when these components are carbonized separately or within the raw precursor.

A kinetic model for the thermal decomposition of coconut lignocellulosic material has been validated. This model can also be applied to the thermal decomposition of plum stones. The model is based on deconvolution of the TGA profile into three contributions corresponding to the three major components: cellulose, hemicellulose, and lignin. The deconvolution method has been checked by comparing the experimental thermal decomposition of the three isolated compounds and their mixing. A particular technique is proposed to reduce drastically the number of adjustable parameters needed in the model from 15 to 3. Using this model, the char formation as a function of temperature during the carbonization step can also be assessed.

The model could be extended to larger-sized particles by taking into account the mass and heat transfer kinetics. The whole model could then be used to optimize the carbonization protocol of lignocellulosic materials with respect to the obtained microporous properties.

REFERENCES

Afrane, G.; Achaw O-W. *Bioresour. Technol.* 2008, 99, 6678-6682.
Antal, M.J. In "Biomass pyrolysis" In K.W. Böer and J.A. Duffie (Eds); Advances in Solar Energy; American Solar Energy Society: New York, 1984; 175-255.

Attia A.A. *Adsorption Science and Technol.* 1997, 15(9), 707-715.

Banerjee, S.K.; Majumdar, S.; Dutta, A.C.; Roy, A.K.; Banerjee, S.C. and Banerjee, D.K. *Ind. J. Technol.* 1976, 14, 45-49.

Bansal, R.C.; Donnet, J.B and Stoeckli, F. In "Active Carbon" Marcel Dekker: New York, 1988.

Brunner PH and Roberts PV. *Carbon* 1980, 18, 217-224.

Byrne, C.E. and Nagle, D.C. *Carbon* 1997, 35(2), 259-266.

Caballero, J.A.; Marcilla, A.; Conesa, J.A. *J. Anal. Appl. Pyrolysis* 1997a, 44, 75-88.

Caballero, J.A.; Conesa, J.A.; Font, R.; Marcilla, A. *J. Anal. Appl. Pyrolysis* 1997b, 42, 159-175.

Cagnon B.; Py X.; Guillot A.; Stoeckli F. *Microporous Mesoporous Mater.* 2003, 57, 273-282.

Cagnon, B.; Py , X.; Guillot, A.; Stoeckli, F.; Chambat, G. *Bioresour. Technol.* 2009, 100, 292-298.

Daud, W.M.A.W.; Ali, W.S.W.; Sulaiman, M.Z. *Carbon* 2000, 38, 1925-1932.

Eickner, H.W. *Forest products Laboratory, Forest service* 1962, 12(4), 194-199.

Fernandez, E.; Centeno, T.A. and Stoeckli, F. *Adsorption Science and Technol.* 2001, 19(8), 645-653.

Font, R.; Marcilla, A.; Verdu, E. and Devesa, J. *J. Anal. Appl. Pyrolysis* 1991, 21, 249-264.

Font, R.; Marcilla, A.; Garcia, A.N.; Caballero, J.A.; Conesa, J.A. *J. Anal. Appl. Pyrolysis* 1995, 32, 29-39.

Gergova, K.; Petrov, N. and Minkova V. *J. Chem. Biotechnol.* 1993, 56, 77-82.

Gergova K., Petrov N., Eser S. *Carbon* 1994, 32, 693-702.

Gonzalez, M.T.; Molina-Sabio, M. and Rodriguez-Reinoso F. *Carbon* 1994, 32, 1407-1413.

Haluk, J.P. "Le bois, matériau d'ingénierie - Composition chimique du bois" Editeur Arbolor : Nancy, France 1994, chapitre II, 53-88.

Haykiri-Acma, H., Yaman, S., Kucukbayrak, S. *Fuel Processing Technol.* 2010, 91, 759-764.

Hillis, W.E. ; Sumimoto, M. "Effect of extractives on pulping" In J.W. Rowe (ed.), Natural Products of Woody Plants II. Springer: Berlin 1989, 880-920.

Jankowska, H. ; Swiatkowski, A. and Choma, J. "Active carbon" Ellis Horwood 1991.

Kaneko, K. ; Ishii, C. ; Ruike, M. and Kuwabara, H. *Carbon* 1992, 30, 1075-1088.

Kastanaki, E.; Vamvuka, D.; Grammelis, P.; Kakaras, E. *Fuel Processing Technol.* 2002, 77-78, 159-166.

Kifani-Sahban, F.; Belkbir, L.; Zoulalian, A. *Thermochim. Acta* 1996, 284, 341-349.

Li, W.; Yang, K.; Peng, J.; Zhang, L.; Guo, S.; Xia, H. *Ind. Crops Prod.* 2008, 28, 190-198.

Lopez-Gonzalez, J de D.; Martinez-Vilchez, F. and Rodriguez-Reinero, F. *Carbon* 1980, 18, 413-418.

Mackay, D.M.; Roberts, P.V. *Carbon* 1982a, 20(2), 105-111.

Mackay DM; Roberts PV. *Carbon* 1982b, 20(2), 87-94.

Marcilla, A.; Garcia-Garcia, S.; Asensio, M.; Conesa, J.A. *Carbon* 2000a, 38(3), 429-440.

Marcilla, A.; Conesa, J.A.; Asensio, M.; Garcia-Garcia, S.M. *Fuel* 2000b, 79, 829-836.

Molina-Sabio, M.; Caturla, F. and Rodriguez-Reinoso, F. *Carbon* 1995, 33(8), 1180-1182.

Mortley, Q.; Mellowes, W.A. and Thomas, S. *Thermochim. Acta* 1988, 129, 173-186.

Muller-Hagedorn, M.; Bockhorn, H.; Krebs, L.; Muller, U. *J. Anal. Appl. Pyrolysis* 2003, 68-69, 231-249.

Orfao, J.J.M.; Antunes, F.J.A; Figueiredo, J.L. *Fuel* 1999, 78, 349-358.

Pollard, S.J.T. ; Fowler, G.D.; Sollars, C.J. and Perry, R. *The Science of the Total Environment* 1992, 116, 31-52.

Roberts, A.F. *Combustion and flame* 1971, 17, 79-86.

Rodriguez-Reinoso, F.; Lopez-Gonzalez, J de D. and Berenguer C. *Carbon* 1982, 20(6), 513-518.

Rodriguez-Reinoso, F.; Lopez-Gonzalez, J de D. and Berenguer C. *Carbon* 1984, 22(1), 13-18.

Rodriguez-Reinoso, F.; Martin-Martinez, J.M.; Molina-Sabio, M.; Perez-Lledo, I. and Prado-Burguete, C. *Carbon* 1985, 23(1), 19-24.

Saeman, J.F.; Moore, W.E.; Mitchell, R.L. and Millett, M.A. *Tappi* 1954, 37(8), 336-343.

Sarwardeker, J.S ; Sloneker, J.H. and Jeanes, A. *Anal.Chem.*1965, 37, 1602-1604.

Shafizadeh F. and Chin P.P.S. "Thermal deterioration of wood" in: Goldstein IS (ed.) Wood Technology : Chemical aspects, *ACS Symposium Series* 43, 1977, 57-81.

Solum, M.S.; Pugmire, R.J.; Jagtoyen, M. and Derbyshire, F. *Carbon* 1995, 33(9), 1247-1254.

Stamm, A.J. *Ind. Eng. Chem.* 1956, 48(3), 413-417.

Suhas, Carrott, P.J.M.; Carrott, M.M.L. *Bioresour. Technol.* 2007, 98, 2301-2312.

Tang, M.M. and Bacon, R. *Carbon* 1964, 2, 211-220.

Wigmans, T. *Carbon* 1989, 27(1), 13-22.

INDEX

C

Q

R

S

T